"Your book is clearly written and very useful, especially as a reference for the roles of transcription factors and genetic defects."

John M. Allman, Ph.D., *Frank P. Hixton Professor of Neurobiology, Caltech*

"This book offers a compelling account of a scientific journey that spans from neuroanatomy to the intricate molecular mechanisms underlying neuronal processing. I highly recommend it to aspiring biomedical students. For those drawn to neuroscience, it introduces bioenergetics as a powerful investigative tool. And for those already interested in bioenergetics, it reveals the brain as a fascinating and rich field of study. A truly enlightening read that bridges two dynamic areas of biomedical science."

Yidong Bai, Ph.D., *Professor, Department of Cell Systems and Anatomy, University of Texas Health San Antonio*

"This book offers valuable insights into neuroscience and bioenergetics, highlighting the role of cytochrome *c* oxidase in cellular respiration and neuronal function. The well-organized chapters provide accessible yet detailed explorations of molecular pathways, linking transcriptional regulation, neuronal activity, and energy metabolism. This authoritative resource is essential for neuroscientists, molecular biologists, and students in these fields."

Shilpa S. Dhar, Ph.D., *Assistant Professor of GI Medical Oncology, University of Texas MD Anderson Cancer Care Center*

"This is great! I learned so much! I think it's safe to say your research has made an impact, and it's truly impressive when it's gathered all in one place like this."

Serena Dudek Sommer, Ph.D.

"This is an extremely impressive body of work that is so well explained with the known details and caveats. The frequent summary interpretations are a big help to the readers. Congratulations on your many stunning findings and this masterful compilation of science."

Carol A. Everson, Ph.D., *Professor of Medicine, Medical College of Wisconsin*

"Your book is an incredible book of great importance."

Francisco Gonzalez-Lima, Ph.D., *George I. Sanchez Centennial Professor of Psychology, The University of Texas at Austin*

"The whole book looks really interesting."

Lawrence I. Grossman, Ph.D., *Henry L. Brasza Professor of Molecular Medicine and Genetics, Director of Center for Molecular Medicine and Genetics, Wayne State University*

"I enjoyed the book very much… I think it is a fantastic contribution to the literature. As always, your writing is so clear and precise that it makes the complex subject material understandable with no loss of detail. I enjoyed reading it all, and especially, catching up on the upstream regulators of COX."

Robert F. Hevner, M.D., Ph.D., *Professor of Neuropathology (retired), University of California, San Diego*

"This book provides a comprehensive overview of cytochrome oxidase and beautifully integrates studies of the enzyme at neurofunctional, developmental, biochemical, genetic, and clinical levels. It is very readable and will provide a valuable resource for the next generation of scientists, putting a trove of useful knowledge all in one place right in their hands. That was your goal and you have achieved it."

Jonathan C. Horton, M.D., Ph.D., *William F. Hoyt Professor of Ophthalmology, Neuroscience Program, UCSF*

"It is a wonderful book, and of course very well illustrated. The preface is excellent…. What a powerful new agent. Chapter 1 is good, and it should come first. Chapter 2 is even better. The Abstract is helpful… chapter 4 …. is very well illustrated. I'm very impressed with what you have done. The chapters are well written and illustrated…. you have produced a wonderful and very impressive book."

Jon Kaas, Ph.D., *Distinguished Centennial Professor of Psychology, Member of the US National Academy of Sciences, Vanderbilt University*

"Overall, a great job! I especially enjoyed much of the EM and COX histochemistry, beautiful work. Congratulations! … Your book is a comprehensive, synthesizing review of cytochrome oxidase. It should be very valuable to anyone working in the area; a treasure trove."

Joseph C. LaManna, Ph.D., *Jeannette M. and Joseph S. Silber Professor Emeritus of Physiology, Case Western Reserve University School of Medicine*

"The first chapter … is a great introduction. You have done an amazing job pulling all this information together."

Margaret I. Lomax, Ph.D., *Research Professor Emerita, Kresge Research Institute, University of Michigan*

"This book is helpful for the Neuroscientists to understand cytochrome oxidase from cell to molecule."

Xuegang Luo, Ph.D., *Professor of Anatomy and Neurobiology (retired), Xiangya School of Medicine, Central South University, China*

"I enjoyed reading your book. I learned a lot that I did not know about COX biology…. This is a well-written, scholarly book that is really two books."

Henry J. Ralston, III, M.D., *Professor and Chair Emeritus, Department of Anatomy, UCSF*

"This wonderful and informative book on Cytochrome *c* Oxidase is written by Margaret Wong-Riley. It contains 18 chapters describing all aspects of this interesting and important enzyme. It also describes how the enzyme was discovered and the subsequent works conducted by the author and many other scientists on the role of the enzyme in the nervous systems. It is a most useful book for established and young scientists to learn about this topic. I am sure this book will inspire the curiosity of neuroscientists to conduct further work on Cytochrome *c* Oxidase."

Kwok-Fai So, Ph.D., *Director, GHM Institute of CNS Regeneration, Member of Chinese Academy of Sciences, Jinan University, China*

"In the late 1970s Margaret Wong-Riley pioneered the histochemical localization of a metabolic enzyme, cytochrome oxidase, in the mammalian cortex. The levels of this enzyme are dramatically affected by the activity of the neurons that express it, integrated over some period of time. This discovery has given rise among other things to an understanding of the compartmentalization of multiple cortical areas, with nearly 300 papers by many authors on cytochrome oxidase patterns in the visual cortex alone. It has also stimulated the discoveries of additional activity-dependent changes. Her new book is a scholarly summary of the literature on cytochrome oxidase together with an interpretation of how neural activity interacts with it. It will be useful for all students of brain organization."

Michael P. Stryker, Ph.D., *William Francis Ganong Professor of Physiology, Kavli Institute for Fundamental Neuroscience, UCSF*

"I would like to congratulate You on this book. It is a very nice contribution to the history of "our enzyme". I am very pleased with the book. … You have presented the facts in a very readable way, and I hope your book will be published soon so that I can cite it."

Sebastian Vogt, M.D., *Professor, Cardiovascular Research Lab for Bioenergetics, Heart Surgery, Philipps University of Marburg, Germany*

Cytochrome *c* Oxidase

This comprehensive reference book on an essential mitochondrial enzyme, cytochrome *c* oxidase (COX), is the culmination of four decades of work. This enzyme's biochemical activity originated and evolved more than a billion years ago. Nevertheless, its contributions to the function of neurons remain relatively obscure. COX is not just a housekeeping enzyme engaged in energy generation, but a treasure trove for discovering numerous neuronal properties chronicled in this book. Various methods of studying COX in neurons are described, including biochemistry, histochemistry, cytochemistry, immunohistochemistry, in situ hybridization, and gene isolation. These methods revealed that COX also plays a critical role in rescuing the metabolic integrity of neurons exposed to toxins. The goal of this book is to raise the curiosity of readers about this invaluable enzyme so researchers will continue the quest to uncover neuronal properties unveiled by this enzyme.

Key Features

- A comprehensive book on cytochrome *c* oxidase in neurons
- Covers all major neuroscientific discoveries of cytochrome *c* oxidase in the mammalian nervous system
- Provides a broad range of themes from systems, through cellular, to molecular
- Contains not only the major discoveries but also various techniques for analyzing cytochrome *c* oxidase, plus a comprehensive list of references
- Describes the mechanisms of energy regulation in neurons and how energy metabolism and neuronal activity are tightly coupled at the cellular and molecular levels
- Documents the role of cytochrome *c* oxidase in uncovering a striking synaptic imbalance during the critical periods of both brain stem respiratory system and visual cortical development

Cytochrome *c* Oxidase

A Treasure Trove in Revealing Neuronal Properties from Systems to Molecules

Margaret T.T. Wong-Riley, PhD

CRC Press
Taylor & Francis Group
Boca Raton London New York

CRC Press is an imprint of the
Taylor & Francis Group, an **informa** business

Designed cover image: Author

First edition published 2026
by CRC Press
2385 NW Executive Center Drive, Suite 320, Boca Raton FL 33431

and by CRC Press
4 Park Square, Milton Park, Abingdon, Oxon, OX14 4RN

CRC Press is an imprint of Taylor & Francis Group, LLC

ISBN: 978-1-032-82592-2 (hbk)
ISBN: 978-1-032-83098-8 (pbk)
ISBN: 978-1-003-50776-5 (ebk)

DOI: 10.1201/9781003507765

Typeset in Times
by SPi Technologies India Pvt Ltd (Straive)

Dedication

To my family, my mentors, and colleagues in my lab

and

To all future neuroscientists who will
carry the field to new heights

Contents

Author Biography

Margaret T.T. Wong-Riley, PhD, Professor Emerita at the Medical College of Wisconsin (MCW), received her BS and MA in Science Education from Teachers College, Columbia University, and her PhD from Stanford University (1970). Her postdoctoral training was at the University of Wisconsin (Dr. Ray Guillery) and National Institutes of Health (NIH). She was an Assistant Professor, then a tenured Associate Professor at UC San Francisco before joining MCW. Her research – funded for four decades by NIH – focused on mammalian visual system ultrastructure and organization, cytochrome *c* oxidase as a metabolic marker of neuronal activity, metabolic plasticity in developing and adult nervous systems, molecular mechanism of bigenomic enzyme regulation, the tight coupling between neuronal activity and energy metabolism, photobiomodulation, and critical periods of respiratory system and visual cortical development. Her publications include 154 articles, 28 invited reviews and book chapters, and she is the editor of the *Neuroscience Secrets* (2000) book. She served on two NIH Study Sections and numerous review panels for NIH and other organizations. She has been on the Scientific Advisory Board of the International Foundation of Optic Nerve Disease (IFOND) since 1998. She served on several editorial boards and was the Founding Editor-in-Chief of Eye and Brain (2010–2023), Dove Medical Press (now part of Taylor & Francis). From 2023 to 2025, she was an Associate Editor of *Frontiers in Neuroscience*. Currently, she is a member of the editorial board of *NeuroSci*. She has won several teaching awards at MCW and was inducted into the Society of Teaching Scholars at MCW in 1996. In 2006, she received the Roy and Sherrington Award from the Gordon Research Conference on "Brain Energy Metabolism and Blood Flow" at Oxford.

Preface

Cytochrome *c* oxidase is one of the most ancient enzymes known, parts of it having evolved more than a billion years ago in bacteria. It is present in almost every cell of our body and is crucial for our survival. Indeed, inactivation of its function, such as by cyanide, azide, or carbon monoxide, is incompatible with life. Yet, its presence in neurons is often taken for granted, its value overlooked, its own properties ignored, and what it can offer neglected. In writing this book, I hope to unlock some of the mysteries of neuronal properties revealed by this vital enzyme. If it should engender some excitement among its readers to further its research, my mission would be accomplished.

I came across this enzyme in neurons rather serendipitously in the mid-1970s. Like many other neuroscientists at the time, I was using horseradish peroxidase (HRP) as a retrograde tracer for neuronal pathways. As my main area of focus was the visual system, my first test of the method was to inject HRP unilaterally in the primary visual cortex of mature monkeys, with the expectation that I would see labeled neurons in the ipsilateral lateral geniculate nucleus (LGN). Not surprisingly, I did see a column of labeled neurons through all the layers of the ipsilateral LGN (Wong-Riley 1974a). However, I *was* surprised to find brown reaction product in neurons of the neighboring globus pallidus bilaterally, as well as in other regions of the brain. Upon closer examination, there was a distinct difference in the appearance of the reaction product. Whereas HRP labeling in the LGN was small, round, and punctate, that in the globus pallidus and elsewhere was heterogeneous in size and shape (Wong-Riley 1976). As the globus pallidus was not known to project directly to the visual cortex, let alone bilaterally, I had to conclude that the labeling was due to some endogenous substance that reacted with the diaminobenzidine and H_2O_2 in the incubation medium. What could that be? Curiosity killed the cat. I had to find out. By running a battery of tests, I concluded that the endogenous labeling was not due to catalase or peroxidase, but to an enzyme that was completely inactivated by potassium cyanide and was localizable under the electron microscope to the inner mitochondrial membrane. That enzyme was cytochrome *c* oxidase (COX) (Wong-Riley 1976). I remember sitting in the Electron Microscope room, feeling a bit disappointed that it was only a common mitochondrial enzyme! A colleague also chided me with, "Why are you spending so much time on false positives of HRP?"

However, my thoughts soon centered on the possibility that, if COX is an essential oxidative enzyme for energy generation, and neurons depend almost entirely on oxidative metabolism in the mitochondria for their energy supply, then COX might serve as an indicator of neuronal activity. I optimized the histochemical method to demonstrate COX's relative activity in neurons (Wong-Riley 1979a). At a social gathering, my colleague Michael Merzenich was bemoaning the fact that he had to euthanize a group of cats whose unilateral cochlear implants with embedded electrodes did not work. A light bulb went off in my head. Although the cochlear implants failed to stimulate the cochlear nerve, they damaged the inner hair cells by their presence. Such damage should deprive the brain stem relay neurons of their excitatory input. Will they show any diminution in their relative COX activity? "Why don't you let me perfuse the cats and look at their brains with cytochrome oxidase histochemistry?" I asked Mike. He agreed. I processed the tissues, examined them under the microscope, and recorded all the ipsi- and contralateral structures in the auditory brain stem that showed reduced COX activity. These findings did not make much sense to me at the time, as I naively thought that all projections were bilateral in the auditory system. However, when I showed the slides to Mike, he said, "That's right.... that's right.... that's right." All the data agreed with the known pathways. It turned out that some projections were dominant on the ipsilateral side, while others were contralateral, and some projections were excitatory, while others had an indirect inhibitory effect. A decrease in COX activity proved to result from the "silencing" of excitatory input (Wong-Riley et al. 1978). Significantly, removal of inhibitory input did not induce much change, indicating that the energy requirement is mainly for repolarizing membrane potentials after depolarization and not hyperpolarization. To determine if changes were reversible, we tested another group of cats whose implants *did* work, and indeed, previously functionally "silenced" neurons regained their COX activity with stimulation (Wong-Riley et al. 1981).

With these encouraging data in the auditory system, I proceeded to test the activity dependence of COX in the cat and monkey visual system and the somatosensory system of mice. In each case, COX levels were reduced in neurons that were deprived of normal excitatory input (Wong-Riley 1979a; Wong-Riley and Welt 1980; Wong-Riley and Riley 1983; Wong-Riley and Carroll 1984a). The case of the primate visual cortex was particularly striking. I discovered a periodic distribution of high COX activity in normal, supragranular layers that assumed a polka-dot-like pattern in tangential sections. I named the COX-rich zones "puffs" because they are globular, reticulated rather than solid, and resemble puffs of cotton. I sent a COX-reacted, histological slide of a normal monkey visual cortex to David Hubel in 1978, but our lab was slow in publishing this finding at the light and EM levels. Once Hubel's lab started using COX histochemistry (Horton and Hubel 1981; Horton 1984; Livingstone and Hubel 1984a), the whole visual community took notice, and soon the method became widely used. The supragranular puffs (which Horton also uncovered and called "patches," but Livingstone and Hubel named them "blobs," and "blobs" quickly became the only commonly used term) opened a whole new field of primate visual cortical research.

Over the years, COX has proven to be a sensitive and reliable metabolic indicator of neuronal activity. It has also uncovered metabolic plasticity in both developing and mature neurons. My lab went on to generate COX antibodies, optimize the detection of its biochemical activity, and localize its subunit genes both with light microscopy and EM (Hevner and Wong-Riley 1989, 1991, 1993; Wong-Riley et al. 1997a).

Another cellular aspect of COX is whether it is involved in photobiomodulation (PBM). PBM reportedly promotes wound healing, but its mechanism of action in neurons was poorly understood. Our lab started analyzing this process in primary cultures of postnatal neurons, and the results indicate that, under specific light conditions, COX plays a critical role in increasing ATP production and in the functional recovery of neurons exposed to various toxins (Wong-Riley et al. 2005).

Concurrently, I became more and more interested in the molecular mechanisms of regulating COX. COX is one of only four proteins in our body that have two genomic sources: nuclear and mitochondrial. How does this singular, bigenomic enzyme get coordinately regulated? This quest propelled my lab into probing the various transcription factors involved in this regulatory process. None of the other three bigenomic proteins (mitochondrial complexes I, III, and V) has yet been analyzed in this fashion. Moreover, we have uncovered the tight coupling between energy metabolism and neuronal activity at the transcriptional level. Again, this vital topic had received very little attention prior to our investigation.

Totally surprising to me was how COX modulation factored significantly in the mechanisms underlying the critical period of postnatal development, not only in the brain stem respiratory system but also in the primary visual cortex of rats. In the case of the respiratory system, our investigation was initiated as a request by a dear, late colleague, Dr. Ralph Franciosi, who was the Professor and Chief of Pediatric Pathology at the Medical College of Wisconsin and the Director of Perinatal and Infant Pathology at the Children's Hospital of Wisconsin. He had published a landmark book on Sudden Infant Death Syndrome (SIDS) and had developed two SIDS Centers to support grieving families: one in Minneapolis and one in Milwaukee. He was troubled by the frequency of SIDS and asked me to study it. I responded that it was outside my field of the visual system. Eventually, however, I consented because one of my sisters-in-law lost her only son to SIDS. Nevertheless, I felt strongly that before I could study SIDS, I needed to understand the *normal* development of the brain stem respiratory system as a solid basis for comparison (the necessary "control"). An initial analysis of postnatal development of the brain stem with COX histochemistry revealed an abrupt, unexpected fall in COX levels on postnatal day 12 in respiratory nuclei of *normal* rats (Liu and Wong-Riley 2002). This was followed by more than a decade of discoveries of concomitant neurochemical and physiological changes during this critical period of normal development. Based on what we have learned about the brain stem respiratory system, I was compelled to re-examine the much-studied critical period of the visual system. Sure enough, we uncovered comparable metabolic, neurochemical, and electrophysiological changes in the primary visual cortex during the critical period of its development in rats (Zhang et al. 2018). Such shared, abrupt alterations in both subcortical and cortical structures strongly suggest a common set of guiding principles governing critical periods in general.

This book is a culmination of four decades of diligent work that has taught us abundantly about neuronal properties, and much of the revelation was contributed by COX. None of the discoveries would have been possible, however, without the dedicated endeavors of all of my talented students, postdocs, technicians, and collaborators. To them I owe eternal gratitude. For myself, what propelled me through the years of successes and failures was my persistent sense of curiosity, my need to find and know the truth, and my desire to learn new things. This curiosity drove me into uncharted territories beyond my expertise at the time, but the reward of discoveries fostered new quests, and so I ploughed on. Indeed, among the many important characteristics of a good scientist: intelligence, diligence, patience, perseverance, skill, knowledge, and curiosity, I believe the most critical one is CURIOSITY.

Acknowledgments

My gratitude is first and foremost to my parents, James E.Y. Wong and Florence Tai Wong, who always encouraged me to do well in school. My emancipated mother attended Shanghai University, a rare privilege, especially for a young woman in her days. She was very happy when I received my Master's degree; however, she did try to discourage me from getting a PhD ("No one will want to marry you!"). I was an obedient daughter, but I did follow my dream. My father, who also attended Shanghai University, was the kindest, most honest and upright, and most generous person I know. I had a happy childhood in Hong Kong with my sister Cissy.

I'm grateful to my husband, Danny A. Riley, Ph.D., who is also my best friend. I cannot ask for a more loving, kind, intelligent, and supportive partner through our many life challenges. Our collaboration on the NASA space research was truly "out of this world." We mourn the premature loss of our dear daughter Michelle, but are deeply thankful that our son Sean is a loving, brainy, thoughtful, creative, and all-around good person, who is also a great husband to Kelly and father to our two wonderful grandchildren, Finn and Kitt.

I thank Drs. Malcolm Carpenter and Charles Noback for opening my eyes and mind to neuroanatomy when I did my Master's degree at Teachers College, Columbia University. My PhD mentor at Stanford University, Henry J. (Peter) Ralston, III, MD, who is kind and generous, taught me electron microscopy and permitted me to work on a system different from his own. His wife Diane was the most internally and externally beautiful person I've ever known. Rainer (Ray) Guillery was an eminent neuroscientist who honed my critical thinking during my postdoctoral training at the University of Wisconsin in Madison. In Arnaldo Lasansky's NIH lab, I learned serial EM sectioning of 200+ consecutive sections that helped me uncover the coexistence of ribbon, regular, and gap junction synapses in bipolar processes of the tiger salamander retina (Wong-Riley 1974b). During my brief sojourn in Michael Fuortes's lab, I learned how to make microelectrodes, fill them with Procion yellow, and record from the turtle retina.

I am deeply indebted to all the talented, conscientious, and diligent individuals who worked tirelessly in my laboratory through the years. Without their invaluable contributions, none of the work that laid the background for this book could have been materialized. I owe a hefty dose of gratitude to all of them, whose names appear in my publications through the years. A few of them deserve additional mention. Glenn Kageyama was my first graduate student, whose immense technical and artistic skills made all the COX EM cytochemistry possible. Edward Carroll did much of the initial work on the monkey visual cortex. Anne Mjaatvedt uncovered the mysteries behind Purkinje cells' transition from early postnatal development to maturity. Dan Hoppe and Wayne Kaboord have exceptional skills in EM cytochemistry. Thomas Trusk analyzed the different forms of visual deprivation in monkeys. Robert Hevner is an exceptional scientist whose sharp mind and magical hands enabled him to excel in biochemistry, antibody generation, histo- and immunohistochemistry, and in situ hybridization of COX in neurons. He also produced, together with Suyan Liu, a comprehensive metabolic map of the rat brain that has no equal. Xuegang Luo made the combined COX histochemistry and IGSS of GABA come to life. Suyan Liu discovered that COX subunit precursor proteins form a precursor pool in dendritic mitochondria poised for processing into mature forms when local energy demand increases. Feng Nie and Zifang Huang are exceptionally skilled in EM immunocytochemistry and EM in situ hybridization of COX and other neurochemicals. Chenyang Zhang perfected primary neuronal cultures and uncovered synthesis and degradation rates of *COX* subunit mRNAs. She and Aili Guo spearheaded NRF-2 research in my lab. Brenda Anderson contributed to the neurochemical analysis. Ying Ying Liu started the SIDS research and continued to excel in EM co-localization of COX and other neurochemicals in the pre-Bötzinger complex in her own lab. Qiuli Liu is the most patient and remarkably skilled individual who spent long hours collecting and processing specimens and performing optical densitometric measurements of individual neurons in the developing respiratory system. He also undertook daily monitoring of ventilatory changes in vast numbers of developing animals. Without his meticulous and diligent work, we would never have uncovered the many neurochemical, physiological, and ventilatory changes during the critical period of respiratory development. Huan Ling Liang is gifted with a pair of golden hands that can perfect any experiment in which she is involved. Sakkapol (Zach) Ongwijitwat is delightful, inquisitive, and has solved many issues about NRF-2's regulation of COX. Shoujing Yang patiently performed EM localization of NRF-1 and NRF-2 in cultured neurons. Hui Meng did the same for PGC-1α. Shilpa Dhar easily conquered all the technically challenging molecular biological experiments and firmly established the role of NRF-1 in regulating all *COX* subunit genes. She perfected the challenging 3C documentation of 13 genomic loci in the same transcription factory. Shilpa Dhar, Kaid Johar, Anusha Priya, and Bindu Nair contributed greatly to the knowledge of the coupling of neuronal activity and energy metabolism at the transcriptional level. Xiu-ping Gao revealed synaptic imbalance during the critical period of respiratory development through her exquisite patch-clamp experiments. Lianwei Mu, Dandan Wang, Dongdong Xia, and especially Hanmeng Zhang worked tirelessly and skillfully in uncovering electrophysiological, neurochemical, and metabolic changes during the critical period of the rat's visual cortical development. I owe everyone mentioned above my deepest appreciation and gratitude.

I thank many of my colleagues: Michael Merzenich offered his deaf cats that started my journey on probing the relationship between COX and neuronal activity; Carol Welt introduced me to barrel field research; Kwok-Fai So is a long-time colleague with a mutual interest in vision research; Thomas Norton collaborated with me on tree shrew research; Jon Kaas taught me extracellular recordings; Edgar (Ted) DeYoe collaborated with me and was instrumental in uncovering spontaneous activity in various compartments of the monkey visual cortex and correlated them with COX; Joseph LaManna and I shared many hours of discussion about energy metabolism in neurons, and we collaborated on a project on prolonged hypoxia; Francisco Gonzalez-Lima shares my interest in cytochrome *c* oxidase over the years, and he honored me with a book on "Cytochrome oxidase in neuronal metabolism and Alzheimer's Disease" edited by him and published in 1998; K.C. Ho helped procure enough Alzheimer's disease brains to make our project possible; Jeffrey Palmer introduced me to the enteric nervous system; Maureen Neitz showed me some basic molecular biological techniques before they became routine; Xandra Breakefield at Mass General enabled me to observe molecular genetics in her lab for a month in 1997; Guiseppe Attardi welcomed me to his Caltech lab for a month in 1997 to observe the cutting-edge research done on the mitochondria and enabled me to form collegial friendship especially with his wife Anne Chomyn and his postdoc Yidong Bai; Margaret Lomax allowed me to spend a month in 1998 in her Kresge lab and, together with Nancy Bachman's help, performed enough molecular biological experiments to result in a joint publication; Richard Scarpulla discovered NRF-1 and NRF-2 and generously provided us with antibodies against these two vital transcription factors; Ralph Franciosi encouraged and supported me in launching a project on the critical period of respiratory system development; Herbert (Bert) Forster enlightened me on the brain stem respiratory system; Harry Whelan and Janis Eells collaborated with me on PBM research; Qing-song Liu generously helped my lab set up and run a patch-clamp recording station; Joseph Besharse enlightened me on the characteristics of KIF17; and Carol Everson shared many light-hearted as well as serious discussions with me, and I thank her most of all for her true friendship through the years.

Appreciation goes to the department chairs who hired me: Henry (Peter) Ralston III at UC San Francisco and David Anderson at the Medical College of Wisconsin (MCW). Thanks also to subsequent chairs at MCW: Roger Markwald, Joseph Besharse, and Jonathan Marchant for their unfailing support.

I greatly appreciate the tens of thousands of investigators whose work on COX and related topics has inspired and propelled me to forge ahead, even when the going was at times rough. I have learned much from their discoveries, and my only regret is that I cannot describe and cite them all, and for this I offer my sincere apology. Indeed, each chapter can easily be expanded into a book of its own. There is much for future curious generations to pursue.

Special thanks go to Drs. John Allman, Yidong Bai, Shilpa Dhar, Carol Everson, Francisco Gonzalez-Lima, Lawrence Grossman, Robert Hevner, Jonathan Horton, Jon Kaas, Joseph LaManna, Margaret Lomax, Xuegang Luo, Henry (Peter) Ralston III, Kwok-Fai So, Serena Dudek Sommer, Michael Stryker, and Sebastian Vogt for their careful and thoughtful reading of this book and for their constructive criticisms.

Finally, I wish to thank C. R. "Chuck" Crumly, PhD, Senior Acquisition Editor of Life Sciences at CRC Press/Taylor & Francis Group, for his gentle guidance and good-humored patience during the entire writing process.

1 Introduction
Cytochrome c Oxidase

ABSTRACT

Cytochrome c oxidase (COX) is one of the most ancient enzymes known; parts of it evolved from primordial bacteria. As the terminal enzyme of the electron transport chain, it is the only one that completes oxidative metabolism for the generation of ATP in the mitochondria. COX is one of only four proteins in mammalian cells that are bigenomic, having 3 subunits encoded by the mitochondrial genome and 10 (or 11) by nuclear DNA. Many of the nuclear-encoded subunits have tissue-specific isoforms. COX forms supercomplexes with other complexes, and its own assembly requires dozens of assembly factors.

When I began my work in 1976, I thought of cytochrome *c* oxidase as just a common, garden-variety enzyme that happens to be in the mitochondria. It does what all enzymes do: catalyzes a reaction. It possesses no obvious mystery, and I never dreamed that it holds the key to unraveling many neuronal properties not realized or anticipated previously. This book chronicles the journey on which this enzyme took me for four decades and the secrets that it revealed to me about our neurons, our prized possession. Before I begin, the biochemical properties of this unique protein are summarized in this first chapter.

Cytochrome *c* oxidase (cytochrome aa_3, ferrocytochrome c: oxygen oxidoreductase, a transmembrane electron transporter, EC 7.1.1.9 [previously EC 1.9.3.1], has been variously abbreviated in the literature as CcO, CCO, CcOX, CO, C.O., COase, COX, CYO, $Cytaa_3$, Cytox, and CytOx; to be consistent, the abbreviation COX is used throughout this book) is one of the most ancient enzymes known; parts of it evolved in purple non-sulfur bacteria (α-proteobacteria) at least 1.5×10^9 years ago, if not much earlier (Castresana et al. 1994). In prokaryotes (aerobic bacteria and archaea), it forms part of the cell membrane, but in eukaryotes (yeast, plant, and animal cells), it is an integral transmembrane protein of the inner mitochondrial membrane (IMM) (Wikström et al. 1981). The cytochrome system was discovered by Keilin only a hundred years ago (1925) as pigmented respiratory proteins in animals, bacteria, yeast, and higher plants. In 1938, Keilin and Hartree named the cytochrome aa_3 bands of oxidative protein cytochrome oxidase, which corresponds to Warburg's iron-containing respiratory enzyme, "atmungsferment" (1924). This is the terminal enzyme, or complex IV, of the electron transport chain, catalyzing the oxidation of its substrate, ferrocytochrome c, and the reduction of dioxygen to water. According to the Chemiosmotic Theory of Peter Mitchell (1961), electron transport is coupled to oxidative phosphorylation for the synthesis of adenosine triphosphate (ATP), the basic currency of energy. In this process, complex I (NADH: ubiquinone oxidoreductase or NADH dehydrogenase), complex III ($CoQH_2$: cytochrome *c* reductase or cytochrome bc1 complex), and complex IV (COX) actively pump protons from the interior of mitochondria, the matrix, across the IMM, to the intermembranous space (IMS). There, an electrochemical proton gradient or the mitochondrial membrane potential ($\Delta\Psi m$) is generated, and the inward flux of protons down this gradient through complex V (ATP synthase or F_oF_1 ATPase) drives the synthesis of ATP from adenosine diphosphate (ADP) and inorganic phosphate (Figure 1.1) (Kanehisa et al. 2025). As this mechanism is dependent on the gradient, complexes I, III, and IV can be considered as integral parts of the energy-generating machinery. COX is particularly important in this process, as it is the only enzyme that can reduce molecular oxygen to water, thereby completing the oxidative metabolic process in the mitochondria. Inactivation of COX by azide, cyanide, or carbon monoxide is incompatible with life.

DOI: 10.1201/9781003507765-1

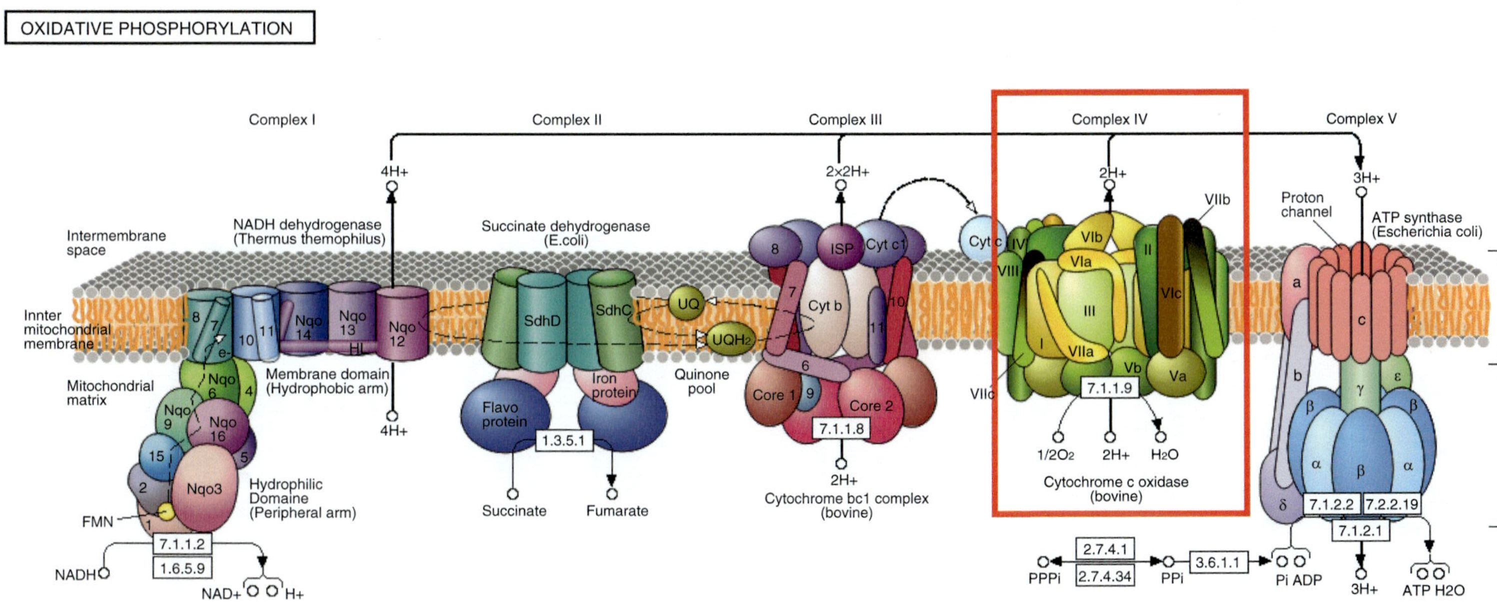

FIGURE 1.1 Electron transport chain (complexes I–IV) and ATP synthase (complex V) of the inner mitochondrial membrane. A rectangle is added to highlight complex IV.

From KEGG Oxidative phosphorylation – Reference pathway. https://www.genome.jp/pathway/map00190. Reproduced with permission from Dr. Kanehisa.

1.1 SUPERCOMPLEXES

Complexes I, III, and IV in yeast and mammalian cells also assemble in various combinations to form supercomplexes or respirasomes, with reported advantages of substrate channeling, enhancement of respiratory chain activity, and reduced reactive oxygen species (ROS) formation (Schägger and Pfeiffer 2000; Kadenbach 2017). In the bovine heart, ~80% of complex I and about two-thirds of complex III are in supercomplexes, but only ~15% of complex IV are in supercomplexes, while the rest are in free, dimeric form (Schägger and Pfeiffer 2001). Notably, the structural and functional integrity of complex I is dependent on assembled COX, as suppressed expressions of *COX4* or *COX1* subunits, which disrupt COX assembly, destabilize complex I, and reduce its activity (Li et al. 2007). Complexes I and III are sites that generate ROS that are both destructive (in damaging proteins, lipids, and DNA, including mitochondrial DNA (mtDNA)) and beneficial (as signaling molecules) (Handy and Loscalzo 2012). More recently, ROS are found to regulate COX assembly and biogenesis (Bourens et al. 2013).

Remarkably, during aging, the longest-lived rodent species, the naked mole-rats, have a significant increase in individual COX and supercomplexes containing complexes I and III or III and IV compared to C57BL/6 mice (Liang et al. 2025). Increased supercomplex assemblies would increase electron transfer efficiency and reduce ROS formation, thereby decreasing oxidative stress and possibly prolonging life. Disassembly and dysregulation of supercomplexes have been proposed as contributing to mitochondrial dysfunction in aging and neurodegenerative diseases (reviewed in Novack et al. 2020). However, the difficulty of interpreting postmortem tissues, especially in diseased brains, and the dearth of comprehensive studies make any conclusion currently premature. Intriguingly, mice with profound depletion of respirasomes apparently preserve oxidative phosphorylation capacity and physiology (Milenkovic et al. 2023).

1.2 BIGENOMIC NATURE OF CYTOCHROME *C* OXIDASE

Cytochrome *c* oxidase is one of only four proteins in mammalian cells that are uniquely bigenomic, that is, they are derived from two different genomes, with some subunits encoded by the nuclear genome, common to all other proteins, while others are specified by the mtDNA. The other three bigenomic proteins are mitochondrial complexes I, III, and V. Complex II (succinate: ubiquinone oxidoreductase or succinate dehydrogenase) is entirely nuclear encoded. The human mtDNA is compact, circular, and double-stranded, comprising only 16,569 base pairs (bp) and no introns (Anderson et al. 1981). It codes for 2 rRNAs, 22 tRNAs, and 13 mRNAs. The 13 polypeptides are as follows: 7 for complex I (ND1, ND2, ND3, ND4, ND4L, ND5, and ND6), 1 for complex III (cytochrome b), 3 for complex IV (COX I, COX II, and COX III), and 2 for complex V (ATPase 6 and 8). All these complexes also have nuclear-encoded subunits (38 for complex I, 10 for complex III, 10 (or 11, see below) for complex IV, and 9 for complex V) (Wallace 2005; Carroll et al. 2006; Mishmar et al. 2006). In recent years, microproteins that are typically less than 100 amino acids long have been documented to be encoded in the mtDNA within small open reading frames (smORFs), translated in the mitochondria, and serving metabolic and signaling functions (reviewed in Yen et al. 2025). Humanin is the first such mitochondrial smORF microprotein discovered with only 24 residues and exhibits antioxidant cytoprotective effects (Paharkova et al. 2015).

The mammalian complex IV (COX), specifically in the bovine heart, was first crystallized by Yoshikawa's group (Tsukihara et al. 1996). Being a very hydrophobic transmembrane protein, COX posed a distinct challenge, but the group succeeded in crystallizing it. They confirmed an earlier study in both heart and liver that used sodium dodecyl sulfate–polyacrylamide gel electrophoresis (SDS-PAGE) (Kadenbach et al. 1983), showing that the enzyme is composed of 13 subunits, with a molecular size of ~200 kDa. The largest three (I, II, and III) that form the catalytic core of the enzyme are, as stated above, encoded by the mitochondrial genome, whereas the other ten (IV, Va, b, VIa, b, c, VIIa, b, c, and VIII) are derived from the nuclear DNA (Kadenbach et al. 1983; Attardi and Schatz 1988), and their primary function is to modulate the activity of the enzyme (Kadenbach et al. 2000). To form the functional holoenzyme, all 13 subunits must be present in a 1:1 stoichiometry. The crystal structure shows that the

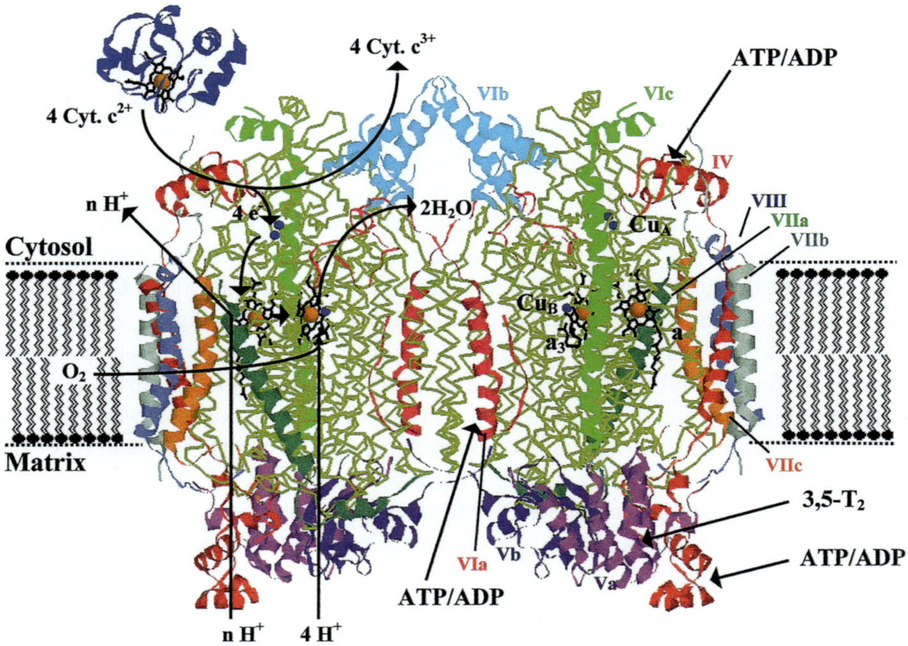

FIGURE 1.2 A crystal structure of dimeric COX from the bovine heart (Kadenbach et al. 2000; based on Tsukihara et al.'s 1996 data). The 3 mitochondrial subunits are in yellowish-green backbone, and the 10 nuclear ones are in different colors. The metallic centers include heme a and heme a_3 (black), iron atoms (orange circles), and Cu_A and Cu_B (blue circles). Chemical reactions are indicated on the left, and the binding sites for the regulatory compounds ATP/ADP and 3,5-diiodothyronine (3,5-T_2) are shown on the right.

enzyme exists as a dimer, with 13 subunits in each monomer (Tsukihara et al. 1996; Kadenbach et al. 2000) (Figure 1.2). Subunits I and II provide the redox center for intramolecular electron transfer via their redox-active heme and copper metal sites. Electrons flow down the electron transport chain to cytochrome c, which binds to subunit II of COX, and within COX, electrons move from the dimeric Cu_A (in II), to heme a (in I), and then to the bimetallic, binuclear heme a_3-Cu_B (in I), which binds and reduces O_2 to water (Winter et al. 1980; Ludwig 1980; Martin et al. 1988; Hall et al. 1988; Hill 1991; Tsukihara et al. 1995). In short, four electrons transferred from cytochrome *c* and four protons from the matrix join to reduce each dioxygen to two molecules of water. Much of the proton-pumping activity in COX is via pathways in subunit I (Shimokata et al. 2007). There is also a binding site for zinc in COX Vb on the matrix side of IMM and a binding site for magnesium near Cu_A at the interface between subunits I and II (Pan et al. 1991b; Tsukihara et al. 1995). An "adventitious copper" (Cu_X) is found to associate with COX III and, even though it is not involved in electron transport, may play a structural role in COX dimerization (Einarsdottir and Caughey 1985; Pan et al. 1991a). A stoichiometry of 5 Cu/4 Fe/2 Zn/2 Mg has been proposed for a dimeric COX (Einarsdottir and Caughey 1985). Subunit III, on the other hand, has neither redox nor known proton-pumping function, but it is involved in proton uptake (before channeling to the proton transfer pathway), in stabilizing the enzyme, and is critical for enzymatic activity (Sharma et al. 2015).

1.2.1 Subunit Isoforms

Among the ten nuclear-encoded COX subunits, at least six have isoforms that are tissue-specific, developmentally induced, or both (COX IV, VIa, VIb, VIIa, VIIb, and VIII; see Table 1.1). Subunit IV has a ubiquitous isoform (COX IV-1; gene *COX4i1*) and a lung-specific one (COX IV-2;

TABLE 1.1
Cytochrome *c* Oxidase Subunits

Protein	Gene	Tissue	References
COX I	*COX1*	Ubiquitous	Anderson et al. 1981; Kuhn-Nentwig & Kadenbach 1985
COX II	*COX2*	Ubiquitous	Anderson et al. 1981; Kuhn-Nentwig & Kadenbach 1985
COX III	*COX3*	Ubiquitous	Anderson et al. 1981; Kuhn-Nentwig & Kadenbach, 1985
COX IV-1	*COX4i1*	Ubiquitous	Kuhn-Nentwig & Kadenbach 1985; Zeviani et al. 1987
COX IV-2	*COX4i2*	Lung	Hüttemann et al. 2001
COX Va	*COX5a*	Ubiquitous	Kuhn-Nentwig & Kadenbach 1985; Rizzuto et al. 1988
COX Vb	*COX5b*	Ubiquitous	Kuhn-Nentwig & Kadenbach 1985; Zeviani et al. 1988a
COX VIa1/COX VIaL	*COX6a1*	Ubiquitous; Liver	Kuhn-Nentwig & Kadenbach 1985; Fabrizi et al. 1989
COX VIa2/COX VIaH	*COX6a2*	Heart & skeletal muscle	Kuhn-Nentwig & Kadenbach 1985; Fabrizi et al. 1992
COX VIb1	*COX6b1*	Somatic	Kuhn-Nentwig & Kadenbach 1985; Taanman et al. 1990
COX VIb2	*COX6b2*	Testis-specific	Hüttemann et al. 2003a
COX VIC	*COX6c*	Ubiquitous	Kuhn-Nentwig & Kadenbach 1985; Otsuka et al. 1988
COX VIIa1/COX VIIaH	*COX7a1*	Heart & skeletal muscle	Kuhn-Nentwig & Kadenbach 1985; Arnaudo et al. 1992
COX VIIa2/COX VIIaL	*COX7a2*	Ubiquitous; Liver	Kuhn-Nentwig & Kadenbach 1985; Hüttemann et al. 2000
COX VIIa3/COX VIIaR/ SIG81	*COX7a3/COX7aR/ SIG81*	Many tissues	Segade et al. 1996; Zhang et al. 2016
COX VIIb/COX VIIb-1	*COX7b* (X-linked)	Ubiquitous	Kuhn-Nentwig & Kadenbach 1985; Sadlock et al. 1993
COX VIIb-2	*COX7b2*	Testis	Kadenbach & Hüttemann 2015
COX VIIc	*COX7c*	Ubiquitous	Kuhn-Nentwig & Kadenbach 1985; Aqua et al. 1991
COX VIII-1/COX VIII-H/ COX VIIIb	*COX8-1/COX8b*	Heart & skeletal muscle (not present in humans)	Kuhn-Nentwig & Kadenbach 1985; Lomax et al. 1995
COX VIII-2/COX VIII-L	*COX8-2/COX8a*	Ubiquitous; Liver	Kuhn-Nentwig & Kadenbach 1985; Rizzuto et al. 1989
COX VIII-3/COX-VIIIc	*COX8-3/COX8c*	Testis, pancreas, placenta	Hüttemann et al. 2003b
NDUFA4 COXFA4	*NDUFA4*	(A 14th subunit of COX?)	Balsa et al. 2012 HUGO Gene Nomenclature Committee, 2025

Note: References for genes are limited mainly, but not exclusively, to humans. Those for proteins are limited to earliest ones. The gene names for humans will be capitalized for *A*, *B*, and *C* isoforms.

gene *COX4i2*), which is induced in the human lung after birth (Hüttemann et al. 2001). COX IV-2 is essential for acute but not chronic pulmonary oxygen sensing (Sommer et al. 2017) and reportedly is induced in hypoxic cells by hypoxia-inducible factor 1 (HIF-1) (Fukuda et al. 2007). Under hypoxia, *COX4i2* transcription can even be induced in cortical astrocytes and in cerebellar granule cells (Horvat et al. 2006). Three subunits (VIa, VIIa, and VIII) in most species have a ubiquitous (liver or L) isoform and a heart- and skeletal muscle-specific (H) isoform expressed after birth (Kuhn-Nentwig and Kadenbach 1985; Kadenbach et al. 1986; Grossman and Lomax 1997; Jaradat et al. 1998). Whereas many mammalian species have separate heart and liver isoforms for subunit VIII, humans and Old World monkeys have only the liver type for all tissues (Rizzuto et al. 1989; Goldberg et al. 2003). In the brain, the isoform is of the liver type. Subunit VIb has a somatic isoform (COX VIb1) and a testes-specific one (COX VIb2) (Hüttemann et al. 2003a). Subunit VIIa also has a homologous gene with high sequence identity named *SIG81* (silica-induced gene, clone 81), *COX7a2L* (*COX7a2*-like), or *COX7aR* (*COX7a2*-related) (Segade et al. 1996; Zhang et al. 2016; Sinkler et al. 2017). It is expressed in all tissues, and its association with supercomplexes containing complexes III and IV indicates that it should be renamed supercomplex assembly factor 1 or SCAF1 (Lapuente-Brun et al. 2013). COX VIIb has a ubiquitous isoform that is X-linked (COX VIIb-1) and another one expressed in testis (COX VIIb-2) (Kadenbach and Hüttemann 2015). The existence of a third isoform of COX VIII, VIII-3 (gene *COX8-3* or *COX8c*), has been proposed at the transcript level in several species, including humans, although its precise tissue expression and function remain unclear (Hüttemann et al. 2003b). These isoforms allow for further refinement of COX activity according to the energy demand of specific tissue types and during specific stages of development. Subunit VIb stabilizes the structure of the dimeric enzyme (Tsukihara et al. 1996), whereas subunit VIc provides a low-affinity binding site for a second molecule of cytochrome *c* (Cooper 1990). A role in COX assembly and CNS development has been inferred for VIIb in two clinical cases of *COX7B* mutation (Indrieri et al. 2012). The function of subunit VIIc is not yet known (Kadenbach et al. 2000; Kadenbach and Hüttemann 2015).

In recent years, a 14th subunit was proposed for COX, namely, NADH dehydrogenase (ubiquinone) 1 alpha subcomplex 4 (NDUFA4), which was originally identified as a subunit of complex I (Mishmar et al. 2006). Based on immunoprecipitation in growing cells as well as genetic and evolutionary studies, Balsa et al. (2012) claimed that NDUFA4 plays a role in COX biogenesis and function, and Pitceathly and Taanman (2018) renamed it COXFA4. However, Kadenbach (2017) did not see the presence of NDUFA4 in non-dividing cells and concluded that it is not the 14th subunit of COX but may serve as an assembly factor for COX or COX supercomplexes in cancer cells or growing cells. In either case, NDUFA4 is peripherally associated with COX and, as it binds to both complexes I and IV, is likely to be involved in the assembly of COX into supercomplexes with complex I (Kadenbach 2017). In a crystal structure of supercomplex $I_1III_2IV_1$, NDUFA4 is assigned to COX, where it lies at the interface of a previously reported crystal structure of COX homodimer, thus precluding COX dimerization (Zong et al. 2018). Because NDUFA4 appears to bind either to complex I or the monomeric COX, and its presence prohibits COX dimerization, Kadenbach and colleagues more recently considered it a loosely bound but not a permanent 14th subunit of COX (Kadenbach 2020; Ramzan et al. 2021). The hypothesis proposed by Kadenbach (2020) is that NDUFA4 toggles its binding between complex I and COX, depending on the ATP requirement and stress level of cells. Under relaxed conditions with low cytosolic calcium, a cAMP-dependent protein kinase A (PKA) phosphorylates both complex I and COX. NDUFA4 binds to phosphorylated complex I, and the dimeric, phosphorylated COX is under allosteric ATP inhibition. With stress-induced increase in cytosolic calcium (>1 μM), both complex I and COX are dephosphorylated by calcium-activated protein phosphatase. This causes the monomerization of COX, relief of its allosteric ATP inhibition, and its binding to NDUFA4, which is dissociated from the dephosphorylated complex I. Further testing is necessary to confirm or refute this hypothesis. NDUFA4 has recently been renamed COXFA4 (cytochrome *c* oxidase–associated subunit FA4) by the HUGO Gene Nomenclature Committee (2025). Variants of NDUFA4, such as NDUFA4L2 (now named COXFA4L2), have been reported to be induced by HIF-1, although they are associated with complex I rather than complex IV (Tello et al. 2011). More research is needed in this regard.

1.3 ROLE OF CYTOCHROME *C* OXIDASE IN REGULATING ENERGY METABOLISM

As the terminal enzyme of the electron transport chain, COX is the pacemaker of respiratory function, and its activity is the rate-limiting step (Villani et al. 1998; Dalmonte et al. 2009; Kadenbach et al. 2013). There is a dynamic interplay among ADP, ATP, and mitochondrial respiration. The binding of ADP stimulates mitochondrial respiration (active, state 3 respiration), and the subsequent conversion of ADP to ATP suppresses respiration (controlled, state 4 respiration) (Chance and Williams 1955). This decreases proton motive force (Δp, which is mostly ΔΨm plus a pH gradient) across IMM and reduces stimulation of proton pumping (Kadenbach et al. 2000; Sinkler et al. 2017). COX plays a significant role in this regulation because ATP binds to its subunits IV and VIa and allosterically inhibits the enzyme (Kadenbach et al. 1997; Napiwotzki and Kadenbach 1998; Acin-Perez et al. 2011). Such inhibition is triggered by a high intramitochondrial ATP/ADP ratio in heart and skeletal muscles (with COX VIaH isoform) but is always present in the liver and other organs (with COX VIaL isoform). The inhibition leads to decreased H^+/e^- stoichiometry and reduced proton pumping (known as "slip") but induces thermogenesis (Kadenbach et al. 2000). The binding of 3,5-diiodothyronine to subunit Va prevents the ATP-induced allosteric inhibition of COX (Arnold et al. 1998), but the binding of the regulatory subunit of PKA to subunit Vb in the bovine heart phosphorylates the subunit and increases the allosteric ATP inhibition of COX (Yang et al. 1998). A cAMP-dependent phosphorylation of Ser-441 of COX I turns on the allosteric ATP inhibition (Ludwig et al. 2001). Likewise, a cAMP-dependent phosphorylation of tyrosine 304 of the same subunit inhibits COX activity (Lee et al. 2005). On the other hand, the phosphorylation of serine 58 in the matrix loop of the ubiquitous COX IV-1 by cAMP-dependent PKA prevents the binding of ATP at that site and switches off the allosteric ATP inhibition (Acin-Perez et al. 2011). The effect of phosphorylation is reversible, however, as dephosphorylation of COX by Ca^{2+}-dependent protein phosphatases switches off the allosteric ATP inhibition (Kadenbach et al. 2000; Kadenbach 2020). Thus, mitochondrial respiration is controlled by ATP and reversible phosphorylation of COX. The efficiency of energy transduction is high when Δp is low, but the reverse is true when Δp is high (Bender and Kadenbach 2000). Under normal conditions, a delicate balance exists between inhibition and stimulation of respiration depending on ATP binding as well as phosphorylation/dephosphorylation of COX (Ramzan et al. 2021).

Another intriguing finding is that the removal of COX VIb1, which is located entirely on the IMS side of the enzyme, induces a twofold increase in COX activity (Weishaupt and Kadenbach 1992). This suggests that the normal function of this subunit may be to downregulate the activity of the enzyme by having the two VIb subunits from the two COX monomers to interact in the dimeric COX and enable allosteric ATP inhibition of cytochrome *c* oxidation (Kadenbach and Hüttemann 2015).

1.4 ASSEMBLY FACTORS, TRANSLOCASES, AND OTHER CONSIDERATIONS

The synthesis and assembly of the bigenomic COX involves more than 30 proteins, and much of the work is based on studies of the baker's yeast *Saccharomyces cerevisiae* and the bread mold *Neurospora crassa*, plus revelations from human diseases (Barrientos et al. 2002; Timón-Gómez et al. 2018; Watson and McStay 2020; Brischigliaro and Zeviani, 2021). Table 1.2 shows some of the key assembly factors in the biogenesis of COX. The listed references are only a representative few and by no means all-inclusive. At least four factors are involved in the delivery of heme (heme a and heme a_3) to COX I, and at least nine factors in the incorporation of copper to COX I (Cu_B) and COX II (Cu_A). The number of participating factors is escalating rapidly, and only a few key points are described below.

TABLE 1.2
Cytochrome *c* Oxidase Assembly Factors

Assembly Factors	Function	References
CMC1/Cmc1p	Twin CX_9C copper chaperone	Horn et al. 2008; Bourens and Barrientos 2017
COA1/MITRAC15/TIMM21	Chaperone in IMS presequence translocase	Chacinska et al. 2005; Mick et al. 2012; Shinde et al. 2021
COA2	Mediates multimerization of COX10 & COX15	Khalimonchuk et al. 2012
COA3/MITRAC12	COX I-specific assembly factor	Mick et al. 2010; Richter-Dennerlein et al. 2016
COA4	Twin CX_9C copper chaperone for Cu_B	Swaminathan et al. 2022
COA5/PET191/C2ORF64	Twin CX_9C COX assembly factor in early assembly	Khalimonchuk et al. 2008; Brischigliaro and Zeviani 2021
COA6	COX II-specific biogenesis factor	Pacheu-Grau et al. 2015
COA7/C1orf163/RESA1	Heme-binding protein in early COX assembly	Martinez Lyons et al. 2016; Formosa et al. 2022
COA8/APOPT1	Intermediate steps of COX assembly, stabilization	Signes et al. 2019
COX10	Heme a farnesyltransferase in COX I biogenesis	Glerum and Tzagoloff 1994; Antonicka et al. 2003
COX11	Metallochaperone for Cu_B insertion in COX I	Nývltová et al. 2022
COX14/MITRACearly/C12ORF62	COX I-specific assembly factor	Mick et al. 2010; Richter-Dennerlein et al. 2016
COX15	Heme a synthase; assist with COX I assembly	Antonicka et al. 2003; Bareth et al. 2013
COX16	Promotes Cu_A-site formation in COX II	Aich et al. 2018
COX17	Twin CX_9C copper chaperone for COX	Palumaa et al. 2004; Bourens et al. 2013
COX18	Translocate COX II C-tail	Gaisne and Bonnefoy 2006
COX19	Assist COX11 to insert copper (Cu_B) in COX I	Bode et al. 2015
COX20/FAM36A	Membrane-bound chaperone for COX II	Hell et al. 2000; Szklarczyk et al. 2013
COX23	Twin CX_9C copper chaperone for COX	Barros et al. 2004
MITRAC7/SMIM20	COX I-specific chaperone and late stability	Dennerlein et al. 2015
MR-1S	Interacts with PET100 & PET117	Vidoni et al. 2017
Mss51	Activates translation & assist in assembly of COX I	Decoster et al. 1990; Perez-Martinez et al. 2003
OXA1L	Insertase for respiratory chain subunits into IMM	Rötig et al. 1997; Koludarova and Battersby 2024
PET100	Late assembly factor	Church et al. 1996; Vidoni et al. 2017
PET117	Interacts with COX15 and PET100	McEwen et al. 1993; Taylor et al. 2017
SCO1	Copper Cu_A metallochaperone for COX II	Leary et al. 2004
SCO2	Copper Cu_A metallochaperone for COX II	Leary et al. 2004
SURF1/Shy1 in yeast	Mutation leads to Leigh; heme a-binding in bacteria	Tiranti et al. 1998; Zhu et al. 1998; Bundschuh et al. 2009
TACO1	Translation regulator of COX I	Weraarpachai et al. 2009
TIM8A/TIMM8A	Member of IMS chaperone network	Kang et al. 2019; Anderson et al. 2023
TIM23/TIMM23	Import nuclear COX precursor subunits	Bauer et al. 1996; Palmer et al. 2021

How is COX assembled in the IMM? Briefly, the largest three subunits (COX I, II, and III) are transcribed and translated in the mitochondria. They associate with mitochondrial ribosomes and require special translocases, such as the OXA1L (oxidase assembly) insertase and cohort proteins, to be co-translationally transported from the matrix and inserted into the IMM (Jia et al. 2003; Szyrach et al. 2003; Pfeffer et al. 2015). Heme a, heme a_3, and Cu_B are synthesized and incorporated into COX I, and Cu_A is added to COX II. Key players include special synthases and chaperone proteins, such as COA7, COX10, COX15, and possibly SURF1 (surfeit locus protein 1) for heme (Glerum and Tzagoloff 1994; Antonicka et al. 2003; Bundschuh et al. 2009; Formosa et al. 2022), and SCO1 (synthesis of cytochrome *c* oxidase 1), SCO2, CMC1 (CX_9C motif containing 1), COA4 (COX assembly factor 4), COX11, COX16, COX17, COX19, and COX23 for copper (Leary et al. 2004; Barros et al. 2004; Palumaa et al. 2004; Bode et al. 2015; Bourens and Barrientos 2017; Aich et al. 2018; Formosa et al. 2022; Nývltová et al. 2022; Swaminathan et al. 2022). The nuclear-derived COX subunits are transcribed in the nucleus, translated in the cytoplasm, and are imported into the mitochondria to be assembled with the mitochondrial subunits in the IMM (Lithgow 2000; Grevel et al. 2019). The import of these subunits encompasses many translocases of the outer and IMMs (TOMs and TIMs), as well as chaperones and partner proteins in the IMS and matrix (see Table 1.2; also reviewed recently in den Brave et al. 2024), driven by the $\Delta\Psi m$. A key translocase in the IMM is the TIM23 complex, which imports most of the nuclear-encoded subunits of the respiratory chain and is known to associate with supercomplexes containing COX (Grevel et al. 2019). A mitochondrial heat shock protein 70 (mtHsp70) is a chaperone in the matrix that associates with a translational activator Mss51 in the biogenesis of COX I and delivery of nuclear-encoded subunits to the assembly line (Decoster et al. 1990; Perez-Martinez et al. 2003; Fontanesi et al. 2011; Böttinger et al. 2013). The binding of assembly factors stabilizes unassembled subunits and prevents them from premature degradation. The assembled intermediates serve as a framework for the further assembly of other subunits in the formation of the final complex (Grevel et al. 2019).

The mitochondria-derived subunit I, whose specific chaperone is MITRAC7 (Dennerlein et al. 2015), is the first to be translocated from the matrix to the IMM (forming assembly intermediate S1). It is stabilized by the twin-CX_9C intermembrane protein CMC1, together with COX14 and MITRAC12 (mitochondrial translation regulation assembly intermediate of cytochrome *c* oxidase or COA3) (Bourens and Barrientos 2017; Clemente et al. 2013; Mick et al. 2012). Heme a biosynthesis and its insertion into COX I require the assembly factors COX10 and COX15 (Barros et al. 2001, 2002). TIM21 is thought to deliver subunit IV to MITRAC before the complete synthesis of COX I, thus promoting the assembly of the enzyme (Mick et al. 2012). The incorporation of subunits IV and Va leads to the assembly intermediate S2. COX IV is essential for assembly of this enzyme and for proper respiratory function (Li et al. 2006). The complex then acquires heme a_3 and Cu_B before incorporating COX II. The assembly protein SURF1 is involved in the early stages of COX assembly, as its absence arrests the process with only subunits I, IV, and Va present (Stiburek et al. 2005). In bacteria, Surf1 protein binds heme a (Bundschuh et al. 2009), but SURF1 probably also plays a role in the regulation of copper homeostasis in COX (Stiburek et al. 2009). The delivery of copper to the Cu_A site of COX II involves metallochaperones SCO1 and SCO2, which are copper-binding proteins that also help maintain copper homeostasis (Leary et al. 2004, 2007). The presentation of these metallochaperones to subunit II is thought to be mediated by an assembly protein, COX20, which helps the incorporation and stabilization of subunit II to the assembly line of COX (Bourens et al. 2014). COX III and most of the other subunits (COX Vb, VIc, VIIa or b, VIIc, and VIII) are then added (assembly intermediate S3), followed by the final addition of the remaining subunits to form the monomeric holocomplex (S4) (Nijtmans et al. 1998; Williams et al. 2004; Stiburek et al. 2005; Fernández-Vizarra et al. 2009).

More recently, however, the COX assembly process is thought to involve a modular, rather than a single-subunit linear, model (Vidoni et al. 2017; Timón-Gómez et al. 2018). In the modular model, which is based mainly on *Saccharomyces cerevisiae* (McStay et al. 2013), each of the three mitochondria-derived COX subunits forms independent modules with their respective chaperones and specific nuclear-encoded subunits. Some preassembled modules are also found in human cells (Bourens and Barrientos 2017; Vidoni et al. 2017). The COX I module includes stabilization and maturation factors such as CMC1, COA1, COA3, COX14, and SURF1. The COX II module contains factors such as COA6, COX18, COX20, SCO1, and SCO2, plus possibly COX Vb, VIc, VIIb, VIIc, and VIII. The COX III module is postulated to include COX VIa, VIb, VIIa, and perhaps NDUFA4. COX IV and COX Va, at least in yeast, most likely also form a module before joining COX I in the assembly (Vidoni et al. 2017; Timón-Gómez et al. 2018). Another candidate assembly factor, myofibrillary-related protein 1 short isoform (MR-1S), is found to interact with PET100 and PET117 chaperones to assist in the assembly of human COX prior to the insertion of the COX III module (Vidoni et al. 2017). Moreover, the hypoxia-inducible gene domain proteins 1A (HIGD1A) and 2A (HIGD2A) not only respond to hypoxia but also participate in COX assembly as well as in the assembly of supercomplexes involving complexes I, III, and IV (Timón-Gómez et al. 2020). Notably, the assembly factor COA8 (APOPT1) is stimulated by ROS to import into the mitochondria and assist with the intermediate steps of COX assembly and stabilization (Signes et al. 2019).

Mutations of various assembly factors cause a few human syndromes and diseases that will be discussed in Chapter 17.

1.5 FACTORS AFFECTING COX ACTIVITY

COX activity is influenced by a variety of factors. Besides tissue-specific isoforms, supercomplexes, allosteric regulation, ATP, ATP/ADP ratio, and phosphorylation/dephosphorylation discussed above, it is also affected by acetylation, substrate concentration (ferrocytochrome *c* and oxygen), pH, ionic strength, membrane potential of IMM ($\Delta\psi m$), hormones, and metal ions (Dalmonte et al. 2009; Kadenbach 2017; Sinkler et al. 2017; Vogt et al. 2021; Pham et al. 2024). More recently, MNRR1 (mitochondrial-nuclear retrograde regulator 1; previously CHCHD2) has been found to be a stress regulator that functions in both the mitochondria (where it binds to COX and stimulates respiration) and the nucleus (where it acts as a transactivator of *COX4i2* and other hypoxia-stimulated genes) during hypoxic stress (reviewed in Grossman et al., 2017). A major focus of this book is how neuronal activity affects COX expression in neurons under normal and experimental conditions. The reverse state, in which defects in COX (including mutations) impact neuronal function, will be discussed in Chapter 17.

2 A Sensitive Indicator of Metabolic Plasticity in Neurons

ABSTRACT

Neurons are highly dependent on oxidative metabolism for their energy supply. As such, cytochrome c oxidase (COX) proves to be a logical, sensitive, and reliable indicator of neurons' oxidative capacity that fuels neuronal activity. The major energy-demanding function in neurons (and in many cell types) is active, ATP-dependent ion transport against both concentration and electrical gradients. A metabolic map of the rat brain using COX histochemistry reveals metabolic heterogeneity among neurons at the regional, laminar, and cellular levels. Significantly, COX uncovers metabolic plasticity in both developing and adult neurons within multiple systems of the mammalian CNS.

The pioneering work of Franz Nissl (1860–1919), Camillo Golgi (1843–1926), and Santiago Ramon y Cajal (1852–1934) has laid a firm foundation for the structural organization of neurons. Less clear were the dynamic properties of neurons until the advent of physiological investigation. Many of these techniques, however, are indirect, costly, and of macroscopic resolution. Is there any way to reveal neuronal activity not only at a systemic, but cellular and subcellular levels? This chapter will describe such an approach.

The brain is known to comprise only 2% of the body's weight, yet it consumes 20% of the total energy and about 25% of the total glucose utilized by the body (Clark and Sokoloff 1999). As neurons depend almost entirely on oxidative metabolism for its energy supply, much of the energy consumed is via the oxidative pathway.

2.1 WAYS OF PROBING BRAIN ACTIVITY

How does one study the activity of the brain? Earlier methods include transcranial electroencephalography (EEG) and visual evoked potentials (VEP), followed by intracranial electrophysiological recordings of neuronal activities. Roy and Sherrington (1890) were among the first to point out the positive relationship between functional brain activity and local cerebral blood flow. This revelation heralded in decades of research using various techniques to assess regional brain activation by monitoring local changes in blood flow and the resultant glucose consumption. Positron emission tomography (PET) was introduced first in the 1950s and refined through subsequent decades (reviewed in Portnow et al. 2013). It utilizes radiotracers, the most popular of which is the radioisotope 2-[^{18}F] fluoro-2-deoxy-D-glucose (FDG), which is delivered intravenously, and its concentration in activated regions of the brain is imaged with a scanner. In 1975, Sokoloff and his colleagues introduced the [^{14}C] deoxyglucose (^{14}C-2DG) method, which measures the rates of glucose consumption in various brain regions correlated with local changes in functional activities (Kennedy et al. 1977; Sokoloff et al. 1977). This method relies on postmortem autoradiographic analysis of signals that must be captured within a brief period (<1 h) after the administration of the radiolabel, as plasma 2DG has a relatively short half-life (Sokoloff 1981). The advent of functional magnetic resonance imaging (fMRI) in the early 1990s opened a whole new and exciting field of imaging activities in the living brain (Kwong et al. 1992; Ogawa et al. 1992; Bandettini et al. 1993). This approach capitalizes on the fact that regional cerebral blood flow and oxygen consumption increase

DOI: 10.1201/9781003507765-2

with neural activation and gives a blood oxygenation-level dependent (BOLD) signal. Another method is diffusion tensor imaging (DTI) (Assaf and Pasternak 2008), which detects changes in microscopic structure in the white matter and reflects the integrity of neural pathways. It is often combined with fMRI. Other methods, such as magnetoencephalography (MEG), which measures magnetic fields produced by electrical brain activities (Cohen 1972), are less frequently used. In recent years, functional near-infrared spectroscopy (fNIRS) has gained much attention, as it is portable, and brain activity inferred by changes in oxygenated (HbO_2) and deoxygenated (HbR) hemoglobin can be monitored even in a freely moving individual (Pinti et al. 2020).

The above methods, except for the ^{14}C-2DG, have the distinct advantage of monitoring activity in the *living* brain. The signal appears in real time and is dynamic. However, all these imaging techniques are based on indirect assessment of neural activity; that is, via blood flow, glucose uptake, or oxygen consumption. Neural activity is inferred, and the resolution is at a macroscopic, regional level.

A simpler, more direct way of revealing neuronal activity would be helpful. Cytochrome *c* oxidase (COX) is an ideal candidate, as it is intrinsic to neurons and directly reflects activity-dependent energy demand within neurons at the regional, laminar, cellular, and even subcellular levels. If captured appropriately, the postmortem signal can reveal a treasure trove of information about neuronal properties, function, and plasticity, as will be detailed in this book.

2.2 CYTOCHROME *c* OXIDASE REVEALS CELLULAR SITES OF HIGHEST ENERGY DEMAND

As an essential component of mitochondria, COX exists in almost every eukaryotic cell, including virtually all cells in the human body except for mature red blood cells that have neither nuclei nor mitochondria. Mitochondria are enriched in highly oxidative organs, such as the heart, kidneys, skeletal muscles, and especially the brain. Oxidative metabolism in the mitochondria has the distinct advantage over anaerobic glycolysis in the cytosol in that 1 mole of glucose generates 32 (or more) moles of ATP via the oxidative pathway, but only 2 moles of ATP via the glycolytic one (Stryer 1975). Whereas the uptake of FDG or 2DG can be for both the glycolytic and the oxidative pathways, COX is strictly a marker of oxidative metabolism. Neurons have limited, if any, capacity for storing glycogen and fat; thus, they rely heavily on ready uptake of glucose to be metabolized mainly via the oxidative pathway. Unlike 2DG, PET, and the other imaging techniques that monitor acute uptake of glucose or changes in blood flow, COX is an integral protein of the inner mitochondrial membrane and needs time for its synthesis, assembly, degradation, and renewal. Its level at any given time, revealed by COX histo-, cyto-, or immunohisto-chemistry, represents a more chronic state (over hours or longer) of cellular metabolic activity, and thus a more stabilized state of neuronal activity. Glial cells differ from neurons in that they rely more on the glycolytic pathway for their energy supply, and thus they can survive under anoxic conditions when neurons cannot (Bolaños et al. 2010).

In an unpublished survey by the Wong-Riley lab of various non-neural organs with cytochrome oxidase histochemistry, cells whose major function is active ion transport were consistently found to have high COX activity (Kageyama and Wong-Riley 1987 Abstract). These include endothelial cells of the cornea, epithelial cells of the choroid plexus, parietal cells of the gastric mucosa, epithelial cells of the small and large intestines, and striated duct cells of the parotid and submandibular glands. In the kidney, the portions with the highest COX level are the thick ascending limbs of the loop of Henle, followed by the distal and then the proximal convoluted tubules. On the other hand, cells that specialize in protein synthesis and secretion, such as the chief cells of the gastric epithelium, Paneth cells of the small intestine, and goblet cells of the gastrointestinal tract, show relatively low COX levels. Thus, the degree of COX activity is directly proportional to *active* (ATP-dependent) ion transport, whereas protein synthesis, contrary to common presumption, consumes very little energy (Bachelard, 1975; Ames et al. 1992). Figure 2.1 illustrates the difference in mitochondrial

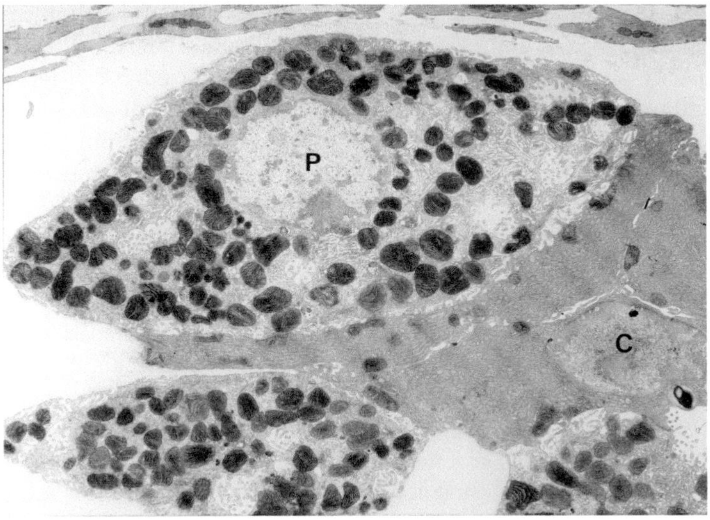

FIGURE 2.1 Gastric chief cells and parietal cells. Within the gastric mucosa, the chief cells (C) are filled with rough endoplasmic reticulum but have very few, small mitochondria. The adjacent parietal cell (P) has numerous large mitochondria, darkly reactive for COX histochemistry.

Kageyama and Wong-Riley, 1987, unpublished observations.

density, hence the energy capacity, between two adjacent cell types in the gastric mucosa. Whereas the chief cells have a rich supply of rough endoplasmic reticulum for the synthesis of pepsinogen, which they also store and secrete, they have very few, small mitochondria with low COX levels. The adjacent parietal cells, however, must actively pump protons against their concentration and electrical gradients by H^+/K^+-ATPase into the gut lumen to form HCl, so they are amply supplied with large mitochondria rich in COX. Clearly, ATP-dependent ion transport is a major energy-demanding function of biological cells.

What is the major determining factor for energy demand in neurons? As will be discussed in Chapter 3, it is not the specific neuronal class, the specific neurotransmitter type, or the size of neurons that is the determining factor, but rather, it is the *specific portions* of any neuronal type that *actively pump ions* against concentration and electrical gradients that matter.

2.3 CYTOCHROME *c* OXIDASE AS A METABOLIC MARKER OF NEURONAL ACTIVITY

The popular Nissl stain shows only the size, location, and groupings of neurons. The classical Golgi stain demonstrates the shapes of neurons, but it is capricious and stains only 1–5% of neurons in any given sample. The myelin stain marks myelinated axons, and neurochemical immunohistochemistry illustrates only their locations. None of these markers reflects the functional demands of neurons. In this regard, COX has much to offer in revealing the metabolic diversity, hence the functional heterogeneity of neurons, and how plastic they are when functionally perturbed.

2.3.1 Advantages of COX as a Marker

Several factors contribute to the choice of COX as a metabolic marker for neurons. First, COX is an oxidative enzyme that assists in generating energy, and neurons depend almost entirely on oxidative metabolism for their energy supply. Second, COX is the terminal enzyme of the mitochondrial electron transport chain, without which oxidative metabolism cannot be carried to completion.

Defects in any complex upstream in the chain can be bypassed, and metabolism can still be completed, but inhibition of COX is almost always fatal to neurons. Thus, COX activity is reflective of the oxidative capacity and metabolic activity of neurons (and of other cell types). Third, COX is quite resilient, and much of its activity withstands a limited range of pH (pH 5.4–8.4 have been tested, with catalytic constant increasing as pH decreases; Thörnström et al. 1988; for COX histochemistry, the Wong-Riley lab uses a pH between 7.4 and 7.45), temperature (20°C down to −80°C, if freezing is done immediately postmortem and the tissue block is ≤ 1 cm^3), and even limited aldehyde fixation (≤4% paraformaldehyde or ≤1.5% glutaraldehyde for <3h at 4°C). This last trait confers a distinct advantage, as neural tissues are extremely fragile; without some fixation, they can easily disintegrate, and the reaction product can be displaced or lost. With limited fixation, COX retains sufficient enzyme activity to oxidize exogenously supplied substrate, cytochrome *c* (which itself is reduced by an electron donor, diaminobenzidine), to accumulate an adequate amount of reaction product on the outer surface of the inner mitochondrial membrane, visible at both the light and EM levels (Wong-Riley 1979a, 1989). Extreme heat, pH, dryness, and prolonged fixation will inactivate the enzyme, and no reaction product can form. Hence, COX histochemistry (and cytochemistry) is an active "reaction" and not a passive "stain," and the term "stain," which has been widely used (including the author's earlier publications), should be replaced by the term "reaction."

2.3.2 Metabolic Map of the Normal Rat Brain Revealed by COX

Whereas the Nissl stain shows the cellular architecture (cytoarchitecture) of the central nervous system and the myelin stain the myeloarchitecture, COX histochemistry reveals the metabolic capacity (or metabo-architecture) of neurons in the brain.

A metabolic map of the normal rat brain was published by Hevner et al. in 1995. The metabolic activity, which reflects functional neuronal activity, is qualitatively demonstrated by COX histochemistry and semi-quantitatively verified by optical densitometry, and the actual enzyme activity is measured by COX biochemistry of tissue homogenates. Linear regression analysis indicates that the optical densitometric values of COX histochemical reactions highly correlate with both the COX activity content (by wet weight) ($r^2 = 0.929$) and the COX-specific activity ($r^2 = 0.946$) monitored biochemically in gray matter and white matter samples (Figure 2.2). Thus, COX reactivity derived histochemically reflects relative COX activity, and these two terms will be used interchangeably in this book. COX activity monitored histochemically also matches its protein amount, both of which represent the metabolic capacity of cells and both are regulated by neuronal activity (Hevner and Wong-Riley 1990).

The metabolic map reveals a very heterogeneous pattern of COX activity among regions, cortical areas, laminae, nuclear groups, and neuronal types (Figure 2.3). The gray matter has 8- to 12-fold higher levels of enzyme activity than does the white matter, consistent with brain activity variation shown by 2DG, PET, MRI, and fMRI. The visual, auditory, vestibular, somatosensory, and somatomotor systems all have relatively high COX activity, whereas associational, limbic, autonomic, and visceral systems, with a few notable exceptions, have much lower levels. In primary sensory cortices, the metabolically most active layer is the thalamic-recipient layer IV. Greater detail of the primate primary visual cortex will be covered in Chapter 4.

Among the thalamic nuclei, the anterodorsal nucleus that receives input from the mammillary bodies has very intense COX activity compared with the adjacent anteroventral and anteromedial nuclei. Equally darkly reactive are the mammillary bodies amidst most of the other hypothalamic nuclei with relatively low COX levels. This indicates that the mammillothalamic pathway, which mediates temporal and contextual memory (Dillingham et al. 2021), is a highly active one. Clinically, the mammillary bodies are severely shrunken in Wernicke–Korsakoff syndrome due to alcohol abuse (Sheedy et al. 1999). Other intensely reactive thalamic nuclei include the visual relay dorsal lateral geniculate nucleus (LGN), the auditory relay medial geniculate nucleus, and the somatosensory relay barreloids (specific to rodents and lagomorphs) in the ventral posterior medial nucleus.

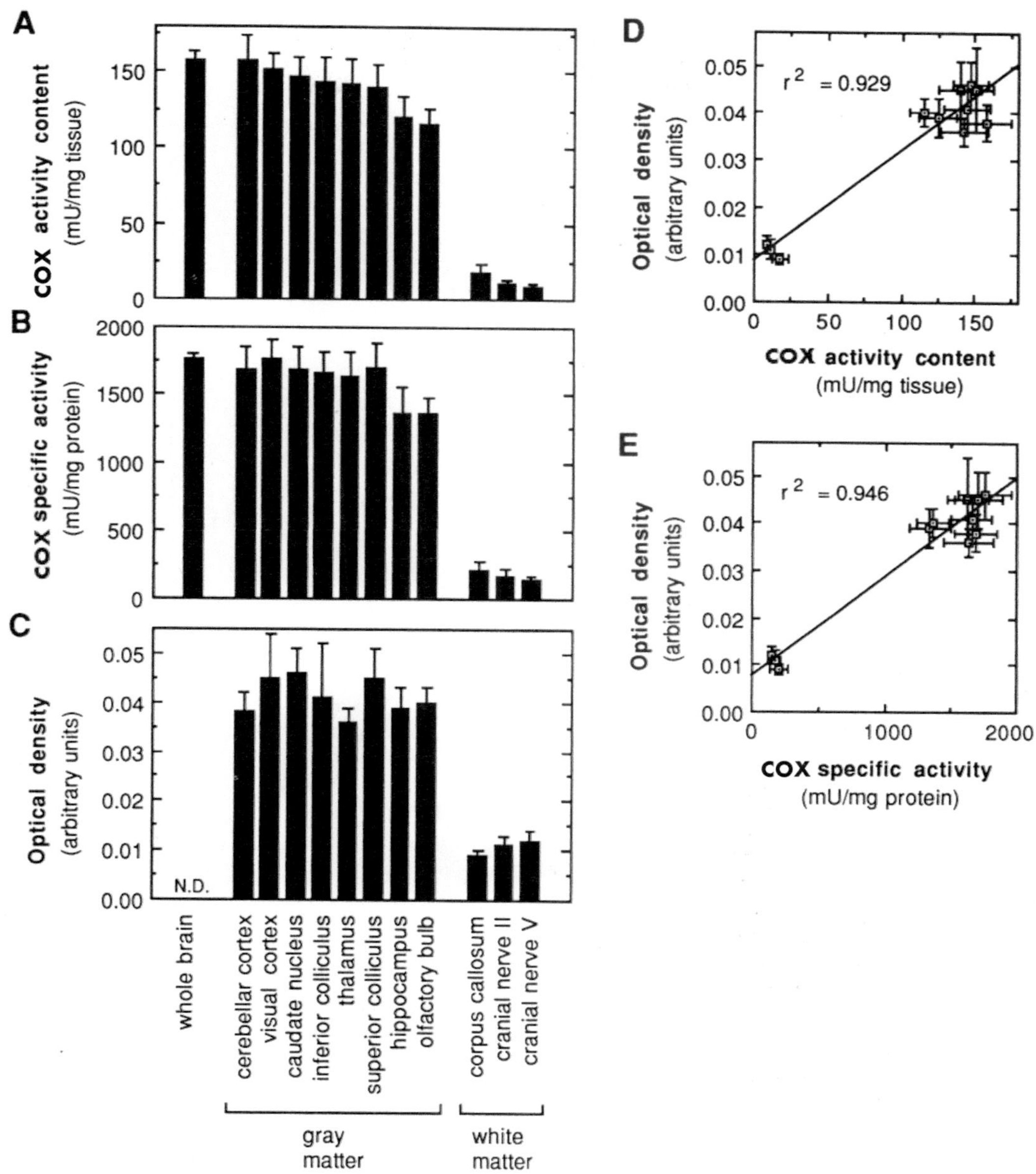

FIGURE 2.2 COX activity content (A) and specific activity (B) measured biochemically in the whole rat brain, eight gray matter areas, and three white matter areas. Optical densitometric values of comparable areas reacted for COX histochemistry are shown in (C). Histochemical values are highly correlated with COX activity content (D) and specific activity (E). All data are mean ± standard error of the mean (SEM). N.D.: not determined.

From Hevner et al. 1995.

Within the basal ganglia, important for motor control and habit formation (Graybiel 2000), the caudate–putamen (striatum) and the nucleus accumbens exhibit much higher COX levels in the neuropil than in neuronal cell bodies, and both have patchy distributions of dark- and moderately reactive COX. In the striatum, most of the COX labeling is in dendritic mitochondria in the neuropil (Difiglia et al. 1987), and the areas of strong COX activity coincide with areas of intense tyrosine hydroxylase and calbindin D-28k-like immunoreactivity, which are markers of striatal matrix

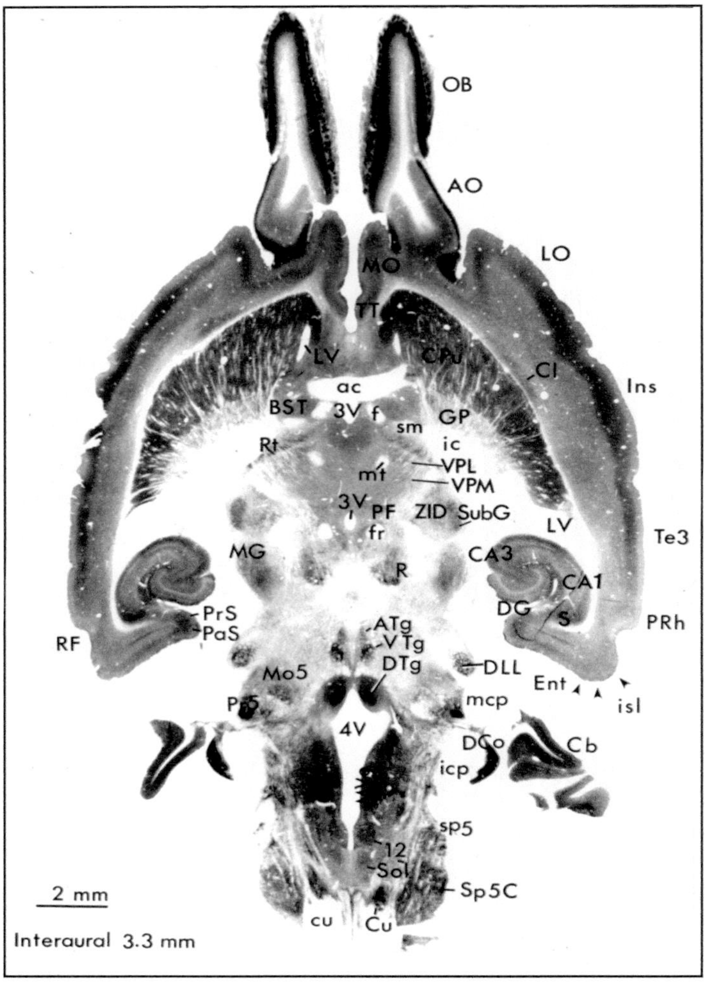

FIGURE 2.3 Horizontal section of the rat brain reacted for COX histochemistry. Note that the intensity of COX labeling varies among different regions and different nuclear groups.

Details and glossary are in Hevner et al. 1995.

compartments, whereas the intervening striosomes exhibit low levels for all three markers (Augood and Emson 1992). Striatal matrix compartments reportedly contain the direct and indirect pathways that are parts of the sensorimotor and associative circuits, whereas the striosomes receive input from the limbic cortex and project to pars compacta neurons of the substantia nigra (Crittenden and Graybiel 2011). In naturally behaving rats, peak metabolic activity measured by ^{14}C-2DG uptake is confined mainly to the matrix compartment of the striatum (Brown et al. 2002), consistent with higher COX activity there. In the midbrain, substantia nigra pars compacta, which projects to the striatum, has moderate levels of COX in the neuropil and low levels in the dopaminergic cell bodies. On the other hand, the major output neurons of the basal ganglia – the GABAergic neurons of substantia nigra pars reticulata and the internal segment of globus pallidus (Feger et al. 1976; Hikosaka 2007) – are darkly reactive for COX, but their neuropil has low COX levels.

The cerebellar cortex will be described in greater detail in Chapter 3 (Section 3.4). Of special note is that the cerebellar vermis contains parasagittal bands of high COX labeling interdigitating with low ones to form a zebra-like pattern. This pattern is complementary to that of the neuropeptide zebrin I (Leclerc et al. 1990). The major efferent neurons of the cerebellum, the deep cerebellar nuclei, are intensely reactive for COX, indicating high functional activity. Two brain stem nuclei – the red nucleus and the inferior olivary nucleus – that directly connect with the cerebellum also have darkly COX-reactive cell bodies.

The hippocampal formation presents a heterogeneous pattern that will be discussed in Chapter 3 (Section 3.3). The superior colliculus in the midbrain has a unique, heterogeneous pattern. Whereas the superficial gray layer (stratum griseum superficiale) that receives excitatory input primarily from the retina (May 2006) is darkly reactive for COX, the stratum opticum, which is composed mainly of fibers from the optic tract, has low levels of COX. The intermediate and deep layers receive information from visual, auditory, and somatosensory inputs important for coordinating eye and head movements toward the target of interest (Hoy and Farrow 2025), but only the intermediate gray layer has patches of dark COX labeling against a background of relatively low enzyme level. This lattice pattern is also present in the intermediate layer of human, monkey, and cat superior colliculus (Wallace 1988). The auditory relay inferior colliculus also has a heterogeneous pattern, with the highest COX level in the caudal nucleus and more moderate levels elsewhere. Within the brain stem, most of the cranial nerve nuclei (including the oculomotor nucleus, trochlear nucleus, mesencephalic trigeminal nucleus, abducens nucleus, facial nucleus, cochlear nucleus, vestibular nucleus, and hypoglossal nucleus), the interpeduncular nucleus, and the ventral tegmental nucleus are darkly reactive for COX. On the other hand, most of the subnuclei of the nucleus solitarius that mediate visceral sensation, and both the Edinger–Westphal nucleus and the dorsal motor vagal nucleus, which are preganglionic parasympathetic, have relatively low levels of COX.

Many of the above findings have been corroborated by publications from other laboratories too numerous to cite here.

2.3.2.1 Modular Organization Revealed by COX Histochemistry

COX histochemistry reveals a modular pattern in several cortical areas. In the somatosensory system of rodents, five major rows of "barrels" have been described with the Nissl stain in layer IV of the primary somatosensory cortex (S1), known as the "posteromedial barrel subfield," that represent five major rows of vibrissae on the contralateral face, with a strictly one-to-one topographical relationship (Woolsey and Van der Loos 1970). The mystacial pads serve an important function of whisking the environment, especially in the dark. The pattern revealed by COX histochemistry, however, is complementary to that of Nissl. Instead of the cell-rich barrel sides demarcated by Nissl, it is the cell-sparse barrel hollows that are highlighted by COX (Wong-Riley and Welt 1980) (Figure 2.4).

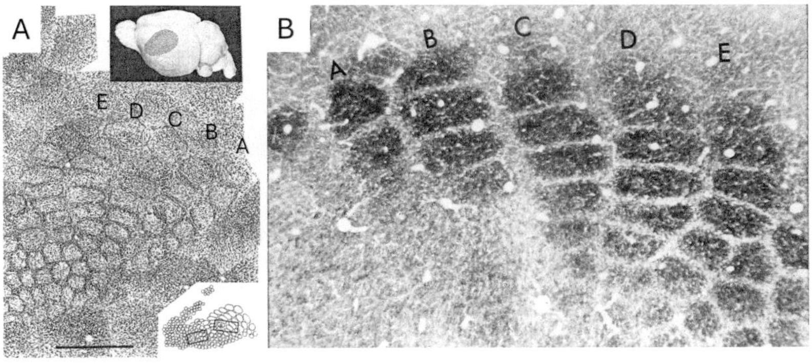

FIGURE 2.4 Mouse barrel cortex. (A) Posteromedial barrel subfield with five rows of barrels (A–E) stained with Nissl and diagramed in the lower inset. The upper inset shows the location of the barrel cortex on the mouse brain (from Woolsey and Van der Loos 1970; reproduced with permission from Elsevier). (B) A section through the mouse barrel cortex reacted for COX histochemistry. Note that it is the barrel hollow rather than the barrel side that is metabolically active.

From Wong-Riley and Welt 1980.

These hollows contain dendrites that receive excitatory input primarily from barreloids in the thalamus (Agmon et al. 1995). With EM, mitochondria darkly reactive for COX are detected mainly in dendrites within the barrel hollows (Wong-Riley and Welt 1980), highlighting metabolically active sites postsynaptic to excitatory inputs. COX-labeled barrel modules exist not only in the rat and mouse but also in the gerbil, degu, hamster, and rabbit (Lauer et al. 2018).

The lateral entorhinal cortex involved in spatial memory consolidation has been noted in monkeys to contain "morphological inhomogeneities" and surface irregularities or verrucae (Amaral et al. 1987). Neuronal clusters in layers II–III are darkly reactive for COX in humans, monkeys, and rats (Hevner and Wong-Riley 1992). In tangential sections of human entorhinal cortex, these COX-rich islands in layers II (sometimes extending into layer I) form a distinct modular pattern that matches remarkably in size, position, and shape to the naturally "dark spots" visible with the naked eye on the surface of fresh, normal parahippocampal gyrus (Hevner and Wong-Riley 1992) (Figure 2.5). The islands have distinct borders, irregular in contours, contain darkly COX-reactive neurons, and are surrounded by interstices with much lighter COX labeling. The patterns are identical between COX histochemistry and immunohistochemistry, although the latter shows a greater intensity in neuronal cell bodies. A similar pattern is also seen with glutamic acid decarboxylase (GAD), the synthesizing enzyme of the inhibitory neurotransmitter gamma-aminobutyric acid (GABA) (Solodkin and Van Hoesen 1996). Cell clusters in layer II of the entorhinal cortex are known to send projections via the

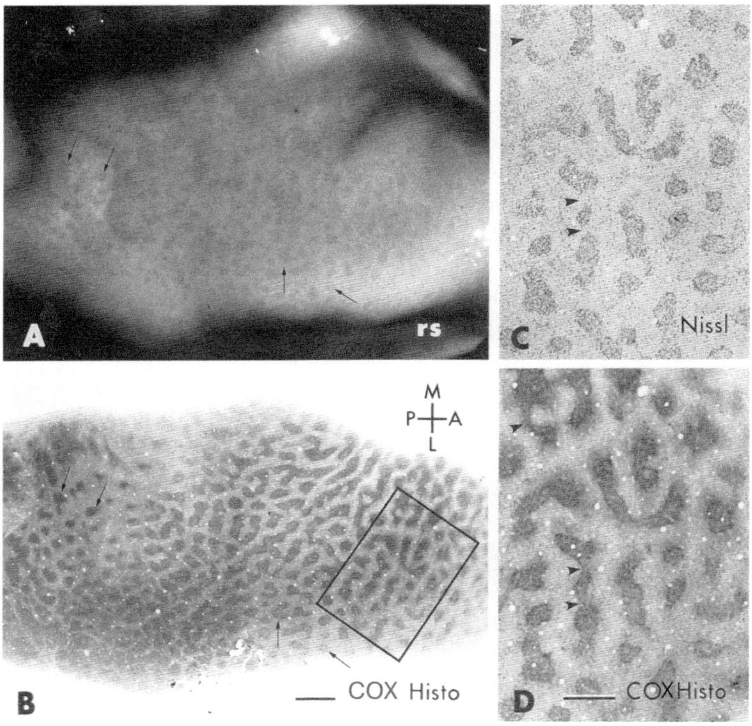

FIGURE 2.5 Entorhinal cortex in the human parahippocampal gyrus. (A) Naturally, dark spots on the unstained brain surface visible to the naked eye. (B) Tangential section through layers I–II of the same cortex reacted for COX shows the same dark spots (arrows). Boxed area in (B) is enlarged in (D). Adjacent section (C) stained with Nissl shows cell clusters corresponding to COX-rich islands. Arrowheads in (C) and (D) denote bridges of dark COX reactivity connecting neuron clusters.

Rs, rhinal sulcus. Scale bar = 1 mm.

From Hevner and Wong-Riley 1992.

perforant pathway to the ipsilateral dentate gyrus and CA2–3 of the hippocampus (Steward and Scoville 1976; Witter and Amaral 1991). Their intense COX labeling indicates that the glutamatergic perforant pathway is highly active. The entorhinal–hippocampal circuit that connects grid cells in the former to place cells in the latter forms the basis for declarative memory formation (Moser et al. 2015). Significantly, the entorhinal cortex and the adjacent perirhinal cortex are the first areas severely affected in Alzheimer's disease (Van Hoesen et al. 2000).

The modular pattern in the primate primary and secondary visual cortical areas will be described in Chapter 4.

2.3.3 Metabolic Plasticity of Neurons Revealed by COX

COX proves to be a reliable indicator of the heterogeneous pattern of energy demands among normal neurons. The logical question is, if neuronal functional activity is vitally dependent on oxidative metabolism, will changes in neuronal activity cause comparable changes in COX even in the adult? That is, is there metabolic plasticity in both the developing and mature nervous systems? What can COX reveal that the widely used Nissl stain cannot? Initial testing in the auditory, visual, and somatosensory systems verifies that COX is a sensitive indicator of metabolic plasticity. These experiments are briefly described below.

2.3.3.1 Plasticity in the Auditory System of Adult Cats

The mammalian auditory system is one of the most active in the CNS. This system was examined in a group of adult cats rendered functionally silenced in one ear with the destruction of hair cells by a silastic cochlear implant in which embedded electrodes failed to function for 18–49 weeks (Wong-Riley et al. 1978) (Figure 2.6A–F). The auditory nerve on the implanted side, however, remained structurally intact. COX histochemistry indicated that those brain stem nuclei normally receiving direct or indirect *excitatory* input from the now deafened ear exhibited reduced COX activity compared to the intact counterparts. The affected nuclei included the ipsilateral anteroventral and posteroventral cochlear nuclei, the ipsilateral ventrolateral periolivary nuclei, the ipsilateral lateral superior olive, the contralateral medial nucleus of the trapezoid body, the contralateral ventral and intermediate nuclei of the lateral lemniscus, and at least the caudal (monaural) portion of the contralateral central nucleus of the inferior colliculus. On the other hand, the medial superior olive and the dorsal nuclei of the lateral lemniscus, both of which normally receive excitatory input from both ears, showed very little change, indicating that input from the normal ear sustained functional and metabolic activity in those nuclei bilaterally. A notable finding is that the contralateral lateral superior olive deprived of its inhibitory input indirectly from the deafened ear remained darkly reactive for COX.

This study revealed that a) the removal of excitatory input one to several synapses away reduces the energy demand, thereby curtailing COX activity; b) the removal of inhibitory input does not significantly impact neurons' energy demand nor their COX activity, as repolarization of membrane potential after inhibitory hyperpolarization is mainly passive; c) adult neurons exhibit a metabolic plasticity simply from functional inactivation without denervation; and d) COX is a sensitive indicator of changes in metabolic activity, which reflect changes in neuronal activity one to several synapses away from the site of perturbation.

Can inactivation be reversible? That is, can neurons that have been functionally inactivated be stimulated back to normal, and will COX be able to demonstrate such a reversal? Another group of adult cats unilaterally deafened for 5–6 months, followed by 1 month of continuous stimulation (biphasic pulses at 20 pps) via electrodes in their cochlear implants, was examined (Wong-Riley et al. 1981). Evoked potentials in the brain stem showed that nearly all cochlear nerve fibers were excited by electrical stimulation. All brain stem auditory relay neurons exhibited COX activity at or close to normal levels (Figure 2.6G–H). Thus, COX is also a sensitive indicator of functional recovery in adult neurons.

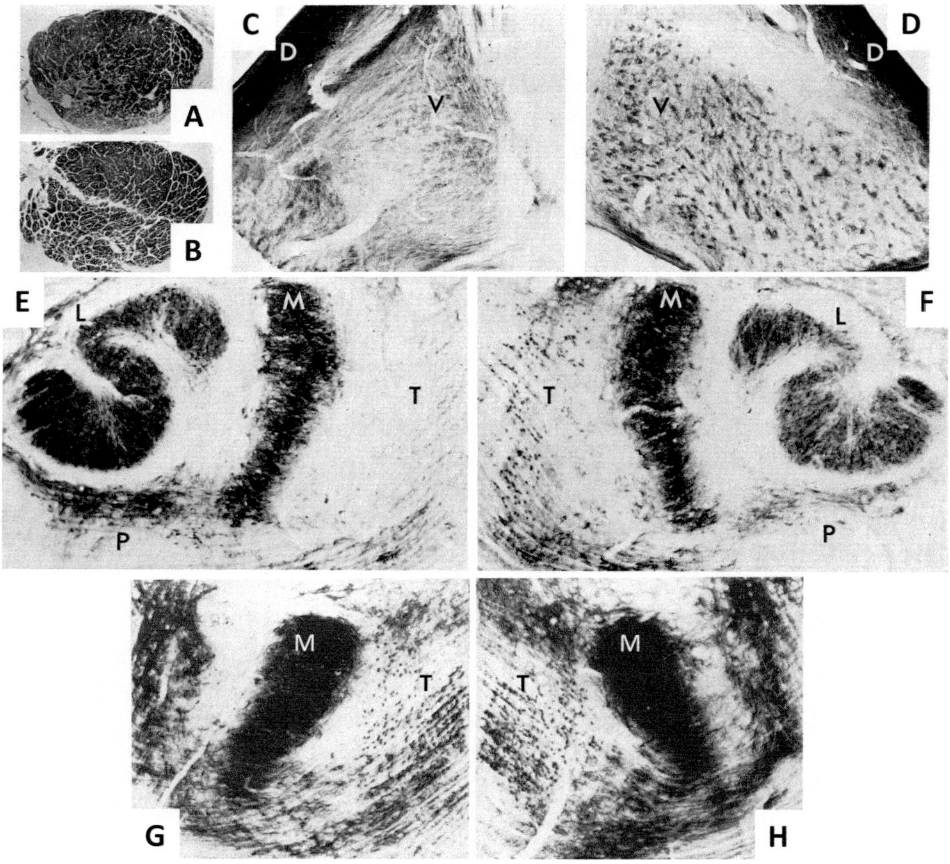

FIGURE 2.6 Metabolic plasticity in the adult cat auditory system with monaural cochlear implant. Cochlear nerve from the implant side (A) is structurally comparable to the normal side (B). The ventral cochlear nucleus (V) on the ipsilateral side (C) has much reduced COX activity compared to the contralateral side (D), whereas the dorsal cochlear nuclei (region D) have comparable activity on both sides (panels C and D). The contralateral medial nucleus of the trapezoid body (T) has much lower COX activity (E) than that of the ipsilateral side (F), but the activity in the ipsilateral lateral superior olive (L) and the ventrolateral periolivary nucleus (P) is much depressed compared to the contralateral side. However, the medial nucleus of the superior olive (M) shows little difference between the two sides. (G and H) Monaural cochlear implant followed by stimulation of the cochlear nerve results in comparable COX activity in relay nuclei bilaterally (compare M and T in panels G and H), indicating full recovery.

A–F: modified from Wong-Riley et al. 1978; G–H: modified from Wong-Riley et al. 1981.

2.3.3.2 Plasticity in the Visual System of Kittens and Adult Cats

The visual system is also conducive to testing. Monocular lid suture from the 1st postnatal week to the 5th month causes a distinct reduction in COX activity in deprived laminae of the binocular portion of the LGN, with the ipsilateral lamina A1 being more affected than the contralateral lamina A (Wong-Riley 1979a) (Figure 2.7A and B).

Monocular lid suture during the 7th postnatal week (i.e., after the critical period; see Chapter 16) for 2.5 weeks does not result in any significant change in COX activity. Monocular enucleation on postnatal day (P) 5, followed by a survival of 54 or 145 days, results in a dramatic decrease in COX levels in deprived LGN laminae that include both the binocular and monocular segments. Notably, COX activity remains high in the border zone of the denervated lamina that faces the "normal" lamina, indicating that neurons in the border zone receive innervation via either translaminar dendrites and/or sprouting axons from the normal lamina. These findings agree with Guillery's

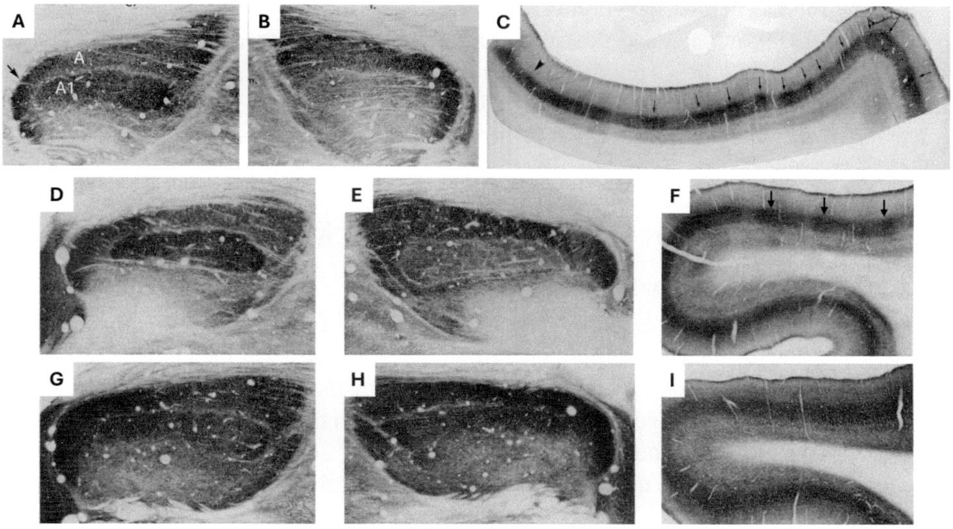

FIGURE 2.7 Metabolic plasticity in the visual system of kittens and adult cats. (A–C) The visual system of a kitten with monocular lid suture from the 1st week to the 5th postnatal month shows decreased COX activity in the contralateral layer A (in panel A) and the ipsilateral layer A1 (in panel B) of its LGN, as well as periodic dark (arrows) and light reactive bands in its visual cortex (C). The contralateral monocular segment of its LGN (arrow in A) retains a high level of enzyme activity. (D–F) The visual system of an adult cat that survived 5.5 weeks of monocular impulse blockade with TTX. The contralateral layer A (D) and the ipsilateral layer A1 (E) exhibit reduced COX activity, and alternating dark (arrows) and light columns are visible in layer IV of its visual cortex (F). (G–I) The visual system of a cat that received 4 weeks of monocular TTX and 6 weeks of recovery has regained its normal level of COX activity in its LGN (G and H) and its visual cortex (I).

A–C: modified from Wong-Riley 1979a. D–I: modified from Wong-Riley and Riley 1983.

original report using the Nissl stain (1971) that border neurons of denervated laminae retain their large size, presumably sustained by innervation from the neighboring intact laminae.

Within the normal visual cortex of cats, COX reveals a pattern not apparent with the cellular Nissl or the myelin stain. Layer IV, which receives input from the LGN, is continuously darkly reactive for COX until the label stops abruptly at the area 17/18 (V1/V2) border (Wong-Riley 1979a. The supragranular layer III and the infragranular layer VI are only moderately reactive. The reaction product is mainly within the neuropil and only within some neuronal cell bodies in layers III, IV, V, and VI. Monocular lid suture for the first 5 postnatal months causes a pattern of alternating dark and light COX-reactive columns in layer IV, often extending into layer III (Wong-Riley 1979a). The dark columns presumably represent non-deprived and the light columns deprived ones (Figure 2.7C). The darker columns are slightly wider than the lighter ones, suggesting that the dominant, open eye outcompetes the weaker, closed one in innervating a larger portion of non-deprived columns. The monocular segment contralateral to the deprived eye exhibits reduced COX activity, but that of the ipsilateral side retains a continuous dark band in layer IV.

The decrease in COX activity in deprived LGN laminae and cortical columns is consistent with the diminished functional activity recorded in deprived LGN cells (Wiesel and Hubel 1963a) and in cortical neurons generally unresponsive to stimulation of the deprived eye (Wiesel and Hubel 1963b). These changes are pronounced when deprivation is induced during the critical period of the cat's visual cortical development (between the start of the 4th week to around the end of the 3rd month), whereas monocular lid suture in adults does not yield detectable physiological effects (Hubel and Wiesel 1970). The topic of the critical period will be discussed in greater detail in Chapter 16.

The adult visual system was tested with unilateral intraocular injections of a voltage-dependent sodium channel blocker, tetrodotoxin (TTX), at a dosage (19 μg in 10 μl of sterile physiological saline or water) tested by Dr. Michael Stryker (UCSF) to block action potentials without blocking axoplasmic transport in retinal ganglion cells. After 1 week of impulse blockade (the earliest time tested), deprived LGN laminae and deprived columns in the visual cortex exhibited reduced COX activity. Changes were progressively more prominent with survivals up to 6 weeks (the longest time examined) (Figure 2.7D–F). When 4 weeks of TTX blockage were followed by 6 weeks of recovery, COX activity in formerly deprived LGN laminae and cortical columns returned to control levels (Figure 2.7G–I). Thus, adult neurons are still responsive to functional deprivation, as well as functional recovery, and, again, COX is a sensitive indicator of such adult plasticity (Wong-Riley and Riley 1983).

Visual deprivation in monkeys will be discussed in greater detail in Chapter 5.

2.3.3.3 Plasticity in the Somatosensory System of Mice

What about the somatosensory system? Vibrissal damage during the 1st postnatal week adversely affects the development of the corresponding cortical barrels, such that they either shrink in size or fail to form, and neighboring barrels expand in size to occupy the vacated space (Woolsey and Wann 1976). However, vibrissal damage or removal on or after P7 in mice results in no detectable change using Nissl staining (Woolsey and Wann 1976).

Shrunken barrels caused by neonatal cauterization of a selected row or rows of contralateral vibrissae also exhibit reduced COX histochemical activity (Figure 2.8) (Wong-Riley and Welt 1980).

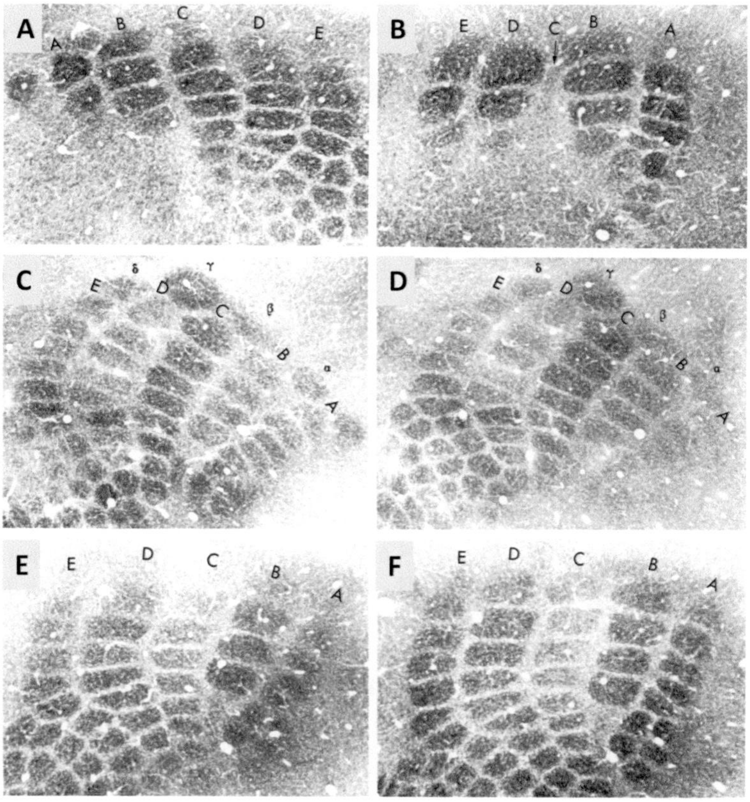

FIGURE 2.8 Metabolic plasticity in the murine somatosensory cortex. (A) Barrel cortex in the normal adult mouse reacted for COX (same image as Figure 2.4B). (B) Row C barrels fail to develop after contralateral row C vibrissae were cauterized at birth, and the animal survived 87 days. Rows B and D barrels enlarge in size in compensation. (C and D) Two consecutive sections showing barrels in rows A, B, D, E, and straddlers α, β, and δ with normal size but reduced COX labeling after comparable rows of contralateral vibrissae were cauterized in the adult, and the animal survived for 62 days. (E and F) Two consecutive sections of barrel cortex exhibiting decreased COX labeling in row C after the comparable row of contralateral vibrissae was repeatedly plucked in the adult for 80 days.

Modified from Wong-Riley and Welt 1980.

Surprisingly, vibrissal removal in the adult likewise induces a significant downregulation of COX in the corresponding sensory-deprived barrels, even though barrel size remains unchanged, and the Nissl stain reveals no abnormality (Wong-Riley and Welt 1980). COX downregulation most likely occurs in postsynaptic dendrites deprived of normal thalamic input. These results indicate that the adult cortex, once thought to be refractory to the effect of peripheral damage (Woolsey and Wann 1976), responds with a significant downregulation of metabolic activity. Even more striking is the fact that the removal of vibrissae by plucking alone and not cauterization in the adult is sufficient to induce changes in cortical neurons at least three synapses away. Adult neurons, therefore, remain highly responsive to sensory deprivation.

These initial sets of experiments performed in the auditory, visual, and somatosensory systems solidify the conclusion that COX can serve as a sensitive and reliable indicator of neuronal metabolic activity in both the developing and adult nervous systems. Increased neuronal activity triggers two coupled processes: (1) energy consumption by Na^+/K^+-ATPase and other ATPases to repolarize the membrane for reactivation, and (2) energy generation by increasing blood flow and glucose utilization to promote cellular respiration for the synthesis of ATP to fuel the ATPases (Figure 2.9). The accumulation of ADP from the hydrolysis of ATP is used to resynthesize ATP and is itself a controlling factor of cellular respiration. As an essential component of the electron transport chain, COX is integral in the energy-generating process. Energy is not produced when energy is not needed, and work dictates the energy consumed (Lowry 1975). The coupling of neuronal activity with energy metabolism renders COX a logical and useful metabolic indicator of neuronal activity, be it a decrease, steady-state, or an increase in energy demand (Wong-Riley 1989).

Metabolic plasticity is likely to be accompanied by microvascular plasticity. Adult rats exposed to chronic hypobaric oxygen (0.5 atm) for up to 3 weeks exhibit overall reduced COX activity

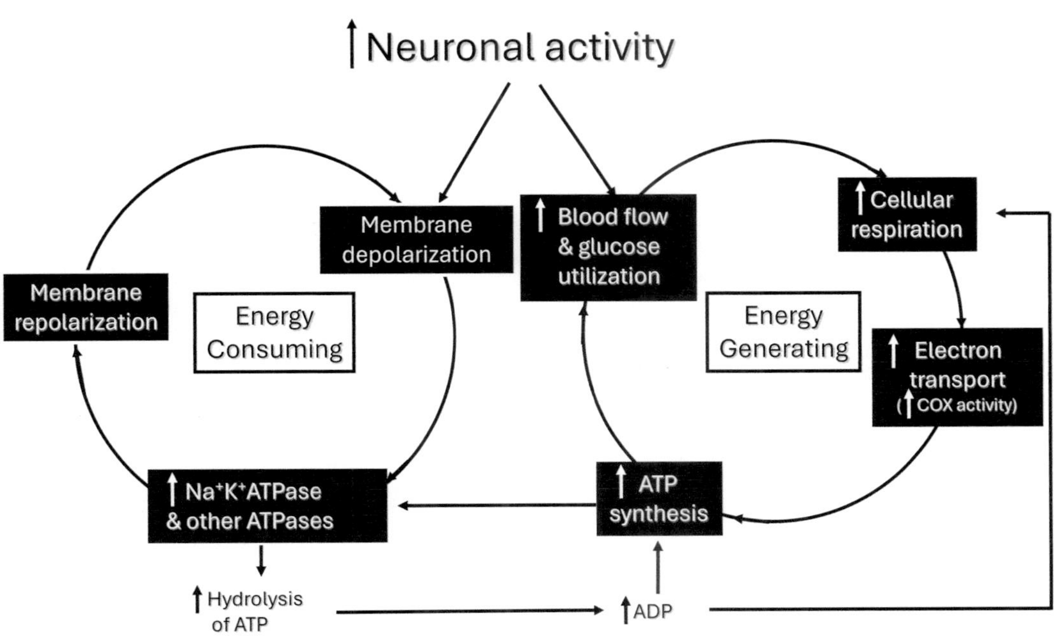

FIGURE 2.9 Schematic diagram of the coupling between neuronal activity and energy metabolism. Increased neuronal activity triggers both energy consumption and energy generation. Energy is consumed by ATPases to repolarize the membrane after depolarization, and energy is generated mainly via the mitochondrial electron transport chain, of which COX is an integral part. The hydrolysis of ATP yields ADP, which is used to re-synthesize ATP and is itself a controlling factor for cellular respiration.

throughout their brains (with some regions affected more than others) (LaManna et al. 1996). Such hypometabolism is accompanied by increased microvessel density by 46–76% in various cortical areas (LaManna et al. 1996).

The original set of experiments laid a firm foundation for numerous studies by many laboratories using COX histochemistry on diverse systems of multiple species in the ensuing decades. They are simply too numerous to cite here.

3 Metabolic Diversity among Cell Types and Segments of the Same Neuron

The Importance of Dendrites

ABSTRACT

Within the same brain region, various neuronal types exhibit distinct metabolic signatures. The metabolic demand of neuronal cell bodies is not determined by their size or by the type of neurotransmitters they synthesize. Rather, it is dictated by the type of somatic synapses they receive. Those that receive exclusively inhibitory synapses generally have low levels of cytochrome c oxidase (COX), whereas those that also receive excitatory ones have abundant mitochondria rich in COX. The entire neuron is also not metabolically homogeneous. Dendrites of almost all neurons have high levels of COX, whereas most axon trunks have low levels, and axon terminals vary in their energy demand depending on the level of tonic activity.

In Chapter 2, levels of cytochrome *c* oxidase (COX) are noted to differ among systems, regions, laminar, and cell types. What is the major factor that contributes to energy demand in neuronal cell bodies? Is the somatic size positively correlated with its energy demand? Is the entire neuron metabolically homogeneous, or does COX activity differ among segments of the same cell? At first glance, a single neuron is minuscule enough that it is easily considered a single metabolic entity. But is it? Electrophysiology records neuronal activity primarily from cell bodies and bundles of axons. Are these the regions where metabolic demands are the highest?

To answer these questions, electron microscopy affords resolution of individual neuronal segments in greater detail, including visualization of a single mitochondrion and the intensity of COX reaction product along the outer surface of IMM. Mitochondria are broadly categorized into darkly (D; when reaction product covers > 50% of IMM), moderately (M; < 50% of IMM), and lightly (L; little or no detectable reaction product) reactive types (Wong-Riley et al. 1989a; Wong-Riley 1989). This degree of resolution renders COX a powerful tool for investigating neuronal metabolic activity beyond the systems, regional, and laminar levels to the cellular and subcellular levels.

3.1 QUANTITATIVE EM ANALYSIS OF MITOCHONDRIAL DISTRIBUTION IN THE MONKEY VISUAL CORTEX

Significant insight was gained from an extensive, quantitative EM analysis of neuronal COX in the adult monkey primary visual cortex (Wong-Riley et al. 1989a, 1989b) with subsequent neurotransmitter analysis (Nie and Wong-Riley 1996a, 1996b). Within globular zones of high COX activity in the supragranular layers (known as puffs, patches, or blobs; see Chapter 4) are three metabolic cell types. Type A neurons are small, glutamatergic, and comprise about 57% of the neuronal population. They have a thin rim of cytoplasm containing small mitochondria of primarily the L type, and receive only symmetric, GABAergic synapses on their somata (Figure 3.1A). Type B cells are medium- and large-glutamatergic pyramidal cells that comprise about 18% of the neuronal population. Their cytoplasm contains a wealth of organelles, including evenly distributed mitochondria of the three metabolic types. Type B cells also receive only symmetric, GABAergic synapses on their

DOI: 10.1201/9781003507765-3

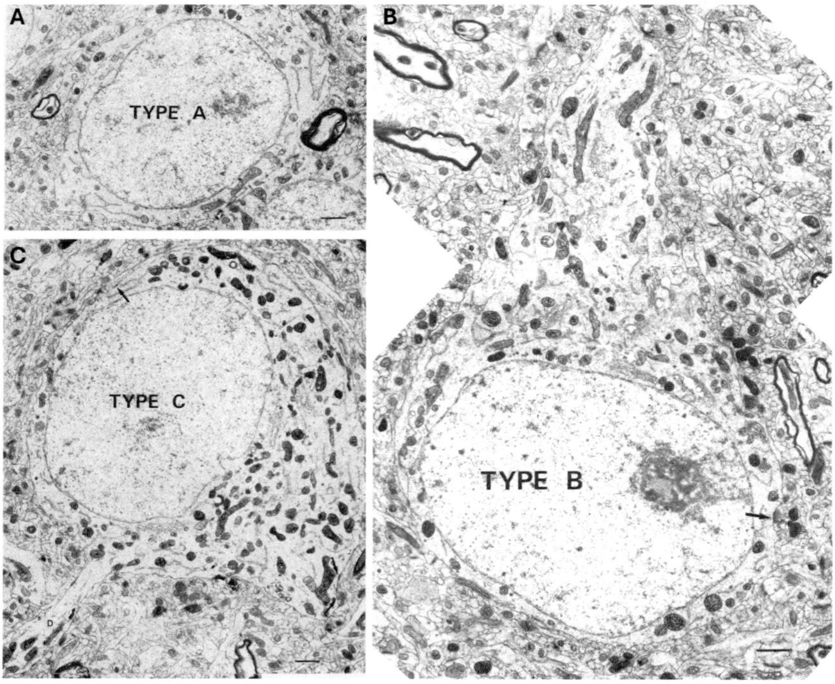

FIGURE 3.1 Three major metabolic neuronal types in supragranular puffs of monkey visual cortex. The small type A cells (A) have lightly COX-reactive mitochondria. The large type B cells (B) are pyramidals with a large apical dendrite and contain a mixture of mostly moderately reactive and some darkly and lightly reactive mitochondria. This cell type, like those of type A cells, receives only symmetric (presumed inhibitory) axo-somatic synapses (arrow). The medium-sized type C cells (C) contain a rich supply of darkly reactive mitochondria and receive both asymmetric (presumed excitatory; arrow) and symmetric axo-somatic synapses.

Scale bar = 1 μm.

Modified from Wong-Riley et al. 1989a.

cell bodies (Figure 3.1B). Type C cells are medium-sized GABAergic neurons comprising about 25% of the neuronal population. This is the only cell type that receives both symmetric GABAergic and asymmetric glutamatergic axo-somatic synapses, and they are metabolically most active with a rich supply of D and M mitochondria (Figure 3.1C). Thus, cell bodies that receive direct excitatory, depolarizing, glutamatergic synapses are functionally more active and have a greater energy demand in repolarizing their membranes than those that receive only inhibitory, GABAergic ones.

In the neuropil of puffs, nearly all dendrites encountered contain D and M mitochondria. This implies that dendrites of all three neuronal types, irrespective of the somatic content of mitochondria, have a strong demand for energy and are amply supplied with COX-rich mitochondria. Considering that dendrites are the major receptive sites for excitatory spiking and graded depolarizing input (in addition to inhibitory ones) for all neurons, their membranes must be repolarized by an ATP-dependent ion-pumping mechanism after each depolarization. Thus, the collective energy utilization of dendritic trees is substantial. A significant 62% of numerical and areal densities of mitochondria in the neuropil reside in dendrites (Figure 3.2A, B, and D) (Wong-Riley et al. 1989b), making dendrites *the major* energy consumers of the brain (Wong-Riley 1989).

Axon terminals in puffs are of two main types: a) glutamatergic, which make asymmetric synapses (termed AA) and are small and contain no or few L-type mitochondria; and b) GABAergic, which form symmetric synapses (termed AS), are larger and typically contain several D-type mitochondria in a single plane of section. AA synapses outnumber the AS ones by almost threefold.

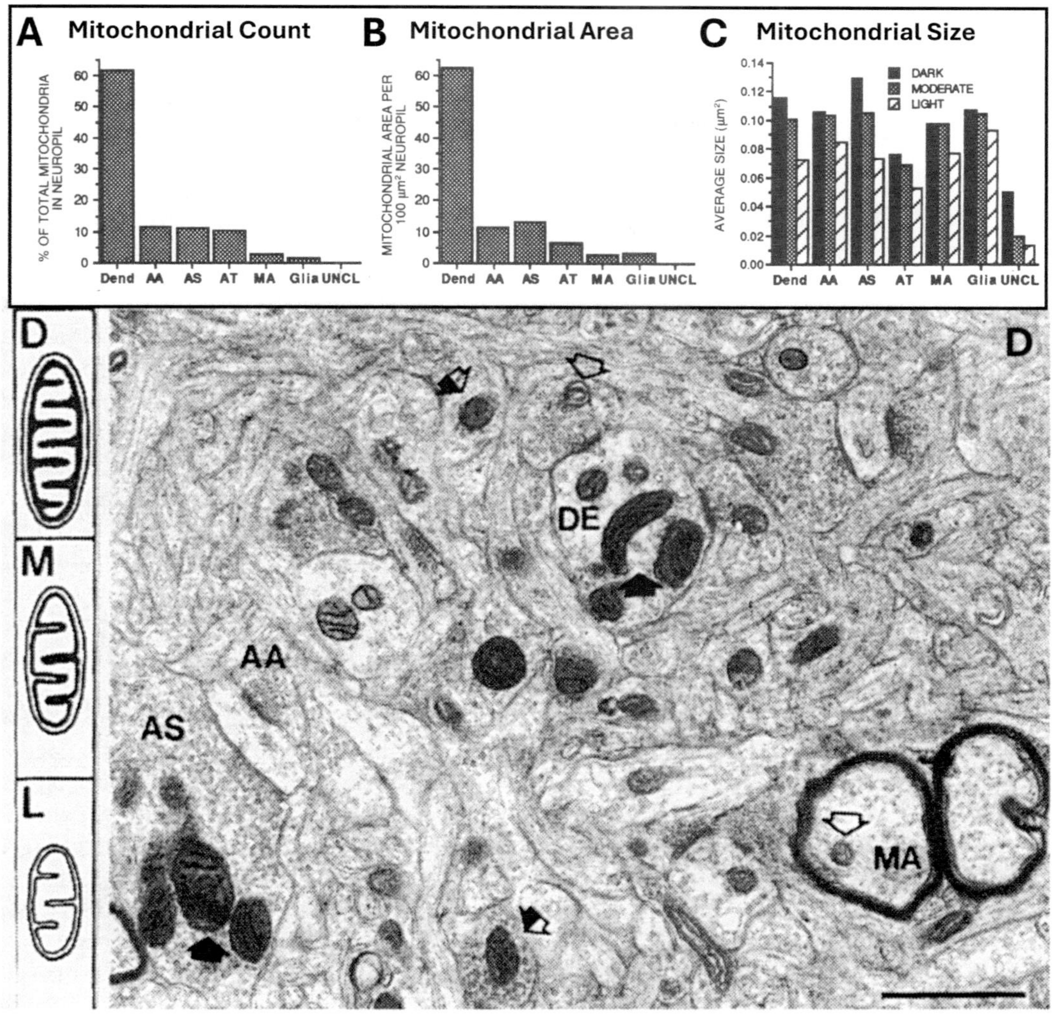

FIGURE 3.2 Mitochondrial distributions in the monkey visual cortex. Percentage distribution of mitochondria by count (A) and area (B) in various profiles in the neuropil. Den, dendrites; AA, axon terminals making asymmetric synapses; AS, axon terminals making symmetric synapses; AT, axon trunks; MA, myelinated axons; UNCL, unclassified. (C) A positive relationship exists between the mean size of mitochondria and their COX reactivity in each type of neuropil profile. (D) An electron micrograph of the neuropil in the monkey visual cortex showing darkly, moderately, and lightly COX-reactive mitochondria (schematically drawn as D, M, and L, respectively, on the left, and as solid, half-solid, and open arrowheads, respectively, in the micrograph) in various neuronal profiles. DE, dendrite.

Modified from Wong-Riley 1989.

About 11% of mitochondria in the neuropil are found in AA terminals, another 11% in AS terminals, and an additional 11% in unmyelinated axon trunks. Only about 3% of the mitochondria in the neuropil reside in myelinated axons, and the remaining 2% are in glia (Figure 3.2A, B, and D) (Wong-Riley et al. 1989b).

A positive relationship exists between the mean size of mitochondria and the level of COX activity, that is, D > M > L. Notably, the mean size of D mitochondria in axon trunks is about the same as the L type in dendrites, indicating that each segment of a neuron regulates the size, packing density, and level of COX activity of its resident mitochondria based on energy utilization (Figure 3.2C and D).

Glial cells, in general, have low levels of COX. Oligodendrocyte cell bodies have very few mitochondria, almost all of which are lightly COX reactive. Astrocyte cell bodies have mainly lightly to moderately COX-reactive mitochondria (Wong-Riley et al. 1989b). In the neuropil, mitochondria in glial processes constitute only 2.14% of the total mitochondrial count and 3.63% of the total mitochondrial area (Figure 3.2A and B) (Wong-Riley et al. 1989b).

3.2 HOW DO CELLULAR LAMINAE AND NEURONAL TYPES IN THE LATERAL GENICULATE NUCLEUS DIFFER IN THEIR COX LEVEL?

The dorsal lateral geniculate nucleus is the major relay station between the retina and the ipsilateral visual cortex, and both its structure and function have been extensively studied in several species, including cats and monkeys (Kaas et al. 1972; Sherman and Spear 1982). Each LGN represents the contralateral hemi-visual field, with input from both eyes forming the binocular segments and from the contralateral eye far periphery forming the monocular segment. Macaque LGN has six laminae, numbered 1–6 ventrodorsally, with laminae 1, 4, and 6 receiving contralateral input, and 2, 3, and 5 ipsilateral ones. Laminae 1 and 2 are magnocellular with large neurons, whereas laminae 3–6 are parvocellular with smaller cells. COX activity is much higher in the magno- than the parvo-cellular layers (except for lamina 6) (Kageyama and Wong-Riley 1984), indicating that the M-pathway is functionally more active than the P-pathway. The "ON" LGN laminae 6 and 5 (responding to lights ON in its receptive field center) are more COX reactive than the "OFF" laminae 3 and 4 in this diurnal animal. The contralateral eye is more dominant, as laminae 1 and 6 (and, less so, 4) are more COX reactive than the ipsilateral laminae 2, 5, and 3.

As discussed in Chapter 2 (Section 2.3.3.2), the LGN in the cat is subdivided into two major laminae (A and A1) dorsally and thinner C bands ventrally, plus the medially located MIN (medial interlaminar nucleus) (Guillery 1969; Sanderson 1974). In the ferret, each of the A laminae is subdivided into A, A' and A1, A1', respectively, such that A and A1 represent the ON pathway and the A' and A1' the OFF one (Sanderson 1974; Linden et al. 1981; Stryker and Zahs 1983). Interestingly, COX highlights the OFF sublaminae A' and A1' much more so than the ON sublaminae A and A1 (Kageyama and Wong-Riley 1984), indicating that these sublaminae are metabolically more active in this nocturnal species.

At the cellular level in the cat LGN, the large Y/Y-like cells are much more COX reactive than the medium-sized X/X-like cells (which often contain cytoplasmic laminated bodies or CLB) and the small W/W-like neurons (Figure 3.3) (Kageyama and Wong-Riley 1984, 1985). This finding is consistent with the maintained firing rates of LGN neurons, being higher in Y cells than X cells, and W cells have the lowest maintained rate (Wilson et al. 1976; Bullier and Norton 1979). Notably, the small, inhibitory LGN interneurons have low somatic COX activity (Kageyama and Wong-Riley 1985), unlike their counterparts in the visual cortex that have high levels (see Section 3.1). On the other hand, terminals making symmetric synapses (presumed inhibitory) are darkly reactive for COX in both the LGN and the visual cortex, indicating a difference in metabolic demands between cell bodies and terminals.

3.3 WHAT CAN THE HIPPOCAMPUS TELL US ABOUT VARIATIONS IN COX DISTRIBUTION?

The hippocampus provides an ideal site where neuronal somata and dendrites, and even proximal versus distal dendrites, are spatially segregated. The pattern of COX labeling is comparable among rats, cats, and monkeys (Kageyama and Wong-Riley 1982). The most intense COX activity is within distal dendrites of pyramidal cells in the stratum moleculare of CA1–3 and the distal dendrites of granule cells in the outer molecular layer of the dentate gyrus, all of which receive excitatory perforant pathway input from the entorhinal cortex (Steward and Scoville 1976; Witter and Amaral 1991)

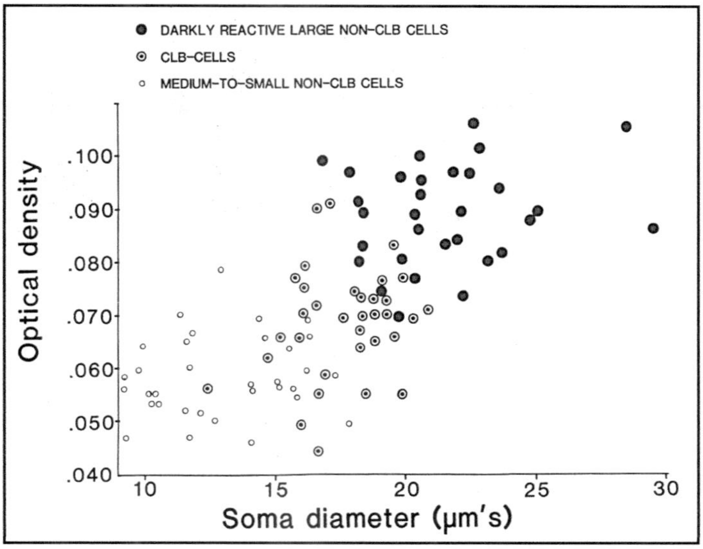

FIGURE 3.3 Distribution of three major cat LGN neuronal types by size and optical density of COX labeling. Darkly reactive large Y/Y-like neurons (non-CLB; solid circles) are distinguishable from medium-sized X/X-like (CLB; circle with a dot) and the small W/W-like neurons (non-CLB; open circles) with some minimal mixing among the classes.

From Kageyama and Wong-Riley 1985.

(Figure 3.4). The stratum radiatum of CA1 contains proximal dendrites of pyramidal neurons that receive excitatory Schaffer collaterals from CA3. However, their level of COX in this region is relatively low, perhaps because CA3 selectively gates specific information to CA1 neurons (Vinogradova 1975; Ahmed and Mehta 2009). The stratum radiatum of CA3, especially the lower half, however, is moderately reactive for COX. This is likely due to the presence of COX-reactive interneurons. The stratum lucidum of CA3 appears lightly reactive for COX. However, the proximal dendrites of CA3 pyramidal neurons that traverse vertically through this region contain darkly COX-reactive mitochondria because they receive massive excitatory mossy fiber input originating from dentate granule cells. The basal dendrites of CA3 pyramidal neurons in stratum oriens receive input from the septo-hippocampal pathway (Raisman et al. 1965), and their mitochondria are darkly COX reactive. The somata of CA3 pyramidal cells and various interneurons are much more COX reactive than those of CA1 pyramidal neurons and dentate granule cells (Kageyama and Wong-Riley 1982). Although grid cells in the entorhinal cortex project to the place fields in both CA1 and CA3 (Ahmed and Mehta 2009; Moser et al. 2015), CA1 place cells only fire when the animal is within a specific spatial location (O'Keefe and Dostrovsky 1971) and not when outside of their place fields, and the majority of them are silent cells (Ahmed and Mehta 2009). CA3 pyramidal cells, on the other hand, maintain interneuronal excitatory synaptic interactions to generate network bursting activity (Traub and Wong 1982). Inhibitory, GABAergic CA1 interneurons with very high firing rates (Ahmed and Mehta 2009) are intensely reactive for COX in both their cell bodies and dendrites, and they contact CA1 pyramidal somata and proximal dendrites, causing these postsynaptic targets to have low COX activity (Kageyama and Wong-Riley 1982).

Thus, COX activity varies among regions, layers, cell types, and even different dendritic segments of the same neuron within the hippocampus and dentate gyrus. Hence, neurons allocate mitochondria with the highest COX levels to sites of greatest energy utilization, and they conserve their mitochondrial supply where energy is less needed.

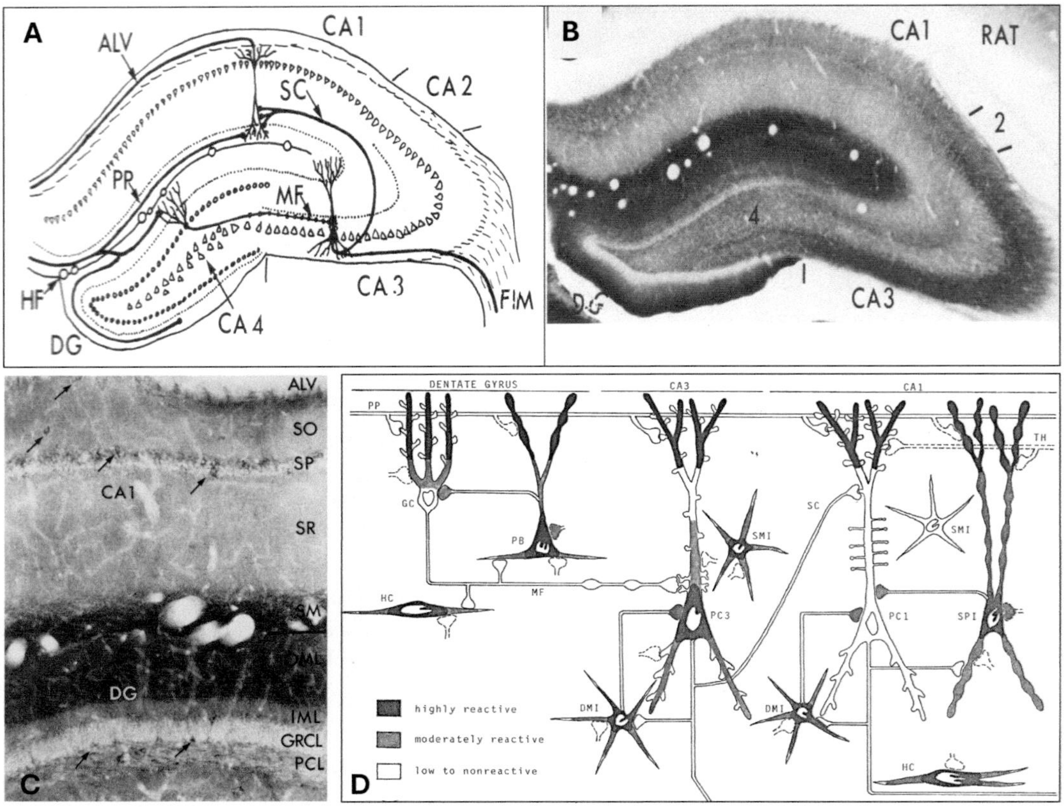

FIGURE 3.4 Mammalian hippocampus. (A) Schematic diagram of the hippocampus with CA1, CA2, CA3, and CA4 subfields and the dentate gyrus (DG). ALV, alveus; Fim, fimbria; HF, hippocampal fissure; MF, mossy fibers; PP, perforant pathway; SC, Schaffer collaterals; small circles, granule cell bodies of DG; small triangles, pyramidal cell bodies of CA1-CA4. (B) A cross-section of the rat hippocampus reacted for COX. Note the intense labeling in the distal dendritic portions of the CA subfields and the DG. Note also that CA3 pyramidal cell bodies are more reactive than those in CA1. (C) A higher magnification of CA1 and the dentate gyrus. Note that the highest levels of COX activity are in the stratum moleculare (SM) of CA1 and the outer molecular layer (OML) of DG. Arrows indicate COX-reactive interneurons in various layers. GCL, granule cell layer; IML, inner molecular layer; PCL, polymorphic cell layer; SO, stratum oriens; SP, stratum pyramidale; SR, stratum radiatum. (D) Schematic diagram of COX labeling in different cell types and different segments of the same cell in the DG, CA3, and CA1. DMI, deep multipolar interneuron; GC, granule cell; HC, horizontal (or fusiform) cell; PB, pyramidal basket cell; PC1, pyramidal cell in CA1; PC3, pyramidal cell in CA3; SMI, superficial multipolar interneuron; SPI, stratum pyramidale interneuron; TH, thalamohippocampal pathway.

Modified from Kageyama and Wong-Riley 1982.

3.4 DEVELOPMENTAL TRANSITIONS IN MITOCHONDRIA AND COX REACTIVITY IN THE CEREBELLUM

The cerebellum is an ideal site for analyzing differences between somatic and dendritic mitochondria within the same cell, as well as the relationship between presynaptic type and postsynaptic energy demand.

The cerebellar cortex is exquisitely organized in layers: densely packed granule cells are in the interior enclosing the innermost afferent and efferent fiber tracts, a single layer of large Purkinje cell bodies lies above the granule cell layer, and the outermost molecular layer is where Purkinje cell dendrites receive direct excitatory climbing fiber input from the contralateral inferior olivary nucleus and indirect excitatory input from mossy fibers via parallel fibers of granule cells. Inhibitory stellate and basket cells are also located in the molecular layer. Purkinje cell axons project to the deep cerebellar nuclei, the output neurons of the cerebellum (Carpenter 1976).

COX cytochemistry reveals, not surprisingly, a rich supply of darkly reactive mitochondria in Purkinje cell dendrites in the molecular layer (Mjaatvedt and Wong-Riley 1988). The Purkinje cell bodies themselves appear darkly reactive at low magnification. However, at higher magnification and under the EM, the strongly COX-reactive mitochondria reside exclusively in basket cell terminals that form a chalice around individual Purkinje cell body (Mjaatvedt and Wong-Riley 1988), the so-called "pinceau" of Ramón y Cajal (1909), indicating that these inhibitory terminals are highly active. Indeed, basket cell terminals are densely supplied with Na^+ channels and Kv1.1 and Kv1.2 potassium channels (Southan and Robertson 1998), and ultra-rapid ephaptic inhibition exists between pinceau and Purkinje cells, especially at the initial segment (Blot and Barbour 2014). Granule cell bodies have very little cytoplasm with a scant supply of lightly reactive mitochondria, but the mossy fiber terminals that synapse with granule cell dendrites contain darkly COX-reactive mitochondria.

What is the major determinant of metabolic activity within neuronal cell bodies? The developing Purkinje cells in postnatal rats offer a model site for investigation (Mjaatvedt and Wong-Riley 1988) (Figure 3.5). At birth, the small cell bodies have no direct synaptic input and no dendritic extensions, and mitochondria are mainly lightly reactive for COX. Between postnatal days P5 and P15, cell bodies enlarge and begin expressing somatic spines (originally described by Cajal, 1909) contacted by excitatory climbing fibers. At this stage, the somata have relatively high COX levels with mitochondria evenly distributed among D, M, and L types. After P15, somatic spines are resorbed, and climbing fibers move up to the developing dendritic branches. COX activity is decreasing in the somata but increasing in dendrites. Once mature, Purkinje cell bodies are fully contacted by highly active, inhibitory basket cell terminals packed with darkly COX-reactive mitochondria. The large cell bodies now have a mixture of moderate and small, lightly COX-reactive mitochondria, but those in the dendritic tree are highly reactive for COX.

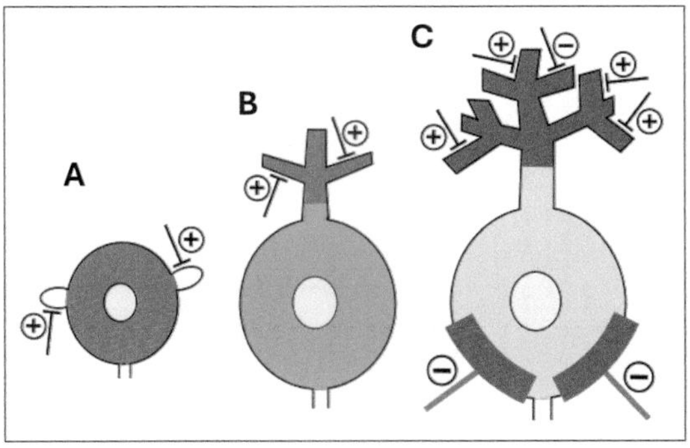

FIGURE 3.5 Schematic diagram of Purkinje cell development in the rat. (A) At postnatal days P5–P15, Purkinje cell bodies express somatic spines (exaggerated in size here) that receive excitatory synapses (plus signs) from climbing fibers, and their level of COX is relatively high. (B) After P15, somatic spines are resorbed, and climbing fibers begin to innervate the newly developed dendritic branches, leaving the cell body at a lower COX level. (C) In the mature state, the much larger cell body is fully innervated by highly active basket cell terminals that are inhibitory (minus signs), thereby significantly reducing somatic COX levels. The dendritic tree, on the other hand, receives the bulk of excitatory inputs (in addition to some inhibitory ones) and its COX level is high. Intensity of labeling indicates the relative level of COX activity.

Based on findings in Mjaatvedt and Wong-Riley 1988.

At first glance, this may appear to be a surface-to-volume ratio issue, with larger cell bodies possibly having lower metabolic rates than smaller ones. However, the fact that type C cells in the primate striate cortex are larger than type A cells yet are metabolically much more active (Section 3.1), and the lack of correlation between somatic size and COX activity (Section 3.5), indicate that surface-to-volume ratio is not universally applicable when considering energy demands and supply in neurons.

In short, it is the type of synaptic input, *not* the cell size, that determines the level of metabolic activity in neuronal cell bodies. When they receive direct excitatory input, their energy demand is high. When the input is strongly inhibitory, the demand is low.

3.5 IS THERE A RELATIONSHIP BETWEEN CELL SIZE AND THE LEVEL OF SOMATIC METABOLIC ACTIVITY? A QUANTITATIVE STUDY OF THE MONKEY SPINAL CORD

The question whether a relationship exists between cell size and somatic COX reactivity was further probed in a quantitative study on ventral horn neurons in the macaque spinal cord (Wong-Riley and Kageyama 1986). When neuronal COX reactivity was plotted against cell size in 1,770 individual neurons of the cervical, thoracic, lumbar, and sacral cords, a rough inverse relationship appears. However, all three cell sizes (small, medium, and large) exist in all three categories of COX reactivity (light, moderate, and dark) (Figure 3.6). Hence, the metabolic demand of neurons does not correlate directly with somatic size.

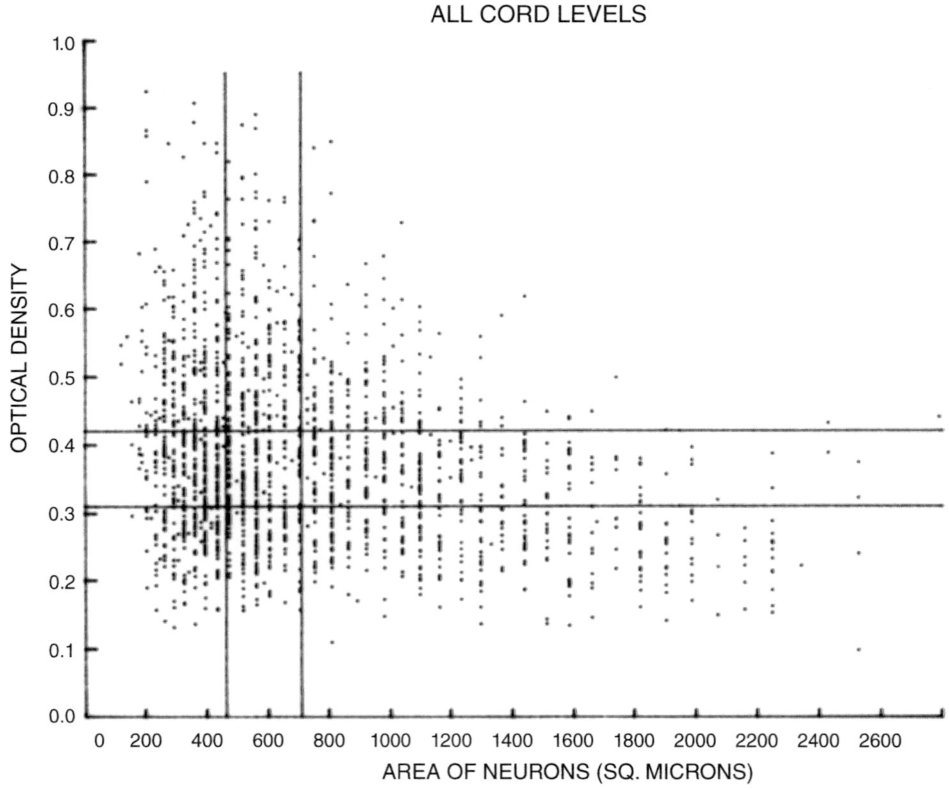

FIGURE 3.6 Scatter plot of 1,770 ventral horn neurons measured for size and optical density of COX reaction product from the cervical, thoracic, lumbar, and sacral levels of the macaque monkey spinal cord. Lines are drawn to arbitrarily divide cell sizes into large, medium, and small, and optical densities into dark, moderate, and light. It is evident that all three size classes have representations in all three reactive groups.

From Wong-Riley and Kageyama 1986.

3.6 VARIATIONS IN ENERGY DEMAND WITHIN THE RETINA

The retina is an exquisitely laminated structure in the CNS. A survey of retinae in the ferret, cat, squirrel monkey, and macaque monkey revealed a consistent pattern (Kageyama and Wong-Riley 1984) (Figure 3.7). The most COX-reactive region encompasses the inner segments of photoreceptor cells. Both rod and cone inner segments are richly populated with D-type mitochondria. Those in cones are larger and more densely packed than those in rods (Figure 3.7G). In tangential sections, the cone inner segments form a mosaic pattern among the sea of rod inner segments (Figure 3.7F). The intense energy demand of photoreceptor cell inner segments serves to maintain the dark current. In the dark, cGMP-gated cation channels in the outer segments open to allow the influx of cations (mainly Na^+), which are actively pumped out from the inner segments by ouabain- and KCN-sensitive Na^+/K^+-ATPase (Penn and Hagins 1969; Zuckerman 1973; Stirling and Lee 1980; Zimmerman et al. 1985). High energy utilization by Na^+/K^+-ATPase is matched by an abundant supply of mitochondria rich in COX. Indeed, a close correlation in COX and Na^+/K^+-ATPase distribution exists in the retina and elsewhere in the brain (Hevner et al. 1992; Wong-Riley et al. 1998a, 1998b).

Cell bodies of rods and cones in the outer nuclear layer contain very few mitochondria, most of which are of the L type. Cone pedicles, on the other hand, are highly reactive for COX, and they form a mosaic pattern in tangential sections (Figure 3.7H and I). The outer plexiform layer, where photoreceptor cell terminals interact synaptically with bipolar cell dendrites and horizontal cell processes, is strongly reactive for COX. Equally reactive is the inner plexiform layer, where bipolar and amacrine cell processes synapse with ganglion cell dendrites. In nocturnal ferrets and cats, sublamina a (OFF) is more COX reactive than sublamina b (ON), but the opposite is true for the diurnal primate (compare IPLa and b in Figure 3.7A–E), comparable to the findings in the LGN (see Section 3.2). This dichotomous pattern is more striking centrally in the macaque parafoveal region than peripherally, and the two sublaminae are separated by a thin band of low COX intensity. Horizontal cells and some retinal ganglion cell bodies are quite reactive for COX, and in the cat, they form a mosaic pattern in tangential sections (Kageyama and Wong-Riley 1984).

A striking finding in the squirrel monkey retina is that unmyelinated ganglion cell axons are darkly COX reactive before they exit the retina. Once they acquire myelin at the lamina cribrosa, their COX activity falls precipitously (Figure 3.7J) (Kageyama and Wong-Riley 1984). In the prelaminar and laminar unmyelinated portions of the optic nerve, an accumulation of mitochondria coincides with an elevated density of voltage-gated Na^+ channels and high COX activity (Barron et al. 2004). The intense energy demand at the optic nerve head renders it exceptionally vulnerable to retinal pathology, including glaucoma (Radius 1987).

3.7 A SINGLE NEURON IS NOT A METABOLICALLY HOMOGENEOUS ENTITY

From the above findings, it is evident that neither neuronal cell size nor transmitter type dictates the level of somatic COX activity. COX-rich neurons exist in all sizes. Inhibitory GABAergic neurons in the monkey visual cortex are highly reactive for COX, yet those in the LGN and Purkinje cells in the cerebellum have low levels of somatic COX. Cell bodies that receive both excitatory (glutamatergic) and inhibitory (GABAergic) inputs have much higher levels of COX than those that receive only inhibitory ones (Wong-Riley et al. 1989a; Nie and Wong-Riley 1996a, 1996b), consistent with the fact that repolarization after depolarization is much more energy demanding than repolarization after hyperpolarization (Figure 3.8).

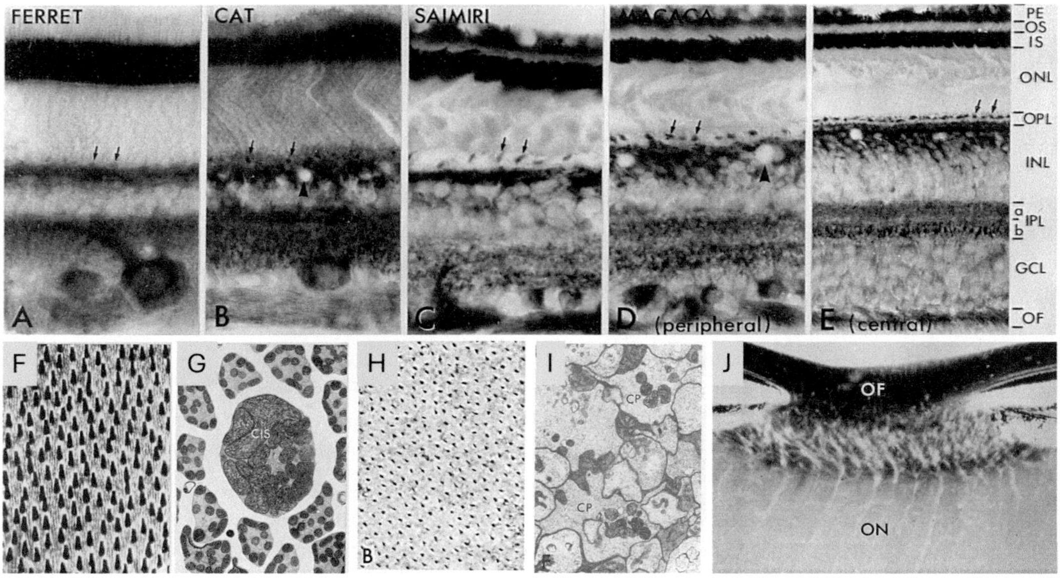

FIGURE 3.7 Cross-sections of retinae reacted with COX histochemistry from the ferret (A), cat (B), squirrel monkey (C), and macaque monkey (D and E). Note that the inner segment (IS) of all species is the most reactive portion of the retina. In tangential sections, the cone inner segments form a mosaic Christmas tree-like pattern against the less reactive rod inner segment (F). At the EM level (G), the cone inner segment (CIS) is packed with large and highly reactive mitochondria, whereas the surrounding rod inner segments are filled with reactive but smaller mitochondria. Arrows in A–E indicate darkly reactive cone pedicles, which exhibit a mosaic pattern in tangential section (H), and at the EM level (I), the cone pedicles (CP) contain darkly reactive mitochondria. The outer plexiform layer (OPL) and the inner plexiform layer (IPL, with a and b sublayers) are highly reactive for COX, as are some large ganglion cells in the ganglion cell layer (GCL) and the optic fiber layer (OF). Note that the outer segments (OS), the outer nuclear layer (ONL), and most of the inner nuclear layer (INL) show very little COX activity. The pigment epithelial layer (PE) is dark because of melanin granules. (J) At the optic disc of the squirrel monkey, the unmyelinated portion of the optic fibers (OF) is darkly reactive for COX. However, once the fibers acquire myelin sheaths at the lamina cribrosa (not labeled) and become the optic nerve (ON), there is a precipitous fall in the level of COX activity.

Modified from Kageyama and Wong-Riley 1984.

Universally, dendrites of all neurons, and especially those segments of dendrites that receive prominent excitatory input, have a rich supply of mitochondria darkly reactive for COX (Figure 3.8). Indeed, the major energy consumers in the brain are dendrites because 1) dendrites are the predominant receptive sites for excitatory synapses; 2) most CNS dendrites mediate slow depolarizing potentials rather than action potentials in response to excitatory input; and 3) the bulk of energy demand in the brain is to repolarize dendritic membranes after slow and fast depolarizing potentials.

Myelinated axon trunks generally have mitochondria spaced widely apart with relatively low COX content. This is because action potential propagation requires relatively little energy (Creutzfeldt 1975) (Figure 3.8). Likewise, the white matter usually exhibits low signals in glucose uptake and oxygen utilization studies, indicating low metabolic rates. Bundles of small unmyelinated axon trunks often have very few mitochondria. However, clear exceptions include the unmyelinated retinal ganglion cell axons before exiting the retina that possess high COX activity (Section 3.6). Axon terminals can have high-to-low levels of COX (Section 3.1), commensurate with the degree of functional activity.

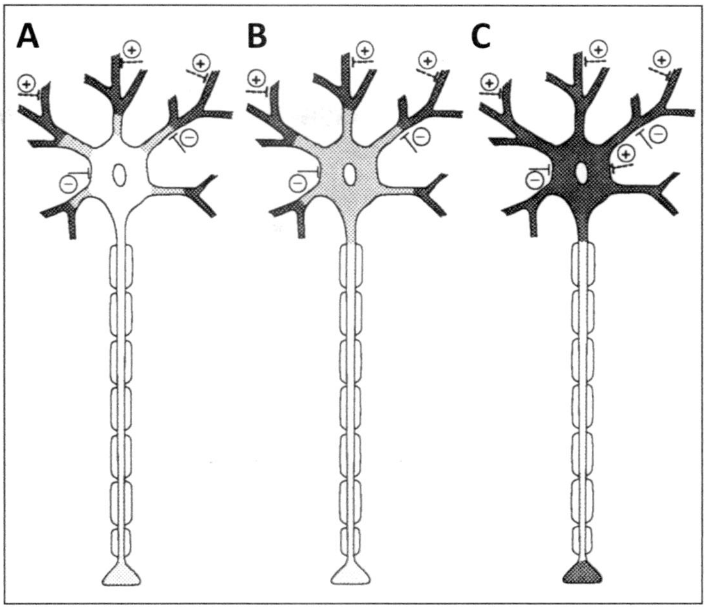

FIGURE 3.8 Diagrammatic rendition of metabolic inhomogeneity among neurons and among different segments of the same neuron. The intensity of shading represents the level of COX activity. Cell bodies may have low (A), moderate (B), or high (C) COX levels depending on their synaptic input (+: excitatory; −: inhibitory) and the degree of spontaneous activity. Dendrites, especially distal portions that receive the bulk of excitatory inputs, are rich in COX, whereas axon trunks, especially myelinated ones, usually have low labeling. The axon terminals may have low, moderate, or high amounts of COX contingent upon activity levels.

From Wong-Riley 1989.

Thus, the segmental metabolic diversity within a single neuron illustrates the exquisite ability of neurons to efficiently couple metabolic capacity locally and tightly to metabolic expenditure. Mitochondria are not simply distributed evenly throughout a neuron, but rather, are more densely allocated in some segments (especially the dendrites) while sparsely in others (such as most axon trunks). Where energy is in low demand, mitochondrial numerical and areal densities are low and COX activity is low. The opposite is true for locales of high energy demand. The control for mitochondrial size, content, density, and COX activity is at an exquisitely local level. Mitochondrial dynamics of fusion and fission are no doubt at play for homeostatic control and responses to functional alterations (reviewed in Zaninello et al. 2024). The high cost of repolarization is not limited to postsynaptic activity, but to active ion pumping in general. The abundance of mitochondria with intense COX activity in the inner segments of retinal photoreceptor cells (Section 3.6) attests to the high energy demand of active ion transport, even independent of direct synaptic transmission.

Neurons must maintain another cation, Ca^{2+}, under homeostatic control. Calcium is involved in multiple biological processes, including signaling for protein synthesis, gene transcription, and remodeling of the actin cytoskeletal network (Jędrzejewska-Szmek et al. 2023). Calcium entry via *N*-methyl-D-aspartate (NMDA) receptors mediates functions such as synaptogenesis, neuronal migration, and differentiation, as well as long-term potentiation in learning and memory (Lau et al. 2009). In dendrites, calcium spikes occur in some neurons (Kitamura and Kano 2013; Manita et al. 2017), and in axon terminals, calcium binding to vesicular synaptotagmin is necessary for the fusion of vesicular and presynaptic membranes before neurotransmitter release (Bennett and Scheller 1994). Excess intracellular Ca^{2+} ions are stored in endoplasmic reticulum and mitochondria and are actively released via two sets of proteins in the plasma membrane: a high-capacity, low-affinity Na^+/Ca^{2+} exchanger (NCX) and a low-capacity, high-affinity Ca^{2+}-ATPase, which has a 1:1 Ca^{2+}:ATP stoichiometry (Brini and Carafoli 2011). A neuron-specific variant of the plasma membrane Ca^{2+}/ATP, PMCA1a, is concentrated in the plasma membrane of neuronal cell bodies, dendrites, and dendritic spines in the rat brain (Kenyon et al. 2010). Their distinct cellular localizations correspond to the energy requirements of synaptic transmission.

4 Uncovering a New Region in the Primate Visual Cortex

Puffs, Patches, Blobs, Dots, or Spots

ABSTRACT

The discovery of COX-rich globular zones forming a polka-dot-like array in the supragranular layers of the primate primary visual cortex opened a new field of visual system research. These zones are variably called puffs, patches, blobs, dots, and spots, and the surrounding COX-poor regions are termed interpuffs, interpatches, or interblobs. Much has been learned about the morphological, physiological, neurochemical, and connectional properties of these regions. COX-rich thick and thin stripes containing globular zones of even higher COX activity (also known as puffs) are found in the primate secondary visual cortex (V2). Connections from striate and prestriate cortices extend to other visual cortical areas.

An entirely unexpected discovery was made in the primate visual cortex using COX histochemistry by the author in 1978. This chapter details the characteristics of these metabolically active zones not previously realized or anticipated.

The primate primary visual cortex (striate cortex, area 17, or V1) has been extensively studied anatomically and physiologically. Anatomically, it contains six cellular layers numbered I–VI (or 1-6) from the surface to the white matter. Layer IV is cell-rich and receives input from the ipsilateral LGN. It is subdivided into IVA, IVB, IVCα, and IVCβ (reviewed in Lund 1988; an alternative naming scheme for the layers was proposed by Hassler in 1966). The supragranular layers receive input primarily from layer IV and project to other cortical areas, such as area 18 (prestriate cortex or V2). The infragranular layer V projects mainly to the superior colliculus, while layer VI sends feedback fibers to the ipsilateral LGN (Lund et al. 1975). Additional layer-specific intralaminar, interlaminar, and long, horizontal clustered connections exist within V1 (Rockland and Lund 1983; Lund 1988; Lund et al. 2003). Moreover, V1 projects to and often receives reciprocal connections from other visual cortical areas (Wong-Riley 1979b; Weller and Kaas 1983). The basic functional organization of the primate V1 was solidly established by the pioneering work of Hubel and Wiesel (1968, 1974, 1977), who described two sets of independent, functional columns extending through the depth of the cortex: a set of left- and right-ocular dominance columns and a set of orientation columns. Cells within the same column interact vertically and share similar functional preferences. Hierarchical and parallel processing streams accomplish the representation of shape, color, depth, and motion of the visual world (reviewed in Van Essen 2005; Nassi and Callaway 2009).

The pattern revealed by COX histochemistry is quite different from that of the Nissl stain (Carroll and Wong-Riley 1984; Horton 1984; Wong-Riley and Carroll 1984a) (Figure 4.1A and B). Whereas Nissl labels only cell bodies of neurons and glia, COX highlights metabolically active neuronal compartments (i.e., mostly dendrites, some axon terminals, and some cell bodies; see Chapter 3). In layer IVC, Nissl emphasizes the very cell-dense IVCβ, but COX is rich throughout the geniculate-recipient layer IV, with stronger labeling in the magnocellular-recipient IVCα than the parvocellular-recipient IVCβ, and a prominent band at the base of IVCβ known as IVCβdark (Wong-Riley 1994). COX also reveals the honeycomb-like pattern in layer IVA, not evident with the Nissl stain. The other geniculate-recipient layer, layer VI, is also moderately reactive for COX.

DOI: 10.1201/9781003507765-4

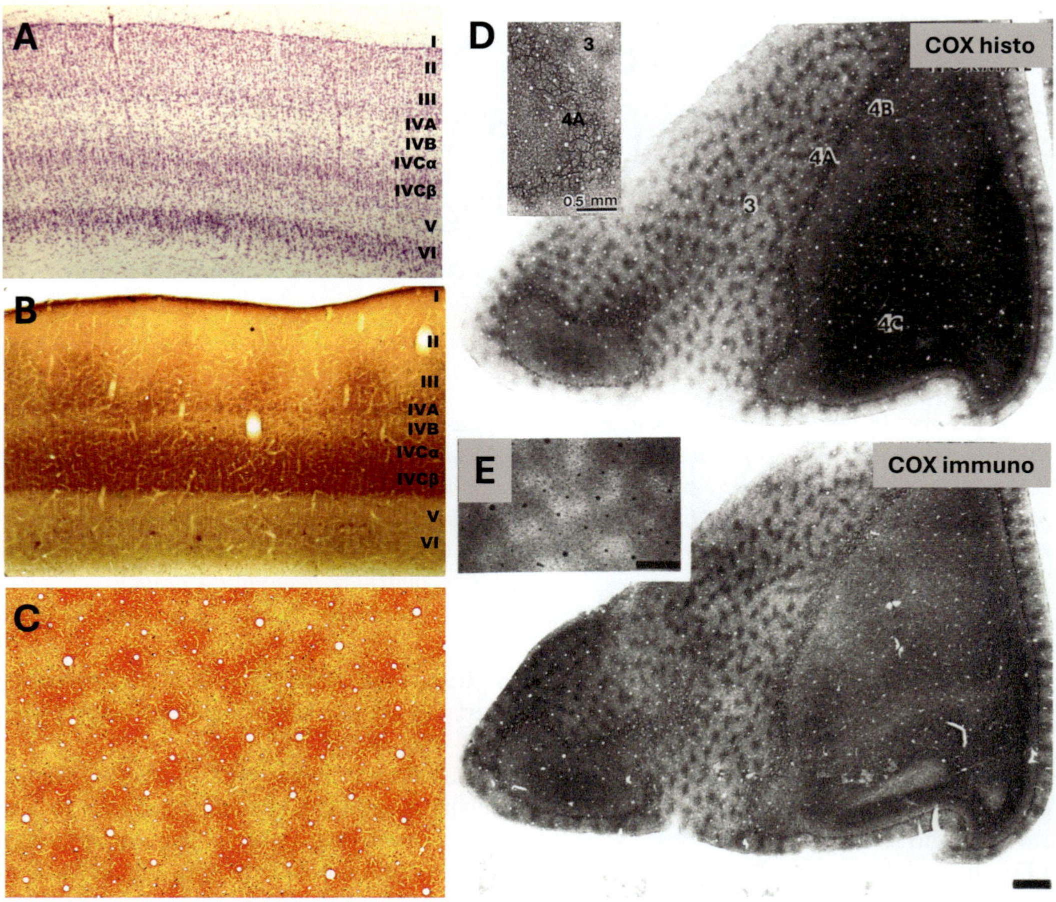

FIGURE 4.1 Macaque monkey visual cortex. Coronal sections of macaque V1 stained for Nissl (A) and reacted for COX histochemistry (B), with layers indicated on the right. (C) Tangential section through layer III showing COX-reactive puffs. (D and E) Adjacent tangential sections of macaque V1 reacted for COX histochemistry (D) or immunohistochemistry (E) with closely matched patterns. Layers are indicated in Arabic. Inset in D shows honeycomb pattern in layer 4A, and inset in E shows indirectly immunofluorescently labeled puffs (light spots).

Scale bar for D and E = 1 mm.

A–C: unpublished observations. D and E are modified from Hevner and Wong-Riley 1990.

4.1 AN ORDERLY ARRAY OF METABOLICALLY ACTIVE ZONES: CYTOCHROME *c* OXIDASE-RICH PUFFS

What came as a total surprise was a polka-dot-like array of globular zones of high COX activity in the supragranular layers of normal striate cortex (Figure 4.1B–E) (discovered but not immediately published by Wong-Riley in 1978; mentioned in Hubel 1982 as a personal communication from Wong-Riley; Horton and Hubel 1981; Carroll and Wong-Riley 1982 Abstract; Horton 1984; Carroll and Wong-Riley 1984; Wong-Riley and Carroll 1984a). These metabolically active zones had not been discerned by existing methods at that time (including electrophysiology, Nissl stain, Golgi stain, and myelin stain). They were named "puffs" by this author because of their three-dimensional, reticulated, and globular appearance, like cotton puffs (this term will also be used in this book). Horton referred to them as "patches," and Livingstone and Hubel (1984a) called them "blobs," but they used the same term for COX-reactive regions in layers IVB, V, and VI. The terms "blobs" and "interblobs" became the predominant ones in the literature. Other names were also introduced, such as "spots" (Tootell et al. 1983) and "dots" (Hendrickson 1985). The COX-poor regions surrounding the puffs are called interpuffs (or "interblobs"). The discovery of puffs opened a whole new dimension of visual neuroscience inquiry.

4.1.1 Puff Dimensions

Puff diameter averages 210 μm in squirrel monkeys (Carroll and Wong-Riley 1984) and ~262 × 377 μm in macaque monkeys (Wong-Riley and Carroll 1984a; however, Horton 1984 reported 150 × 250 μm). Their mean cross-sectional area (0.040 mm^2) remains quite constant, as is their mean volume of 0.0127 ± 0.0027 mm^3 (Trusk et al. 1990). Three-dimensional reconstruction indicates that puffs are roughly pear-shaped with a larger base in layer IIIB, where the area ranges between 0.046 and 0.050 mm^2, but gradually decreases toward upper IIIA and II (Trusk et al. 1990). In squirrel monkeys, where ocular dominance columns are poorly delineated, the center-to-center spacing of puffs is around 500 μm (Carroll and Wong-Riley 1984). In macaque monkeys, puffs are centered on ODCs, and they form rows that radiate perpendicularly from the area 17/18 border (representing the vertical meridian), in register with ODCs in layer IVC. Their center-to-center spacing within rows is ~450 μm and between rows is ~353 μm (Wong-Riley and Carroll 1984a; Horton in 1984 reported a within row spacing of ~550 μm). The spacing within rows tends to vary with eccentricity (Livingstone and Hubel 1984a), but the spacing between rows tends to remain quite constant, as ODCs have relatively constant widths from 0° to 15° eccentricity in the macaque (Hubel and Wiesel 1972). The center-to-center distance of adjacent puffs at the foveal representation is equivalent to a visual angle of ~4′ of arc, but at 8° eccentricity, this represents ½° of arc (Tootell et al. 1988b).

COX-rich puffs also exist in human primary visual cortex (Horton and Hedley-Whyte 1984; Wong-Riley 1993; Wong-Riley et al. 1993; Marcondes et al. 2019). Their reported size varies from 250 × 400 μm (Horton and Hedley-Whyte 1984) to ~230 × 450 μm (Wong-Riley et al. 1993), with a mean area of 0.2–0.8 mm^2 (Marcondes et al. 2019). The center-to-center spacing within ocular dominance rows is 650 μm and between rows is ~800 μm (Horton and Hedley-Whyte 1984; Wong-Riley et al. 1993). Whereas puffs develop weeks before birth in macaque monkeys, they are not well-organized at birth in humans but become clearly formed at postnatal day 24 and fully developed by the 4th postnatal month (Wong-Riley et al. 1993) (Figure 4.2).

Puffs are present throughout the primary visual cortex of humans and non-human primates. The density is about one puff per 0.20 mm^2 in the macaque monkey and one per 0.22 mm^2 in the squirrel monkeys (Horton 1984). The density in the foveal representation in humans is only about a third of that in the macaque, or ~1 puff per 0.6–0.8 mm^2 (Horton and Hedley-Whyte 1984), but ~0.6/mm^2 at ~60° eccentricity (Marcondes et al. 2019). The average puff density ranges from 2 to 3 puffs/mm^2, with a mean of 2.35 puffs/mm^2 (Wong-Riley et al. 1993). Within 1 mm^2 of the macaque cortex, ~5.6 puffs exist near the optic disc representation, ~9.0 puffs near the monocular crescent, and ~5.5 puffs within the crescent (Horton 1984). As the surface area of the macaque striate cortex is estimated to be ~1200–1400 mm^2, ~6000–9000 puffs exist from the central to more peripheral portions of the binocular cortex (Horton 1984). In humans, a comparable number of puffs should exist for a surface area of 3000–3600 mm^2, as puff density is only half of that of macaques (Horton and Hedley-Whyte 1984). This value, however, may vary according to the actual dimension of the human striate cortex.

4.1.2 Cellular and Synaptic Differences Between Puffs and Interpuffs

What are the major differences between puffs and interpuffs at the cellular level? The overall density of neurons within puffs is equivalent to that in interpuffs, the three metabolic neuronal types (A, B, and C) found in puffs (see Chapter 3, Section 3.1) are qualitatively the same in interpuffs, and the numerical densities of mitochondria in the neuropil are also comparable between the two regions (Carroll and Wong-Riley 1984; Trusk et al. 1990). However, type B and C neurons are much less COX-reactive in interpuffs than in puffs, containing only 2/3 to 3/4 the number of D mitochondria, as their counterparts in puffs (Wong-Riley et al. 1994). The COX-reactive, GABAergic type C cells

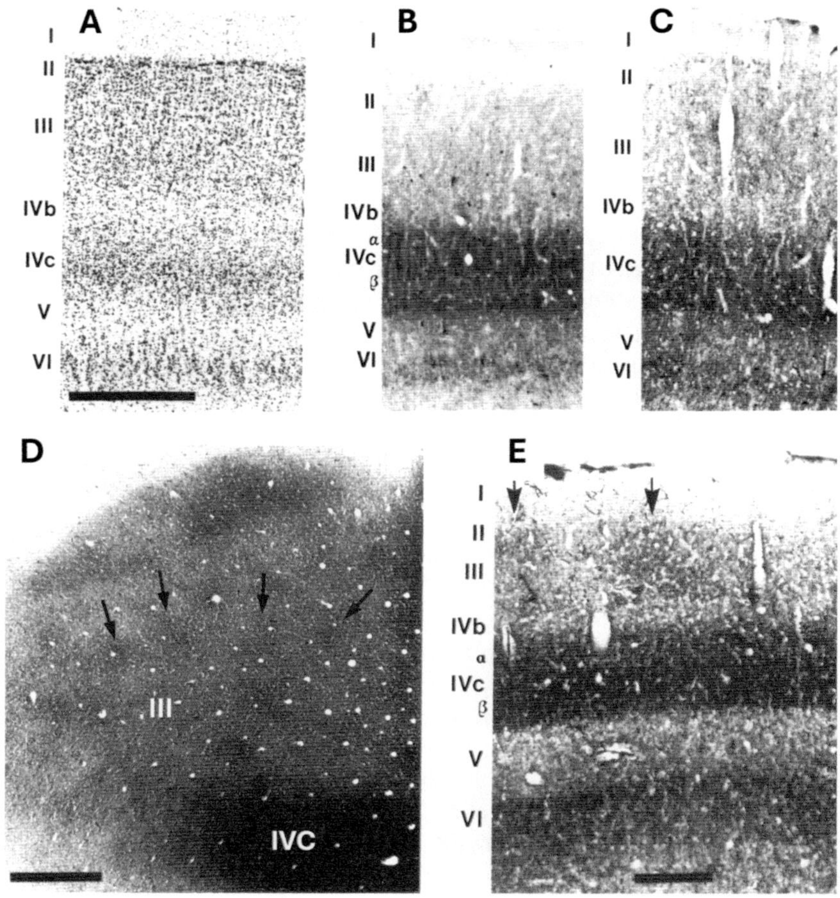

FIGURE 4.2 Development of puffs in humans. At birth, the cellular laminae in the human striate cortex are well demarcated by the Nissl stain (A), and layers IVC and VI show moderate-to-high levels of COX, revealed histochemically (B) or immunohistochemically (C). However, there are no discernible puffs. At postnatal day 24, supragranular puffs are clearly visible (arrows in D), and they are well organized at 4 months of age (arrows in E).

Scale bar for A–C = 0.5 mm; D = 1 mm; E = 0.5 mm.

Modified from Wong-Riley et al. 1993.

are 13% smaller than those in puffs and contain only 45–60% of mitochondria in their somata, as those in puffs. The type B pyramidal cells are also smaller in interpuffs and have 30% fewer somatic mitochondria than those in puffs. The small type A cells, on the other hand, have about the same number of somatic mitochondria in both regions, but the prevalence of D mitochondria is even lower in interpuffs (Wong-Riley et al. 1994). The ratio of non-GABAergic to GABAergic axo-somatic synapses is higher in puffs than interpuffs (2.3:1 vs. 1.6:1; $P < 0.01$) (Nie and Wong-Riley 1995).

In the puff neuropil, most mitochondria, especially the D type, consistently reside in dendrites (Wong-Riley et al. 1989b; also see Section 3.1). Although not quantified, dendritic mitochondria in interpuffs are less COX-reactive than those in puffs. GABAergic axon terminals that make symmetric synapses are larger and have twice as many D mitochondria in puffs than in interpuffs (Nie and Wong-Riley 1995). Overall, there are significantly more synapses of both GABAergic and non-GABAergic types in puffs than in interpuffs, and the ratio of non-GABAergic (presumed excitatory) to GABAergic synapses is significantly higher in the neuropil of puffs than of interpuffs (2.86:1 vs. 2.08:1; $P < 0.01$) (Nie and Wong-Riley 1995).

Thus, puffs have more excitatory synapses requiring more energy for repolarization than interpuffs. Such excitation is not only on dendrites of both glutamatergic, excitatory neurons (types A and B), but also on dendrites and cell bodies of GABAergic, inhibitory type C neurons. It is not known whether GABAergic, type C neurons in puffs exert inhibitory influence also on interpuffs or whether type C cells in interpuffs confine their output solely to binocular interpuffs and not monocular puff neurons.

4.1.3 Neurochemical Differences Between Puffs and Interpuffs

From the neurochemical standpoint, puffs are synaptically and metabolically much more active than interpuffs. Synaptically, puffs have a denser presence of both excitatory (glutamate, NMDA receptor 1 [NMDAR1], neuronal nitric oxide synthase [more distinctly labeled by NADPH-diaphorase] associated with NMDA receptors, vesicular glutamate transporter 2, and the neuropeptide tachykinin) and inhibitory (GAD, GABA, and glycine) neurochemicals than interpuffs (Hendrickson et al. 1981; Hendry et al. 1988; Carder and Hendry 1994; Wong-Riley et al. 1998a; Rockoff et al. 2014). Metabolically, both oxidative (COX and succinate dehydrogenase or SDH) and glycolytic (lactate dehydrogenase or LDH) enzymes, the energy-consuming enzyme Na^+/K^+-ATPase, and the uptake of 2DG are all higher in puffs than in interpuffs (Horton and Hubel 1981; Humphrey and Hendrickson 1983; Horton 1984; Wong-Riley et al. 1998a). Acetylcholinesterase and myelin are also more densely distributed in puffs than in interpuffs (Horton 1984). On the other hand, vitamin D-induced calcium-binding protein calbindin-D-28K and neuropeptide Y have distributions complementary to puffs, that is, higher in interpuffs (Celio et al. 1986; Kuljis and Rakic 1989). The rationale is not entirely clear, but an association with extra-geniculocortical connections has been proposed (Celio et al. 1986).

4.1.4 Physiological Properties of Puff and Interpuff Neurons

What are the physiological properties that distinguish puffs from the surrounding COX-poor interpuffs? Why are these zones metabolically more active? Several functional studies had been performed. Livingstone and Hubel (1984a) spearheaded the investigation of macaque monkeys. They found that cells within puffs lacked orientation selectivity but tended to be monocular, color-selective, and specialized in maintaining color constancy. These neurons had concentric, circular symmetric receptive fields and were classified into three main types: the most common were double-opponent (either the more prevalent red-green, such as R+G− center and R−G+ surround, or the blue-yellow variety); a quarter of them were broad-band center-surround, which gave vigorous responses at all wavelengths; and about 13% were color opponent center-only (COCO) cells. A small percentage of neurons in puffs had broad-band center and tonic red surround, and they exhibited rhythmic, bursting firing. The double-opponent cells are apparently unique to puffs. The broad-band and COCO cells are comparable to type III and type II cells, respectively, in the parvocellular layers of the LGN, whereas the last, rare type resembles type IV cells in the magnocellular layers. Spontaneous activity was high within puffs but low in interpuffs. Cells within interpuffs were highly selective for orientation, and many of them were also color-coded.

Subsequent studies showed variations in the above scheme. Ts'o and Gilbert (1988) reported very few true double-color-opponent cells in puffs. They also found neurons with parvocellular, LGN-like receptive fields, including center-surround spatial and chromatic opponency and center-only chromatic opponency. In addition, there were color-selective cells with no geniculate counterpart that exhibited center color opponency with a non-chromatic opponent surround antagonism, which they named modified type II cells. Individual puffs appeared to process a single-color opponency, and the R-G ones were three times more prevalent than the B-Y ones. Moreover, puff neurons with similar receptive field type, ocular dominance, and color opponency interacted with each other via horizontal connections. Cells in the interpuffs were primarily orientation-selective, and those with the same orientation and eye preference also interconnected horizontally. Yoshioka and Dow (1996) further noted that puff neurons preferred end-spectral colors (e.g., blue and red) and were non-oriented, whereas interpuff neurons were oriented and preferred midspectral colors (e.g., yellow and green).

Johnson et al. (2008) did find double-opponent cells, but most of them were orientation-selective to color or achromatic stimuli. Unfortunately, no COX histochemistry was performed, so the exact locations are not clear. Callaway and colleagues (Garg et al. 2019) used a fluorescent calcium indicator GCaMP6f and two-photon imaging to monitor a large number of neurons. They noted that unoriented, color-preferring neurons were located mainly in COX-rich regions (presumably puffs), whereas strongly color-preferring, orientation-tuned neurons were located in regions with lower COX labeling (presumably interpuffs). However, 11.6% of their neurons were strongly color-preferring and orientation-tuned. On the other hand, neurons that responded most strongly to green tended to be where COX labeling was the densest, and they were the least orientation-selective. Using an acutely implanted 100-electrode array, Economides et al. (2011) found that, whereas puff neurons were less orientation-selective than interpuff ones, the difference is very subtle, suggesting that the processing of form and color is not strictly segregated in V1 as Livingstone and Hubel originally proposed.

A series of studies by the Tootell lab (1988a, 1988b, 1988c, 1988d, 1988e) uncovered a number of other functional properties that distinguished puffs from interpuffs. Using the 2DG method, they confirmed that puffs were centered on ODCs and contained neurons that were selective for color (such as double opponency) but non-selective for stimulus orientation. They also documented that puff neurons responded to stimuli of low spatial frequencies, whereas those in interpuffs preferred gratings of high spatial frequencies.

Is there a distinct physiological border between puffs and interpuffs? Kaplan's group described a continuum from puff centers to the interpuffs (Edwards et al. 1995). Neurons with high contrast sensitivity, comparable to that of magnocellular LGN cells, were clustered near the puff centers, whereas neurons with poor contrast sensitivity, like those of parvocellular LGN cells, were detected throughout layers II/III. These findings are consistent with puff neurons receiving both magno- and parvocellular LGN input (indirectly via layer IV), whereas interpuffs receive input mainly from parvocellular LGN. Neurons in puff centers preferred stimuli of low spatial frequencies, while those in interpuffs responded to spatial frequencies over a wider range.

In sum, color coding and orientation specificity appear to coexist in both puffs and interpuffs, but puffs tend to have more non-oriented ones. Puff neurons are likely to be monocularly driven, respond to stimuli of low spatial frequency gratings, and maintain a high level of spontaneous activity (Livingstone and Hubel 1984a; Tootell et al. 1983, 1988a-e; Ts'o and Gilbert 1988; Silverman et al. 1989; Ts'o et al. 1990; DeYoe et al. 1995; Economides et al. 2011). These properties, except for color, all contribute to a higher energy demand and hence higher metabolic activity. Color processing, per se, is not likely to be the main driving force for greater metabolic activity for two main reasons. First, interpuff neurons also process color; and second, puffs are present in nocturnal animals, such as the owl monkeys (Horton 1984; Tootell et al. 1985). Those color neurons that have a high firing rate (see above), however, would have a greater metabolic demand. Regarding the other properties, a higher level of spontaneous activity in puffs would definitely require more energy. Being monocular means that there is less competition and less interference from the other eye. A low level of orientation specificity indicates that cells may respond to stimuli of all orientations, and, therefore, are more frequently stimulated. Low spatial frequency preference implies broader spatial frequency and orientation bandwidths than those with high spatial frequencies, and cells probably receive more chronic stimulation (Silverman et al. 1989). An enigma with the owl monkeys is that O'Keefe et al. (1998) did not find any difference in eye dominance, orientation, contrast sensitivity, spatial and temporal frequency tuning between puffs and interpuffs, but they did find fewer direction-selective cells in puffs than in interpuffs. Overall, levels of spontaneous and synaptically driven activities likely play the most important role in the greater energy demand of puffs.

4.1.5 Afferent Inputs to Puffs and Interpuffs

Puffs receive input from the ipsilateral LGN, either directly or indirectly via layer IV of the visual cortex (Livingstone and Hubel 1982). In squirrel monkeys, a direct input exists from intercalated neurons surrounding the magnocellular laminae of the LGN to puffs (Fitzpatrick et al. 1983).

Likewise, puffs (but not interpuffs) in the prosimian *Galago crassicaudatus* receive direct input from the small, koniocellular laminae of the LGN, establishing the K pathway (Lachica and Casagrande 1992). In macaque monkeys, however, this source has been questioned (Tootell et al. 1988c). Instead, puffs likely receive a strong input from the parvocellular LGN layers via layers IVCβ and IVA, as well as a weaker magnocellular input via IVCα (Tootell et al. 1988c, 1988d). Using iontophoretic injections of biocytin in specific cortical layers, Yoshioka et al. (1994) reported that both puffs and interpuffs received both parvocellular (P) and magnocellular (M) inputs. Interpuffs obtain them directly from mid-layer IVC, whereas puffs get them indirectly from layers IVB and IVA, with the M pathway relayed from IVCα to IVB, and the P pathway from IVCβ to IVA. Puff neurons are intrinsically connected with neurons in neighboring puffs (but not with interpuffs). Likewise, interpuff neurons are connected mainly with other interpuff neurons, avoiding the puffs (Livingstone and Hubel 1984b)

4.1.6 Are Puffs Specific to Primates?

Horton (1984) examined seven primate species (cynomolgus, owl, rhesus, and squirrel monkeys, bushbaby, baboon, and human), all of which have COX-rich "patches." He also looked at seven non-primate species (mouse, rat, ground squirrel, cat, rabbit, mink, and tree shrew), and none of them exhibits any non-homogeneous COX labeling in the supragranular layers. More detailed studies of various primate species have also been done over the years, and they have all shown puffs (Olavarria et al. 2022). These metabolic hot spots, then, have been accepted as a distinct feature of the primate striate cortex. However, COX-rich puffs also reportedly exist in the cat (Murphy et al. 1995). As the labeling is much weaker in cats than in monkeys, Shoham et al. (1997) generated contrast-enhanced images of COX and demonstrated that COX-rich "blobs" coincided with domains that processed low spatial and high temporal frequency stimuli, whereas the surrounding COX-poor "interblobs" corresponded to high-spatial and low temporal frequency domains. These attributes are very similar to those found in primates.

4.2 COX LABELING OF GRANULAR AND INFRAGRANULAR LAYERS

Layer IV (or 4) is the major geniculocortical-recipient layer in the striate cortex and is highlighted by COX. In tangential sections, IVA takes on a COX-reactive, honeycomb-like appearance (Figure 4.1D inset). Layer IVB is only lightly reactive for COX. Layer IVCα, which receives magnocellular LGN input, is highly sensitive to changes in contrast and temporal content but has low sensitivity to wavelength and spatial resolution. It is slightly more COX-reactive than IVCβ, which receives parvocellular input, has low contrast sensitivity, high spatial resolution, and responds to color and high spatial frequency gratings (Tootell et al. 1988d, 1988e). The base of IVCβ, however, is again darkly reactive for COX and has been named "IVCβdark" (Wong-Riley 1994). This pattern in IVC is reminiscent of that in the LGN, where the magnocellular layers are more darkly reactive for COX than the parvo ones, but lamina 6 is quite reactive (Kageyama and Wong-Riley 1984; see Chapter 3, Section 3.2). It is not known if IVCβ dark receives a major input from the ipsilateral geniculate lamina 6.

The infragranular layer VI is moderately reactive for COX, while layer V is less reactive. Layer VI also receives some geniculate input, and it has magnocellular characteristics like those of IVCα and IVB. Layer V, on the other hand, shares parvocellular properties with IVA, IVCβ, and supragranular layers (Tootell et al. 1988d). Neurons in those portions of infragranular layers and IVB that align in register with the supragranular puffs also prefer stimuli of low spatial frequency, whereas those that align with interpuffs respond best to stimuli of high spatial frequency (Tootell et al. 1988e).

In cross sections, a pillar of higher COX activity is seen from layer I to VI and is centered on puffs. This gives a periodicity not only to puffs but also to layers IVB, V, and VI. Livingstone and Hubel (1984a) have also referred to these parts as "blobs." A schematic diagram of COX labeling in the macaque striate cortex is shown in Figure 4.3 (Wong-Riley 2004).

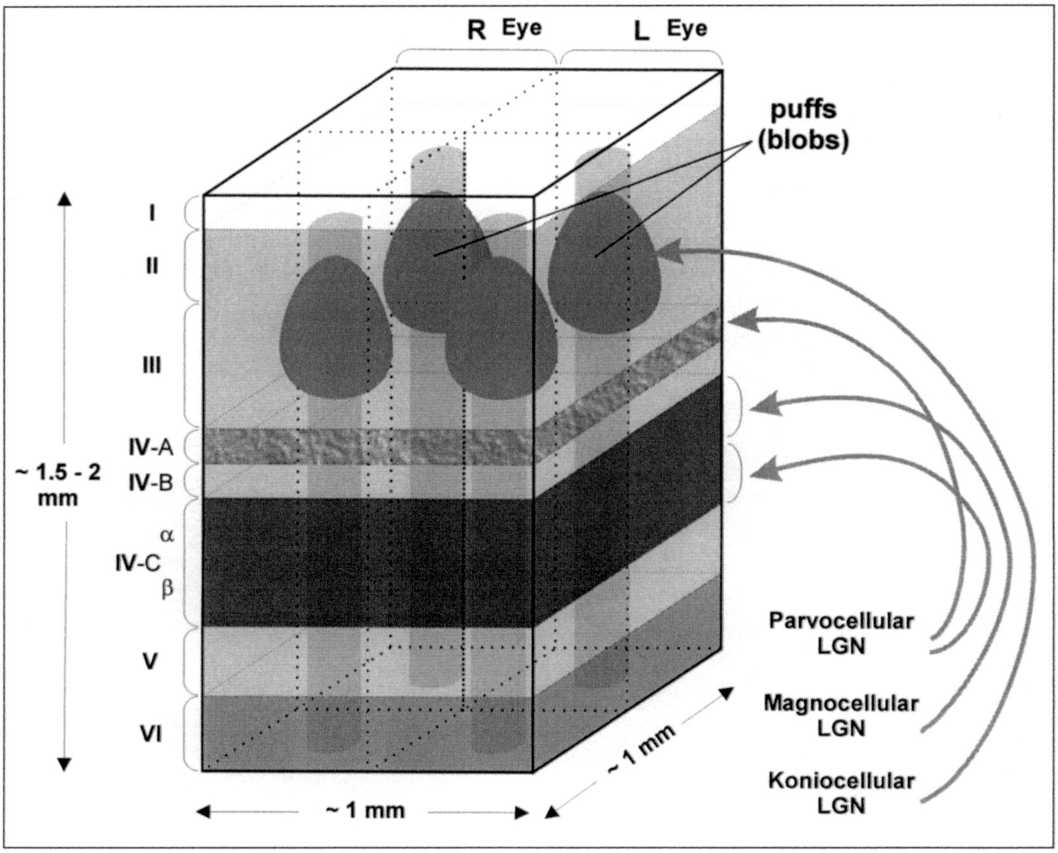

FIGURE 4.3 Schematic diagram of the pattern of COX labeling in the macaque striate cortex. Layers are indicated on the left, and inputs from different layers of the LGN are shown on the right. Layers II/III puffs are centered on ocular dominance columns (R eye and L eye). The term "blobs" refers to puffs and COX-reactive portions of layers IVB, V, and VI below the puffs. Layer IVA is reticulated, layer IVC is intensely labeled, whereas layer VI is more moderately labeled. Only major projections from various LGN laminae to the striate cortex are indicated here.

From Wong-Riley 2004.

Notably, the microvascular density of the macaque primary visual cortex is significantly correlated not with cell density, but with the steady-state metabolic demand indicated by COX activity (Weber et al. 2008).

4.3 PRESTRIATE CORTEX, AREA 18, OR V2

Beyond the striate cortex and directly adjacent to it is the prestriate cortex, area 18, or V2. It has a distinct metabolic pattern of its own, highlighted by COX. In the squirrel monkey, this region, about 6 mm wide, forms, in tangential sections, alternating thin and thick stripes extending orthogonally from the V1/V2 border, with pale or interstripes in between (Livingstone and Hubel 1982; Tootell et al. 1983; Horton 1984; Wong-Riley and Carroll 1984b). Within each thin and thick stripe are globular zones of higher COX activity known as "V2 puffs" (Wong-Riley and Carroll 1984b) (Figure 4.4). V2 puffs in thick stripes are larger than those in thin ones (700–1,100 μm vs. 400–890 μm in diameter, respectively), and puffs in thin stripes next to the V1/V2 border are even smaller (370–480 μm in diameter). The thick

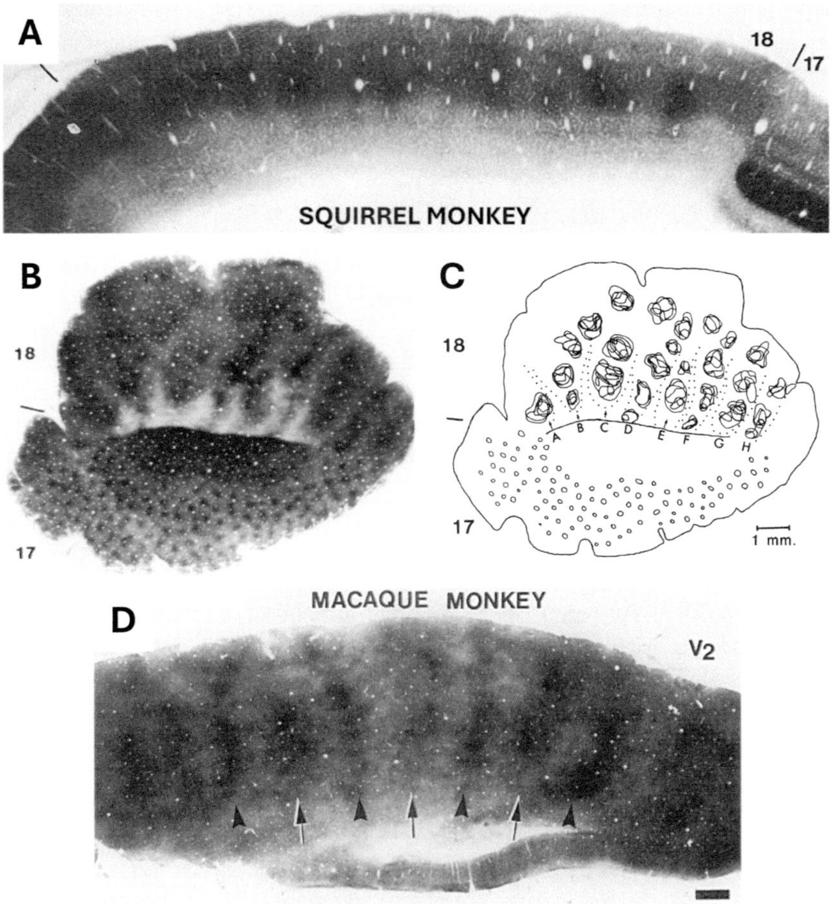

FIGURE 4.4 Monkey V2 (area 18) reacted for COX histochemistry. (A) Horizontal section of squirrel monkey visual cortex showing the abrupt transition of COX labeling from the striate cortex (area 17 or V1) on the right to the adjacent area 18 (V2). Note the globular zones of intense COX labeling (also called puffs) in layers II/III and upper IV of V2. (B) Tangential section through squirrel monkey V1 (17) and V2 (18). Note the polka dot-like pattern of puffs in V1 and the larger puffs in V2 that are within rows or stripes perpendicular to the 17/18 border. (C) A schematic tracing of the same V2 puffs as in (B) through seven sections spaced 120 µm apart, indicating that the puffs remain globular with some irregular edges and are within alternating wide (A, C, E, G) and narrow (B, D, F, H) bands, known as thick and thin stripes, respectively. Puffs in the thin stripes are closer to the 17/18 border. The light bands in between the thick and thin stripes are known as pale stripes. (D) Tangential section through macaque V2 showing COX-rich puffs within alternating thick (arrows) and thin (arrowheads) stripes, with pale stripes in between.

Scale bar: 1 mm.

A–C: modified from Wong-Riley and Carroll 1984b. D: from Wong-Riley 1994.

stripes range between 1,000 and 1,370 µm in width, whereas the thin ones are 750–870 µm. The pattern is more regular in squirrel monkeys than in macaques. In cross-sections, COX-rich puffs extend from lower layer II to upper layer IV, and a thin band of moderate COX activity can be seen between layers V and VI.

The metabolic cell types in V2 are very similar to those in V1. The most COX-reactive ones in V2 puffs are the medium-sized non-pyramidals and the medium-to-large pyramidals, whereas the small neurons are lightly reactive for COX (Wong-Riley and Carroll 1984b). In the neuropil, most of the mitochondria, especially the D and M ones, reside in dendrites and axon terminals, making symmetric (presumed inhibitory) synapses, whereas those making asymmetric (presumed excitatory) synapses are less COX-reactive. COX-poor regions (non-puffs) in V2 have qualitatively similar neuronal and neuropil composition as in puffs, but they decidedly have fewer D-type mitochondria in both cell bodies and neuropil and fewer asymmetric synapses than COX-rich puffs. As in V1, greater excitatory postsynaptic activity demands a higher level of energy metabolism.

4.3.1 Functional Characteristics of V2

Neurons in thin, thick, and pale (inter) stripes have different functional features. Those in thin stripes tend to be unoriented and wavelength-selective; those in thick stripes are orientation-selective, often binocular disparity-sensitive, and directionally selective; and those in pale stripes are also orientation-selective, and many of them are end-stopped (Hubel and Livingstone 1985, 1987; Shipp and Zeki 1985; DeYoe and Van Essen 1985, 1988; Livingstone and Hubel 1988). The fact that within each thin and thick stripe there are globular zones (puffs) with even higher COX activity suggests that further functional subdivisions exist within each stripe.

4.3.2 Specific Connections Between V1 and V2

Remarkably, connections between V1 and V2 appear to be functionally precise. Thin stripes in V2 receive input from puffs in V1, pale stripes from interpuffs in V1, and thick stripes from layer IVB of V1 (Hubel and Livingstone 1987; Maunsell and Newsome 1987; Livingstone and Hubel 1988; Zeki and Shipp 1988). Thus, thin and pale stripes represent the parvocellular system selective for form and color, and thick stripes the magnocellular system specialized in stereopsis (depth perception) and movement. An alternative arrangement was found by Horton and his colleagues (Sincich and Horton 2002; Sincich et al. 2010). Whereas the bulk of projections to thin stripes comes from puffs, both pale stripes and thick stripes receive their major input from interpuffs. Additionally, all three stripes receive minor inputs from layers IVA, IVB, and V/VI, with thick stripes receiving more contributions from IVB than do pale or thin stripes (23% of their striate input vs. 10% for pale and 3% for thin stripes). Thus, there is a mixing of parvo- and magnocellular streams. Regardless, cross-talk between visual channels is evident (Levitt et al. 1994).

4.4 BEYOND V2

Projections from V2 are functionally precise. Thin stripes and interstripes project to V4, which specializes in color and other attributes of object recognition, whereas clusters of neurons in thick stripes project to the middle temporal area (MT) or V5, which processes motion and is directionally selective for spatial guidance (Shipp and Zeki 1985; DeYoe and Van Essen 1985, 1988). These two are the ventral and the dorsal streams, comparable to the What and Where streams, respectively, of Ungerleider and Mishkin (1982). In the Cebus monkey, an additional visual information-processing stream has been reported: a dorsomedial stream from thick stripes and interstripes to parietooccipital areas that represents mainly the binocular and monocular periphery of the visual field (Nascimento-Silva et al. 2003). These authors refer to the MT-projecting pathway as the dorsolateral stream. All three stripes in V2 reportedly also project to the ventral posterior area (VP) or V3v, indicating that VP may process color, form, space, and disparity (Nakamura et al. 2004). Direct projections from the foveal representation of V1 (both puffs and interpuffs) to V4 and from V2 (both thin stripes and interstripes) to the temporooccipital area TEO suggest that coarse information about form and color may rapidly reach the temporal lobe (Nakamura et al. 1993). Many other connections have also been reported, indicating multistream processing (Felleman and Van Essen 1991; DeYoe et al. 1994; Lyon and Kaas 2002). Visual areas beyond V2, such as MT and the mediolateral face perception region of the inferotemporal cortex highlighted by COX, are found in multiple primate species (Krubitzer and Kaas 1990; Oishi et al. 2024). A representation of COX labeling in V1, V2, and extrastriate cortical areas in humans is shown in Figure 4.5 (Wong-Riley et al. 1993).

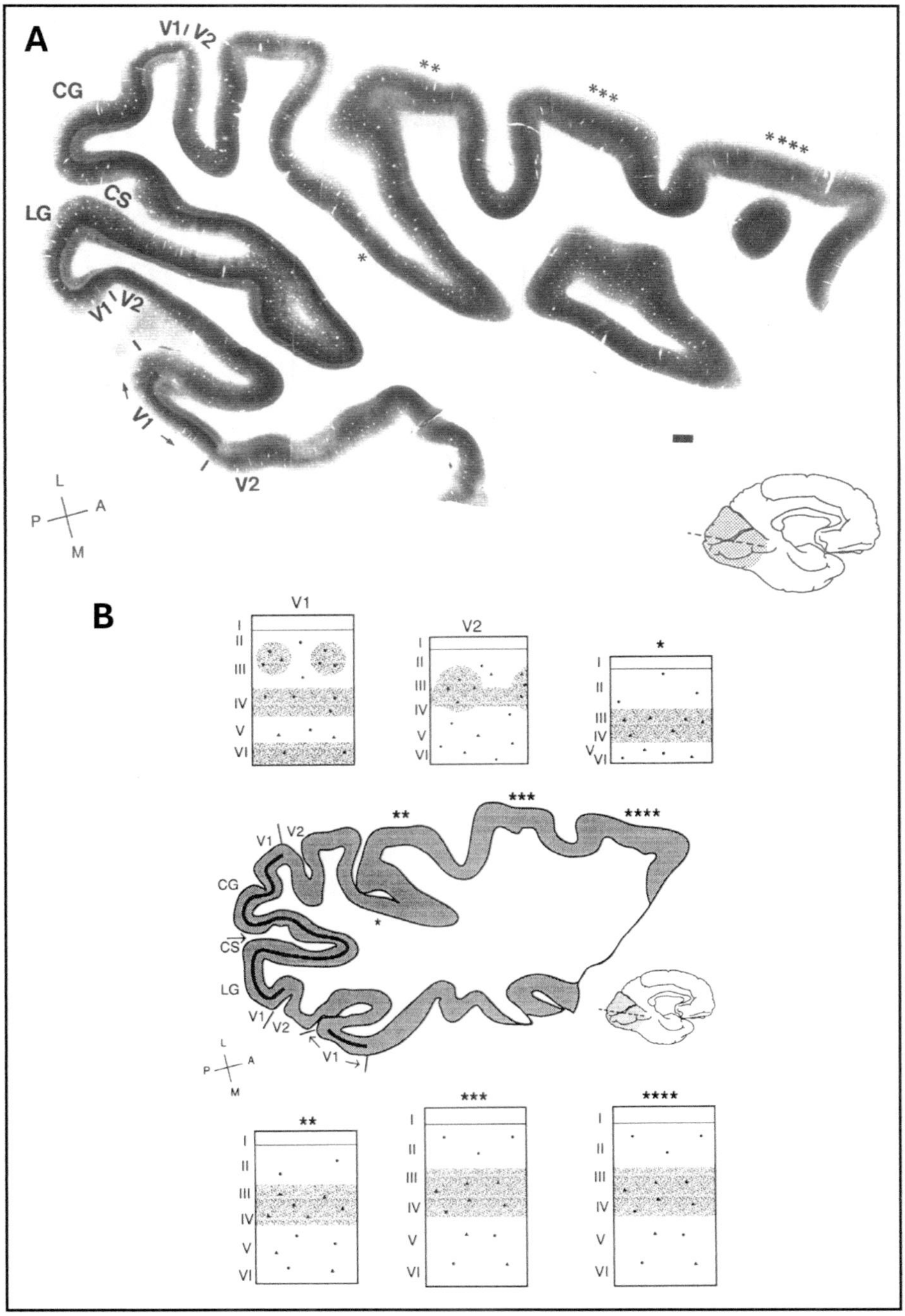

FIGURE 4.5 Human visual cortex reacted for COX histochemistry. (A) Horizontal section through the human occipital cortex (shaded area diagrammed in the lower right), showing V1 extending from the V1/V2 border in the cuneus gyrus (CG), through the calcarine sulcus (CS) to the lingual gyrus (LG), with two more V1/V2 borders ventrally. The pattern of labeling (stippled) is schematically represented in (B). Note that beyond V2, COX highlights mainly layers III and IV, with moderate labeling in layer VI. Black dots indicate the distribution of COX-reactive neurons. The single asterisk denotes the anterior base of the gyrus next to CG, and areas marked with one, two, three, or four asterisks all belong to the extrastriate cortex.

Scale bar in A = 1 mm.

Modified from Wong-Riley et al. 1993.

5 Metabolic Plasticity in the Adult Primate Visual Cortex

Not All Adult Neurons Respond Alike to the Same Functional Insult

ABSTRACT

Adult macaque monkeys remain responsive to visual deprivation by downregulating their COX levels in deprived ocular dominance columns, supragranular puffs, and interpuffs within the primary visual cortex. The effect is least severe with monocular lid suture and most severe with monocular enucleation, which is irreversible. Monocular impulse blockade, on the other hand, induces swift but reversible changes in deprived cortical columns and puffs. Metabolic plasticity in the visual cortex is also evident with retinal laser lesions and in humans with macular degeneration. Not all mature neurons respond alike to the same functional insult. The metabolically most active neurons incur the greatest loss of COX from functional deprivation.

Functional disruption during the critical period of postnatal visual system development can have dire structural and physiological consequences (see Chapter 16). However, visual perturbations in the adult appear to be virtually innocuous (Wiesel and Hubel 1963a). Monocular lid sutures in adult monkeys for 6 ¼ months do not yield detectable electrophysiological alterations in the deprived visual cortex (Blakemore et al. 1978). Long-term monocular enucleation in cynomolgus or macaque monkeys (but not in squirrel monkeys) does cause cellular shrinkage in deprived ocular dominance columns, shown by the Nissl stain (Haseltine et al. 1979; Hendrickson and Tigges 1985), but less attention was paid to functional alterations.

The sensitive marker COX has revealed metabolic changes in the adult visual system not previously recognized. This chapter summarizes some of these findings.

5.1 MONOCULAR LID SUTURE IN ADULT MACAQUE MONKEYS

Eyelid suture in adult monkeys reduces the entrance of light, allowing < 5% of blue light (450 nm), <10% of the green range (500–580 nm), and no more than 15–25% of the longer wavelengths to pass (Crawford and Marc 1976). The loss is mainly form vision, as many retinal ganglion cells remain active in dim light and in darkness, and structural changes are not noted in the LGN (Horton 1984). However, a faint banding pattern is detectable with COX histochemistry in monocularly deprived layer IVC of the visual cortex (Horton 1984; Trusk et al. 1990; Wong-Riley 1994), indicating that deprived cortical neurons are functionally less active. The pattern is more striking with prolonged deprivation (1–3 years), but interanimal variation exists, indicating that some functional recovery is possible (Trusk et al. 1990). Figure 5.1 shows the effect of monocular lid suture on cortical COX activity in a juvenile macaque deprived since birth for 11 weeks, an adult deprived for 11 weeks, and an adult deprived for 48 weeks. The severity is clearly age- and duration-dependent.

Monocular lid suture appears to induce a curious wider pale column and thinner dark column in layer IVC, demonstrable with COX and measured by Horton (1984) and his colleague (Horton and Hocking 1998). They explained that each ODC is made up of a central monocular core zone flanked

DOI: 10.1201/9781003507765-5

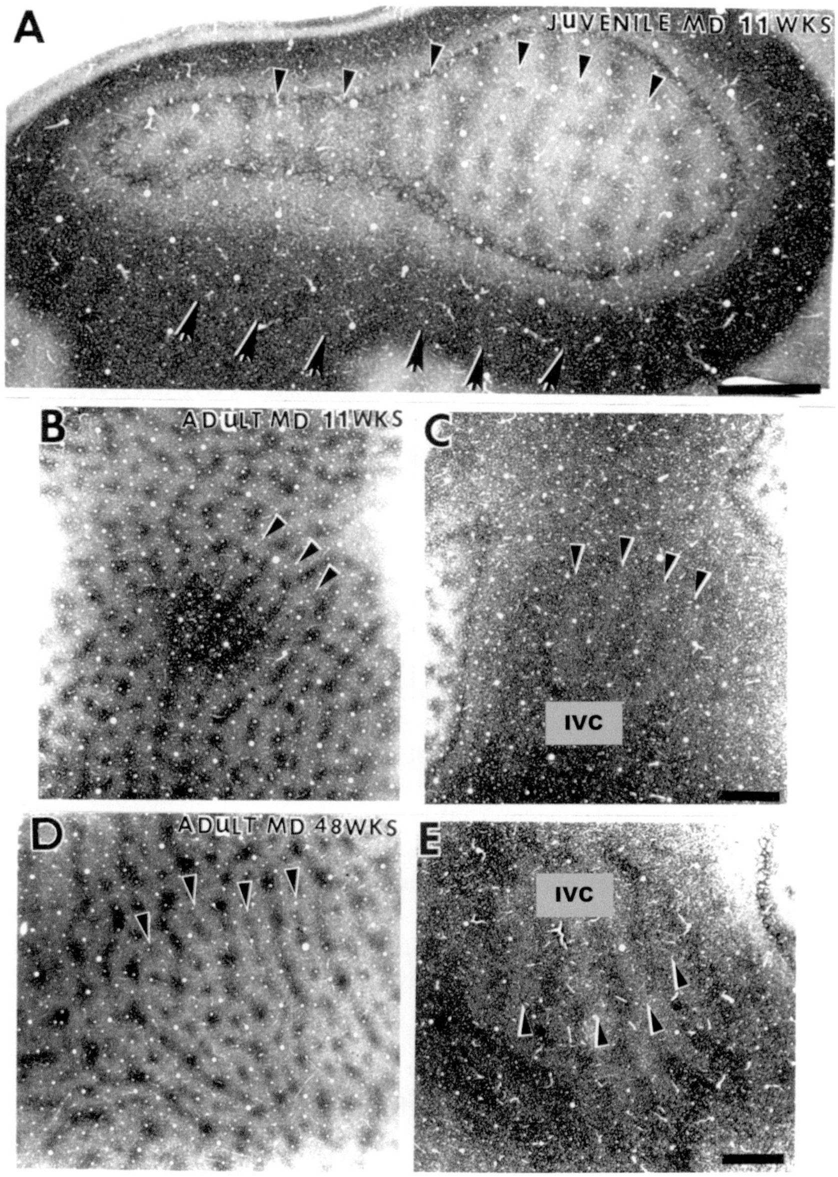

FIGURE 5.1 Effect of monocular deprivation on COX activity in the macaque visual cortex. (A) Tangential section through V1 of a juvenile monkey monocularly lid sutured (MD) for 11 weeks since birth. Note the alternating rows of normal and shrunken puffs (upper small arrowheads) and the faint ocular dominance columns in layer IVC (lower large arrowheads). Similar treatment in the adult has a much milder effect, with indistinct pale puffs (arrowheads in B) and pale bands in layer IVC (arrowheads in C). Monocular deprivation for 48 weeks in the adult results in more distinct alternating rows of normal puffs and pale, shrunken puffs (arrowheads in D) and alternating dark and light bands (arrowheads in E) in layer IVC.

Scale bar = 1 mm.

Modified from Wong-Riley 1994.

by a binocular border zone. Monocular lid suture appears to affect not only both zones in deprived ODCs but also the border zone of nondeprived ODC. However, they did not see a difference in ODC widths in their monocular enucleated or laser-lesioned monkeys.

In the supragranular layers, alternate rows of puffs with reduced COX activity are in register with paler COX bands in layer IVC (Horton 1984; Trusk et al. 1990; Wong-Riley 1994). Puff volume progressively decreases with longer monocular lid suture: 33% after 11 weeks and 53% after 48 weeks to 2½ years ($P < 0.001$ for both) (Trusk et al. 1990; Wong-Riley 1994). A faint pattern of light and dark COX-bands is discernible in the infragranular layers V and VI in register with those in the granular and supragranular layers (Horton 1984; Trusk et al. 1990), indicating that cells in all layers within deprived ODCs are affected.

5.2 MONOCULAR ENUCLEATION IN ADULT MACAQUE MONKEYS

With monocular enucleation in adult monkeys, a much more striking pattern than lid suture appears with COX histochemistry (Figure 5.2A–C). Alternating bands of high and low COX labeling are evident in layer IVC, in register with dark and pale puffs in the supragranular layers and corresponding fainter bands in the infragranular layers, consistent with puffs being centered on ocular dominance columns (Horton and Hubel 1981; Horton 1984; Trusk et al. 1990). Within IVC, the deprived parvocellular-recipient IVCβ has lower COX activity than the magnocellular-recipient IVCα (Horton 1984), and the darkly reactive, non-deprived ODCs in IVCβ are significantly wider than the pale, deprived ones (Trusk et al. 1990; Wong-Riley 1994) (Figure 5.2D), indicating that the parvocellular pathway may be more severely affected than the magnocellular one. The difference in ODC size may reflect a shrinkage in deprived ODCs verified morphologically (Haseltine et al. 1979; Hendrickson and Tigges 1985), a contralateral dominance in ODC size, or inherent variability in ODC periodicity that exists in normal macaques (Horton and Hocking 1996).

Deprived puffs decrease in size and in COX activity, the severity being greater in deeper layer IIIB than in layers IIIA and II. Puff volume is reduced by 34% after 2 weeks and by 53% after 60 weeks of enucleation, at which time, the shrunken ones are difficult to discern (Figure 5.2B) ($P < 0.001$ for both) (Trusk et al. 1990). In comparison, non-deprived puffs are larger, more confluent, and are darkly reactive for COX (Horton 1984; Trusk et al. 1990). A light and dark COX banding pattern exists in layer IVA and the infragranular layers V and VI, in register with the ODCs in layer IVC (Horton 1984).

5.3 MONOCULAR IMPULSE BLOCKADE IN ADULT MACAQUE MONKEYS

Enucleation is an irreversible process. To decipher the effect of functional deprivation alone without denervation or cell death, reversible intraocular impulse blockade with TTX is the method of choice (see Chapter 2, Section 2.3.3.2). The minute volume of TTX (19 μg in 10 μl) permeates the vitreous humor and impedes all impulse activities without causing structural damage (Wong-Riley and Carroll 1984a). Blockage is verifiable by the cessation of pupillary light reflex commencing ~90 min following injection and lasting at least 4 days in adult monkeys (and in adult cats). Maintaining blockage longer than 3 days requires injections twice a week.

5.3.1 Effect on the Retina

All retinal spiking neurons (ganglion cells and some amacrine cells) in the injected eye remain structurally intact, and all non-spiking neurons (photoreceptor, bipolar, and horizontal cells) are also functionally intact. The inner segments of photoreceptor cells, the cone pedicles, and the outer plexiform layer retain high levels of COX. The spiking neurons are functionally silenced and downregulate their COX activity (Figure 5.3). The inner plexiform and nerve fiber layers also show significant reductions in enzyme activity ($P < 0.001$) (Wong-Riley et al. 1998b).

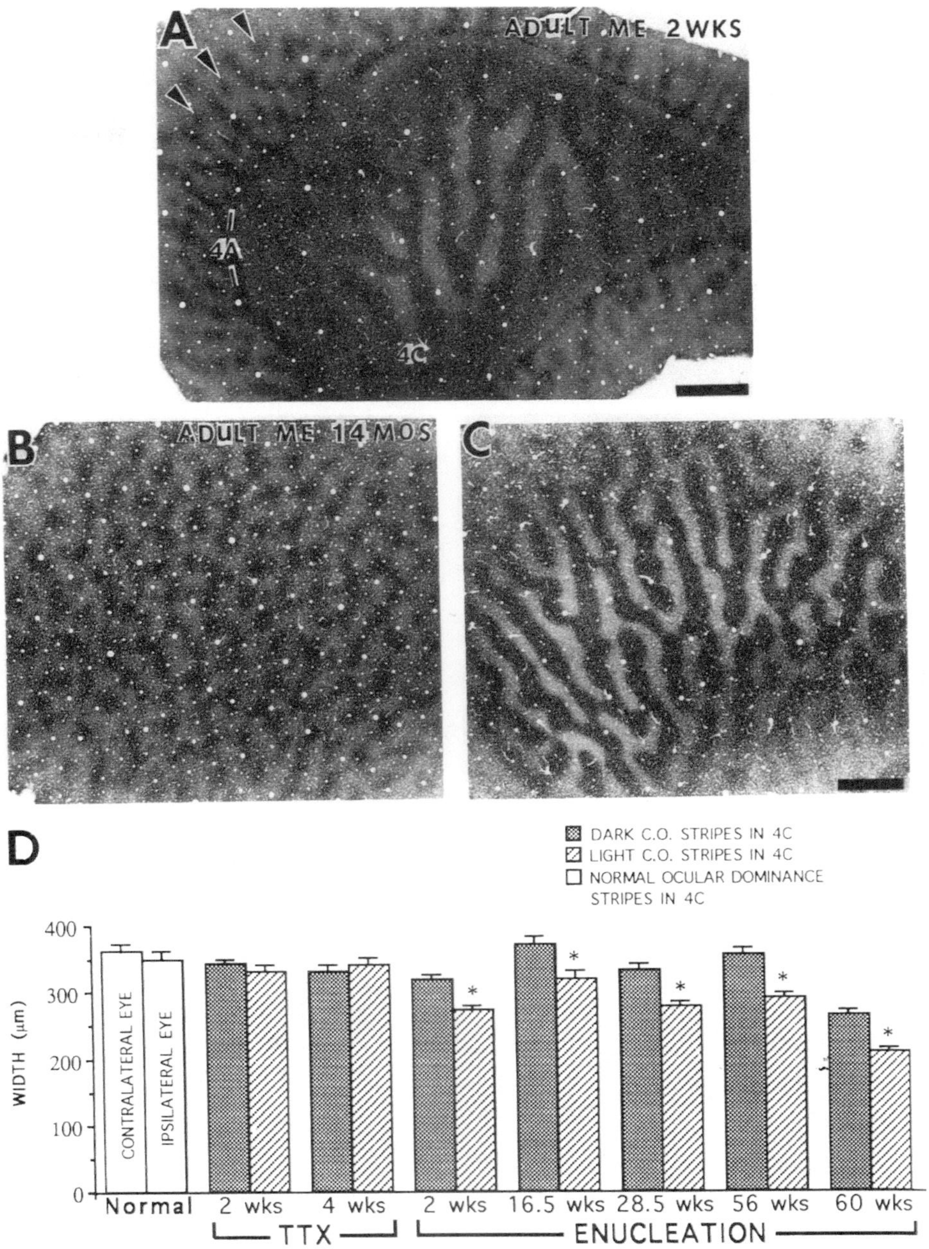

FIGURE 5.2 Effect of monocular enucleation on COX activity in the macaque visual cortex. Tangential sections through V1 of adult monkeys monocularly enucleated (ME) for 2 weeks (A) or 14 months (B and C). Note the banding pattern in layer IVC (4C) (A and C) and the alternating large, dark puffs and shrunken, pale puffs in the supragranular layers. The shrunken puffs are more prominent after 2 weeks (arrowheads in A), but difficult to discern after 14 months (B). Presumably, long-term ME induces a severe shrinkage of deprived puffs and an enlargement of the non-deprived puffs. (D) The mean (± SE) widths of COX-labeled ocular dominance columns in layer IVCβ (labeled as C.O. stripes in 4C) of monkeys either monocularly injected with TTX or survived monocular enucleation for varying periods. The normal monkey received a unilateral injection of ^{3}H-proline that was transneuronally transported to the visual cortex. Note that the widths of light columns are significantly narrower than those of dark columns in enucleated but not TTX or normal animals.

***: *P* < 0.01. Scale bars = 1 mm.**

From Wong-Riley 1994.

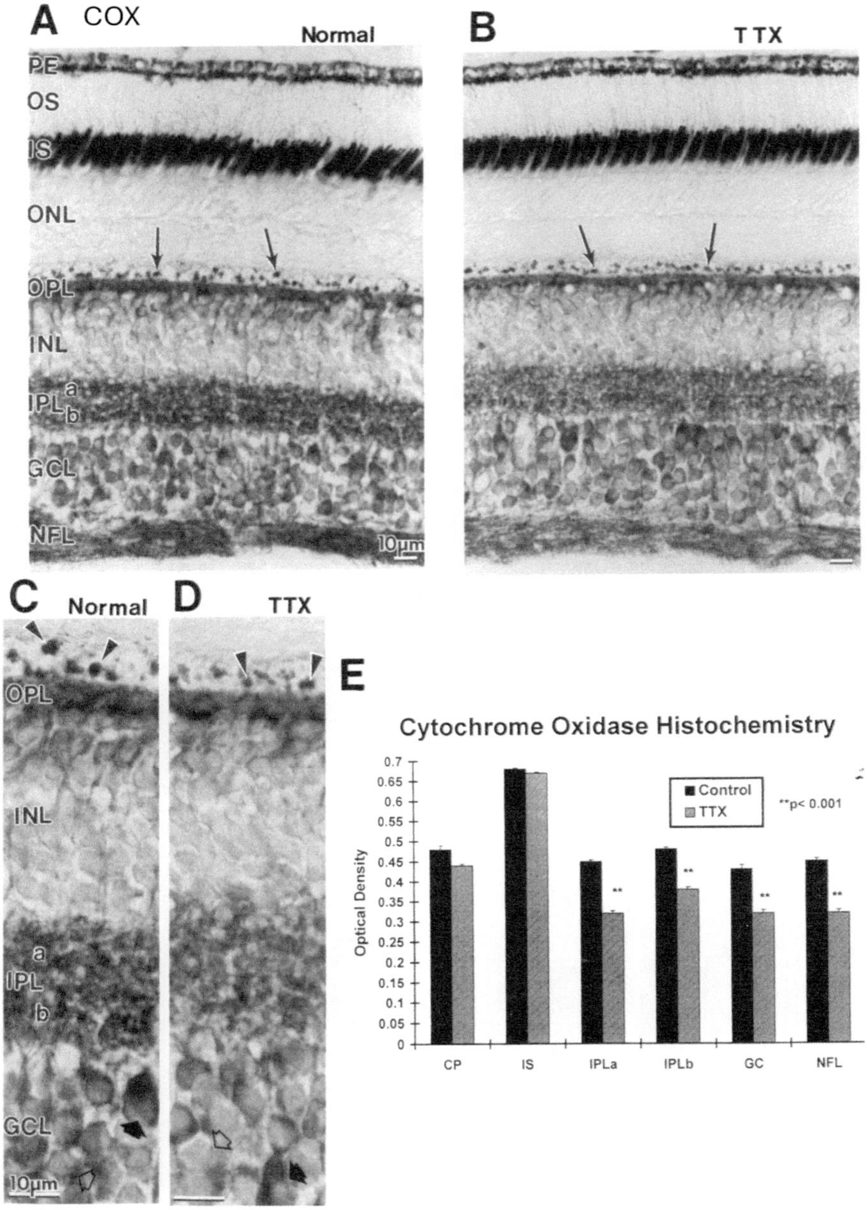

FIGURE 5.3 Effect of monocular TTX on COX activity in the macaque retina. Comparison of COX labeling in normal retina (A) and retina after 4 weeks of TTX (B) (enlarged respectively in C and D). The widths of retinal layers are comparable between the two specimens. Note that the inner segment (IS) of photoreceptors and the cone pedicles in the outer plexiform layer (OPL) (arrows in A and B; arrowheads in C and D) remain highly reactive for COX after TTX, but the a and b layers of the inner plexiform layer (IPLa, b), the ganglion cell layer (GCL), and the nerve fiber layer (NFL) show significantly reduced enzyme activity, which are verified by optical densitometry (E). Higher magnifications of retinal ganglion cells show that those in the TTX-injected eye (D) are less COX-reactive than their counterparts in the normal eye (C) (open arrowheads for lightly reactive and solid arrowheads for darkly reactive cells in each). INL, inner nuclear layer; ONL, outer nuclear layer; OS, outer segments of photoreceptors; PE, pigment epithelium

Modified from Wong-Riley et al. 1998b.

5.3.2 Effect on the LGN

In the LGN, deprived laminae (contralateral 1, 4, and 6 and ipsilateral 2, 3, and 5) downregulate their COX activity, their mtDNA copy number, as well mRNAs of *COX1*, *COX4*, and *COX8* subunit genes (Wong-Riley and Carroll 1984a; Hevner and Wong-Riley 1993) (Figure 5.4).

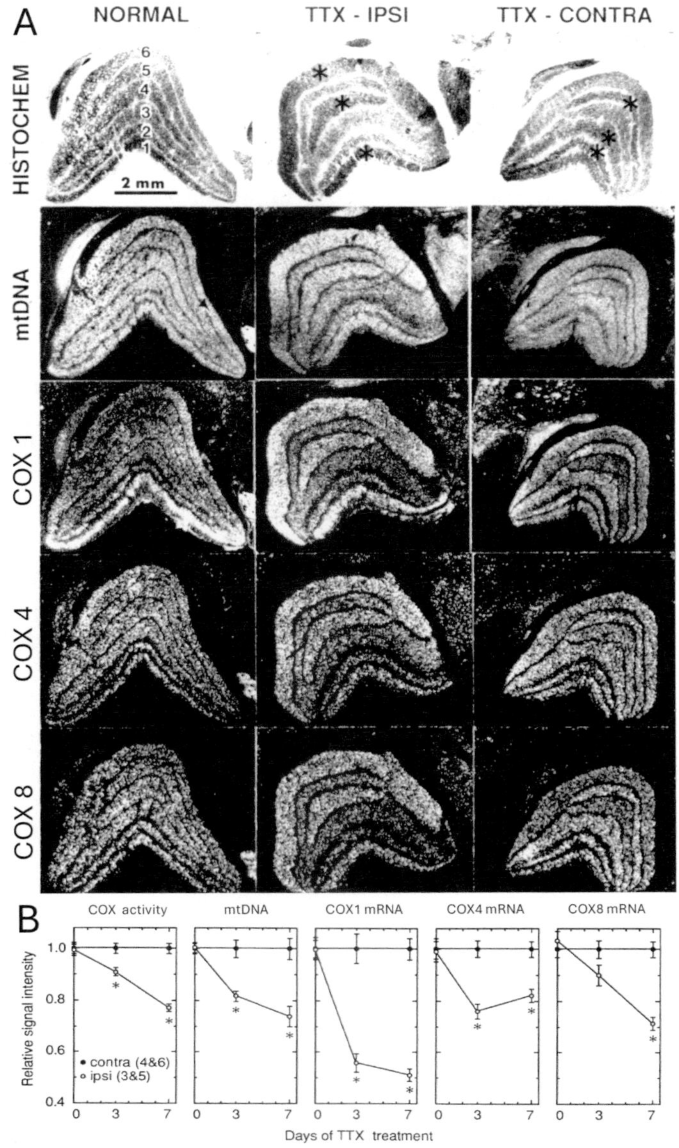

FIGURE 5.4 Effect of monocular TTX on COX activity, mtDNA, and *COX* subunit mRNAs in the macaque LGN. (A) Under normal conditions (left column), COX activity revealed with histochemistry shows the highest levels in laminae 1, 2, and 6 of the LGN, a pattern that is comparable to that of mtDNA and mRNAs of *COX1*, *COX4*, and *COX8* shown with *in situ* hybridization. After 7 days of intraocular TTX, laminae 2, 3, and 5 on the ipsilateral side (middle column) and laminae 1, 4, and 6 on the contralateral side (right column) show reduced labeling of all five markers. Asterisks indicate LGN layers receiving input from the untreated eye. (B) Relative signal intensity of the five markers in individual neurons of deprived layers with reference to those in non-deprived layers (normalized to a unit of 1.0 for each marker) after 0, 3, and 7 days of monocular TTX. Note that *COX1* falls 1.5- to 4-fold more than *COX4* or *COX8*. Values are mean ± SEM of 44 measurements for each.

***$P < 0.001$ as compared to the respective control.**

Modified from Hevner and Wong-Riley 1993.

Significantly, mitochondrial-encoded *COX1* message falls 1.5- to 4-fold greater than those of the nuclear-encoded *COX4* and *COX8*. This indicates that COX activity over days is controlled mainly by mitochondrial genes encoding the catalytic core of the enzyme.

5.3.3 Effect on the Visual Cortex

Mature neurons in the adult visual cortex remain highly sensitive to the cessation of excitatory input from the LGN. Shortly after an intravitreal injection of TTX, there is a brief initial increase in background activity (presumably due to a transient decrease in intrinsic inhibition) followed by a profound depression of both visually evoked and ongoing background spiking activity in the visual cortex (DeYoe et al. 1995). Depressed responses are recordable in layers II/III puffs, interpuffs, and strikingly so in IVB and layers IVC, the last one of which normally has the highest level of spontaneous activity (Figure 5.5). In layer IVC, the average background multiunit spike rate for the TTX-inactivated ODCs is only 8% of that in the normal eye column. Background activity in interpuffs is reduced threefold in < 1 h after TTX injection, but stabilizes within 3–4 h to a level only 53% of that before injection (DeYoe et al. 1995). Puff numbers recorded post-TTX were too few for statistical analysis.

As early as 14 hours (the earliest time examined) after a single, monocular injection of TTX, faint alternating pale and dark COX-labeled bands are discernible in tangential sections of layer IVC in the adult monkey visual cortex (Figure 5.6) (Wong-Riley 1994; DeYoe et al. 1995). No changes are detectable in supra- or infragranular layers at that time. The much earlier TTX-induced functional inactivation preceding reduction in COX confirms that neuronal activity governs energy expenditure revealed by COX.

Longer exposures to TTX yield an increasingly striking, zebra-like pattern of COX activity in layer IVC, in register with dark and light rows of layers II/III puffs (Figure 5.7) and fainter bands in layers IVA and VI (Wong-Riley and Carroll 1984a; Carroll and Wong-Riley 1984; Trusk et al. 1990). Changes in COX activity are accompanied by a parallel decrease in COX protein content revealed by COX immunohistochemistry (Hevner and Wong-Riley 1990). Unlike monocular enucleation, the widths of dark and light ODCs remain the same up to 4 weeks of monocular impulse blockade (Figure 5.2D) (Trusk et al. 1990). In the supragranular layers, deprived puffs lose 17% of their volume after 2 weeks of TTX ($P < 0.001$). This is less severe, though still significant, than the loss after a comparable period of monocular enucleation (see Section 5.2). However, after 4 weeks of monocular TTX, deprived puffs lose 57% of their average volume ($P < 0.001$), a reduction comparable to that after 60 weeks of monocular enucleation. The decrease extends through the depth of layers II–III, but is greatest in layer IIIB (Trusk et al. 1990). As puff size longer than 4 weeks of enucleation is not known, it is possible that maximal reduction in COX activity is achieved after 1 month of retinal input removal.

The labeling pattern of the energy-consuming enzyme Na^+/K^+-ATPase closely parallels that of COX in the retina and visual cortex under normal and TTX-inactivated conditions (Hevner et al. 1992; Wong-Riley et al. 1998a, 1998b). With monocular TTX, dark and light bands of this enzyme in layer IVC and alternating rows of large dark and small pale globular zones of ATPase activity in supragranular layers of V1 match precisely those of COX (Figure 5.8).

Reduced $GABA_A$ receptor immunoreactivity and reduced binding of ligands to $GABA_A$ receptors (^{3}H-muscimol and ^{3}H-flunitrazepam) are noted in deprived layer IVCβ columns after monocular TTX and enucleation in adult macaque monkeys (Hendry et al. 1990). Likewise, immunoreactivity for microtubule-associated protein 2 (MAP-2) is decreased in deprived layer IVC and puffs of adult monkeys after monocular TTX blockade (Hendry and Bhandari 1992). Notably, however,

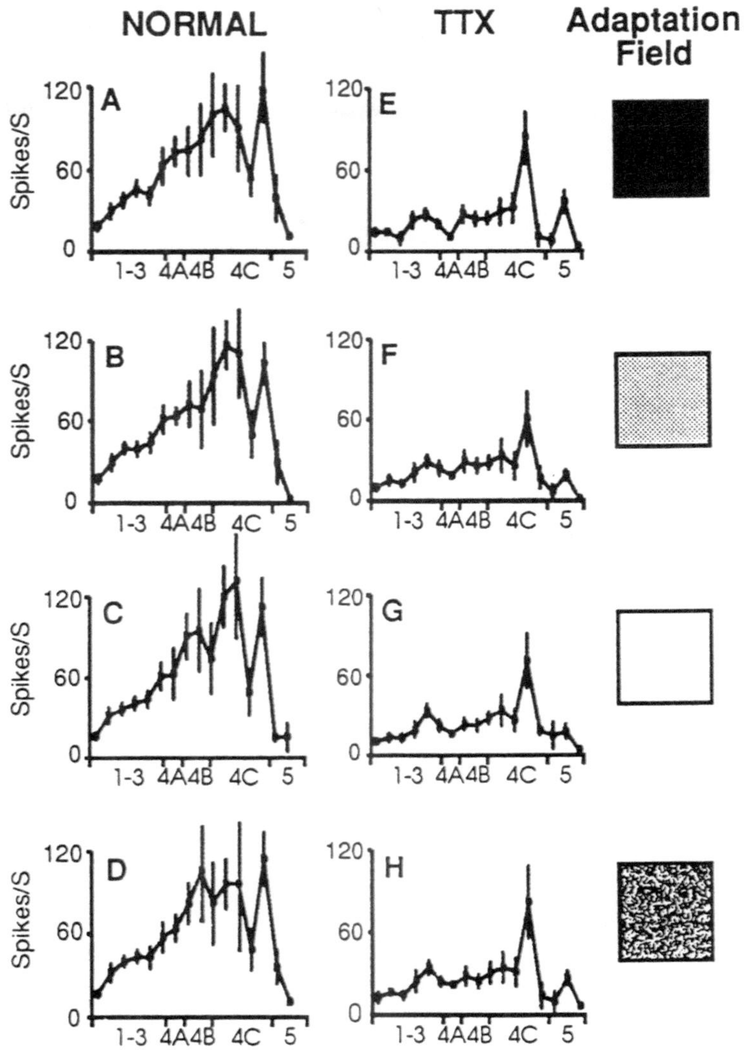

FIGURE 5.5 Spontaneous background activity recorded from different V1 layers of normal (pre-TTX; A–D) and post-TTX (E–H) macaque monkeys viewing four different adaptive fields: Black (<0.1 cd/m^2), uniform gray (20 cd/m^2), white (60 cd/m^2), and random dot (average luminance 20 cd/m^2). Each profile is the average of data from eight penetrations of each layer, and each point is the mean ± SEM. The patterns are similar among the four adaptation conditions. Note that under normal conditions, the spiking rate increases from layers 1–3 to peak in layer IVC (labeled in Arabic numbers), with a dip between the upper half and lower half of this layer and a distinct fall in layer V. Values from puffs and interpuffs are averaged here. Intraocular TTX causes a significant reduction in the background activity of all layers, especially in layer IVC.

From DeYoe et al. 1995.

the transcript level of calcium/calmodulin-dependent protein kinase II-α (CaMK II-α) is upregulated in deprived cortical columns and reduced in nondeprived ones (Benson et al. 1991). The reason for this is unclear, especially when the expression of the beta subunit (CaMK II-β) is downregulated in deprived ODCs (Tighilet et al. 1998).

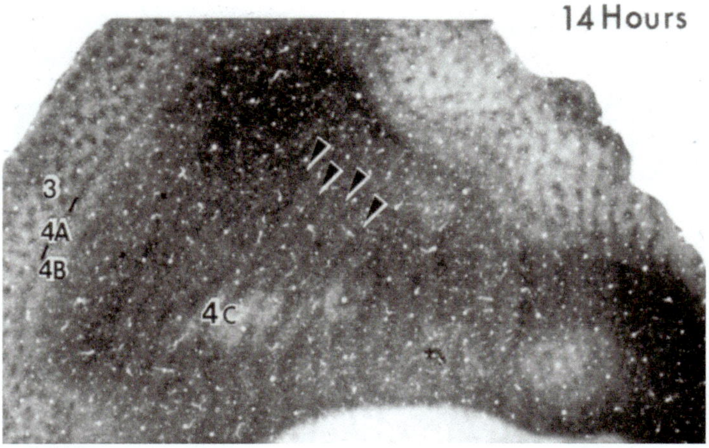

FIGURE 5.6 As early as 14 hours after a single monocular injection, faint alternating bands of dark (arrowheads) and light COX labeling can be discerned in layer IVC, whereas labeling in the other layers, such as puffs in II/III, and honeycomb pattern in IVA appears relatively normal (layer numbering in Arabic)

From Wong-Riley 1994.

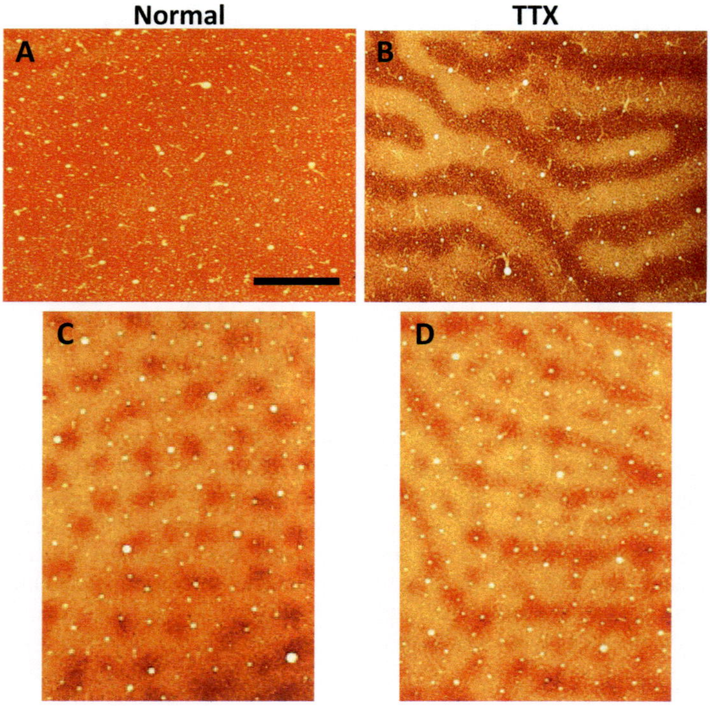

FIGURE 5.7 Effect of monocular TTX on COX activity in the macaque V1. Under normal conditions, COX labeling in tangential sections is homogeneous in layer IVC (A), and puffs in layers II/III are homogeneously dispersed in a polka dot-like pattern (C). After monocular TTX for 4 weeks, an alternating dark and light banding pattern is seen in layer IVC (B), and alternating bands of normal puffs and shrunken paler puffs are visible in layers II/III (D), centering on ocular dominance columns in IVC.

Scale bar = 1 mm.

Colored figures similar to those in Hevner and Wong-Riley 1990.

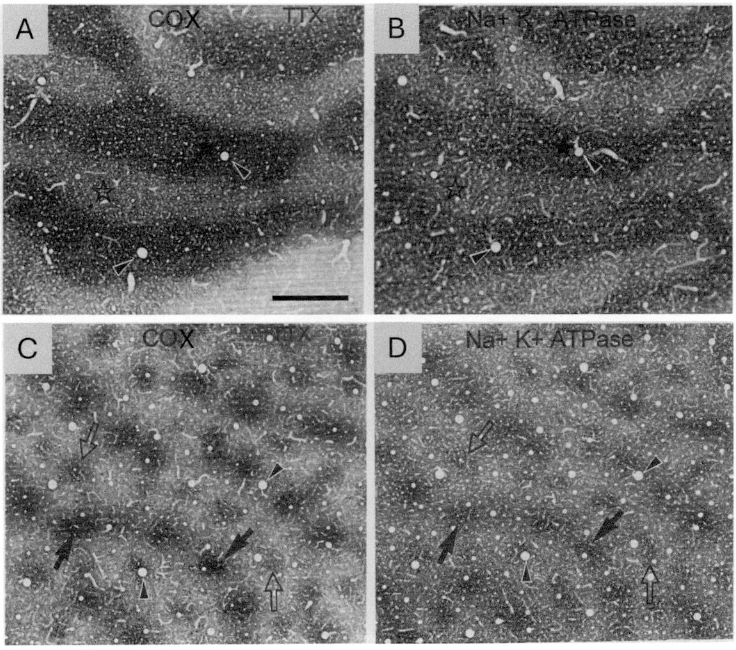

FIGURE 5.8 Changes in Na^+/K^+-ATPase match those of COX in macaque V1 after 4 weeks of monocular TTX. Adjacent tangential sections reacted for COX (A and C) or immuno-labeled for Na^+/K^+-ATPase (B and D) show comparable patterns of dark (solid stars) and light (open stars) bands in layer IVC (A and B; arrowheads indicate blood vessels for alignment) and alternating rows of large dark puffs (solid arrows) and small pale puffs (open arrow) in supragranular layers (C and D; arrowheads indicate blood vessels for alignment). The set in C and D is not aligned with that in A and B.

Scale bar = 0.5 mm.

From Wong-Riley et al. 1998a.

5.3.4 Recovery in the Visual Cortex from Monocular Impulse Blockade

A major advantage of intraocular TTX is that it is an effective but reversible functional blocker. The pupillary light reflex returns within a week after the last injection, and COX activity eventually returns to control levels in the visual cortex (Figure 5.9A and B). The length of recovery depends on the duration of inactivation. For example, after 1 week of TTX treatment, full recovery of COX labeling in deprived layer IVC takes about 6 weeks. With 2 weeks of TTX, full recovery of COX activity and areal size of deprived puffs takes 9 weeks (Figure 5.9C and D) (Wong-Riley 1994). The cortical response is at least two synapses away from the inactivated and recovering retina. Time is needed for new synthesis and assembly of COX subunits from both the mitochondrial and nuclear genomes to generate functional holoenzymes and for reactivating existing synapses and/or establishing new synapses at each stage of visual processing. Ultimately, the system will return to its pre-inactivated state.

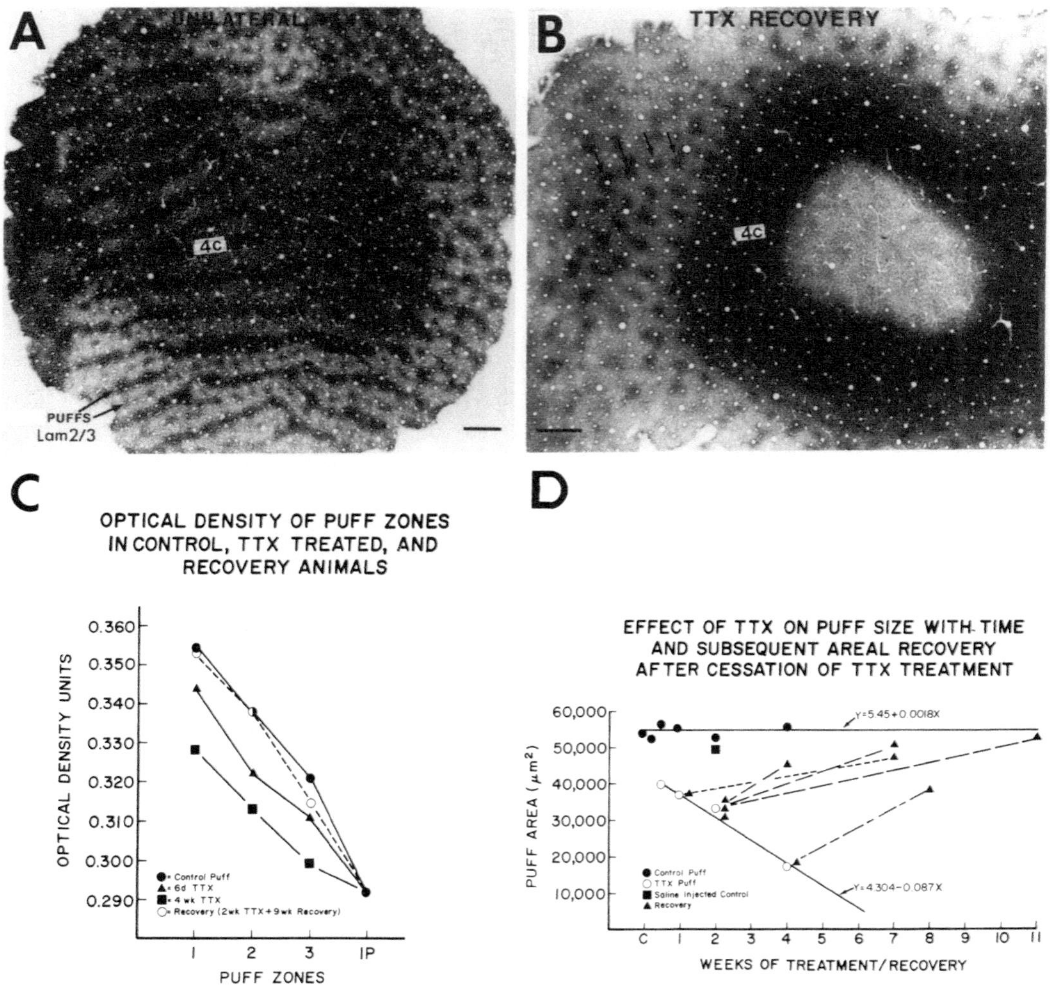

FIGURE 5.9 TTX recovery in the monkey V1. (A) Tangential section of a macaque striate cortex after 4 weeks of monocular TTX. Note the zebra-like banding pattern of COX labeling in layer IVC (4C) and the alternating rows of dark and pale puffs in layer II/III (arrows; Lam 2/3). (B) Tangential section through several layers of area V1 in a macaque that recovered 6 weeks after 1 week of TTX. The COX labeling pattern is indistinguishable from that of normal monkeys, indicating a complete metabolic recovery from the effect of TTX at the cortical level. Layer IVC (4C) appears as a homogeneous ring around the central, lightly reactive layer V (not labeled). The supragranular puffs (arrows) are of uniform shape and size. (C) Optical density values of COX labeling are highest in the center (zone 1) of puffs and decrease toward the periphery (zone 3), with the lowest in the interpuffs (IP). Solid circles represent control puffs. After 2 weeks (solid triangles) and 4 weeks (solid squares) of TTX, the labeling decreases uniformly in the three zones and is more severe with time ($P < 0.01$ for all). However, 2 weeks of TTX followed by 9 weeks of recovery (open circles) bring the values back to control. (D) The time course of areal recovery of puffs from monocular TTX. Controls are taken from three groups of macaques: normal (solid circles), saline-injected for 2 weeks (solid squares), and TTX control puffs (also solid circles). There is no significant difference among the controls (straight line at the top). Values for TTX-affected puffs (open circles) gradually decrease with longer inactivation. Recovery periods (solid triangles) are as follows: 1 week TTX/6 weeks recovery; 2 weeks/2 weeks; 2 weeks/5 weeks; 2 weeks/9 weeks; and 4 weeks/4 weeks. They all show recovery, but only the 2-week/9-week one reaches control values.

From Wong-Riley 1994; A and B are also from Wong-Riley 1989.

5.4 DISCRETE LASER RETINAL LESIONS

Laser lesions are commonly used clinically to treat retinal detachments and proliferative diabetic retinopathy. What is the effect of such lesions on the visual cortex? Focal retinal lesions were made with a blue–green argon laser (150–400 mW; 600 μm spot size, centered on the vertical midline above but not including the fovea) in adult macaque monkeys (Wong-Riley 1994). Such lesions destroy photoreceptor cells, but the retinal ganglion cells remain intact with normal levels of COX. However, in the laser-damaged representation of the striate cortex (scotoma), alternate rows of dark and light puffs can be seen in register with high and low COX-reactive ODCs in layer IVC (Figure 5.10). A banding pattern is also present in infragranular layers. In the transitional zones

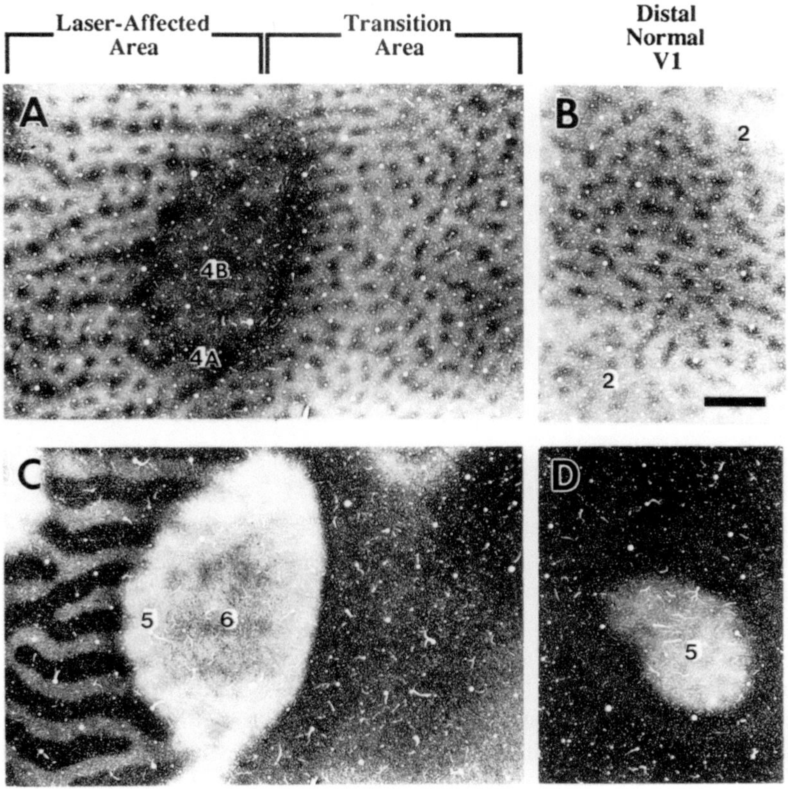

FIGURE 5.10 The effect of focal retinal laser lesion on COX activity in the macaque V1. Tangential sections through V1 of a macaque monkey that has received a focal retinal lesion on the vertical meridian above but not including the fovea. The 600-μm spot lesion was made with a 150–400 mW blue–green argon laser that destroyed the photoreceptors but left the ganglion cells intact. Note the alternating rows of large dark puffs and small pale puffs in the laser-affected area (left half of A) that are in exact register with the dark and light banding pattern in layer IVC (left half of C). A hint of banding is also seen in layer VI (labeled 6 in C). In the transition area, the puffs appear normal (right half of A) and layer IVC is homogeneous (right half of C), but their COX activity is slightly lower than that in an isoeccentric region approximately 10 mm away (B and D). Note also that the interpuffs in the dark rows of puffs in the laser-affected area (left half of A) have a higher level of COX activity than those in the transition area (right half of A) and the distal normal V1 (B). Layers are numbered in Arabic. The majority of puffs in B are in layer III (not labeled), while those that graze through the upper layer II (2; upper right and lower part of B) appear less COX-reactive.

Scale bar = 1 mm.

From Wong-Riley 1994.

(between the laser-affected and normal areas), no banding pattern of puffs or layer IVC occurs, but COX levels are slightly reduced. Interestingly, interpuffs in non-deprived rows of laser-affected areas show heightened COX activity. These results indicate that the destruction of photoreceptors alone, sparing retinal ganglion cells, in adult macaque monkeys severely downregulates the functional activity of visual cortical neurons deprived of normal input. This has important clinical implications when focal laser lesions result from therapeutic procedures or accidental exposures to lasers.

When focal laser lesions destroyed all retinal layers in adult monkeys, the dark and light COX banding pattern in the corresponding cortical areas persisted for months (Horton and Hocking 1998). Non-deprived puffs reportedly "expand" in size (Farias et al. 2019). However, it remains to be determined with electrophysiology whether true "expansion" of puffs occurs or simply an increase in COX activity in the surrounding non-deprived interpuffs.

Takahata et al. (2018) raised macaque monkeys with monocular laser burns restricted to the macula or the peripheral retina at 3 weeks of age and form deprivation in the same eye by wearing a diffuser lens during the critical period (3 weeks to 5 months), followed by 5 years of normal visual environment. They found with COX histochemistry that the laser projection zones (LPZs) in V1 had significant widening of dark ODCs corresponding to the intact eye and shrunken, pale columns representing the lesioned eye (Takahata et al. 2018). This pattern resembles the effect of monocular enucleation (see Section 5.2; Trusk et al. 1990). However, outside the LPZ, the pale columns were wider than the dark ones. It appears that, without binocular competition in the LPZ in infancy, geniculocortical afferents representing the intact eye occupy a larger territory or simply fail to retract. On the other hand, binocular competition remains outside of the LPZ, and, as in monocular suture (Horton 1984; Horton and Hocking 1998), form deprivation affects the border zone of the intact eye column as well, resulting in apparent wider pale columns.

5.5 RETINAL LESIONS ALTER HUMAN VISUAL CORTEX

In a human with verified macular degeneration, COX reveals a low-contrast dark and light banding pattern representing ODCs in layer IV of the macular representation in the visual cortex (Horton and Hocking 1998). In another patient with a long-standing monocular blindness (with severely shrunken optic nerves and transsynaptic neuronal degeneration in the affected LGN layers), the visual cortex displays alternating rows of dark and light COX-reactive ODCs in layer IVC and puffs in supragranular layers, as well as alternate bands of reduced neurofilament (SMI-32) immunoreactivity aligning with the pale COX columns (Duffy et al. 2007). Non-deprived puffs reportedly enlarge in size in humans with retinal lesions (Marcondes et al. 2024).

Thus, adult neurons remain responsive to functional perturbations by adjusting their metabolic levels, even before detectable structural alterations, that is, they exhibit metabolic plasticity. COX is a sensitive indicator of such metabolic adjustments.

5.6 DO ALL ADULT NEURONS RESPOND ALIKE TO THE SAME FUNCTIONAL INSULT?

It is reasonable to assume that all neurons within a restricted region, such as puffs, would be affected by functional insult to the same degree, yet this is not the case. Which neuronal type is most vulnerable?

In normal puffs, three metabolic neuronal types exist (see Section 3.1). Briefly, 57% are small, glutamatergic type A neurons with mostly lightly COX-reactive (L) mitochondria; 18% belong to medium and large, glutamatergic type B pyramidal neurons with mainly moderately reactive (M) mitochondria; and 25% are the medium-sized, GABAergic type C neurons densely filled with darkly reactive (D) mitochondria. Both type A and type B neurons receive only inhibitory, GABAergic synapses on their cell bodies, and type C exclusively receive both GABAergic and excitatory glutamatergic axo-somatic synapses.

How do they respond to monocular TTX? Type A neurons display little change in cell size or percent distribution (Figure 5.11A). Their limited numbers of D and M mitochondria are further and significantly reduced in both numerical and areal densities after 2 weeks of TTX ($P < 0.001$ for D, and < 0.05–0.01 for M) but return to normal after 4 weeks, whereas the count of L type is significantly increased after 4 weeks ($P < 0.01$). Type B neurons also exhibit no change in percent distribution, and their slightly smaller cell size after 4 weeks of TTX is not significantly different from controls (Figure 5.11B). Their mitochondrial population, however, undergoes a notable reduction in numerical densities of D and M types and areal density of the D type after 2 weeks of TTX ($P < 0.01$ for all). However, these return to normal levels after 4 weeks (Wong-Riley et al. 1989a).

By far, the most remarkable change is in type C cells. Although retaining the same percent distribution, indicating no cell loss, this is the only cell type that exhibits significant shrinkage in cell size after 2 and 4 weeks of TTX ($P < 0.05$ for both) (Figure 5.11C). Their normally abundant D mitochondria are dramatically reduced in numerical and areal densities at 2 and 4 weeks ($P < 0.05$) with a concomitant increase in L-type mitochondria. Overall areal densities of mitochondria are severely reduced at the two survival time points ($P < 0.001$ for both). All three categories of mitochondria exhibit slight reductions in mean size but retain the positive relationship between mean size and COX level. This relationship holds for all three metabolic cell types under normal and TTX conditions (Wong-Riley et al. 1989a).

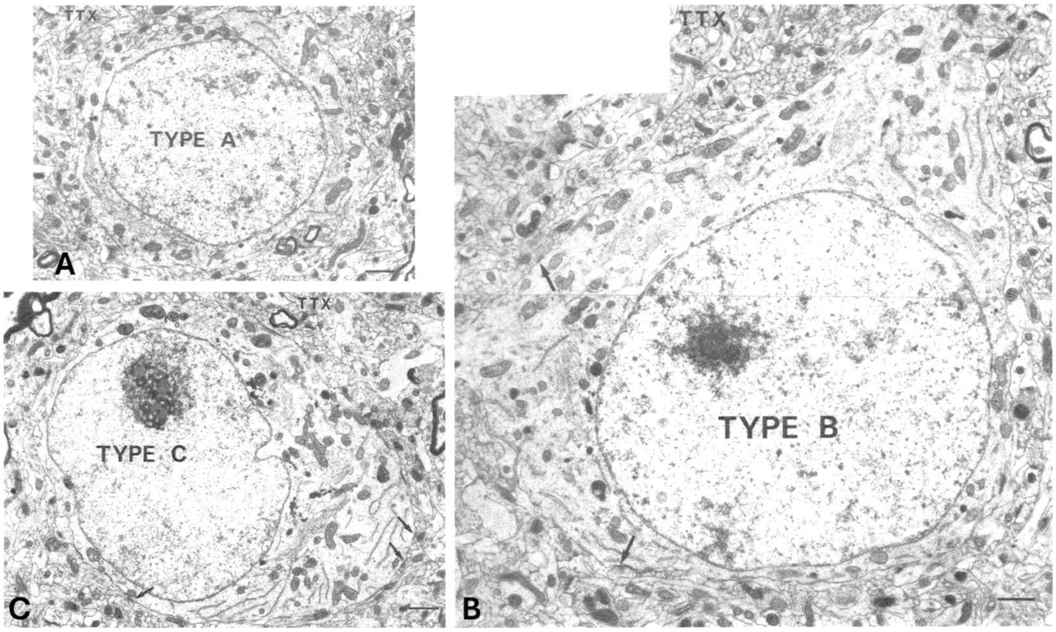

FIGURE 5.11 Electron micrographs of the effect of 4 weeks of monocular TTX on the three major neuronal types in V1 puffs. Compare this figure with Figure 3.1, which shows the cell types under normal conditions. (A) A small type A cell in deprived puff shows very little change, and its mitochondria remain lightly reactive for COX. (B) A large type B cell in deprived puff contains mainly lightly reactive mitochondria, but it is structurally intact and still receives symmetric synapses (arrows) on its cell body. (C) A medium-sized type C cell in deprived puff contains mostly moderate to lightly COX-reactive mitochondria. It is the only cell type that shows a significant shrinkage in cell size, but it still receives both asymmetric (arrows on the right) and symmetric (arrow on the left) synapses on its cell body.

Scale bars = 1 μm.

Modified from Wong-Riley et al. 1989a.

Therefore, not all mature neurons within a small subregion of the visual cortex respond alike to the same functional insult. The metabolically most active neurons (type C) are the most perturbed by functional inactivation alone, without any denervation or cell death. Shrinkage of cell size results presumably from disuse, leading to a severe downregulation of metabolic activity, converting mitochondria to a less active form with diminished packing density. The metabolically least active neurons (type A) are the least changed, though the few D mitochondria are also converted to the L type. An analogy is that bed rest affects trained athletes' muscles much more than those less trained in the general population (Spiering et al. 2023).

5.7 NEUROPIL ALTERATIONS IN DEPRIVED PUFFS AFTER MONOCULAR IMPULSE BLOCKADE

Monocular TTX for 2 or 4 weeks does not result in significant changes in either numerical or areal densities of dendrites, but their mean size is significantly reduced at 4 weeks ($P < 0.001$) (Wong-Riley et al. 1989b). As described in Chapter 3 (Section 3.1), at least 60% of mitochondria reside in dendrites of normal puffs, and most are the D and M types. Monocular impulse blockade progressively decreases both the numerical and areal densities of D and M mitochondria and increases the L-type in dendrites. Likewise, TTX converts 11% of mitochondria residing in axon terminals making symmetric (presumed inhibitory) synapses from D and M to the L type. The 11% of mitochondria in axon terminals making asymmetric (presumed excitatory) synapses remain lightly reactive, with the scant D and M ones all converting to the L type. A further 11% of mitochondria in the neuropil are in axon trunks, and they retain their L-type mitochondria after TTX (Wong-Riley et al. 1989b).

Thus, segments of neurons in the neuropil with the highest metabolic activity under normal conditions (dendrites of all neurons and axon terminals making symmetric synapses) are the most affected by the elimination of excitatory functional input.

After 4 weeks of TTX, quantitative analysis indicates that the outermost concentric zone (zone 3) of deprived puffs exhibits significant reductions in numerical density and proportion of asymmetric synapses and an increase in symmetric ones ($P < 0.01$). This may account for the greater reduction in COX activity and the apparent shrinkage in puff size in that zone. It further suggests possible synaptic reorganization with retraction of excitatory synapses and expansion of inhibitory ones (Wong-Riley et al. 1989b).

5.8 HOW DO MATURE NEURONS IN THE INTERPUFFS RESPOND TO MONOCULAR IMPULSE BLOCKADE?

As described in Chapter 4 (Section 4.1.2), interpuffs normally contain the same three metabolic cell types with comparable percent distribution as in puffs, i.e., types A, B, and C, except that types B and C neurons are smaller and less COX-reactive in interpuffs than in puffs. How do these adult neurons respond to 2 weeks of monocular TTX? Notably, both deprived and non-deprived interpuff neurons respond, but in opposite directions (Wong-Riley et al. 1994). These two categories of interpuffs are defined according to whether they are in register with the underlying deprived or non-deprived ODCs in layer IVC.

5.8.1 Neurons in Deprived Interpuffs (DI)

Overall, there is no cell loss in deprived interpuffs. Type A neurons in DI retain normal cell size and percent distribution, but the few D and M somatic mitochondria are all converted to the L type, resulting in reduced areal density ($P < 0.05$) but not the numerical density of mitochondria. Type B neurons also maintain their size after 2 weeks of monocular TTX. However, their D and M mitochondria are all converted to the L type, causing a significant decrease in the areal density ($P < 0.05$), but only a slight decrease in the numerical density of mitochondria. Type C neurons exhibit the greatest conversion from mainly D- to L-type mitochondria, with a significant loss of D and a significant gain of L ($P < 0.01$ for both) and a loss of total areal density of mitochondria. Unlike type C cells in deprived puffs, those in DI do not decrease significantly in cell size (Wong-Riley et al. 1994). All three cell types in DI retain their normal composition of axosomatic synapses.

The fact that type C neurons in DI do not exhibit a shrinkage in their cell size as their counterparts do in deprived puffs may be due to two main reasons: a) their normal level of metabolic activity is lower than that of type C neurons in puffs, so further reduction in functional activity does not induce a significant change in cell size; and b) whereas puff neurons are largely monocular, those in interpuffs receive binocular input (Livingstone and Hubel 1984a; Ts'o and Gilbert 1988). DI neurons may receive a weak, excitatory input from the non-dominant, non-inactivated eye. Nonetheless, the loss of excitatory input from the dominant and inactivated eye causes a distinct conversion of D and M mitochondria to the L type in all three metabolic cell types.

5.8.2 Neurons in Non-Deprived Interpuffs (NDI)

Unlike neurons in DI, those in NDI transform in the opposite direction, even though their numbers and percent distribution are comparable to those in normal interpuffs (NI). Type A cells increase in their mean size by 7% ($P < 0.05$), and their mitochondria are converted from the L to the D type ($P < 0.05$–0.01). Both type B and type C neurons also increase their mean size (B by 7% and C by 17%; $P < 0.05$ for B and $P < 0.01$ for C), but their D mitochondrial population falls ($P < 0.05$ for B and $P < 0.01$ for C) while those of M and L remain within the normal range. An increase in cell size without a change in mitochondrial numerical density suggests that each cell type has a net gain in mitochondria (Wong-Riley et al. 1994). All three cell types retain a normal composition of axosomatic synapses. At the light microscopic level, increased COX activity in NDI results in apparent "bridges" between adjacent non-deprived puffs not observed normally (Wong-Riley et al. 1994).

The difference in response between DI and NDI neurons strongly suggests that interpuff neurons normally receive stronger excitatory input from the dominant eye and a weaker excitatory input (mainly to B and C cells) as well as an indirect inhibitory input from the non-dominant eye. The removal of the dominant eye input causes decreased activity and energy demand in DI neurons, but type C cells do not shrink in size because they are sustained by the weak excitatory input from the non-deprived eye. NDI neurons, on the other hand, are advantaged because they retain excitatory input from the dominant eye and lose indirect inhibition from the non-dominant TTX eye. All three cell types enlarge in size and gain additional mitochondria to maintain a normal mitochondrial numerical density. Type A neurons are at a greater competitive advantage because their excitatory input likely comes only from the non-deprived eye. Types B and C cells, however, lose the weak excitatory input from the TTX eye and downregulate their D mitochondria (Figure 5.12). Binocular interaction and competition, therefore, play a more important role in regulating functional and metabolic activities in interpuff neurons than visual input alone.

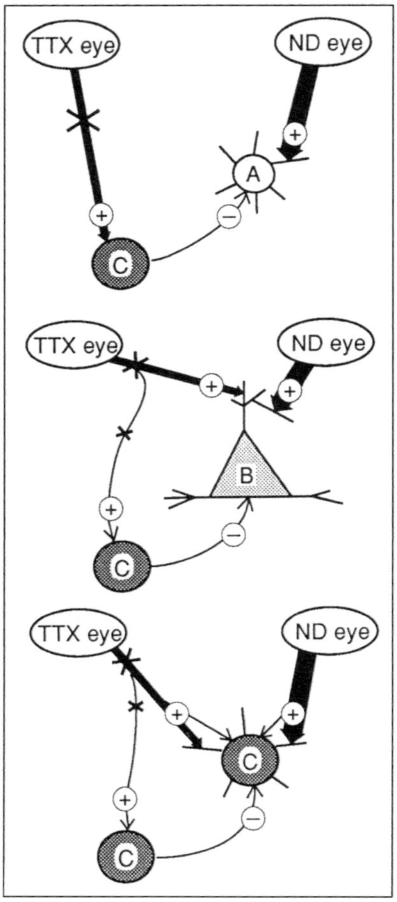

FIGURE 5.12 A working model of the effect of monocular TTX on the three major types of neurons in non-deprived interpuffs, all of which receive excitatory input from the non-deprived (ND) eye and all receive inhibitory input that, at least in part, originates from the TTX-injected eye via inhibitory type C cells. The removal of such inhibitory input induces all three cell types to increase in somal size and absolute number of mitochondria. Type A cells (top panel) also have a net gain of D mitochondria and a loss of L ones, indicating a conversion of L to the D form. Type B (middle panel) and type C (bottom panel) cells, on the other hand, normally may receive excitatory input also from the other (TTX) eye, and the removal of that excitation may cause some of their D mitochondria to be converted to the M and L classes. Overall, by 2 weeks after TTX, non-deprived interpuffs show an increase in COX activity.

From Wong-Riley et al. 1994.

6 Role of Cytochrome *c* Oxidase in Photobiomodulation

ABSTRACT

Light in the far red to near-infrared range (R-NIR) has been used for decades to promote wound healing and many other therapeutic applications known as photobiomodulation (PBM) or low-level light therapy (LLLT). Both lasers and light-emitting diodes (LEDs) have been used at various wavelengths and energy fluences in vivo and in vitro. The therapeutic mechanisms are unclear, but COX has been proposed as an important photoacceptor in the R-NIR range due to its heme and copper centers. Experiments in primary neuronal cultures exposed to various toxins attest to the benefit of PBM via the upregulation of both COX activity and ATP production. Questions remain, however, regarding true mediators of PBM and optimal wavelength/fluence for peak biomedical benefit.

Light is essential for almost all living organisms on Earth. Plants utilize light energy for photosynthesis, and animals depend on sunlight for energy, growth, circadian rhythm, and various biological processes. The solar electromagnetic spectrum extends from radio waves to gamma rays, and within it, the ultraviolet (10–400 nm), through the visible (400–700 nm), to the infrared (700 nm–1 mm) light (Haynes 2011) are of special interest (see below). Near-infrared (NIR) is at the short wavelength portion of infrared, but its defined range varies in the literature, such as 700–2,000 nm (Weidner et al. 1977) and a range that includes 650–980 nm (Mason et al. 2014). Phototherapy, especially in the far red to NIR (R-NIR) range, has received increasing attention in recent decades, and COX is implicated in playing a critical role. This chapter examines the relationship between the two.

6.1 ABSORPTION SPECTRUM OF CYTOCHROME *c* OXIDASE

The absorption spectrum of COX has long been examined. Depending on the cell and tissue type, in vitro or in vivo, experimental conditions, and the redox state of the enzyme, the absorption wavelengths vary. The method of purifying COX is paramount because the catalytic activity of the enzyme can be severely and adversely affected during isolation (Vanneste et al. 1974). Chance and Williams (1956) described a large Soret band at 445 nm and a smaller one at ~600 nm. Likewise, Vanneste (1966) noted a fully reduced form of COX with a major peak at 443 nm and a minor one at 603 nm, whereas the fully oxidized one has a major peak at 421 nm and a minor one at 598 nm. The Soret band in the blue range is named after the Swiss chemist Jacques-Louis Soret and is typical of porphyrin or heme-containing compounds, including the cytochrome heme proteins of the electron transport chain (cytochrome bc_1 or complex III, cytochrome c, and COX), myoglobin, and hemoglobin. The 603–605 nm band is the α band that is attributable to heme a (Chance and Williams 1956). In *Thermus thermophilus*, heme a reportedly has absorbances at 446 nm and 605 nm, while heme a_3 has absorbances at 441 nm, 611 nm, and 656 nm (Hellwig et al. 1999). Wharton and Tzagoloff (1964) described a broad NIR band of purified, oxidized COX from 710 nm to 930 nm, with the broad maximum near 830 nm, which they attributed to copper. Later studies indicated that the broad band around 830 nm is due to Cu_A (Ferrari et al. 1990; Cooper and Springett 1997; Cooper et al. 1999). Beef heart Cu_A has a maximum absorption at 835 nm (Mason et al. 2014). In figure 2 of Griffiths and Wharton's 1961 paper, the absorption spectrum of oxidized COX between

DOI: 10.1201/9781003507765-6

580 nm and 650 nm is much greater than that at 830 nm, and the one around 680 nm is similar to that at 830 nm. In the beef heart, the 655 nm band is assigned to the fully oxidized heme a_3/Cu_B binuclear center, but a "perturbed" binuclear center (when modified by ligands) reportedly also contributes to a broad 712–920 nm band, even though it is due mainly to Cu_A (Mason et al. 2014). Figure 6.1 shows the absorption spectra of oxidized and reduced COX in the beef heart (Mason et al. 2014).

Jöbsis (1977) recognized the relative transparency of biological tissues to wavelengths in the near-infrared range between 700 and 1300 nm, and the longer the wavelength, the less scattering and the deeper the penetration. Below 700 nm, hemoglobin (Hb) has an intense absorption band, and above 1300 nm, water absorbs all photons. He noted that cytochrome *c* oxidase has a weak absorption band in the 780–870 nm region, with a broad maximum from 820 to 840 nm. His work forms the basis of the near-infrared spectroscopy (NIRS) for measuring oxyhemoglobin, deoxyhemoglobin, and COX in vivo, still used today, including the imaging of human brain function (Villringer and Chance 1997). As the absorption signal of COX in the NIR range in vivo is only one-tenth that of hemoglobin, Wray and her colleagues (1988) replaced the blood of rats with a fluorocarbon substitute so that they could more clearly detect the COX spectrum. Their figure 2 shows the average cytochrome aa_3 difference spectrum from six rats after transfusion. From a moderately high starting point at 700 nm, the absorption falls to a low point around 750 nm, followed by a prominent broad rise that peaks around 810–830 nm before falling again at ~880 nm. When they compared the absorption spectra (between 650 nm and 1050 nm) across a rat's head with that of 3.5 cm of pure water, they found that water essentially showed little or no signal between 650 nm and 930 nm, but it did have a sizable peak at 965 nm (their figure 4).

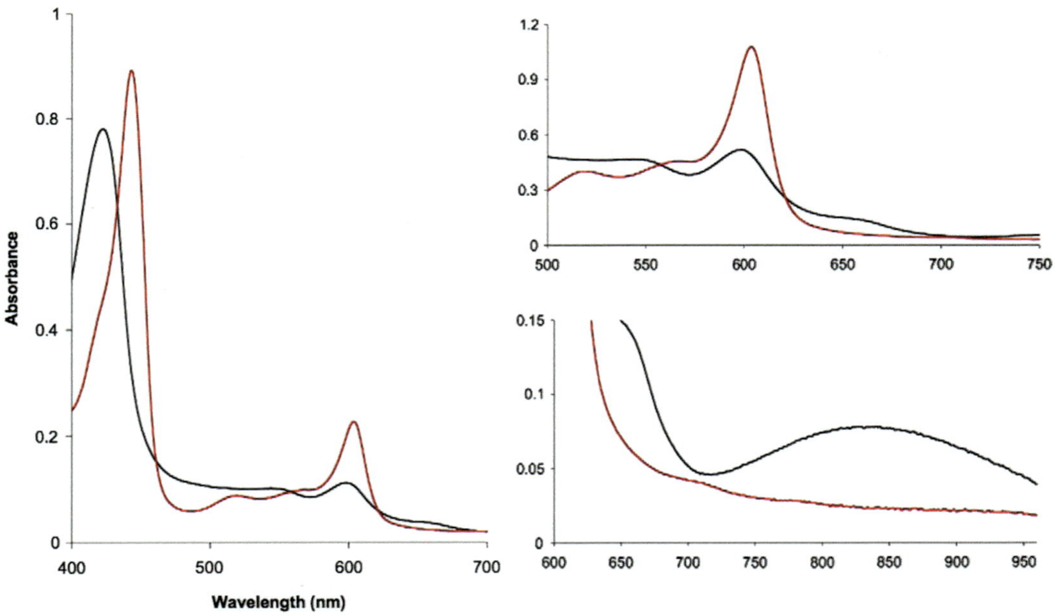

FIGURE 6.1 Absorption spectra of oxidized (black) and fully reduced (red) COX in the beef heart obtained in the Soret, visible, and NIR regions. The enzyme is in 0.1 M sodium borate buffer, pH 8.75, and 0.1% lauryl maltoside at 22°C, and reduction is with a small aliquot of 10% sodium dithionite for 5 min. The left panel shows the enzyme at 4.9 μM concentration, whereas the right ones are at a fivefold higher concentration.

From Mason et al. 2014; open access.

6.2 BACKGROUND OF LIGHT THERAPY

Light as a therapy is thought to date back to ancient Greek and Chinese culture, but it is difficult to track down the original documentation. In the Western world, the Danish physician and scientist Niels Ryberg Finsen may be considered the father of phototherapy. He won the Nobel Prize in Medicine and Physiology (1903) for his use of light radiation in the treatment of lupus vulgaris or skin tuberculosis. He also reported on the effectiveness of red light in the treatment of smallpox. For that condition, his main goal was to block out "chemical rays" in "daylight" that exacerbated the condition. When patients were exposed to red light continuously from the 3rd to the 12th day of the disease, they reportedly were able to reduce or avoid suppuration and pockmarks from the smallpox (Finsen 1895).

The Hungarian physician Endre Mester was one of the pioneers in using laser rays as a therapy for wound healing. His group discovered that argon ion gas laser between 488 and 501 nm or He-Ne laser at 632 nm was effective in promoting the healing of wounds, presumably via a suppressive effect of some adverse immune response (Mester et al. 1971, 1978). Soon, the benefit of low-energy laser as well as light-emitting diode (LED) irradiation for wound healing gained greater popularity (Conlan et al. 1996; Whelan et al. 2001). The approach is known as "low-level laser therapy, LLLT," "low-level light therapy, LLLT," "photobiomodulation, PBM," or "photobiomodulation therapy, PBMT."

A series of papers published by Tiina Karu in Russia directed the attention of the laser/LED community to cytochrome *c* oxidase as the "photoacceptor" or a "nonspecialized chromophore" (as opposed to specialized photoreceptors, such as rhodopsin, phytochrome, bacteriorhodopsin, and chlorophyll) in the R-NIR range (Karu 1999). Karu used lasers at various wavelengths, including 632.8 nm, 820, 1066, and 1286 nm, and noted stimulatory activity on different cell types (e.g., *E. coli*, HeLa, bone marrow, blood, fibroblasts, and human lymphocytes), resulting in increased ATP, cell growth, cell adherence, and DNA synthesis. However, she also reported the generation of singlet oxygen, increased superoxide anion production, and localized transient heating of absorbing chromophores (Karu et al. 1997; Karu 1999). She proposed COX as a key mediator of photobiomodulation and attributed four action spectrum peaks for increased DNA and RNA synthesis in HeLa cells to copper centers in COX: Cu_A reduced at 620 nm, Cu_B oxidized at 680 nm, Cu_B reduced at 760 nm, and Cu_A oxidized at 820 nm (Karu 1999). However, it is unclear how she derived those values and whether she performed absorption spectra experiments to verify the assignment of specific copper centers. Purified and air-oxidized COX is also found to absorb light at 632.8 nm from a He-Ne laser (Pastore et al. 2000). There is significant activation from the light at low cytochrome c/COX ratios (5–20), but inhibition at high ratios (200–800). The activity of neither cytochrome *c* nor cytochrome bc1 complex (complex III) is affected by the He-Ne laser at 632.8 nm. The authors concluded that "COX must be considered a molecular target of the He-Ne laser light, thus acting as a mitochondrial photo-acceptor" (Pastore et al. 2000).

6.3 EXPERIMENTS ON PRIMARY NEURONAL CULTURES WITH VARIOUS TOXINS AND LED

What is the effect of R-NIR on neurons, and what is the role of COX? Rather than using an in vivo model that involves many cell types, blood elements, and oxy- and deoxyhemoglobin, and rather than using immortal cell lines that may have functional and survival properties not possessed by normal cells, our lab conducted experiments on cultured primary neurons taken from postnatal rat cortex and grown facing but separated from a glial feeder layer. More than 95% of cells in the culture were neurons, and they remained viable for at least 16 days in culture (Wong-Riley et al. 2001). Instead of using a laser, which emits a monochromatic, focused beam of light that may generate some heat with adverse effects, we used an LED prototype developed by Quantum Devices, Inc. (Barneveld, WI) originally for NASA plant growth experiments in space. It consists of gallium aluminum arsenide (GaAlAs) LED of 670 nm, bandwidth of 25–30 nm at power intensity of 50 mW/cm^2, and energy density or fluence of 4 joules (J)/cm^2 when applied for 80 s. This dosage is beneficial for wound healing (Whelan et al. 2001). With a built-in fan, the device emits virtually no heat, and the wide bandwidth has the advantage of irradiating a larger area than the pinpoint source of lasers.

Single layers of primary neurons yield too few cells to perform biochemical assays; thus, relative COX activity was monitored with optical densitometry of histochemically reacted single neurons categorized into three metabolic cell types: darkly (D), moderately (M), or lightly (L) reactive for COX, based on reaction product density in the cytoplasm (Figure 6.2A).

6.3.1 TTX Blockade

When cultured neurons are bathed in 0.4 μM of the voltage-dependent sodium channel blocker TTX for 6 days (days 11–16 of culture), COX activity is significantly reduced in all three metabolic cell types (D, M, and L) ($P < 0.001$ for all) (Figure 6.2B and E). The size, shape, and viability of neurons are not altered. However, in the presence of TTX, as little as 80 s exposure per day to LED for the last 5 of the 6 days returns COX levels to those of controls (Wong-Riley et al. 2001) (Figure 6.2C and E). In the absence of TTX, LED once a day for 5 days significantly increases COX activity above control levels in all metabolic types (Wong-Riley et al. 2001) (Figure 6.2D and E). Under the microscope, these neurons appear healthy with branching dendrites. Thus, LED likely triggers a cascade of events that upregulates the metabolic activity of neurons, even when their spiking activity (though not their graded depolarizing activity) is inhibited by TTX.

6.3.2 Inhibitors of Cytochrome *c* Oxidase

If COX plays any role in the therapeutic effect of PBM, then inhibition of COX with potassium cyanide (KCN, a potent, specific inhibitor of COX) should reduce the light's beneficial effect. This was tested in sister cultures of primary neurons exposed to 10 μM, 100 μM, 1 mM, 10 mM, or 100 mM KCN for 5 days. Without LED, all three metabolic cell types exhibit significant reductions in COX activity measured optical densitometrically within single neurons ($P < 0.001$) even at the lowest concentration of KCN (Wong-Riley et al. 2005a) (Figure 6.3A). With increasing concentrations of KCN, progressive declines in enzyme activity and cell viability are evident. LED treatment twice a day (80 s per treatment) for each of the 5 days in 10–100 μM KCN partially but significantly restores COX activity ($P < 0.05$ to $P < 0.001$). With increasing KCN concentrations (1–100 mM), LED benefit decreases while neuronal death increases. Exposure of sister cultures to various concentrations of another COX inhibitor, sodium azide (NaN_3), with or without LED yields results comparable to those of KCN, although a higher concentration of NaN_3 is needed to attain a similar degree of inhibition. LED treatment reduces the equilibrium constant for both KCN and NaN_3, indicating the light's action in rescuing COX (Wong-Riley et al. 2005a). However, LED is not able to restore COX activity fully, even at the lowest concentration of KCN tested (10 μM). On the other hand, LED can completely reverse the detrimental effect of TTX, which *indirectly* and not directly downregulates COX.

Cellular ATP content is significantly reduced by KCN. LED partially reverses the effect at 10 μM to a level not significantly different from controls, but LED has little or no effect at 100 μM or higher concentrations of KCN (Figure 6.3B). Pretreatment of sister cultures with LED twice a day for 3 or

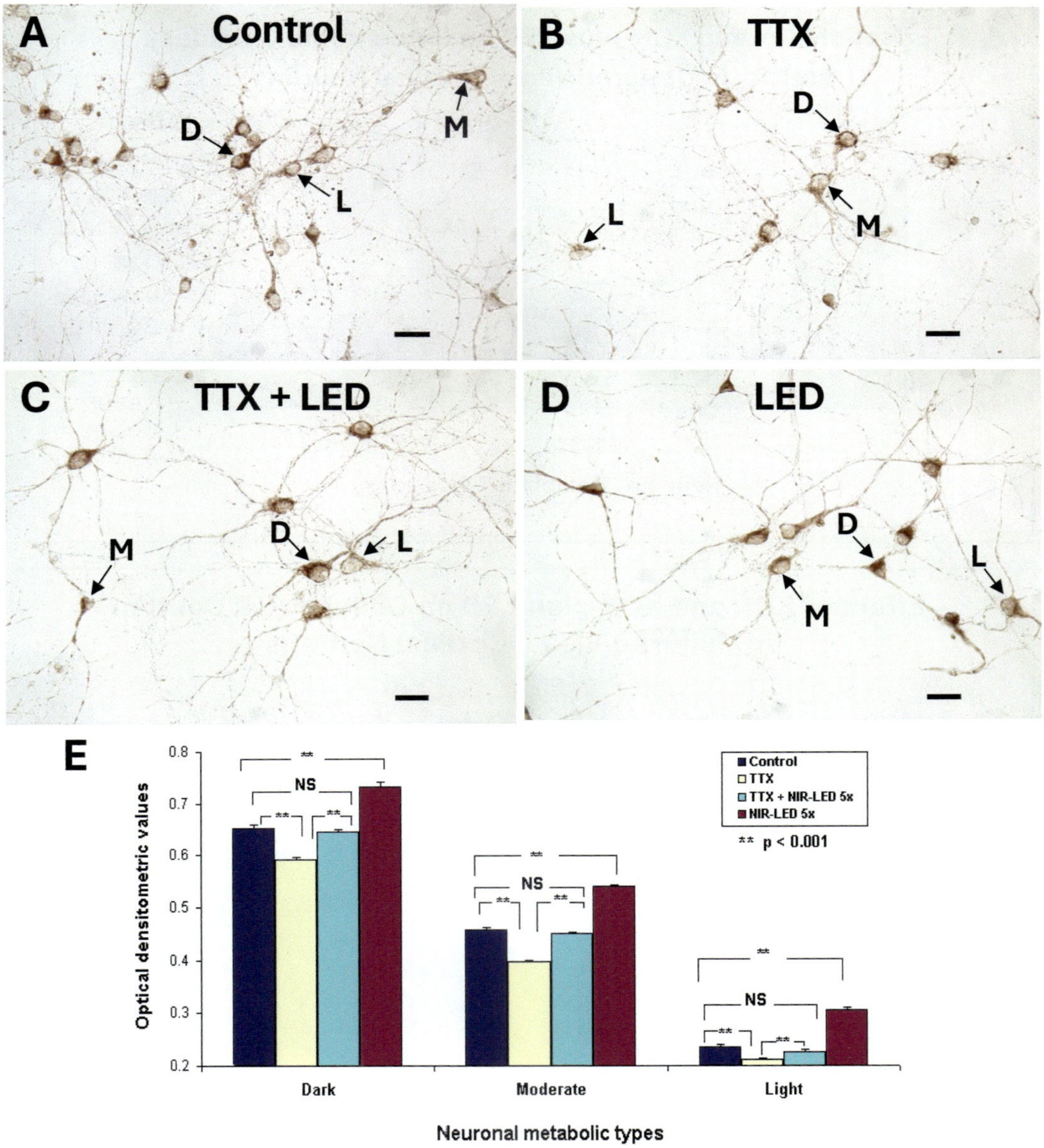

FIGURE 6.2 Effects of TTX and LED light on COX activity in primary cultures of postnatal visual cortical neurons. (A) Control neurons cultured for 16 days, showing examples of the three metabolic cell types: dark (D), moderate (M), and lightly (L) reactive for COX. (B) Neurons exposed to 0.4 µM TTX on days 11–16 of culture have reduced COX activity in all three metabolic cell types. (C) In the presence of TTX, neurons were treated with 670 nm LED (see text) for 80 s per day on days 12–16 of culture. COX levels of all three metabolic cell types are not significantly (NS) different from those of controls. (D) Normal neurons treated with LED for 80 s per day on days 12–16 of culture. COX levels for all cells are higher than those of controls. (E) Optical density measurements of COX reaction product in the three metabolic cell types under the above four conditions. Between 150 and 300 neurons were measured from each group.

$^{**}P < 0.001$.

Modified from Wong-Riley et al. 2001.

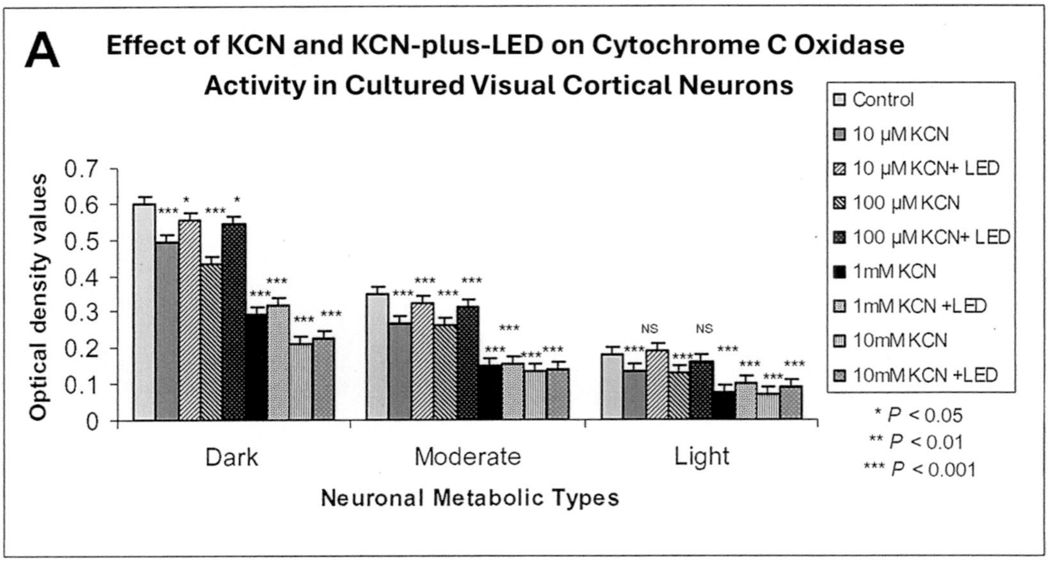

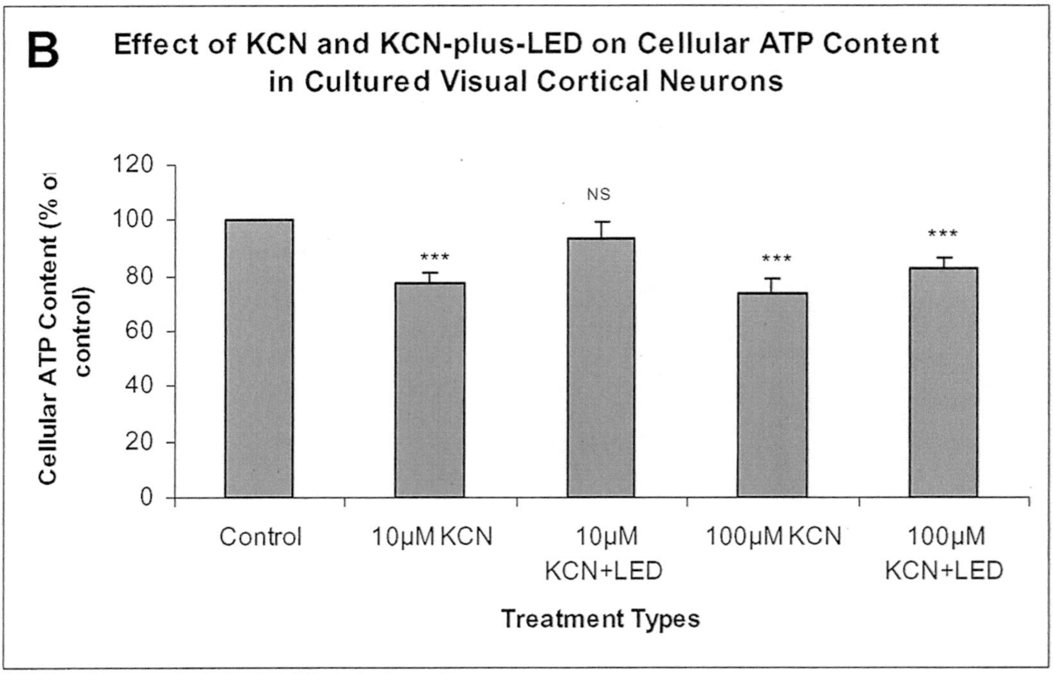

FIGURE 6.3 Effects of various concentrations of KCN and KCN-plus-LED on COX activity and ATP content in cultured visual cortical neurons. (A) Note that the three metabolic cell types have comparable COX responses to each treatment. All *P* values are comparisons with controls in each metabolic category. With increasing concentrations of KCN, COX levels progressively decrease. LED partially but significantly reverses the reduction caused by 10 and 100 μM KCN, but it has a minimal effect on higher concentrations of KCN. (B) Cellular ATP content is significantly reduced by KCN. LED partially reverses the effect at 10 μM to a level not significantly different from controls, but LED has little or no effect at 100 μM or higher concentrations of KCN.

All *P* values are compared to controls.

From Wong-Riley et al. 2005a.

5 days before exposure to 10 μM or 100 μM KCN confers additional protection to neurons. Nevertheless, enhanced COX activity, though significant, does not achieve control levels (Wong-Riley et al. 2005a).

6.3.2.1 Pretreatment with LED

Can pretreatment with LED arrest cell death? About 83.6% of cultured neurons exposed to 300 μM KCN for 28 h contain condensed chromatin (propidium iodide DNA labeling), indicative of cell death. Ten minutes of LED pretreatment (equivalent to 30 J/cm^2 of energy density) reduces cell death to 43.47% ($P < 0.001$). Comparable results are obtained when KCN is replaced with 1 mM of NaN_3 (Wong-Riley et al. 2005a). The mode of cell death from 100 μM or 300 μM KCN is determined by Hoechst 33258 stain, single-stranded DNA, Bax, active caspase 3, and EM to be apoptotic rather than necrotic (Liang et al. 2006a). Applying caspase inhibitor I to neurons for 30 min before exposure to 28 h of 300 μM KCN significantly reduces the number of apoptotic neurons ($P < 0.01$). LED pretreatment markedly reverses KCN-induced increase in the expressions of pro-apoptotic caspase 3 and Bax, as well as decrease in anti-apoptotic Bcl-2 (Figure 6.4A–F). LED pretreatment also reduces ROS expression caused by KCN, but it does not change ROS levels in control neurons in the absence of KCN (Liang et al. 2006a) (Figure 6.4G).

6.3.2.2 Optimal Regimen of LED Treatment Against Cell Death

If pretreatments are not possible, and if LED treatment once a day is beneficial, does increasing the frequency of treatments per day prove more beneficial? To address this question, primary neurons were exposed to 300 μM KCN for 24 h, and sister cultures were treated with LED (670 nm, 4 J/cm^2, 50 mW/cm^2, 80 s per application) 1, 2, 3, or 4 times within the first 8 h of the 24-h period, and all cells were harvested at the end of the period. LED once a day reduces the number of Hoechst-positive apoptotic neurons by 22.8% ($P < 0.05$), and LED twice a day reduces the number by 33.8% ($P < 0.001$). Significantly, LED treatments 3 or 4 times a day are not beneficial (Liang et al. 2008). LED twice a day is the most effective in decreasing ROS and reactive nitrogen species induced by KCN ($P < 0.001$). In this case, 3 or 4 times a day are also beneficial, but not as effective as twice a day, and once a day is not helpful with this toxin (Liang et al. 2008).

LED twice a day is also the most effective in increasing COX activity and ATP content in neurons exposed to KCN. For control neurons not exposed to any toxins, LED twice a day significantly upregulates ATP content above baseline ($P < 0.001$). The increase is 2.63-. 3.08-, and 4.08-fold after 1, 3, and 5 days of LED treatment, respectively ($P < 0.001$ for all) (Liang et al. 2008).

6.3.3 Varying R-NIR Wavelengths

Does LED benefit increase with increasing wavelengths between 670 nm and 880 nm? To determine this, sister cultures of primary neurons were exposed to 0.4 μM of TTX for 6 days and treated once a ay (80 s) for the last 5 of the 6 days in TTX with one of the five wavelengths (670, 728, 770, 830, and 880 nm), each at a fluence of 4 J/cm^2 and power intensity of 50 mW/cm^2 (Wong-Riley et al. 2005a). Significantly, the effectiveness is not incremental with wavelengths (Figure 6.5); 670 nm and 830 nm are equally effective in fully reversing the deleterious effect of TTX on COX activity to non-TTX control levels; 880 nm induces a 98.4% recovery of COX activity, and 770 nm, a 67.2% recovery. On the other hand, 728 nm is the least effective, with only 27.5% recovery. Six days of TTX reduces neuronal ATP content to 82.5% of controls. All wavelengths of LED tested, except for 728 nm, completely reverse the TTX effect to control levels. Clearly, 728 nm is the least effective wavelength.

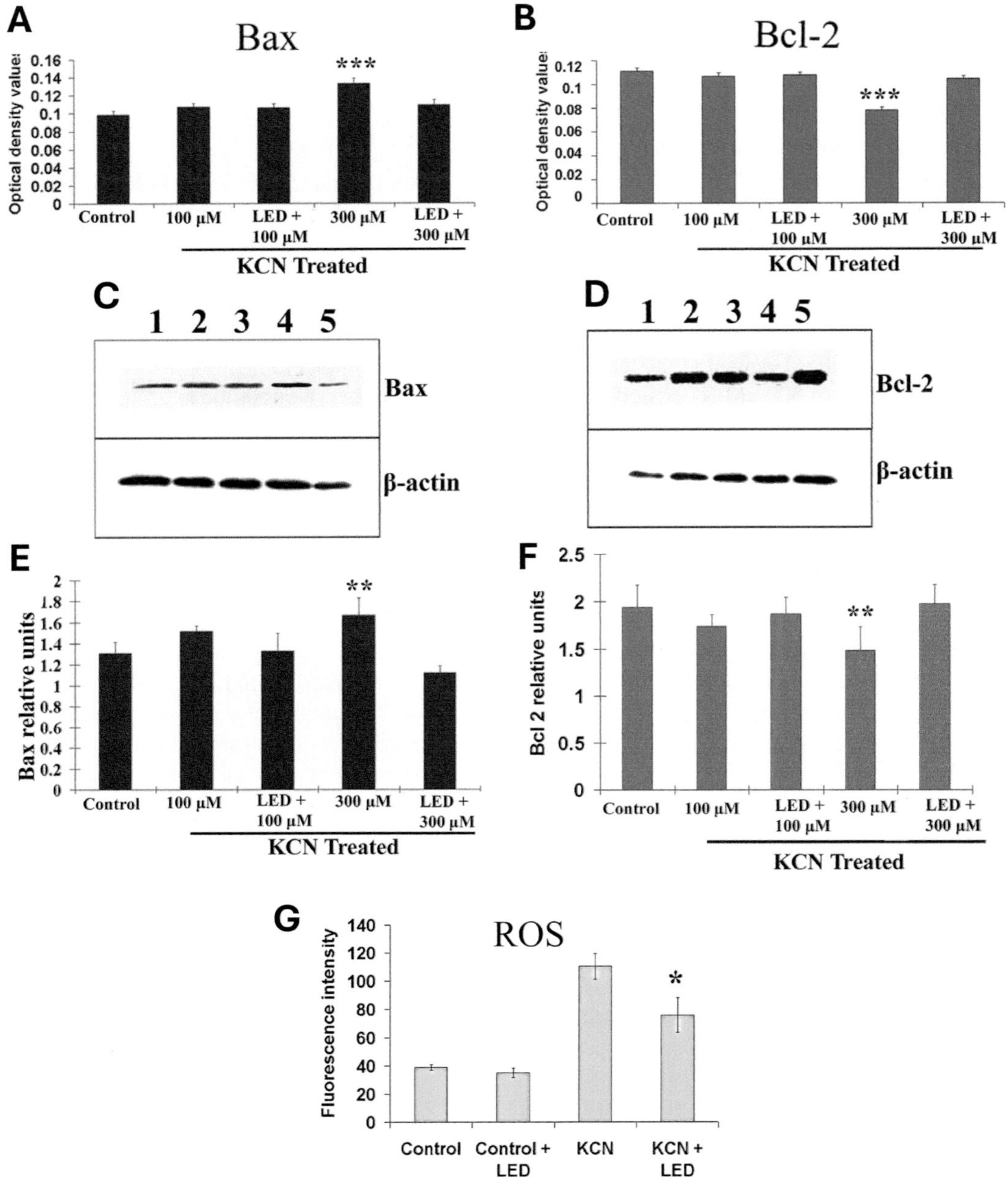

FIGURE 6.4 The effect of KCN on Bax, Bcl-2, and ROS in cultured visual cortical neurons with and without LED pretreatment. Exposure to 300 µM (but not 100 µM) KCN for 28 h induces an upregulation of BAX (A) and a downregulation of Bcl-2 (B), but 10 min of LED pretreatment reverses these changes to control values. Each of the five experiments for Bax and for Bcl-2 is analyzed with western blots (C and D) and quantified (E and F), respectively. All P values (** $P < 0.01$; *** $P < 0.001$) are compared to controls. (G) Exposure to 300 µM KCN for 28 h induces an increase in reactive oxygen species (ROS) revealed by CM-H_2DCFDA in neurons, but LED pretreatment significantly reduces such an increase. LED for 10 min does not affect ROS levels in normal neurons. * $P < 0.05$ compared to KCN alone.

Modified from Liang et al. 2006a.

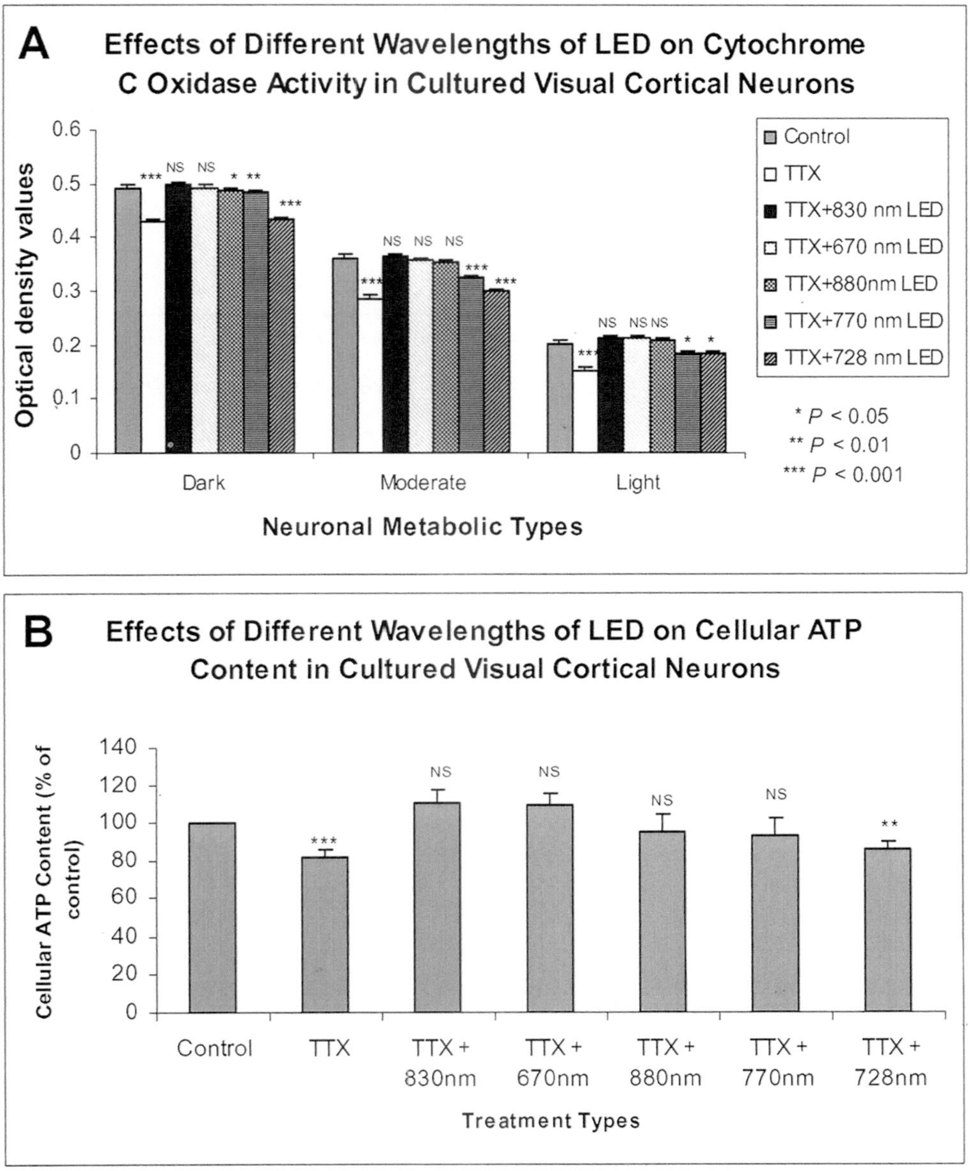

FIGURE 6.5 Effects of various wavelengths of LED on COX activity (A) and ATP content (B) of cultured visual cortical neurons exposed to TTX for 6 days. All *P*-values are compared to controls. Note that 670 nm, 830 nm, and 880 nm are all effective, and 770 nm is slightly less effective in reversing COX and ATP values downregulated by TTX to or close to control levels, whereas 728 nm is the least effective. NS, nonsignificant.

From Wong-Riley et al. 2005a.

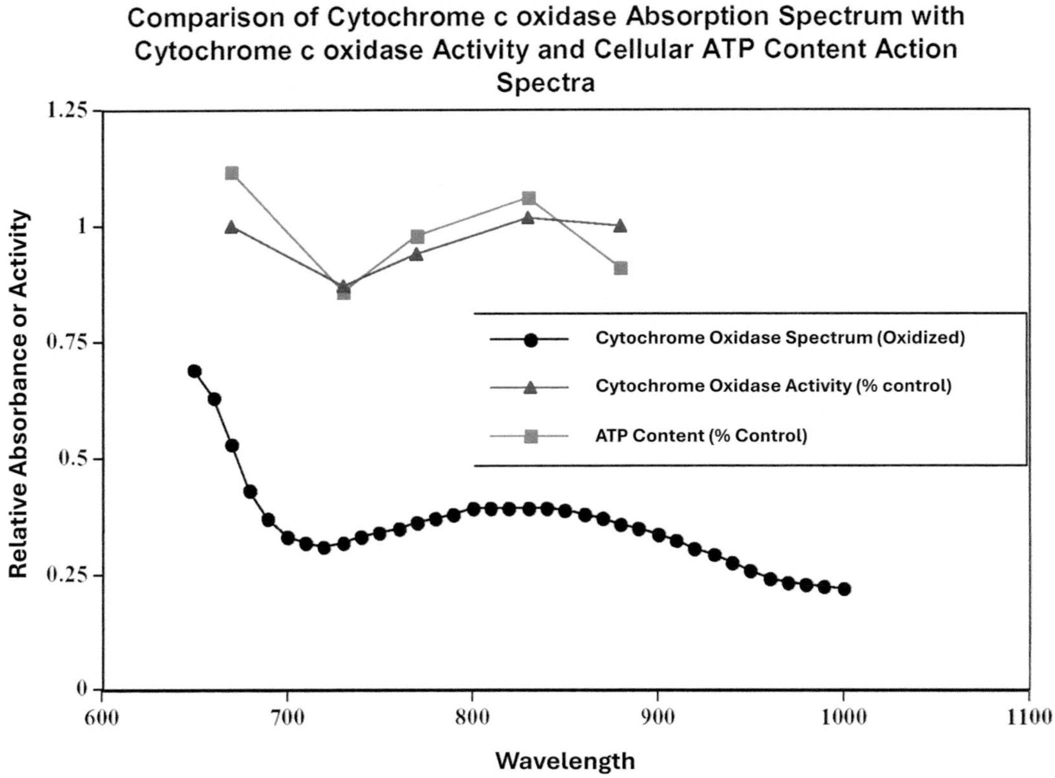

FIGURE 6.6 Comparison of oxidized COX absorption spectrum (circles; derived from Wharton and Tzagoloff 1964 and Mason et al. 2014) with action spectrum of relative COX activity (triangles) and relative ATP content (squares) in TTX-treated neurons exposed to various wavelengths of LED (see Figure 6.5). Note the correspondence of effective wavelengths (especially 670 nm and 830 nm) with the absorption spectrum of COX and the non-correspondence of the least effective 728 nm.

From Wong-Riley et al. 2005a. Note that the reference to the absorption spectrum of COX in that article should have been Wharton and Tzagoloff 1964, not Cooper and Springett 1997.

The action spectra of COX activity and ATP content, being most effective at 670 nm and 830 nm, correspond well with the reported absorption spectrum for oxidized COX within the R-NIR range (Wharton and Tzagoloff 1964; Mason et al. 2014), whereas the least effective 728 nm does not (Wong-Riley et al. 2005a) (Figure 6.6).

6.3.4 Rotenone and MPP^+

Does PBM have any beneficial effect on neurons exposed to toxins implicated in Parkinson's disease (PD)? Mitochondrial complex I inhibitors rotenone and MPP^+ (1-methyl-4-phenylpyridinium ion, the ultimate metabolic product of 1-methyl-4-phenyl-1,2,3,6-tetrahydropyridine or MPTP) have been used in animal models of PD (Schmidt and Ferger 2001; Sherer et al. 2003). Primary cultures of striatal and cortical neurons exposed to these toxins and treated with LED (670 nm, power intensity 50 mW/cm^2, fluence 4 J/cm^2, 80 s per treatment, twice daily for 2 days) significantly increase ATP content (Figure 6.7A–D), decrease cell death (Figure 6.7E,F), and reduce the expression of reactive oxygen and nitrogen species compared to toxin-exposed neurons without LED (Liang et al. 2008) (Figure 6.7).

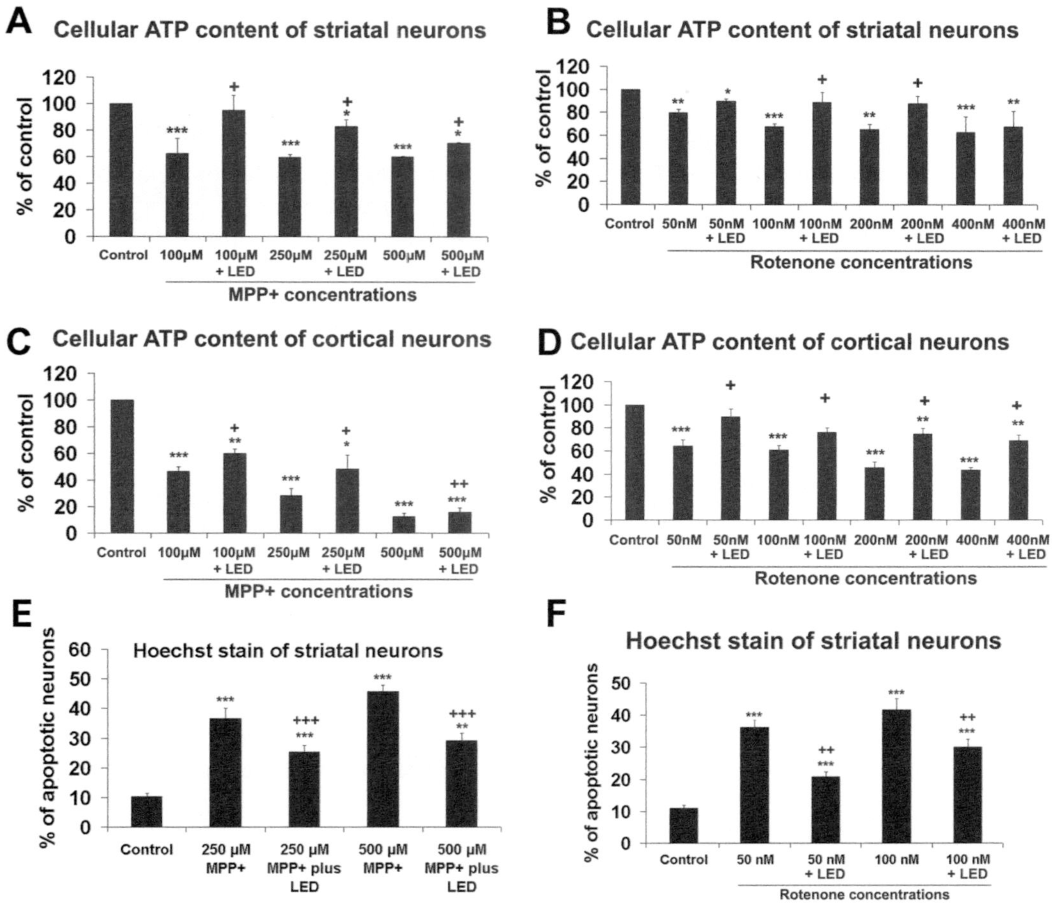

FIGURE 6.7 Beneficial effects of LED twice a day on ATP content of striatal and cortical neurons exposed to various concentrations of MPP^+ (A and C) or rotenone (B and D). LED is able to increase ATP content in all conditions, though not reaching control levels. LED twice a day also reduces the percentage of apoptotic neurons shown with the Hoechst stain in striatal neurons exposed to different concentrations of MPP^+ (E) or rotenone (F).

$^*P < 0.05$; $^{**}P < 0.01$; $^{***}P < 0.001$. All $^*P$-values are compared with controls. All ^+P-values are compared either to MPP^+ alone or rotenone alone.

From Liang et al. 2008.

Thus, cultured primary neurons exposed to various toxins benefit from R-NIR light treatment. The absence of oxy-/deoxy-hemoglobin and myoglobin in these cultures and the knowledge that other heme proteins absorb light only in the blue–green range point to COX as the only known protein dually responding to R-NIR light and directly involved in the generation of ATP. Irrespective of whether COX is the direct or indirect agent in PBM, its activity and cellular ATP content are consistently upregulated by R-NIR in intoxicated and irradiated neurons. Inhibitors of COX have a dose-dependent effect in offsetting the benefit of the light, consistent with COX's role in R-NIR light therapy. The long-lasting effect of LED indicates that the activation of COX and ATP triggers a cascade of events involving many players that ultimately enable neurons to improve health and attenuate their toxin-induced propensity toward death.

6.4 IN VIVO EXPERIMENTS WITH LED

6.4.1 Methanol Toxicity

An in vivo model of methanol poisoning was tested for the effect of PBM (Eells et al. 2003). Methanol causes metabolic acidosis, blindness, coma, and even death (Tephly 1991). A rat model was created to mimic the disorder (Eells et al. 2000). Formate, a metabolite of methanol, inhibits COX (Nicholls 1975) and is thought to be the basis for the ocular disorder.

Rats treated with N_2O/O_2 and given methanol i.p. experience a prolonged state of formic acidemia and visual dysfunction (Eells et al. 2003). Retinal abnormalities include a profound decrease in retinal sensitivity to light, an attenuation of maximal, flicker, and UV cone ERG response amplitudes, retinal edema, swelling of photoreceptor inner segments, and fragmentation of the outer nuclear layer. Three brief LED treatments (670 nm, 4 J/cm^2, 2 min 24 s per treatment) given at 5, 25, and 50 h of methanol intoxication result in a significant recovery of rod- and cone-mediated function and protection from detrimental retinal histopathologic changes (Eells et al. 2003).

6.5 LOW-LEVEL LIGHT THERAPY (LLLT) WITH LASERS

Lasers have been widely used over the years at various wavelengths and energy densities on different tissue types, in vivo and in vitro, for a variety of applications, including wound healing, bone repair, peripheral nerve repair, pain relief, improvement of muscle performance, and the treatment of arthritis, radiculopathy, depression, retinal diseases, traumatic brain injury, acute ischemic stroke, and even neurodegenerative diseases, such as Parkinson's disease and Alzheimer's disease (AD; reviewed in Conlan et al. 1996; Hamblin et al. 2017; Bathini et al. 2022). More recently, violet/blue (400–470 nm) light has been proposed as an antimicrobial treatment for the coronavirus (Enwemeka et al. 2020).

Does LLLT affect the biochemical activity of cytochrome *c* oxidase?

6.5.1 660 nm Laser Activates Electron Transport Chain Enzymes in Normal Rat Liver Mitochondria

Biochemical assays have been performed on isolated mitochondria from normal rat liver irradiated with 660 nm argon dye laser at power intensity of 10 mW/cm^2 and various energy densities (Yu et al. 1997). At a fluence of 0.6 J/cm^2 and 1.2 J/cm^2, oxygen consumption is significantly increased. At 1.8 J/cm^2 and 2.4 J/cm^2, phosphate potential and energy charge of mitochondria are augmented ($P < 0.05$ for all). At 0.06, 1.2, 2.4, and 4.8 J/cm^2, activities of complexes I, III, and IV of these mitochondria are enhanced ($P < 0.05$). Complex II, ATPase, and lactate dehydrogenase, however, are not affected.

Hence, 660 nm laser light at certain fluences positively modifies the oxidative metabolism and biochemical activities of respiratory enzymes in normal rat liver mitochondria. The effect, as indicated by the authors, lasts only a "short period." However, they have not examined longer treatment times.

6.5.2 660 nm Laser Stimulates COX in Stressed Human Fibroblast Mitochondria

Another direct demonstration of COX activation was obtained on mitochondria isolated from normal, diabetic, and ischemic human skin fibroblasts in vitro and irradiated with 660 nm diode laser at 5 or 15 J/cm^2 fluence (Houreld et al. 2012). Complex IV (COX) exhibits a significant increase in activity in normal ($P < 0.01$) and diabetic ($P < 0.05$) mitochondria irradiated at either 5 or 15 J/cm^2. Likewise, ATP is significantly increased in normal cells irradiated at 5 J/cm^2 ($P < 0.05$) and in diabetic cells irradiated at 15 J/cm^2 ($P < 0.01$). Irradiated cells also have a higher accumulation of active mitochondria than non-irradiated ones. On the other hand, complex III activity is significantly decreased in diabetic mitochondria irradiated with a fluence of 15 J/cm^2 ($P < 0.05$). The rationale for this is unclear, and complexes I and II show no significant effect.

6.5.3 Transcranial LLLT in Humans

Transcranial infrared (with a deep penetrating 1064 nm) laser stimulation in healthy human subjects increases cerebral concentrations of oxidized COX, oxyhemoglobin, and total hemoglobin monitored with a broadband near-infrared spectroscopy, indicating improved metabolic–hemodynamic coupling (Wang et al. 2017). This approach also increases EEG spectral powers at the α (8–13 Hz) and β (13–30 Hz) bands and enhances cognitive capabilities in humans (reviewed in Gonzalez-Lima (2021). Transcranial laser or LED PBM shows some promise as a therapy in stroke, traumatic brain injury, depression, and neurological diseases, though it still faces technical challenges of insufficient tissue penetration (reviewed in Salehpour et al. 2018; Nairuz and Sangwoo-Cho 2024).

6.5.4 Laser Light Induces Structural Changes in COX

In a study not directed at LLLT/PBM *per se* but at the effect of light on COX, Hallén and Brzezinski (1994) observed that Nd-YAG pulsed laser (10–200 mJ fluence, 10 ns pulse width) alters absorbance in one-electron reduced COX at different wavelengths. A rapid increase in absorbance occurs at 445 nm, consistent with heme reduction, before decreasing to pre-illumination level. The change is less when COX is further reduced or oxidized, but no change occurs in fully reduced or oxidized COX. A smaller absorbance shift at 830 nm is attributed to heme a_3 (but not Cu_B). Cyanide blocks absorbance changes, likely due to its binding to the heme a_3-Cu_B center of COX (Wikström et al. 1981). Azide, which binds to the same binuclear center, further implicates heme a_3's involvement in the light-induced absorbance shift. As the time constants of the relatively small structural changes in the binuclear site are comparable to those of dioxygen reduction by reduced COX, the authors proposed that binuclear changes control the rates of dioxygen reduction and the internal electron- and proton-transfer reactions.

Thus, different laser wavelengths act on COX and induce changes in the structure, absorption spectra, and activity of this enzyme.

6.6 CONTROVERSIES REGARDING THE ROLE OF COX IN LLLT/PBM

With such wide-ranging applications, it is not surprising that LLLT/PBM has gained a great deal of popularity, attention, as well as controversy. For those not directly involved, the approach is often viewed as bordering on phenomenology and incredulity. A major hurdle is the question of the basic mechanism of LLLT/PBM. What is the key chromophore? What is the mechanism of healing?

6.6.1 Cell Lines Lacking COX

Although COX has been widely accepted as playing an important role in the beneficial effect of LLLT/PBM, there have been challenges in recent years that it is not the "major photoacceptor" in light therapy. In two cell lines lacking COX assembly and function (a mouse cell line with a knockout of COX assembly factor, COX10, and a human cell line with mtDNA mutation in the tRNA lysine gene, resulting in impaired synthesis of COX I–III), PBM at 660 nm still increases cell proliferation and ATP levels (Lima et al. 2019). In these cases, COX appears unnecessary. The exact candidate is unclear. Nevertheless, it is possible that these immortalized cell lines channel the light to activate downstream mechanisms, such as anaerobic production of ATP and ROS, which trigger cellular proliferation. By itself, this study cannot rule out that LLLT/PBM activates COX in normal cells to help generate ATP.

6.6.2 Inhibitory Action of Certain NIR Wavelengths

Another paper reports that NIR via LED at 750 nm and 950 nm, unlike the stimulatory effect at 810 nm, decreases activity of isolated COX, decreases mitochondrial respiration in isolated mitochondria, and decreases mitochondrial membrane potential ($\Delta\Psi_m$) in an immortalized hippocampal cell line (HT22) (Sanderson et al. 2018). The authors proposed that this inhibitory effect reduces ROS production and neuronal death in oxidatively stressed cells after glutamate toxicity or reperfusion following ischemia. They suggest that NIR at these two wavelengths is a potential therapy for global brain ischemia. It remains to be seen whether this approach yields clinical benefit. Application of 808 nm diode laser 6 hours after global cerebral ischemia protects against neuronal death, preserves healthy mitochondrial dynamics, reduces mitochondrial oxidative damage, and suppresses mitochondrial fragmentation in hippocampal neurons (Wang et al. 2019). In their in vivo ischemic model, these authors validate that COX is a key PBM target for mitochondrial functional preservation and neuroprotection.

6.6.3 Role of Nanoscopic Interfacial Water

Yet another report condemns the claim that COX is the main photoacceptor for R-NIR photons (Sommer et al. 2020). One strong objection was that the paper (Cooper and Springett 1997) referenced in figure 8 of Wong-Riley et al. 2005a does not contain a diagram of COX absorption spectrum in the R-NIR range, even though the main focus of that paper is that CuA in COX absorbs light in the NIR range and forms the basis of NIRS. Indeed, a better choice for a diagram of the COX spectrum would have been Wharton and Tzagoloff (1964) or several papers by Cooper's group, such as Mason et al. 2014 (see Figure 6.1).

Instead of COX, Sommer et al. proposed that R-NIR light promotes ATP synthesis by reducing the viscosity of the nanoscopic interfacial water layer within and around the mitochondrial rotary motor (ATP synthase or complex V). They suggest that the inhibitory effects of 750 nm and 950 nm observed by Sanderson et al. (2018) represent a disturbance of electron transfer from cytochrome *c* to COX across the nanoscopic interfacial water layer. The presence of three water layers between cytochrome *c* and COX is found with X-ray crystallography by Shimada et al. (2017). These authors regard the protein–protein interaction "soft and specific," contributing to the rapid association/dissociation of the cytochrome c-COX complex that facilitates the sequential transfer of four electrons for oxygen reduction. Whether the nanoscopic interfacial water is sufficient to activate the synthesis of ATP, as suggested by Sommer's group, awaits future experimentation. Water as a photoacceptor has been proposed to facilitate COX energy transfer for increased efficiency in ATP production (Santana-Blank et al. 2017). It is unknown whether the water interface behaves differently at different wavelengths; for example, 728 nm is much less effective in activating COX and increasing cellular ATP than 670 nm and 830 nm (Wong-Riley et al. 2005a). Additional questions also await investigation. Is COX dispensable in the PBM process for normal cells, especially neurons? Can mitochondrial ATP be produced in PBM without the involvement of COX? Does the known absorption of R-NIR light by oxidized Cu_A in COX have absolutely no effect on the generation of ATP in the mitochondria?

6.7 OTHER AGENTS

6.7.1 Role of Nitric Oxide

Nitric oxide (NO) has been proposed as a signaling molecule in LLLT/PBM (Poyton and Ball 2011). Besides the known inhibitory effect of NO on COX, nitrite is reduced to NO by COX, and such synthesis reportedly is enhanced by LLLT (Poyton and Ball 2011). On the other hand, adding NO donors before irradiating HeLa cells with 600–860 nm monochromatic light eliminates the light-induced increase in cell attachment (Karu et al. 2005). Irradiation with 590 nm light increases

NO production, and under hypoxic conditions, NO production is optimal. This led Quirk and Whelan (2020) to postulate that photorelease of NO from inhibiting COX may be integral to PBM. However, when isolated COX was tested with NO donors, NO-induced COX inhibition is not relieved by irradiation with either 670 nm or 830 nm light, calling into question the NO inhibition relief model of PBM (Quirk and Whelan 2021).

Endogenous NO can act in various ways with COX (Cooper 2002). The oxygen-binding site of COX is an iron/copper (heme a_3/Cu_B) binuclear center. At high substrate (ferrocytochrome c) concentration, NO binds reversibly to reduced iron in heme a_3 in competition with oxygen. This inhibition is relieved by dissociating NO from reduced iron. At low substrate concentration, NO binds to oxidized copper, and such inhibition is relieved by the oxidation of the bound NO, followed by the dissociation of nitrite from COX. Hence, NO serves as a substrate, effector, or inhibitor of COX (Cooper 2002). Inhibition of oxygen consumption by NO is also dependent on the concentration of oxygen, and NO can be readily metabolized by the oxidative processes within COX. Thus, COX is a possible physiological detoxifier of NO (Cooper and Brown 2008).

NO has been considered a neurotoxin in stroke models. Nitric oxide synthase (NOS) activity increases during the first week of experimentally induced cerebral ischemia in rats (Leung et al. 2002). A low-energy 660 nm GalAlAs laser irradiation of the transiently ischemic cerebrum of rats decreases NOS activity, but increases the level of neuroprotective TGFβ1 (Leung et al. 2002).

6.7.2 Carbon Monoxide

Although not used in PBM, carbon monoxide is another reversible inhibitor of COX worth mentioning. In his Nobel Lecture in 1931, Otto Warburg described how carbon monoxide binds to the "respiratory oxygen-transferring ferment" (later named cytochromes) in the dark and inhibits cellular respiration. In the presence of ultraviolet or visible light, however, carbon monoxide "splits off from the iron" (later found to be heme a_3 of the heme a_3-Cu_B binuclear center in COX) and respiration is restored (Warburg 1931). Carbon monoxide is produced endogenously by heme oxygenase during heme degradation. Like NO, carbon monoxide in small amounts is a signaling molecule that acts as a gaseous neurotransmitter (Snyder et al. 1998) and a therapeutic agent, in that it modulates inflammation, proliferation, and cell death (reviewed in Bauer et al. 2025).

6.7.3 Other Chromophores

Additional chromophores may be considered. Many absorb light in the blue and green range, such as opsin, flavoproteins, cryptochrome, and channel rhodopsin. Others, such as transient receptor potential vanilloid 1 (TRPV1) and TRPV4 channels, respond to 2,780 nm and 1875 nm light, respectively (reviewed in Hamblin 2019).

6.8 WHAT IS THE BASIC MECHANISM OF LLLT/PBM?

Currently, a universally accepted basic mechanism of LLLT/PBM is still lacking. The primary "photoacceptor" is debated. A major confounder is that numerous approaches have been used with common therapeutic goals. Many laboratories are engaged in LLLT/PBM research, and the number of publications increases exponentially. The approaches vary in methods of isolating enzymes and mitochondria, in cell types, organs, organisms, in vivo or in vitro, wavelengths, fluence, experimental conditions, disease type, severity of disease, and many other parameters. A universal protocol is theoretically more desirable. However, the optimal therapeutic PBM protocol may need to be tailored to specific maladies or even to specific patients. In cultured neurons, LED treatments twice a day are more beneficial than one, three, or four times a day (Section 6.3.2.2). Likewise, ROS generation in small amounts is beneficial as signaling molecules, but in large amounts, it becomes detrimental.

The bulk of data collected thus far from laboratories around the world points to COX as a key mediator of LLLT/PBM, primarily because it contains heme and copper metallic centers that absorb light in the Soret, red, and near-infrared spectra. Near-infrared spectroscopy detects the redox state of COX, although the signal is weak compared to oxy- and deoxyhemoglobin (Mason et al. 2014; Bale et al. 2016). Another major reason for implicating COX is that it is the only enzyme that can complete the oxidative metabolic pathway for the synthesis of ATP in the mitochondria. Whereas ATP can be synthesized via anaerobic glycolysis in the cytosol, a candidate chromophore/photoacceptor remains to be identified to trigger such synthesis. Regardless of the candidate(s), it is only the first step in a cascade of events mediating multiple regulatory mechanisms that ultimately result in the healing of cells, organs, and organisms. Many players have been proposed, including signaling molecules (such as ROS), transient receptor potential vanilloid (TrpV) channels, brain-derived neurotrophic factor (BDNF), and transcription factors (Albert et al. 2012; Hamblin et al. 2017; Hamblin 2019). The escalating list awaits dedicated experimentation.

With a healthy dose of informed skepticism, a focus on well-designed, carefully executed studies that are reproducible is still critical in advancing the field. LLLT/PBM should not be regarded as *the* cure for any disease, but it likely serves as a beneficial adjunct therapy.

7 Molecular Mechanisms of Regulating the Bigenomic Cytochrome *c* Oxidase in Neurons

ABSTRACT

Being one of only four proteins that have two genomic sources, nuclear and mitochondrial, COX poses a distinct challenge in unraveling the molecular mechanisms of its regulation. Are all the subunits from the two sources coordinately or independently regulated? We found that depolarizing stimulation upregulates all COX subunit transcripts in unison. However, impulse blockade downregulates the mitochondrial mRNAs earlier and to a greater extent than the nuclear ones, indicating that the mitochondrial subunits play a more dominant role. The mitochondrial messages are regulated mainly at the transcriptional level, but the nuclear transcripts are regulated by both synthesis and degradation. The expressions of all subunits, nevertheless, are controlled by neuronal activity. With chromosome conformation capture (3C), all ten nuclear COX genes on nine different chromosomes and three transcription factor genes responsible for the transcription of the three mitochondrial COX genes interact in the same transcription factory. Finally, representative nuclear and mitochondrial COX subunit genes were isolated from brain cDNA libraries and their transcripts identified with in situ hybridization in both normal and functionally perturbed neurons. These findings lay the foundation for exploring transcription factors that regulate all 13 COX subunits from the two genomes detailed in the subsequent three chapters.

A most remarkable property of cytochrome *c* oxidase is that it is one of only four proteins out of tens of thousands in mammalian cells that are bigenomic, that is, they are derived from two different genomes: nuclear and mitochondrial. As discussed in Chapter 1, COX has three of its largest subunits (I, II, and III) encoded by the mitochondrial genome and the other ten (IV, Va, Vb, VIa, VIb, VIc, VIIa, VIIb, VIIc, and VIII) by the nuclear DNA (Kadenbach et al. 1983). These 13 subunits are the major ones investigated in neurons. The role of NDUFA4 (see Chapter 1, Section 1.2.1) in neurons is unclear and will not be discussed further in this book.

The bigenomic nature of COX poses a distinct challenge in understanding its regulation. What mechanisms coordinate the two genomes to generate all 13 subunits in a 1:1 stoichiometry? Are there common transcription factors that orchestrate expression from the two genomes? Are all 13 subunits synthesized and degraded at equal rates? Do they respond alike to changes in neuronal activity? These and other questions are addressed in this and the subsequent three chapters. As none of the other three bigenomic proteins (mitochondrial complexes I, III, and V) has been probed with similar queries, novel findings in COX will serve as a springboard for future investigations into this special class of proteins.

7.1 ARE ALL 13 SUBUNITS REGULATED COORDINATELY BY NEURONAL ACTIVITY?

Under normal, non-pathological conditions, neuronal activity regulates energy metabolism and not vice versa (Lowry 1975). Are all 13 COX subunits coordinately regulated by alterations in neuronal activity? This question was investigated both in vitro and in vivo. Cultured primary neurons were

DOI: 10.1201/9781003507765-7

stimulated with 20 mM potassium chloride for 5–24 h. This dosage causes mild depolarization (del Rio et al. 1999), and cultured neurons remain healthy in appearance (Zhang and Wong-Riley 1999). Real-time quantitative polymerase chain reaction (RT-qPCR) demonstrates that all 13 COX subunit mRNAs are significantly upregulated at 5 h (Figure 7.1), but COX I and COX IV proteins measured by western blots are not increased until 24 h later (Liang et al. 2006b).

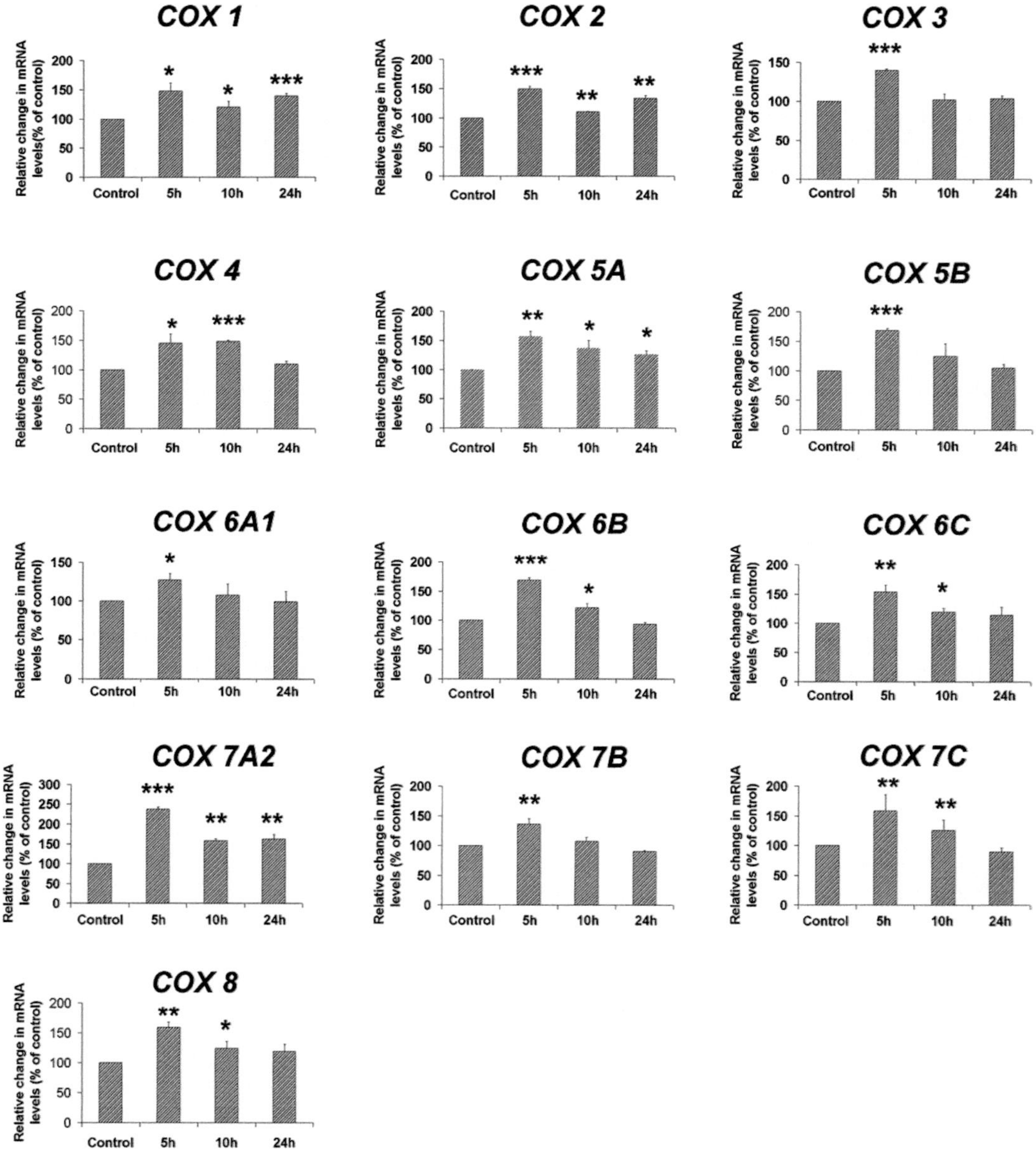

FIGURE 7.1 All 13 *COX* subunit mRNAs are upregulated by depolarizing stimulation in primary cultures of neurons. Responses after 5, 10, and 24 h of exposure to 20 mM KCl are analyzed with RT-qPCR. ($N = 6$ cultures for each group). All comparisons are with the control of each subunit. $*P < 0.05$, $**P < 0.01$, $***P < 0.001$.

From Liang et al. 2006b.

When cultured neurons are exposed to 0.4 μM TTX for 2–6 days (a dosage that causes impulse blockade but neurons remain structurally intact; see Chapter 6, Section 6.3.1), the three mitochondrial-encoded subunit mRNAs are downregulated earlier (at 2 days) than the ten nuclear ones (at 4 days). Likewise, in vivo monocular enucleation causes a downregulation of the mitochondrial-encoded transcripts before the nuclear ones in the deprived visual cortex (4 days vs. 1 week) (Liang et al. 2006b) (Figure 7.2). COX activity and protein levels are reduced in parallel after 4 days of deprivation in vitro and 1 week in vivo.

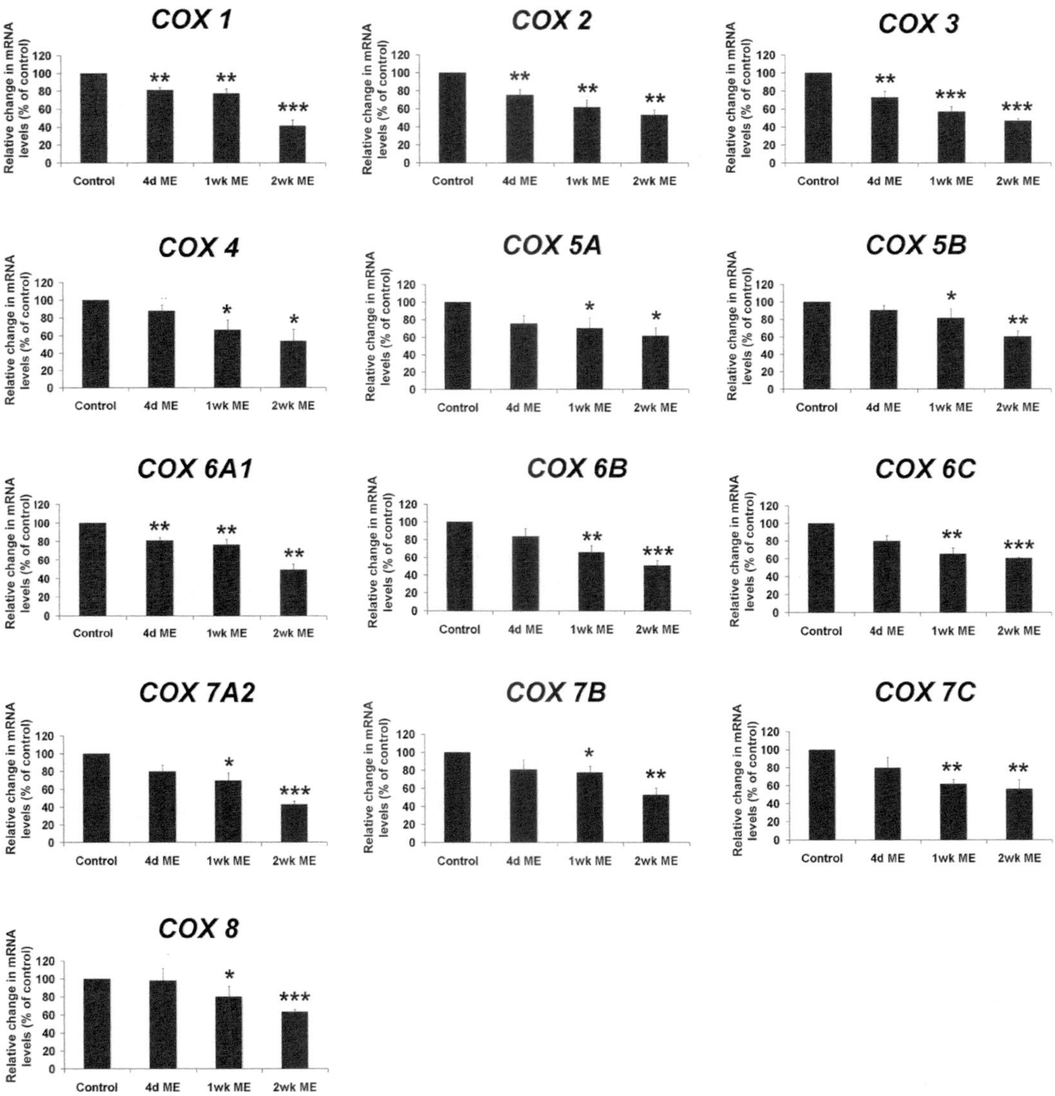

FIGURE 7.2 All 13 *COX* subunit mRNAs are downregulated in the deprived visual cortex after monocular enucleation (ME) in the rat. Note that the mitochondrial subunits (*COX 1*, *2*, and *3*) are downregulated earlier (after 4 days) than the nuclear ones (except for *COX6A1*). By 1 and 2 weeks after enucleation, all 13 subunits are significantly reduced. Measurements are done with RT-qPCR ($N = 6$ for each group). All comparisons are with the control of each subunit. $*P < 0.05$, $**P < 0.01$, $***P < 0.001$.

From Liang et al. 2006b.

The findings in the rat visual cortex agree with those in the monkey lateral geniculate nucleus, where representative mitochondrial and nuclear *COX* subunits were analyzed with in situ hybridization and grain counting. With monocular impulse blockade, mitochondrial *COX1* message in deprived LGN layers downregulates earlier (3 days for *COX1* vs. 7 days for *COX8*) and to a greater extent (by 1.5- to 4-fold more) than nuclear ones (Hevner and Wong-Riley 1993). The reason for this difference is not entirely clear. Several factors can be considered: 1) Mitochondrial-encoded transcripts are synthesized and stored within mitochondria distributed throughout the neuron and are especially enriched in dendrites, the major receptive sites of excitatory synapses. Changes in neuronal activity are sensed locally in dendrites, where adjustments in transcription and translation are made readily and efficiently. 2) Nuclear-encoded *COX* subunit transcripts are synthesized exclusively in the nucleus and translated only in cell bodies and not in dendrites or axons (Hevner and Wong-Riley 1991; Wong-Riley et al. 1997a, 1998c). A signal to downregulate transcription in response to reduced synaptic activity would be sensed in the dendritic tree before the cell body; hence, the adjustment of nuclear transcripts would lag that of mitochondrial ones. 3) The distance between the soma and distal dendrites necessitates neurons to synthesize excess nuclear transcripts for an adequate supply of precursor proteins to be imported into somatic mitochondria and transported to distal processes, where they form a precursor pool to be readily converted to mature proteins when the energy demand increases (Liu and Wong-Riley 1994). An upregulation is in line with the active process, but a significant downregulation to a detectable level requires more time. 4) The more severe downregulation of mitochondrial subunits indicates that they are the major determinant of COX responses to local changes in energy demand.

Overall, neurons achieve a coordinated regulation of all ten nuclear-encoded and three mitochondrial COX subunits in line with changing neuronal activity (Wong-Riley et al. 2008). The response to increased energy demand is in relative unison, whereas that to reduced need varies slightly in rates. In general, mitochondrial COX subunits that form the catalytic core of the enzyme play a dominant role in sensing activity-dependent metabolic demands of neurons and adjusting COX protein amount and activity accordingly. COX activity generally relates most closely to levels of mitochondrial DNA and mitochondrial *COX* subunit transcripts (Hevner and Wong-Riley 1993).

7.2 SYNTHESIS VERSUS DEGRADATION RATES

What is the mechanism of regulating the levels of mitochondrial- and nuclear-encoded *COX* subunit transcripts? RNA species decay with first-order kinetics (Hunt 1974), so their levels can be deduced by the ratio of synthesis-to-degradation rates. Synthesis and degradation also control the degree of mRNA accumulation.

To determine if alterations in *COX* transcript levels were caused by RNA synthesis rate, stability, or both, RNase protection assays and ^{3}H-uridine pulse-chase labeling of *COX* transcripts were conducted on cultured primary neurons subjected to varying periods of 20 mM KCl. This dosage provides gentle depolarizing stimulation without affecting the viability of cultured neurons (Zhang and Wong-Riley 2000a). The synthesis rates of both mitochondrial (*COX2*) and nuclear (*COX4*) subunit mRNAs are elevated significantly after KCl treatment. The half-life of *COX2* mRNA at 84 min remains unchanged with KCl, but that of *COX4* is increased from 50 min to 102 min. Thus, the mitochondrial subunit is regulated mainly at the transcriptional level, but the nuclear one is regulated by both synthesis and degradation. Both subunits, however, are regulated by neuronal activity (Zhang and Wong-Riley 2000a).

The half-lives of *COX* subunit transcripts in primary neuronal cultures may be much shorter than those in whole brains that contain different types of glial cells and blood elements in addition to neurons. Future investigations will shed more light on this issue.

7.3 CHROMOSOME CONFORMATION CAPTURE (3C) OF ALL 13 GENOMIC LOCI

Whereas the largest three subunits of COX are encoded by the mitochondrial genome, the ten nuclear ones are on nine different chromosomes. Transcriptional regulation of this multisubunit, multichromosomal, and bigenomic enzyme is mechanistically challenging. Are all ten nuclear subunits transcribed in unison? If so, how do all nine chromosomes synchronize their transcription of *COX* subunits? How do mitochondrial subunits coordinate transcription with those of nuclear ones? How does one probe such a complicated process? What tools are available?

One possible mechanism of coordinating the transcription of multiple genes is to assemble all relevant chromatin loci in close proximity to one another. Such discrete and dynamic sites within the nucleus have been proposed as "transcription factories," containing several RNA polymerase II molecules surrounded by transcription factors and loops of active chromatin for coordinated transcription of different genes (Jackson et al. 1998; Osborne et al. 2004). The looping of chromosomes brings distant loci on the same chromosome or loci from different chromosomes in close contact for intrachromosomal and interchromosomal interactions. An effective method is the high-throughput, chromosome conformation capture (3C) that detects the frequency of interactions between two genomic loci joined by common restriction sites, yielding products demonstrable with PCR (Dekker et al. 2002; Miele and Dekker 2009). Using this method, dynamic intra- and interchromosomal interactions within the same or different chromosomes have been found in specialized nuclear sites, the presumed transcription factories (Spilianakis et al. 2005; Ling et al. 2006; Lomvardas et al. 2006).

Can 3C, previously used for only two genomic loci, determine whether all ten nuclear *COX* genes and nuclear genes encoding transcription factors that mediate mtDNA transcription (and hence the three mitochondrial *COX* genes) are transcribed in the same transcription factory? Besides the nine chromosomes for *COX* genes, the three known transcription factors responsible for the transcription and replication of mtDNA – transcription factors A, B1, and B2 of mitochondria (gene names *Tfam, Tfb1m,* and *Tfb2m*, respectively) (Clayton 2000; Falkenberg et al. 2002) – are on three different chromosomes. The large number of 12 chromosomes presented a technical challenge not previously undertaken. By systematically pairing members of the 13 genes, the Wong-Riley lab demonstrated that, indeed, all 10 nuclear *COX* subunit genes, plus *Tfam, Tfb1m,* and *Tfb2m*, interact within a common transcription factory in murine neuronal nuclei (Dhar et al. 2009a) (Figure 7.3A and B). Control genes for another mitochondrial enzyme (citrate synthase) and a non-mitochondrial protein (calreticulin) do not interact with *COX* genes. Likewise, no interactions exist between β-actin and any of the *COX* subunit genes tested, and no interaction occurs if the crosslinking agent or ligase is withheld (Figure 7.3C and D).

Are these interactions perturbed by altered neuronal activity? Primary neurons stimulated with 20 mM KCl depolarization for 5 h increase the frequency of *COX* gene interactions by 219–259% ($P < 0.05$ for both). On the other hand, exposure to 0.4 μM TTX for 3 days reduces the cross-linking frequency by 55–65% ($P < 0.001$ for both) (Dhar et al. 2009a) (Figure 7.3E and F). Hence, neuronal activity directly dictates the rate of interactions of genomic loci involved in the transcription of *COX* subunit genes.

These results indicate that a bigenomic, multisubunit, and multichromosomal enzyme possesses the mechanism for coordinately regulating transcription of all ten nuclear *COX* subunit genes and all three mitochondrial *Tf* genes. The proximity of these genes from 12 different chromosomes within the same transcription factory greatly facilitates their efficient and synchronized transcription for the ultimate formation of a functional holoenzyme (Figure 7.4).

7.4 ISOLATION OF NEURONAL NUCLEAR *COX* SUBUNIT GENES

Nuclear-encoded *COX* subunit genes have been isolated from mammalian non-neuronal tissues, mainly heart, liver, and kidney (Lomax et al. 1984; Suske et al. 1987; Zeviani et al. 1987, 1988a; Rizzuto et al. 1988; Schlerf et al. 1988; Lightowlers et al. 1989; Taanman et al. 1990; Aqua et al. 1991; Seelan and Grossman 1992). However, no study has focused on brain or neuronal sources prior to 1997.

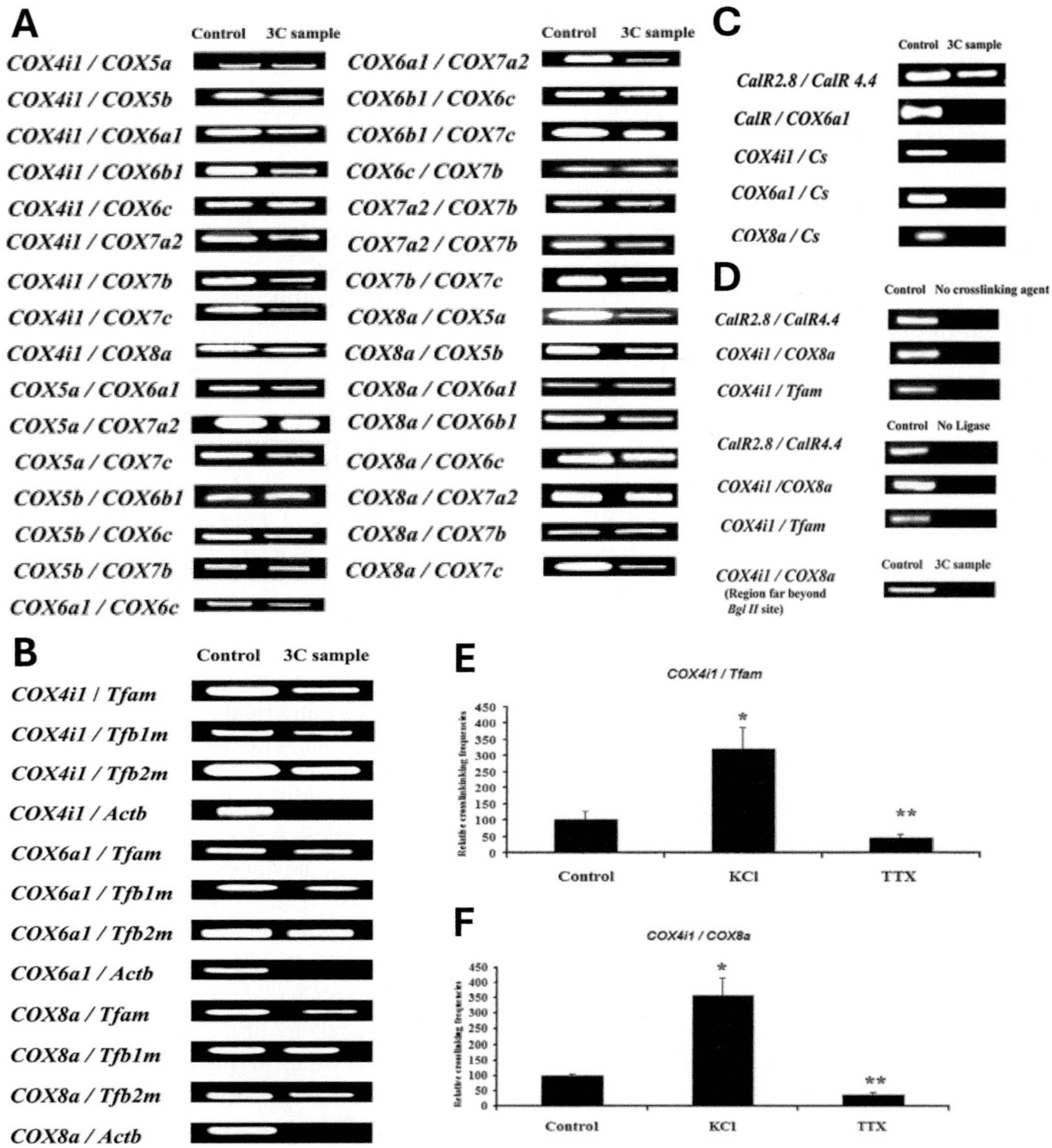

FIGURE 7.3 Chromosome conformation capture (3C) of all 13 genomic loci of COX in neurons. (A) All possible interactive pairs among the 10 nuclear-encoded COX subunit genes test positive for 3C as compared to control templates. (B) Nuclear-encoded COX subunit genes also interact with *Tfam*, *Tfb1m*, and *Tfb2m*, respectively. However, they do not interact with β-actin (*Actb*) gene. (C) A known interaction between two calreticulin genes, *CalR2.8* and *CalR4.4*, serves as a positive control. However, *CalR* does not interact with *COX6a1*, nor does the mitochondrial enzyme citrate synthase (*Cs*) interact with any *COX* subunit genes. (D) In the absence of a crosslinking agent (formaldehyde) or T4 ligase, no interaction occurs between known interactive pairs. *COX4i1* and *COX8a* also do not interact far beyond the *BglII* site tested. All control samples in A–D result from artificially ligating two genomic loci across a *BglII* site. All experiments are performed in triplicate. (E and F) The frequency of interactions between a nuclear *COX* gene and *Tfam*, and between two *COX* genes, is significantly increased by 5 h of KCl depolarizing stimulation and significantly reduced by 3 days of TTX impulse blockade. $^{*}P < 0.05$, $^{**}P < 0.001$ when compared with controls.

From Dhar et al. 2009a.

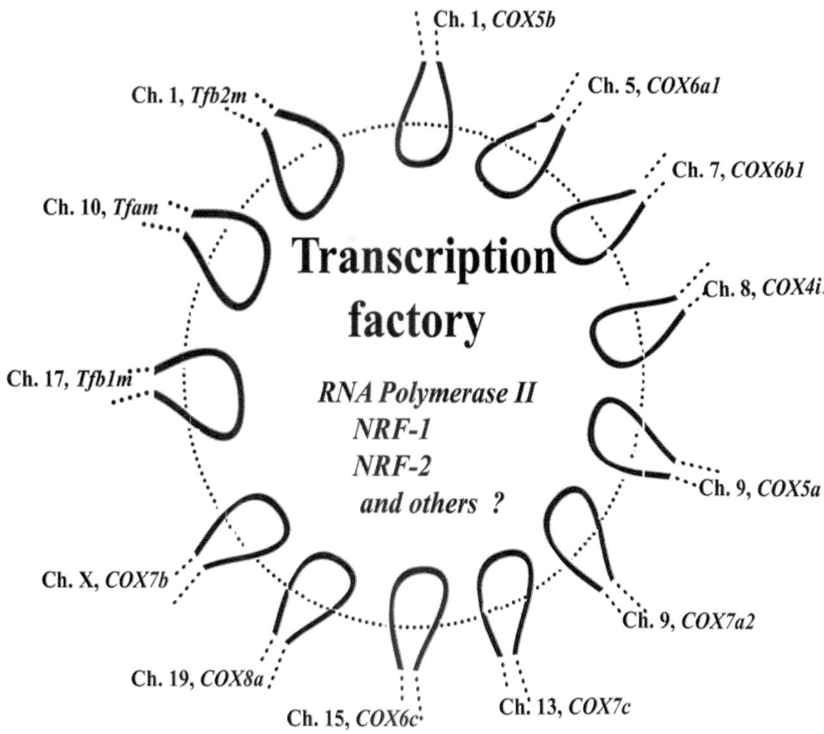

FIGURE 7.4 Schematic diagram of the co-localization of loops of active chromatin loci containing the ten nuclear-encoded *COX* subunit genes and mitochondrial transcription factor genes (*Tfam*, *Tfb1m*, and *Tfb2m*), plus RNA polymerase II and the relevant transcription factors (NRF-1, NRF-2, and others), in the same presumed transcription factory in neuronal nuclei.

From Dhar et al. 2009a.

7.4.1 Isolation of *COX3* and *COX6a* Genes from Murine Brain cDNA Library

To characterize the two genomic sources of *COX* subunit genes from brains, mitochondrial *COX3* and nuclear *COX6a* cDNAs were isolated from a murine brain cDNA library, subcloned, and sequenced. Sense and antisense riboprobes were synthesized for in situ hybridization, and Northern (RNA) hybridization was performed (Wong-Riley et al. 1997a).

Results indicate that brain *COX3* shares 100% identity with that of murine L cells (Bibb et al. 1981), whereas brain *COX6a* is only 93% homologous to that of the rat liver, the only rodent sequence known at the time (Schlerf et al. 1988). Northern blots confirm that the antisense transcripts hybridize specifically to a single band of murine RNA at the appropriate position for *COX3* (~830 bases) and *COX6a* (~355 bases) (Wong-Riley et al. 1997a).

7.4.1.1 In Situ Hybridization at the Light and Electron Microscopic Levels

At the light microscopic level, the distribution of *COX* mRNAs is similar, but not identical, to that of COX activity revealed histochemically. As described in Chapter 2, COX activity is intense in specific, but not all, neuronal cell bodies and in specific processes in the neuropil (mainly dendrites and some axon terminals). All *COX* transcripts reside mainly in cell bodies. The neuropil has some mitochondrial (*COX3*) transcripts but none of the nuclear (*COX6a*) ones (Wong-Riley et al. 1997a).

With EM in situ hybridization, *COX3* mRNAs are found exclusively in mitochondria within the cell bodies, throughout the entire extent of dendrites, and in some axonal terminals (Figure 7.5). *COX6a* transcripts, on the other hand, are localized to the extramitochondrial compartment confined to the cell bodies and absent in neuronal processes. In the somatic cytoplasm, *COX6a* mRNAs

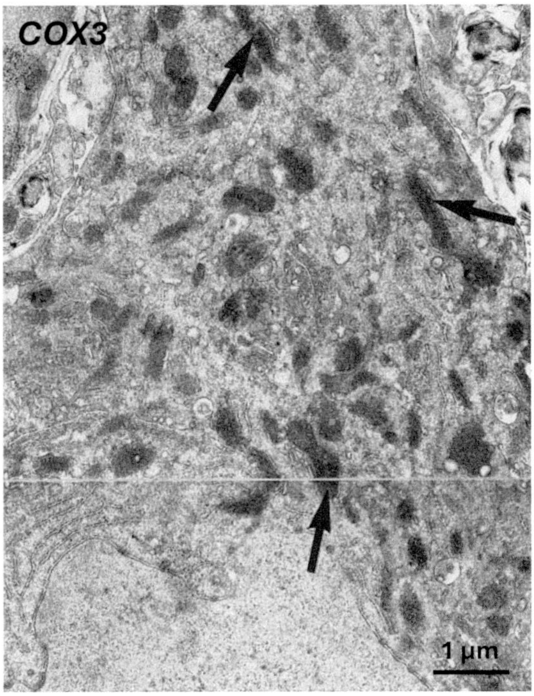

FIGURE 7.5 EM in situ hybridization of mitochondrial-encoded *COX3* mRNAs in a Purkinje cell with antisense riboprobes labeled with digoxigenin-11-UTP. Note that the electron-dense labeling is confined to mitochondria (arrows).

Modified from Wong-Riley et al. 1997a.

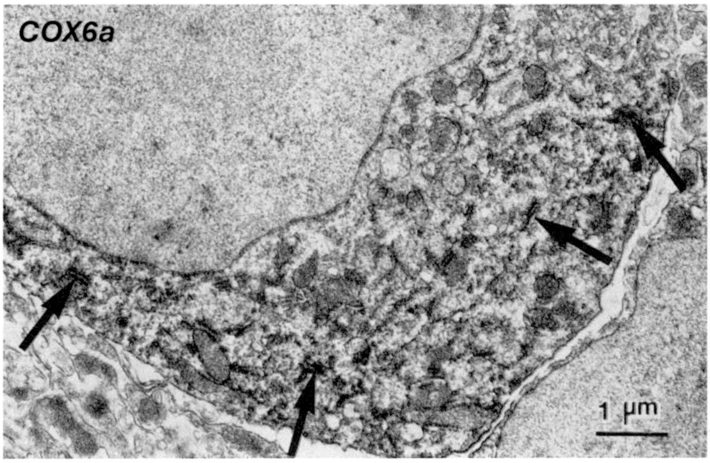

FIGURE 7.6 EM in situ hybridization of nuclear-encoded *COX6a* mRNAs in a Purkinje cell with antisense riboprobes labeled with digoxigenin-11-UTP. Note that the labeling is in the extramitochondrial compartment, primarily associated with the cytoplasmic side of the rough endoplasmic reticulum (arrows) and the outer nuclear membrane.

Modified from Wong-Riley et al. 1997a.

associate primarily with the cytoplasmic side of the rough endoplasmic reticulum (RER) and the outer membrane of the nuclear envelope (Figure 7.6) (Wong-Riley et al. 1997a). This is somewhat surprising, as nuclear-encoded mitochondrial proteins are thought to be synthesized on free ribosomes (Neupert and Herrmann 2007).

7.4.1.2 A Working Model of COX Subunit Synthesis and Assembly in Neurons

Based on the data obtained thus far, a working model of COX subunit synthesis and assembly is proposed (Figure 7.7). Within the cell body, the nuclear-encoded COX subunit precursors are transcribed in the nucleus, translated in the cytoplasm, imported into the mitochondria, and inserted

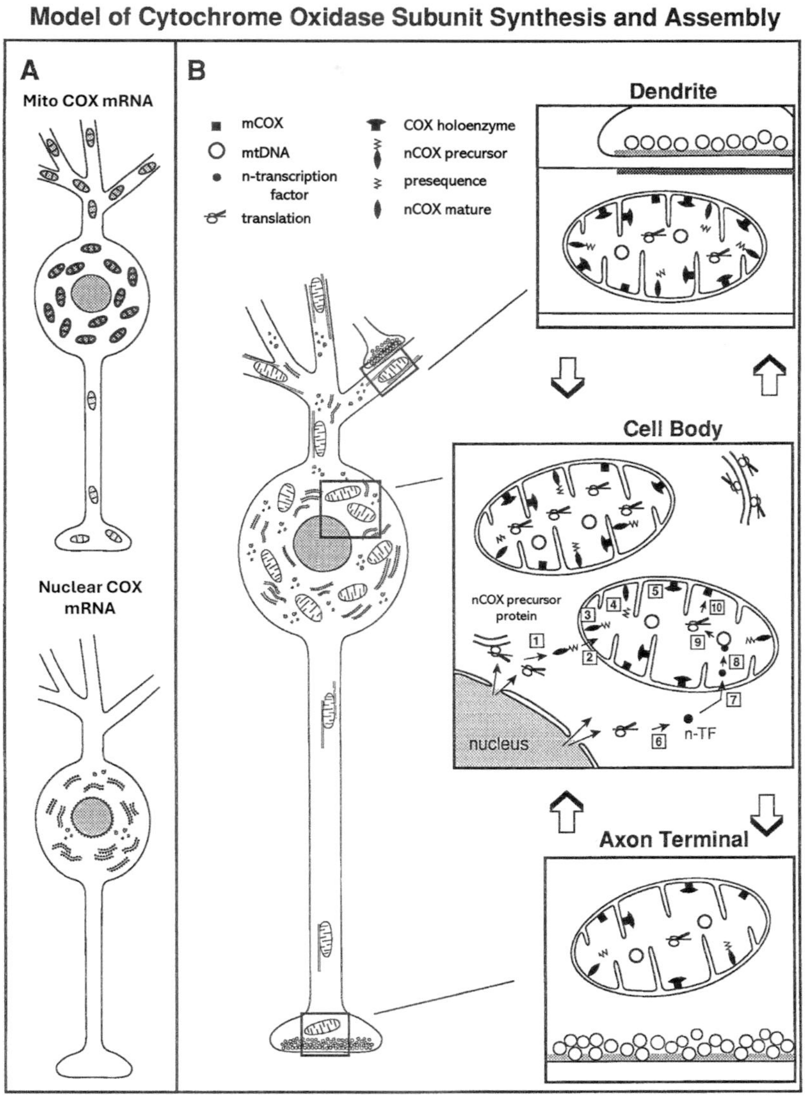

FIGURE 7.7 A working model of COX subunit synthesis and assembly in neurons. (A) Mitochondrial-encoded *COX* subunit mRNAs are localized within mitochondria throughout the neuron, whereas nuclear-encoded *COX* mRNAs are found only in the cell body in association with rough endoplasmic reticulum. (B) In the cell body (middle panel), nuclear *COX* subunit (nCOX) precursors are transcribed in the nucleus, translated in the cytoplasm (1), imported into the mitochondria (2), and inserted into the IMM (3) either with their presequence attached or cleaved (4) for the assembly into COX holoenzyme (5). Meanwhile, nuclear-encoded transcription factors (n-TF) are also transcribed in the nucleus, translated in the cytoplasm (6), imported into the mitochondria (7), and bind to mtDNA (8) to activate the transcription of the mitochondrial *COX* subunits (mCOX) (9). mCOX is translated in the mitochondria and inserted into the IMM to form the catalytic core of the COX holoenzyme (10). This process for mCOX also continues in dendrites (upper panel) and some axon terminals (lower panel). In dendrites that are actively driven by excitatory input, some nCOX precursor proteins may be processed into mature forms to generate additional holoenzymes.

Modified from Wong-Riley et al. 1997a.

into the IMM with or without the cleavage of their presequences. The mature subunits are assembled with the other subunits to form the holoenzyme. Some precursor proteins remain as such within the mitochondria and are transported to the dendritic tree, where they await increasing demands of energy before being processed into their mature forms. Meanwhile, nuclear-encoded mitochondrial transcription factors (*Tfam*, *Tfb1m*, and *Tfb2m*) are also transcribed in the nucleus, translated in the cytoplasm, imported into the mitochondria, and bind to mtDNA to activate transcription of mitochondrial *COX* subunits, which are translated and incorporated into IMM to form the catalytic core of the enzyme. The synthesis and degradation of the enzyme are activity dependent in neurons.

7.4.1.3 Effect of Monocular Enucleation on *COX* Subunit Transcripts in the Rodent LGN

Monocular enucleation for 1–18 days in adult mice significantly downregulates both *COX3* and *COX6a* mRNAs in deprived, contralateral LGN layers ($P < 0.001$ for all) measured with reflectance densitometry of silver grains in in situ hybridized sections (Wong-Riley et al. 1997a). However, the reduction is more severe for the mitochondrial transcripts than the nuclear ones at every post-enucleation time point examined (1, 2, 4, 8, 12, and 18 days). These findings are consistent with those in the monkey visual system (Hevner and Wong-Riley 1993; see Section 7.1).

7.5 ISOLATION OF *COX3* AND *COX4* GENES FROM HUMAN BRAIN CDNA LIBRARY

Partial mitochondrial *COX3* and nuclear *COX4* genes were isolated from a human brain cDNA library (Wong-Riley et al. 1998b). *COX3* measures ~731 base pairs, and *COX4* ~504 bp. *COX3* cDNAs share 99.7% identity with the known human genomic *COX3* (Anderson et al. 1981). A 353G>C substitution, converting arginine to proline, and a 492T>C silent substitution for leucine are noted. Whether these substitutions are technique-induced artifacts or are brain-specific is yet unknown (Wong-Riley et al. 1998b). *COX4*, on the other hand, shares 100% sequence homology with that of the human liver (Zeviani et al. 1987).

7.5.1 Effect of Monocular TTX on *COX* Subunit mRNAs in the Macaque Retina

Intravitreal injections of TTX in adult monkeys block the activity of spiking neurons in the retina (retinal ganglion cells and amacrine cells) but do not block graded potentials in non-spiking neurons (photoreceptors, bipolars, and horizontal cells). The expressions of *COX3* and *COX4* transcripts (revealed by riboprobes in in situ hybridized retinal sections) are significantly downregulated in deprived retinal ganglion cells after 2 weeks and 4 weeks of impulse blockade ($P < 0.001$ for both subunits and at both survival times). The other cell types are unchanged (Wong-Riley et al. 1998b). Thus, neuronal activity plays a significant role in adjusting the expression of *COX* subunits from the two genomes.

7.6 TRANSCRIPTION FACTORS INVOLVED IN THE REGULATION OF ALL 13 *COX* SUBUNIT GENES IN NEURONS

Transcription factors are proteins that bind to promoters or enhancers of their target genes and regulate the transcription of these genes, either positively or negatively. The bigenomic nature of COX presents a unique challenge when one considers its transcriptional regulation. Are the ten nuclear *COX* genes and the three mitochondrial *Tf* genes regulated by different or the same transcription factor(s)? Can a single transcription factor regulate all 13 *COX* subunit genes from the two genomes?

The 5' regulatory or promoter regions of a few nuclear *COX* genes have been analyzed in non-neuronal cells, and subunits *COX4* and *COX5b* were the best characterized (Virbasius and Scarpulla

1991; Carter et al. 1992; Basu et al. 1993; Grossman and Lomax 1997). Like most housekeeping genes (Smale and Baltimore 1989), the basal promoters 5' to the transcription start site (TSS), often with multiple initiation sites, of *COX* subunit genes tend to be compact, lack the TATA box for transcriptional initiation, and are GC-rich in CpG islands. They have binding motifs (sequence of nucleotides) for a few transcription factor candidates. These include nuclear respiratory factor 1 (NRF-1), nuclear respiratory factor 2 (NRF-2), specificity protein 1 (Sp1), and yin yang 1 (YY1). Comparable binding motifs exist in some of the brain-derived *COX* subunit genes (Wong-Riley et al. 2000; Ongwijitwat and Wong-Riley 2004). The following three chapters will detail the roles that NRF-1, NRF-2, and Sp1/Sp4 play in coordinating the transcription of all 13 *COX* subunit genes from the two genomes in neurons.

8 Nuclear Respiratory Factor 1 (NRF-1)

A Transcriptional Regulator of All 13 COX Subunit and Na^+/K^+-ATPase Genes in Neurons

ABSTRACT

Is there any transcription factor that can coordinate the expression of all 13 COX subunit genes from the two genomes? A viable candidate is nuclear respiratory factor 1 (NRF-1), discovered by the Scarpulla lab as a transcriptional regulator of cytochrome c, the substrate of COX. In non-neuronal cells, NRF-1 regulates the three transcription factor genes (Tfam, Tfb1m, and Tfb2m) that control the replication and transcription of mtDNA (and hence the three mitochondrial COX genes). NRF-1 binding motifs were found on some COX subunit genes, but not fully characterized. By means of multiple molecular biological approaches, we found that NRF-1 directly regulates all ten nuclear COX genes as well as Tfam, Tfb1m, and Tfb2m in neurons. Thus, it is a master regulator of all 13 COX subunit genes, and itself is controlled by neuronal activity. Moreover, NRF-1 positively regulates the expression of the β1 subunit but negatively controls that of the α1 subunit gene of Na^+/K^+-ATPase, a major energy-consuming protein in neurons. In this way, NRF-1 also coordinates the expression of genes involved in energy production and in energy consumption.

How can the three mitochondrial-encoded and the ten nuclear-encoded COX subunit genes be coordinately regulated in neurons? It would be most efficient if they were all controlled by the *same* transcription factor. Such a transcription factor will directly regulate the ten nuclear COX genes and indirectly control the expression of the three mitochondrial COX subunits by regulating a set of mitochondrial transcription factors A, B1, and B2 (*Tfam*, *Tfb1m*, and *Tfb2m*) important for mtDNA transcription and replication (Figure 8.1). Can NRF-1 be such a factor?

Nuclear respiratory 1 (NRF-1) is a redox-responsive, homodimeric transcription factor originally discovered by Evans and Scarpulla (1989) in rat somatic cells to activate the gene for cytochrome c, the substrate of COX. They named it NRF-1 because similar sequence motifs for its binding are found in the upstream promoter regions of several nuclear genes encoding mitochondrial proteins required for cellular respiration, including the murine mitochondrial RNA processing (MRP) endonuclease gene and the rat *COX6c* gene. In a subsequent study, they confirmed that these genes indeed are functionally activated by NRF-1 (Evans and Scarpulla 1990). Other mitochondrial proteins, such as subunit 8 of complex I, subunits B, C, and D of complex II, ubiquinone binding protein of complex III, COX IV, Vb, VIaL, and VIIc, the γ-subunit of ATP synthase (complex V), 5-aminolevulinate synthase for heme biosynthesis, Tfam, Tfb1m, Tfb2m, SURF1, voltage-dependent anion channel (VDAC) in the outer mitochondrial membrane, and translocase of the outer mitochondrial membrane 20 (TOM20) also have their genes regulated by NRF-1 (Evans and Scarpulla 1990; Chau et al. 1992; Virbasius and Scarpulla 1994; Scarpulla 1997, 2002; Au and Scheffler 1998; Wong-Riley et al. 2000; Ongwijitwat and Wong-Riley 2004; Gleyzer et al. 2005). In addition, NRF-1 binding sites are present in many other genes important for cell growth, chromosome maintenance, and cell cycle regulation, which are not directly involved in cellular respiration (Scarpulla 1997).

DOI: 10.1201/9781003507765-8

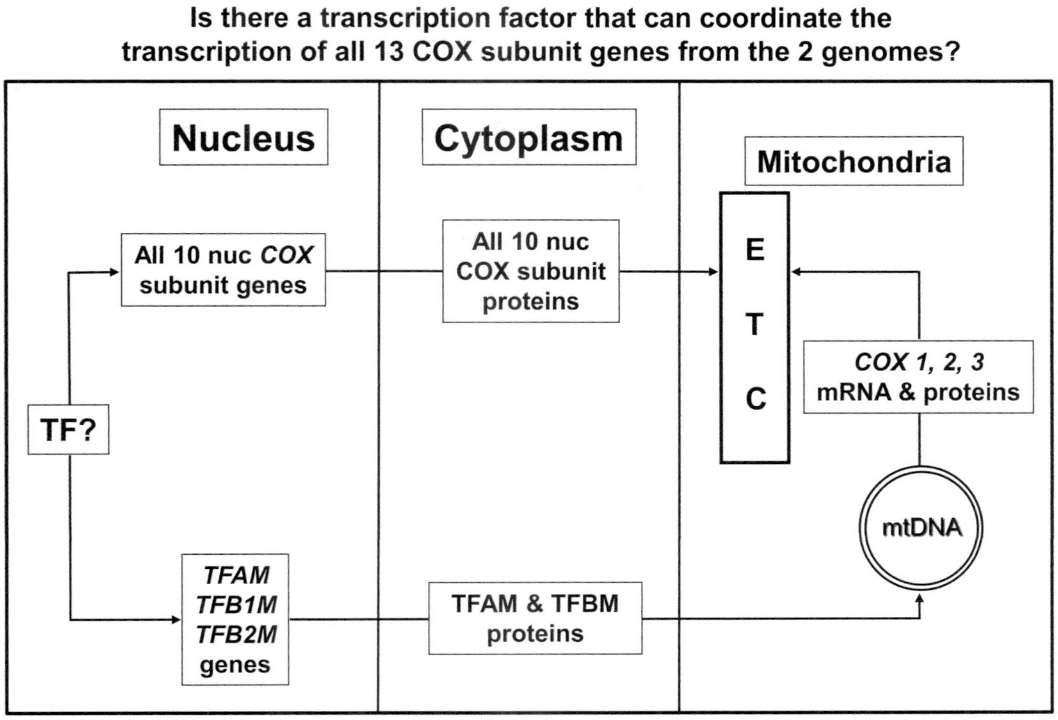

FIGURE 8.1 Schematic diagram of a possible transcription factor (TF) that directly regulates ten nuclear-encoded (nuc) *COX* subunit genes that are transcribed in the nucleus, translated in the cytoplasm, and whose proteins are imported into the mitochondria to form COX holoenzyme in the electron transport chain (ETC). The same TF also indirectly regulates the mitochondrial-encoded *COX 1*, *2*, and *3* by controlling the expressions of mitochondrial transcription factors A, B1, and B2 (*Tfam*, *Tfb1m*, and *Tfb2m*), whose genes are translated in the cytoplasm and whose proteins are imported into the mitochondria to control the replication and transcription of mtDNA. The resultant mitochondrial COX subunits form the catalytic core of COX holoenzyme in the ETC.

Upstream of the DNA-binding domain of NRF-1 is the amino (N)-terminus, which contains a nuclear localization signal (NLS) with clusters of basic amino acids and serine residues, the phosphorylation of which enhances DNA binding (Gugneja et al. 1996; Scarpulla 1997). The DNA-binding locus has high-sequence homology with two developmental regulatory factors, sea urchin P3A2 and the *Drosophila* erect wing gene product (EWG) (Virbasius CA et al. 1993; Scarpulla 1997). It recognizes a GC-rich sequence (see Section 8.1.1) in the *cis*-acting promoter elements in target genes. The *trans*-activation domain of NRF-1 is in the carboxy (C)-terminus of the protein, and transactivation is dependent on clusters of hydrophobic residues, each containing a glutamine. In zebrafish, a gene called "*not really finished*" has high-sequence homology with NRF-1 and is critical for photoreceptor development (Becker et al. 1998). In humans, the single-copy, 68-kb NRF-1 gene has been localized to chromosome 7q31 (Gopalakrishnan and Scarpulla 1995). Knocking out *NRF-1* is embryonically lethal (Huo and Scarpulla 2001).

The abbreviation NRF-1 should not be confused with a similar abbreviation (Nrf1, sometimes also named NRF-1) for an antioxidant transcription factor known as nuclear factor erythroid 2-related factor 1 (Chan et al. 1993; Venugopal and Jaiswal 1996).

The fact that NRF-1 regulates *Tfam*, *Tfb1m*, and *Tfb2m* genes indicates that it indirectly governs the transcription of the three mitochondrial-encoded *COX* subunit genes (*COX1*, *2*, and *3*). Does NRF-1 also regulate all ten nuclear-encoded *COX* subunit genes in neurons? To address this possibility, multiple approaches were used to validate the functional role of NRF-1 in neurons (Dhar et al. 2008).

8.1 DOES NRF-1 REGULATE ALL TEN NUCLEAR-ENCODED *COX* SUBUNIT GENES?

Many molecular biological approaches are available to identify specific transcription factors for specific genes. Any single approach is not sufficient, but a combination of techniques strengthens the conclusion. The methods will be briefly described to help readers appreciate the processes involved.

8.1.1 In Silico Analysis of NRF-1 Binding Motifs in Murine *COX* Subunit Promoters

DNA sequences 1 kb 5′ upstream and up to 1 kb downstream (excluding protein-coding sequences) of the TSS of each of the ten nuclear *COX* genes were analyzed in silico. Sequences were based on the published murine genome database in GenBank™. The GC-rich putative NRF-1 core binding motifs "GCGCAT/CGC" or "GCGCAG/CGC" (Chau et al. 1992) were used as a reference for computer-assisted search of each promoter.

Nuclear-encoded murine *COX 5b*, *6a1*, *6c*, and *8a* promoters all have typical sequence G/T**GCGCATG/CC/T**G/C/T, but *COX 4i1*, *5a*, *6b*, *7a2*, *7b*, and *7c* have atypical sequences **C**/(A/**T**)/**G**/(**C/A**)**C**/(G/**A**)***GCA*****T**/G**G**/(C/A)**C**/G**G**/**T**/C. The **GCA** core, however, is invariant (Dhar et al. 2008).

8.1.2 In Vitro Electrophoretic Mobility Shift and Supershift Assays

To determine possible interactions of NRF-1 with putative binding sites on the ten nuclear *COX* subunit genes, an in vitro electrophoretic mobility shift assay (EMSA) was performed (Dhar et al. 2008). Oligonucleotides of promoter sequences containing NRF-1 binding motifs on each murine *COX* gene were synthesized, annealed, and labeled by a Klenow fragment fill-in reaction with [α-^{32}P]dATP. Each labeled probe was incubated with calf thymus DNA and HeLa nuclear extract and processed for EMSA. Supershift assays were performed with antibodies specific to NRF-1. A positive control was the rat cytochrome *c* promoter known to bind to NRF-1 (Evans and Scarpulla 1989). Two types of competition were used: (a) a 100-fold excess of unlabeled (cold) probes was added to the nuclear extract before applying labeled probes to outcompete the hot ones; and (b) an excess of unlabeled probes containing mutant NRF-1 site was added before the hot ones. Other negative controls included labeled probes with NRF-1 antibodies only and labeled probes with a mutant NRF-1 site.

The positive control cytochrome *c* shows a specific DNA/NRF-1 shift and a supershift complex. Likewise, each of the ten labeled *COX* subunit oligonucleotides exhibits a specific DNA–protein shift band and a DNA–NRF-1–antibody supershift complex. The shift band is outcompeted by an excess of unlabeled probes, but an excess of unlabeled probes containing mutant NRF-1 site is not able to compete. Exposing the labeled oligonucleotides to NRF-1 antibodies without the nuclear extract does not yield any shift or supershift band, indicating an absence of non-specific probe–antibody interaction. Finally, labeled probes containing mutant NRF-1 site do not yield shift or supershift bands (Dhar et al. 2008) (Figure 8.2).

EMSA studies have also been done on the bovine liver isoform of *COX7a2* (Seelan et al. 1996). Thus, all ten nuclear *COX* subunit basal promoters contain sites that bind to NRF-1 in vitro.

8.1.3 In Vivo Chromatin Immunoprecipitation (ChIP) Assays

To ensure that interactions between NRF-1 and each of the ten *COX* subunit promoters exist in vivo, chromatin immunoprecipitation (ChIP) assays were performed using neuroblastoma N2a cells (Dhar et al. 2008). After a brief formaldehyde fixation to immobilize existing DNA–transcription factor interactions, cells were suspended in a buffer containing protease inhibitors to prevent protein degradation. Cells were then homogenized, and nuclei were isolated by centrifugation and

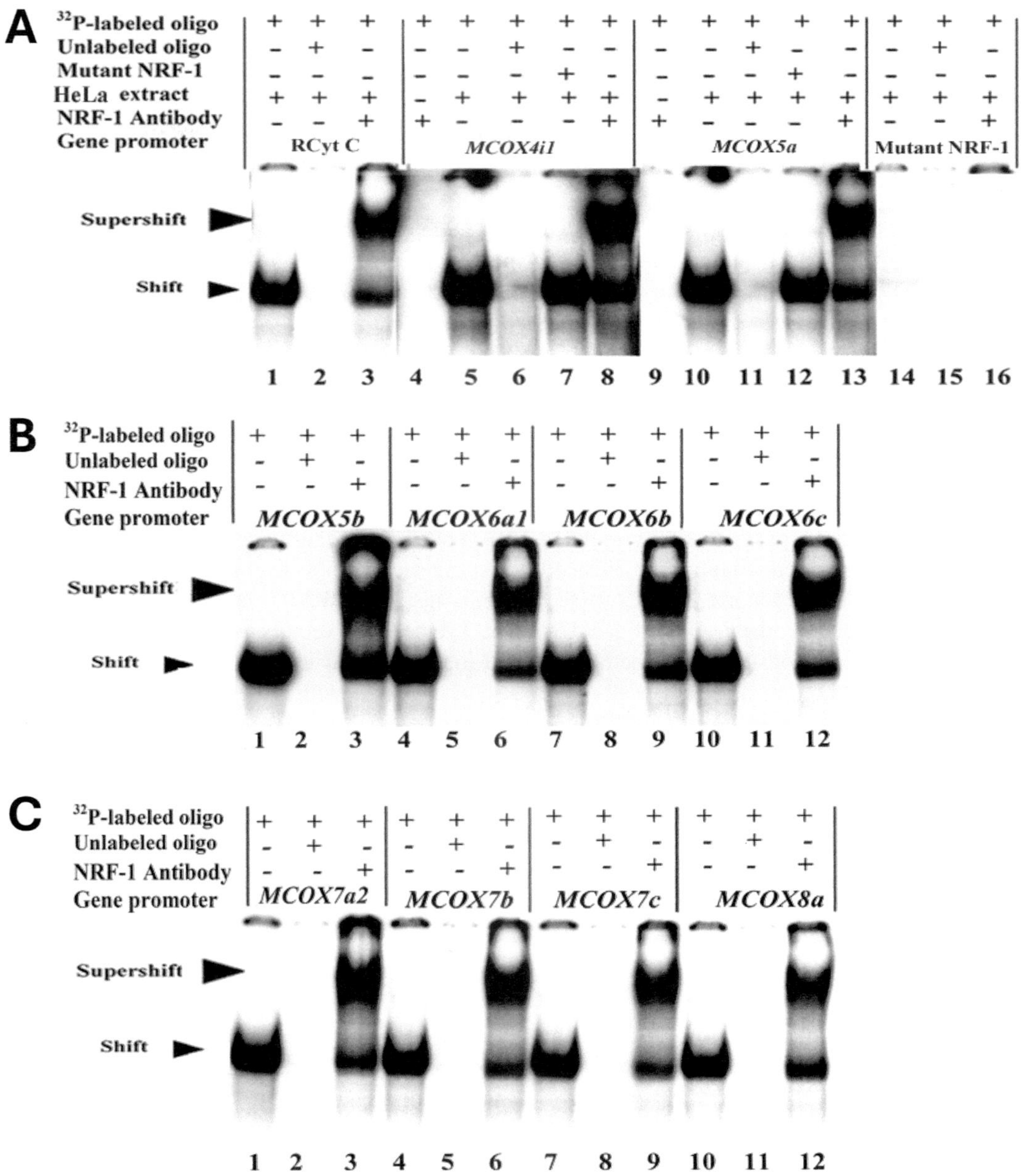

FIGURE 8.2 EMSA and supershift assays of NRF-1 binding to putative sites on all ten nuclear-encoded murine *COX* promoters. (A) The positive control, rat cytochrome *c* promoter, shows a shift band (small arrowhead, lane 1) when the ^{32}P-labeled oligonucleotide is combined with HeLa extract (+ signs for both). In the presence of excess unlabeled oligonucleotide (cold competitor), no band is formed (lane 2). When the NRF-1 antibody is added, both a supershift band (large arrowhead) and a shift band are seen (lane 3). This pattern is repeated for *MCOX4i1* and *MCOX5a* (lanes 5 and 10 for shift, 6 and 11 for cold competitor, and 8 and 13 for supershift and shift). In addition, no band is formed when the ^{32}P-labeled oligonucleotide is added to the antibody without the HeLa extract (lanes 4 and 9). However, when the NRF-1 site is mutated in the unlabeled cold competitor, it can no longer compete and a shift band is seen (lanes 7 and 12). Mutating the NRF-1 binding site on ^{32}P-labeled probe of *COX4i1* yields no shift or supershift bands (lanes 14–16). (B and C) Sets of three lanes (shift, cold competitor, and supershift plus shift) are shown for each of the other eight *COX* subunits.

From Dhar et al. 2008.

sonication. The sonicated lysate was then immunoprecipitated with NRF-1 antibodies, and the precipitated chromatin fragments were revealed by semiquantitative PCR. *Tfb2m* promoter known to bind to NRF-1 (Gleyzer et al. 2005) was used as a positive control, and exon 5 of β-actin not known to interact with NRF-1 was chosen as a negative control. For a negative immunoprecipitation control, antibodies to nerve growth factor receptor p75 (NGFR) were used. A 0.5 and 0.1% dilution of input chromatin (before immunoprecipitation) served as a standard to indicate the efficiency of PCRs.

Results indicate that the proximal promoters of all ten nuclear *COX* subunit genes (*4i1*, *5a*, *5b*, *6a1*, *6b*, *6c*, *7a2*, *7b*, *7c*, and *8a*) coimmunoprecipitate with NRF-1 antibodies and are amplified by PCR. Anti-NGFR does not yield any signal above background. The positive control *Tfb2m* shows a clear band, whereas the negative control β-actin does not (Dhar et al. 2008) (Figure 8.3A). Hence, functional interactions between NRF-1 and promoters of each *COX* subunit gene are confirmed in vivo.

8.1.4 Promoter Mutational Analysis

Another way to test the specificity of NRF-1 binding is to perform promoter mutational analysis. The ten *COX* subunit promoters with mutated NRF-1 sites were first made by PCR cloning using genomic DNA prepared from murine N2a cells, then digested with restriction enzymes and ligated directly into luciferase reporter vectors (pGL3 basic in this case). Wild-type *COX* subunit promoter constructs were also synthesized and ligated into pGL3 vectors. All vectors were verified by sequencing. Each promoter construct was then transfected into N2a cells using Lipofectamine 2000. To control for transfection efficiency, cells were also transfected with a vector that constitutively expressed β-galactosidase. Forty-eight hours after transfection, cells were harvested, lysed, and luciferase activity assayed and normalized with β-galactosidase activity (Dhar et al. 2008).

These site-directed mutagenesis assays show a significant reduction in promoter activity when NRF-1 sites are mutated in each of the ten *COX* subunit promoters. The reductions range from 40% to 80% of the wild-type promoters among the *COX* genes examined (Dhar et al. 2008) (Figure 8.3B). The regulatory role of NRF-1 on the ten *COX* genes is validated.

8.1.5 Silencing of NRF-1 with RNA Interference

What happens to target gene expression if NRF-1 is silenced? To determine this, short hairpin RNAs (shRNAs) against four target sequences in the murine *NRF-1* mRNA were cloned into expression vectors containing the U6 promoters (pLL3.7) with puromycin resistance. Empty vectors served as negative controls. A reporter vector with green fluorescent protein (GFP) (PLVTHM) was also used to identify N2a cells transfected with *NRF-1* shRNAs or empty vectors via Lipofectamine 2000. Puromycin helped to select for purely transfected cells, and transfection efficiency was monitored by the GFP fluorescence (Dhar et al. 2008).

RNA interference (RNAi)-mediated silencing of NRF-1 results in an 80–90% decrease in *NRF-1* mRNA relative to empty vector controls as measured by RT-qPCR and a 75–85% reduction in NRF-1 protein as determined by western blots. The levels of the ten nuclear *COX* subunit mRNAs decline by 40–75% in N2a cells transfected with *NRF-1* shRNA as compared to those transfected with empty vectors ($P < 0.05$–0.001). As a negative control, the expression of another transcription factor, *NRF-2α*, is not changed (Dhar et al. 2008) (Figure 8.4A–C).

In addition, *Tfam*, *Tfb1m*, *Tfb2m*, *Surf1*, *Vdac1*, and *Tom20*, all known target genes of NRF-1, also exhibit significant decreases in the expression of their transcripts in N2a cells transfected with *NRF-1* shRNA. The decrease ranges from ~35% to 70% ($P < 0.05$–0.01) (Dhar et al. 2008) (Figure 8.4D).

These studies strengthen the conclusion that NRF-1 regulates all ten *COX* subunit genes in neurons.

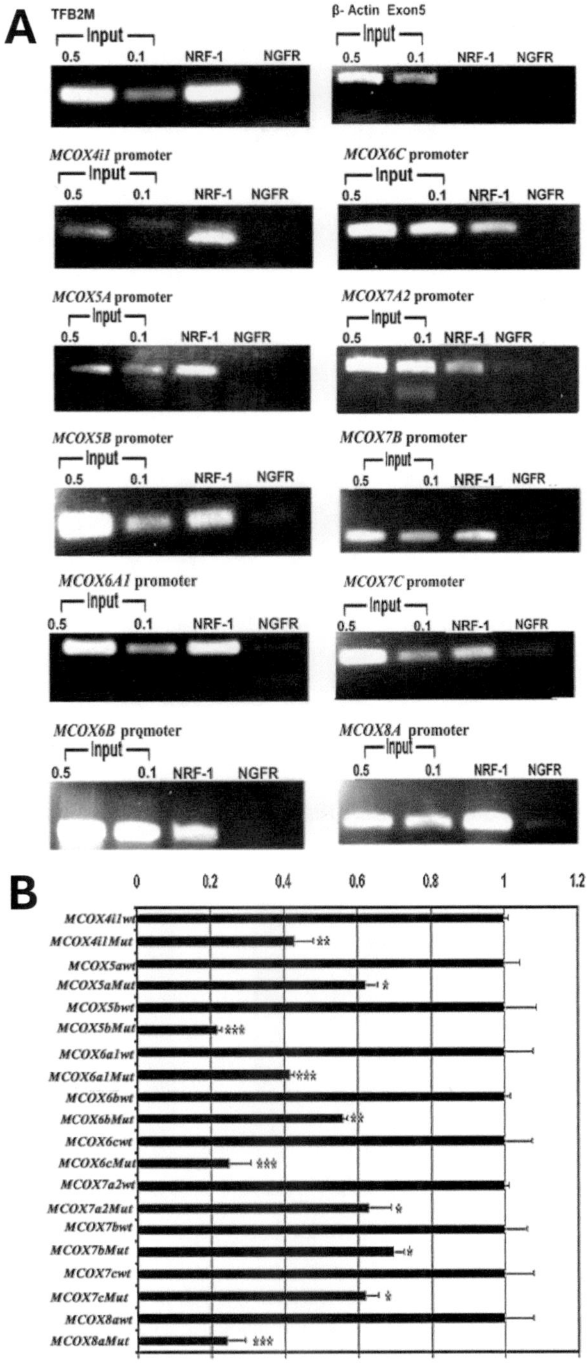

FIGURE 8.3 In vivo ChIP assays of NRF-1 and promoter mutational analysis of all ten nuclear-encoded *COX* subunit promoters. (A) PCR reactions on N2a cell chromatin precipitated with anti-NRF-1 antibodies (NRF-1 lanes) or anti-nerve growth factor receptor p75 antibodies (negative control, NGFR lanes). Control reactions are performed with 0.5% and 0.1% of input chromatin (Input lanes). *TFB2M* serves as a positive control, and *β-actin* exon 5 as a negative control. All ten *COX* subunit promoters are coimmunoprecipitated with NRF-1. (B) Relative luciferase activity of wild type and site-directed mutation of NRF-1 binding sites on all ten nuclear-encoded COX promoters indicates significant reductions in luciferase activity in all ten mutants ($N = 6$ for each construct). $*P < 0.05$, $**P < 0.01$, $***P < 0.001$ when compared to respective wild type.

From Dhar et al. 2008.

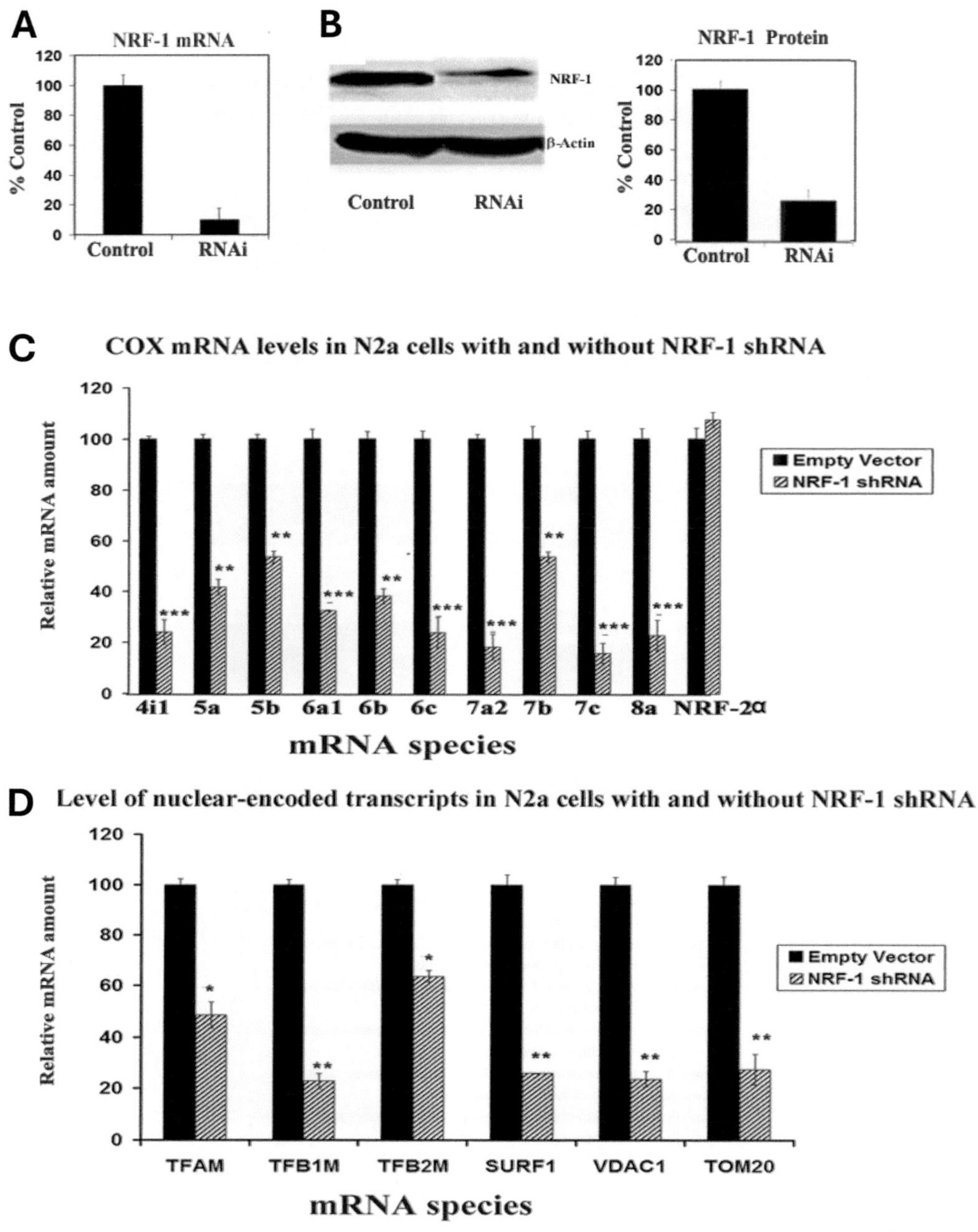

FIGURE 8.4 Silencing of NRF-1 with RNA interference (RNAi) in N2a cells. Such silencing downregulates *NRF-1* message (A) and protein (B), mRNAs of all ten nuclear-encoded *COX* genes (C), as well as those of *Tfam*, *Tfb1m*, *Tfb2m*, *Surf1*, *Vdac1*, and *Tom20* (D). On the other hand, *NRF-2α* message is not affected (C). All comparisons are made with empty vector controls. $^*P < 0.05$, $^{**}P < 0.01$, $^{***}P < 0.001$.

From Dhar et al. 2008.

8.1.6 Conservation of NRF-1 Binding Sites Among Mouse, Rat, and Human *COX* Promoters

A sequence alignment of the ten nuclear *COX* subunit promoter regions containing NRF-1 binding sites from human, mouse, and rat indicates a high degree of homology (60–100%) among the three species. Hence, NRF-1 binding sites are highly conserved.

8.2 NRF-1 ITSELF IS REGULATED BY NEURONAL ACTIVITY

Is NRF-1 itself regulated by neuronal activity? Cultures of primary neurons from postnatal rat visual cortex were exposed to 20 mM KCl depolarizing treatment for 1, 3, 5, and 7 h, or to 7 h KCl followed by withdrawal for 1, 3, 5, and 7 h. The expression of NRF-1 was analyzed by immunoblots, immunocytochemistry at light and EM levels, RT-qPCR, and in situ hybridization (Yang et al. 2006).

The basal level of NRF-1 protein in cultured neurons is relatively low and is present in both the nucleus and the cytoplasm. In the nucleus, NRF-1 is associated mainly with euchromatin, indicating active transcription. In the cytoplasm, it is localized to free ribosomes and infrequently to the Golgi apparatus and the outer nuclear membrane. Depolarizing stimulation with 20 mM KCl induces a time-dependent upregulation of both NRF-1 protein and mRNA in the nucleus and the cytoplasm (Figure 8.5A and C). After 1 and 3 h of stimulation, *NRF-1* message is increased more in the nucleus than in the cytoplasm, indicating *de novo* synthesis of transcripts. It remains high at 5 h before falling off in the nucleus at 7 h. The steady increase of *NRF-1* mRNA in the cytoplasm confirms active translation. The progressive increase of NRF-1 protein in both the nucleus and cytoplasm (from 1 to 7 h of stimulation) suggests an active translocation from the cytoplasm to the nucleus, where it initiates transcription of its target genes. After KCl withdrawal, a progressive downregulation of both message and protein occurs in the nucleus and cytoplasm until they reach close to baseline level by 7 h (Figure 8.5B and D).

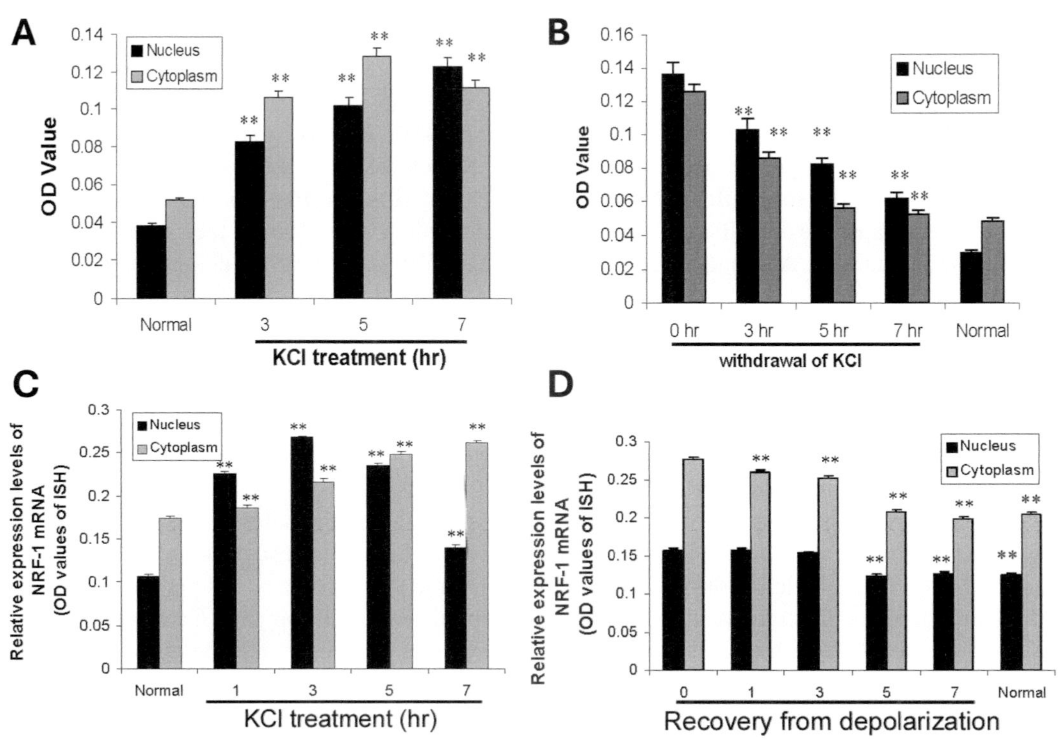

FIGURE 8.5 Neuronal activity regulates NRF-1 protein and message expression in primary neurons in culture. Depolarizing stimulation with 20 mM KCl progressively upregulates NRF-1 protein (A) and mRNA (C) expressions in both the nucleus and cytoplasm from 1 to 7 h. The withdrawal of KCl induces a progressive downregulation of NRF-1 protein (B) and message (D; numbers indicate hours) in the nucleus and cytoplasm until they reach close to baseline level at 7 h. All comparisons are with respective normal controls. $^{**}P < 0.001$ for A and B, $^{**}P < 0.01$ for C and D.

Modified from Yang et al. 2006.

Together, the above experiments provide solid evidence that NRF-1 is an important bigenomic coordinator in orchestrating the transcription of the three mitochondrial *COX* subunit genes via *Tfam*, *Tfb1m*, and *Tfb2m*, as well as the ten nuclear *COX* subunit genes in neurons. A combination of approaches is necessary because no single set of experiments is definitive. In silico analysis reveals a putative binding site, but does not indicate if the site is functional (and often they are not). EMSA alone, even with supershift, shows specific binding in vitro but does not guarantee in vivo binding. The ChIP assay is more definitive, although it is technically more challenging and is dependent on the quality of the antibodies. The promoter mutational analysis and RNAi experiments add weight to the significance of the transcription factor, and functional assays validate that neuronal activity ultimately dictates the transcriptional regulation of *COX* via the specific transcription factor.

8.3 NRF-1 REGULATES SOME SUBUNITS OF Na^+/K^+-ATPASE IN NEURONS

If NRF-1 regulates all 13 subunit genes of *COX*, an important energy-generating enzyme, is it possible that it also regulates the transcription of a critical energy-consuming enzyme, Na^+/K^+-ATPase, in neurons? As discussed in Chapter 3, the expression of Na^+/K^+-ATPase closely parallels that of COX, even though one is in the plasma membrane and the other is in the mitochondria. Where there is a high demand for energy to pump Na^+ out of the cell against their concentration and electrical gradients by Na^+/K^+-ATPase, such as the inner segments of photoreceptor cells, it is met by a rich supply of COX (Chapter 3, Section 3.6). Under normal conditions, Na^+/K^+-ATPase accounts for almost half of the total energy consumption in the brain (Shrivastava et al. 2020).

Na^+/K^+-ATPase was discovered in 1957 by a Danish biochemist, Jens Christian Skou, who studied the leg nerve of a shore crab (*Carcinus maenas*) and found that the enzyme acts like a pump and transports ions across the membrane (Skou 1957). He won a Nobel Prize in 1998 for characterizing this enzyme, which actively pumps 3 Na^+ out of the cell in exchange for 2 K^+ into the cell for every ATP molecule hydrolyzed (Skou 1998). This enzyme is present in all mammalian cell membranes, as all cells must maintain a stable ionic gradient across their membranes (the membrane potential or $\Delta\Psi$) and sustain an osmotic equilibrium critical for cell homeostasis, cell volume, and survival (Skou 1998; Kay and Blaustein 2019; Contreras et al. 2024). The Na^+ gradient is also the energy source for the transport of ions and solutes across the plasma membrane. In neurons, it is the driving force for cell–cell communication via excitatory membrane depolarization and action potentials.

Na^+/K^+-ATPase belongs to the P-type ATPase family with a larger α and a heavily glycosylated β subunit, and in some tissues, an additional smaller γ or FXYD subunit (reviewed in Kaplan 2002; Clausen et al. 2017; Contreras et al. 2024). The α subunit is about 110 kDa, which catalyzes the hydrolysis of ATP and transports the cations. It provides ATP-binding sites on the intracellular or cytoplasmic domain and ouabain binding sites on its extracellular end. The β subunit has a molecular mass of ~55 kDa, is rich in oligosaccharides on its extracellular domain, and plays a critical role in the structural and functional maturation as well as regulation of the enzyme (Geering 2008). The α and β subunits assemble in the endoplasmic reticulum, the β subunits being necessary for the stabilization of the newly synthesized α subunit, and both subunits are dependent on each other for their exit as a holoenzyme out of the organelle and insertion into the plasma membrane (Ackermann and Geering 1990). There are four isoforms for the α subunit (α1–4), each encoded by a different gene. Subunits α1, α2, and α3 are found in the brain (Clausen et al. 2017), with the *Atp1a1* and *Atp1a3* mRNAs expressed in neurons, and that of *Atp1a2* expressed almost exclusively in glial cells (Murata et al. 2020). The α4 isoform is localized to the testis (Kaplan 2002). The β subunit has four isoforms (β1–4). β1 is ubiquitous, β2 is in the heart and skeletal muscles, β3 is in the testis and CNS, and β4 is in nonmammalian species (Armesto et al. 2015). The distribution of

Atp1b1 mRNA is heterogeneous across brain regions and is found at relatively high levels within parvalbumin-positive neurons, where the expressions of the other two isoforms, *Atp1b2* and *Atp1b3*, are very low (Murata et al. 2020).

8.3.1 In Silico Analysis of Promoters of Na+/K+-ATPase Subunit Genes

DNA sequences 1 kb 5′ upstream and 1 kb 3′ downstream of the TSS of *Atp1a1*, *Atp1a3*, and *Atp1b1* genes were analyzed in silico. They all have putative NRF-1 binding motifs with the invariant GCA core and flanking GC-rich regions (Johar et al. 2012).

8.3.2 In Vitro Binding of NRF-1 to Na+/K+-ATPase Subunit Promoters

EMSA studies indicate that *Atp1a1* and *Atp1b1* promoters each forms a specific DNA–protein shift band in the presence of HeLa nuclear extract and a supershift complex with antibodies against NRF-1 (Johar et al. 2012). Despite the fact that *Atp1a3* promoter has an NRF-1 binding motif, it does not display any shift or supershift band. The rat cytochrome *c* promoter is the positive control, and mutating its NRF-1 binding site eliminates any shift or supershift band (Figure 8.6A).

8.3.3 In Vivo Interaction of NRF-1 with Na+/K+-ATPase Subunit Promoters

In vivo ChIP assays using N2a cell chromatin and anti-NRF-1 antibodies demonstrate that NRF-1 interacts with promoters of *Atp1a1* and *Atp1b1* genes, but not with that of the *Atp1a3* gene (Johar et al. 2012). The *Tfb2m* promoter serves as the positive control, and the negative control, β-actin promoter, does not yield any band. Immunoprecipitation with nerve growth factor receptor (NGFR) antibodies also fails to show any band and serves as another negative control (Figure 8.6B).

8.3.4 Mutation of NRF-1 Binding Sites on Promoter Activity

Site-directed mutagenesis of NRF-1 binding sites on *Atp1a1*, *Atp1b1*, and the control *COX6b* promoters leads to a significant decrease in promoter (luciferase) activity of *Atp1b1* and *COX6b* ($P < 0.001$ for both). However, the activity of the *Atp1a1* promoter is significantly increased ($P < 0.05$), indicating that NRF-1 negatively regulates the expression of this gene (Johar et al. 2012) (Figure 8.6C).

8.3.5 Effect of Silencing NRF-1 by RNA Interference

When NRF-1 is silenced in neurons with shRNA against four of its target sequences, its protein is reduced by 75–85% and its mRNA by 65–70% ($P < 0.001$ for both) (Figure 8.7). The mRNA expressions of three positive controls are also reduced: *COX2* by 63% ($P < 0.001$), *COX7c* by 50% ($P < 0.001$), and *Tfb2m* by 22% ($P < 0.01$). Likewise, *Atp1b1* message is reduced by 35% ($P < 0.001$), but that of *Atp1a1* is increased by 12–35% ($P < 0.05$), and that of *Atp1a3* remains unchanged. The protein level of Atp1b1 is decreased by 20–30% ($P < 0.001$) and that of Atp1a1 is increased by 12–30% ($P < 0.05$), consistent with changes in their transcripts (Johar et al. 2012).

When neurons are exposed to 20 mM KCl depolarizing stimulation for 5 h, mRNA levels of *NRF-1*, *COX2*, *COX7c*, *Tfb2m*, *Atp1b1*, *Atp1a1*, and *Atp1a3* are significantly upregulated ($P < 0.001$ for all). In the presence of *NRF-1* shRNA, however, the first five transcripts no longer respond to KCl. On the other hand, *Atp1a1* transcripts are increased ($P < 0.001$), and those of *Atp1a3* are unaffected. *NRF-1*, *Atp1b1*, and *Atp1a1* proteins respond the same way as their respective mRNAs (Johar et al. 2012).

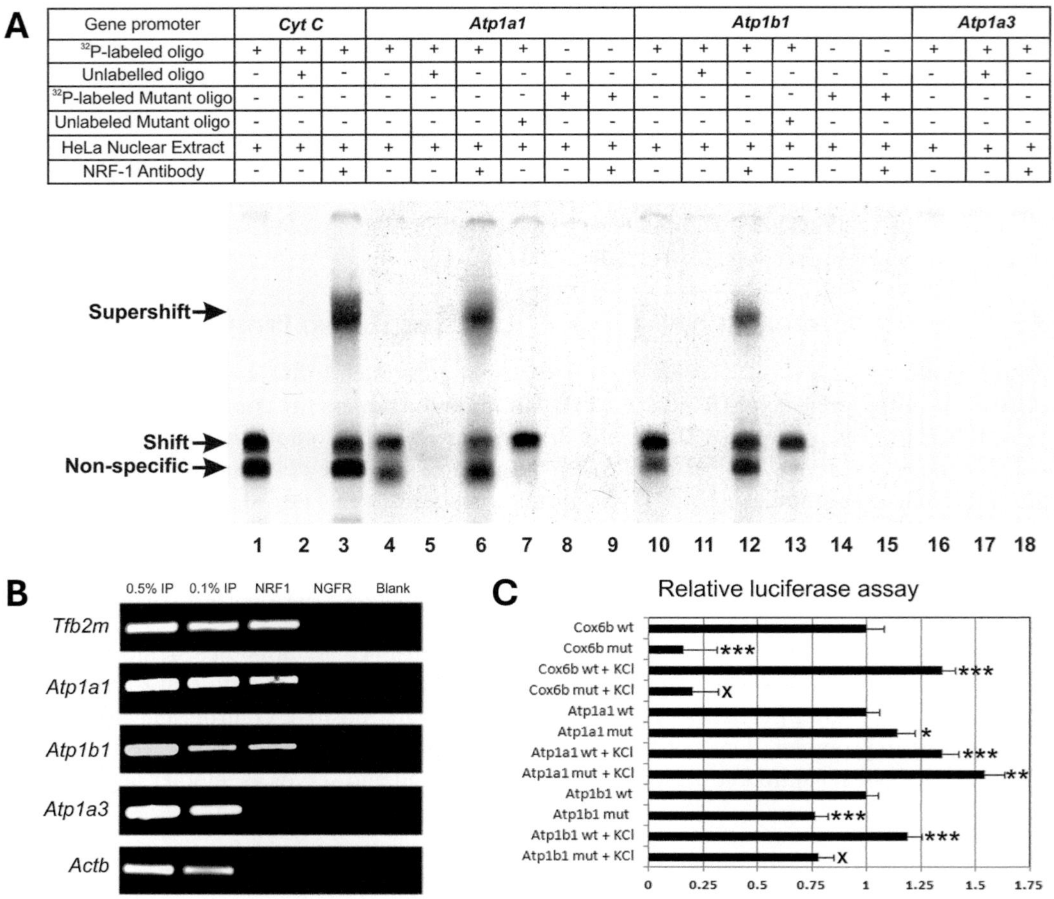

FIGURE 8.6 NRF-1 regulates Na^+/K^+-ATPase subunit genes. (A) EMSA testing *in vitro* binding of NRF-1 to ^{32}P-labeled oligonucleotide containing NRF-1 binding site on subunit *Atp1a1*, *Atp1b1*, and *Atp1a3* genes of Na^+/K^+-ATPase. The components in the assay are shown in the left column, and their individual presence is indicated by a (+) sign. Cytochrome *c* (Cyt C) serves as the positive control, with shift, cold competitor, and supershift bands shown (arrows for lanes 1, 2, and 3, respectively). *Atp1a1* and *Atp1b1* also reveal the same labeling (lanes 4–6 and 10–12, respectively), but *Atp1a3* does not (lanes 16–18). Mutating the NRF-1 site in the unlabeled oligonucleotide disables its competition with the labeled one, and shift bands are present (lanes 7 and 13). However, mutating the site in the labeled probe fails to elicit shift (lanes 8 and 14) or supershift (lanes 9 and 15) bands. (B) In vivo ChIP assays indicate binding of NRF-1 to *Atp1a1* and *Atp1b1* immunoprecipitated by NRF-1 antibodies, but not to *Atp1a3*. *Tfb2m* serves as the positive control, and β-actin (*Actb*) as the negative control. Antibodies against nerve growth factor receptor (NGFR) or no antibody (Blank) fail to generate any band. (C) Luciferase assays of promoter activity of *COX6b*, *Atp1a1*, and *Atp1b1* in the wild type (wt; normalized to 1) and their activation by 20 mM KCl stimulation for 5 h. Mutating the NRF-1 binding site significantly reduces the promoter activity of *COX6b* and *Atp1b1* in the wt and KCl-exposed mutants (mut) but significantly increases that of *Atp1a1*. $N = 6$ for each construct. All mutants and wt + KCl are compared with wt, and all mutants + KCl are compared with mutants. $^*P < 0.05$, $^{**}P < 0.01$, $^{***}P < 0.001$, X = non-significant. Error bars = ± SE.

From Johar et al. 2012.

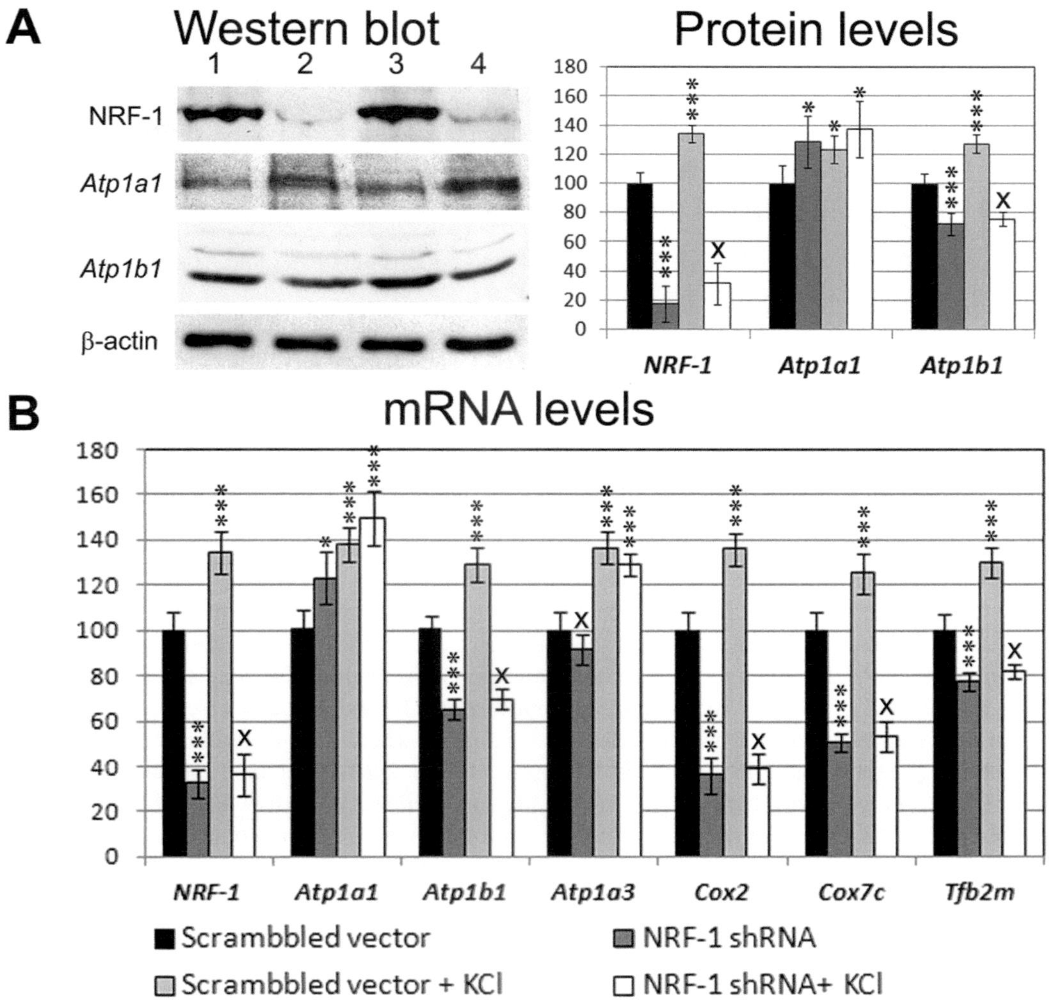

FIGURE 8.7 Effect of NRF-1 silencing with RNA interference on the expression of Na^+/K^+-ATPase genes in N2a cells. (A) Western blots of protein levels of control (lane 1; transfected with scrambled vectors), *NRF-1* shRNA (lane 2), control + KCl (lane 3), and *NRF-1* shRNA + KCl (lane 4) for *NRF-1*, *Atp1a1*, and *Atp1b1*, with β-actin as the loading control. Quantitative analysis indicates that KCl upregulates the expression of all three proteins. Silencing NRF-1 with shRNA suppresses *NRF-1* and *Atp1b1* proteins and prevents them from being upregulated by KCl. However, the opposite is true for *Atp1a1*, whose protein is increased by *NRF-1* shRNA, with and without KCl. (B) Comparable effects of *NRF-1* shRNA are found for mRNA levels of *NRF-1*, *Atp1a1*, and *Atp1b1*. In addition, *NRF-1* shRNA adversely affects the positive controls *COX2*, *COX7C*, and *Tfb2m*, but not *Atp1a3*. $N = 10$ for each data point. All *NRF-1* shRNA and scrambled vector + KCl are compared with scrambled vector controls, and all *NRF-1* shRNA + KCl are compared with *NRF-1* shRNA. $*P < 0.05$, $***P < 0.001$, X = non-significant. Error bars = ± SE.

From Johar et al. 2012.

8.3.6 Effect of Overexpressing NRF-1

A reverse experiment was performed on N2a cells transfected with vectors overexpressing NRF-1. This results in a 29-fold increase in *NRF-1* transcripts and a concomitant increase in the transcript levels of *COX2* (220%), *COX7c* (160%), *Tfb2m* (140%), and *Atp1b1* (165%) ($P < 0.001$ for all). On the other hand, *Atp1a1* transcripts are decreased by 65% ($P < 0.001$), but that of *Atp1a3* is not affected (Johar et al. 2012).

Exposing N2a neurons transfected with empty vectors to 0.4 μM of the impulse blocker TTX for 3 days leads to a significant reduction in the transcript levels of *NRF-1*, *COX2*, *COX7c*, *Tfb2m*, *Atp1b1*, *Atp1a1*, and *Atp1a3* ($P < 0.001$ for all except $P < 0.01$ for *Tfb2m*). NRF-1 overexpression partially rescues the first five transcripts from the detrimental effect of TTX. However, it does not rescue the expression of *Atp1a1* and has no impact on that of *Atp1a3* (Johar et al. 2012).

8.3.7 Homology of NRF-1 Binding Sites

Functional NRF-1 binding sites on the 5′-untranslated regions of *Atp1a1* and *Atp1b1* are conserved among mice, rats, and humans (Johar et al. 2012).

Together, the above experiments illustrate that NRF-1 positively regulates the *b1* subunit gene and negatively controls the *a1* subunit gene of Na^+/K^+-ATPase in neurons. However, the expressions of all three subunits (*a1*, *a3*, and *b1*) are dictated by neuronal activity. NRF-1, therefore, not only regulates the genes of an essential energy-generating enzyme, COX, but also major subunit genes of a key energy-consuming enzyme, Na^+/K^+-ATPase.

The opposing action of NRF-1 on the *Atp1b1* and *Atp1a1* subunit genes is intriguing. It has been reported that the β subunit, being the regulatory one, may be more closely involved in the active ion transport mechanism and thus with the functional activity of the enzyme (Kaplan 2002). The level of β1 mRNA is significantly lower than that of the α subunit in N2a cells and many other tissues (Sweadner 1989; Johar et al. 2012), indicating that the synthesis of the β subunit is tightly controlled and is one of the determining factors for the abundance of the holoenzyme. The negative regulation of *Atp1a1* by NRF-1 may prove to be advantageous for some neurons. In comparison to the α3 subunit, α1 has a greater affinity for Na^+ and a lower affinity for K^+ and ATP, and α1-containing enzymes reportedly operate under normal physiological conditions, whereas α3-containing enzymes restore membrane potential faster in highly active neurons (Jewell and Lingrel 1991; Habiba et al. 2000). By negatively regulating the α1 subunit, NRF-1 may indirectly facilitate α3 when repeated firing and repolarization are in greater demand in specific neurons under obligatory conditions. However, it is highly likely that the various subunits of Na^+/K^+-ATPase are under the control of multiple transcription factors and upstream signaling mechanisms, and that NRF-1 functions in concert with other transcription factors, such as specificity protein (Sp) factors, whose binding sites have been noted on the *Atp1a1* promoter (Suzuki-Yagawa et al. 1992). The role of Sp factors in regulating Na^+/K^+-ATPase will be addressed in Chapter 10, Section 10.3.

9 Nuclear Respiratory Factor 2 (NRF-2)

A Transcriptional Regulator of All 13 COX Subunit Genes in Neurons

ABSTRACT

In addition to NRF-1, nuclear respiratory factor 2 (NRF-2) is another transcription factor that can potentially coordinate the expression of all 13 COX subunit genes in neurons. NRF-2 was discovered by the Scarpulla lab to activate several respiratory genes. It is the human homolog of the GA binding protein (GABP) that has a DNA-binding α subunit and a transactivating β subunit. NRF-2 is special in that its distribution pattern in the monkey visual cortex virtually matches that of COX, including an enrichment in supragranular puffs and type C cells. NRF-2 itself responds to changes in neuronal activity both in primary neuronal cultures and in vivo. Experiments combining multiple molecular biological methods solidify the conclusion that NRF-2 is another master regulator of all nuclear-encoded COX genes, as well as the three mitochondrial COX genes, by controlling the expressions of Tfam, Tfb1m, and Tfb2m. Thus, NRF-2 works in parallel with but independently of NRF-1.

NRF-1 is not likely to function alone. The promoters of several *COX* subunit genes have binding motifs for other transcription factors. NRF-2 is one such factor with structure and binding motifs different from those of NRF-1. This chapter discusses the significance of NRF-2 in neurons and its role in regulating all *COX* subunit genes.

In 1991, Virbasius and Scarpulla reported that the promoter region of the rat *COX4* subunit gene has a tandem-repeat pair of *GGAA/T* binding motifs for a nuclear factor characteristic of the Ets (erythroblast transformation-specific) domain family of proteins, originally identified in the E26 avian erythroblastosis virus (Karim et al. 1990). The Ets domain proteins belong to the superfamily of winged helix-turn-helix transcription factors that recognize the *GGAA/T* core element on the DNA. Scarpulla's group named their factor NRF-2 (nuclear respiratory factor 2) because they found the same motif in the β subunit of the human ATP synthase (complex V) gene, and it was the second redox-responsive transcription factor (after NRF-1) for mitochondrial respiratory genes (however, NRF-2 is structurally and genetically not related to NRF-1). Subsequently, they and others have identified NRF-2 binding sites in the murine *COX5b*, rat *COX6a1*, bovine *COX7a2*, human *TFAM, TFB1M, TFB2M*, human iron–sulfur *SDH2* of complex II, murine uroporphyrinogen III synthase for heme biogenesis, and the human *TOM20* genes (Virbasius JV et al. 1993; Sucharov et al. 1995; Seelan et al. 1996; Scarpulla 1997; Au and Scheffler 1998; Scarpulla 2002; Ongwijitwat and Wong-Riley 2004; Gleyzer et al. 2005).

NRF-2 is the human homolog of the rodent GA-binding protein or GABP, first identified by the McKnight's group to be an activator of herpes simplex virus immediate early genes (Thompson et al. 1991; LaMarco et al. 1991). GABP has two major subunits: the 51.3 kDa DNA-binding α subunit with the Ets domain of protooncogenes, and the transactivating β subunit that contains four 33-amino acid Notch-ankyrin repeats close to its N-terminus that interact with the α subunit. DNA binding will not occur unless both subunits are present. The β subunit is further subdivided into the 41.3 kDa β1 and the 37 kDa β2 subunits that likely represent alternatively spliced variants, with β1 being capable of forming a stable homodimer while β2 is not. A model was proposed for a

DOI: 10.1201/9781003507765-9

heterotetramer, with two α subunits, each binding to a purine-rich hexanucleotide (*CGGAAA/G*) and each associating with the N-terminus of β subunit, while the leucine zipper carboxy ends of the two β subunits (β1) form a homodimer (Thompson et al. 1991). The β1 subunit also contains the nuclear localizing signal, which enables the translocation of the β subunit and influences the transport efficiency of the α subunit into the nucleus (Sawa et al. 1996). Knocking out GABPα is embryonically lethal (Ristevski et al. 2004).

Virbasius JV et al. (1993) purified NRF-2 from HeLa cells and noted that it contains five polypeptides, which they named α (56 kDa) with the Ets DNA-binding domain, β1 (48 kDa), β2 (46 kDa), γ1 (41 kDa), and γ2 (39 kDa), respectively. The β subunits, when complexed with the α subunits, confer high-affinity binding to the tandem *cis* element on the target genes. The β and γ subunits of human NRF-2 appear to correspond to the β1 and β2 subunits of GABP Virbasius JV et al. (1993). However, in cloning and expressing the five subunits of NRF-2, Gugneja et al. (1995) uncovered the fact that β1 and γ1 of human NRF-2 are variants of the β1 subunit of rodent GABP and share the same C-terminus as the rodent β1, but with a 36-bp serine-rich insertion encoding 12 amino acids that are missing in GABP (and in rodent NRF-2). In contrast, the β2 and γ2 of NRF-2 are variants of the β2 subunit of GABP and share a common C-terminus. There is a slight difference in the molecular mass of the five subunits of NRF-2: α (58 kDa), β1 (48 kDa), β2 (46 kDa), γ1 (38 kDa), and γ2 (37 kDa) (Gugneja et al. 1995). All four β and γ subunits have equal potency in binding to the α subunit, and all can activate transcription via their glutamine-rich transactivation domain. The relative amount of different subunits may vary among tissues, thereby conferring tissue specificity.

The abbreviation NRF-2 should not be confused with a similar abbreviation (Nrf2, sometimes also named NRF-2) for an antioxidant transcription factor known as nuclear factor erythroid 2-related factor 2 (Moi et al. 1994; Venugopal and Jaiswal 1996).

Almost nothing was known about the role of NRF-2 in neurons prior to 1999. What is its distribution in neural tissues? Does it regulate *COX* subunit genes in neurons? Does it respond to neuronal activity? Some key experiments probing these questions are outlined below.

9.1 DOES THE DISTRIBUTION OF NRF-2 CORRELATE WITH THAT OF COX IN NEURONS?

9.1.1 Generation of Polyclonal Antibodies Against NRF-2α Subunit Proteins

To determine if the distribution of NRF-2 correlates with that of COX, polyclonal antibodies were generated against synthetic peptides (Nie and Wong-Riley 1999) bearing a sequence of 16 amino acids (residues 301–316) of the human NRF-2α reported by the Scarpulla lab (Virbasius and Scarpulla 1991; Virbasius et al. 1993): RISGEDRSSPGNRTGN. This sequence shares 100% homology with that of the α subunit of GABP. The synthetic peptides were coupled to the carrier protein keyhole limpet hemocyanin (KLH) with a glutaraldehyde link and were injected into New Zealand White rabbits to produce polyclonal antibodies. Immunoglobulins (IgGs) against this sequence were affinity-purified and determined to be specific by enzyme-linked immunosorbent assay (ELISA), SDS-PAGE, and western blots. The antibodies react with the NRF-2α subunit from crude monkey brain tissues and display a single band of the correct size, 56 kDa, in immunoblots, whereas preimmune sera or antisera preadsorbed with synthetic peptides yield no band (Nie and Wong-Riley 1999). Purified IgG was used in immunohistochemistry.

9.1.2 Immunohistochemical Distribution of NRF-2α in the Monkey Primary Visual Cortex and Its Correlation with COX

The pattern of NRF-2α immunohistochemical distribution is strikingly similar to that of COX histochemistry in the monkey primary visual cortex, a feature not seen with NRF-1. Like COX, NRF-2α is densely labeled in IVC, IVA, and II/III puffs, moderately so in layer VI, and lightly labeled in layers V, IVB, and interpuffs (Nie and Wong-Riley 1999) (Figure 9.1A–D). The overlap between the two labels in supragranular puffs is unmistakable in tangential sections, and NRF-2α-immunoreactive neurons within puffs are more intensely labeled than those in interpuffs, just as COX-rich neurons are more plentiful in puffs than in interpuffs (see Chapter 4).

9.1.3 Effect of Retinal Impulse Blockage on NRF-2α Immunoreactivity

Is the expression of NRF-2α influenced by neuronal activity? This was tested in adult macaque monkeys monocularly inactivated with TTX and survived for 1, 3, or 7 days. After 1 day of retinal impulse blockade, NRF-2α immunoreactivity is visibly reduced in deprived ocular dominance columns in layer IVC, in perfect register with decreased COX activity in adjacent sections. Longer survivals lead to further reductions in both NRF-2α and COX in deprived columns in layers IVC, IVA, V, VI, and supragranular puffs (Figure 9.1E and F). Decreases in both NRF-2α and COX in deprived ODCs are statistically significant at all three survival time points tested ($P < 0.001$ for all) (Nie and Wong-Riley 1999).

Clearly, visual input regulates NRF-2α expression in the adult monkey visual cortex. The nearly identical distributions of NRF-2α and COX in both normal and functionally inactivated states strongly support NRF-2α's robust influence on COX expression in neurons, and the regulation is activity dependent at a very local level.

9.2 DISTRIBUTION OF NRF-2α mRNA IN NORMAL AND MONOCULARLY DEPRIVED MACAQUE VISUAL SYSTEM

To determine if the distribution of NRF-2 and its regulation by neuronal activity was at the transcriptional level, a partial cDNA of NRF-2α was isolated from a human brain cDNA library, subcloned, sequenced for specificity, riboprobes generated, and the distribution of mRNA in the adult monkey visual system assessed with in situ hybridization (Guo et al. 2000).

Under normal conditions, the expression pattern of *NRF-2α* mRNA in the monkey striate cortex parallels closely that of COX activity, being densest in layer IVC, moderate in supragranular puffs and layer VI, and light in layers I, II/III interpuffs, IVB, and V. Whereas COX activity is high in the neuropil (mainly dendrites) and in some neuronal cell bodies, *NRF-2α* mRNA understandably resides primarily in the cytoplasm of neuronal somata. Monocular impulse blockade with TTX leads to a decrease in *NRF-2α* transcripts in deprived ocular dominance columns, not only most visible in layer IVC, but also detectable in layers II/III puffs and layer VI (Figure 9.1G and H). Such a downregulation is seen even 1 day after TTX treatment, but more striking with longer survivals up to 2 weeks ($P < 0.001$ for all). Likewise, *NRF-2α* mRNA is reduced in deprived laminae of the lateral geniculate nucleus, in parallel with decreased COX activity there as early as 1 day after monocular TTX, but becomes more prominent after 1 week ($P < 0.001$) (Guo et al. 2000). Hence, NRF-2α senses decreases in neuronal activity and downregulates its transcript levels accordingly.

9.3 HOW DO NRF-2/GABP α AND β SUBUNITS RESPOND TO INCREASED NEURONAL ACTIVITY?

9.3.1 Response of NRF-2/GABP α and β Subunits and *COX* Transcripts to Depolarizing Stimulation in Cultured Primary Neurons

To ensure that responses of NRF-2/GABP are confined mainly to neurons, primary cultures of neurons from postnatal rat cortex (see Chapter 6, Section 6.3) were exposed to 20 mM KCl for varying periods. Expressions of NRF-2α (GABPα) and NRF-2β (GABPβ) proteins were determined immunohistochemically with antibodies from the McKnight's lab, and a monoclonal antibody against neuronal nuclear protein, NeuN, was the negative control. COX activity was performed histochemically, and *COX* subunit

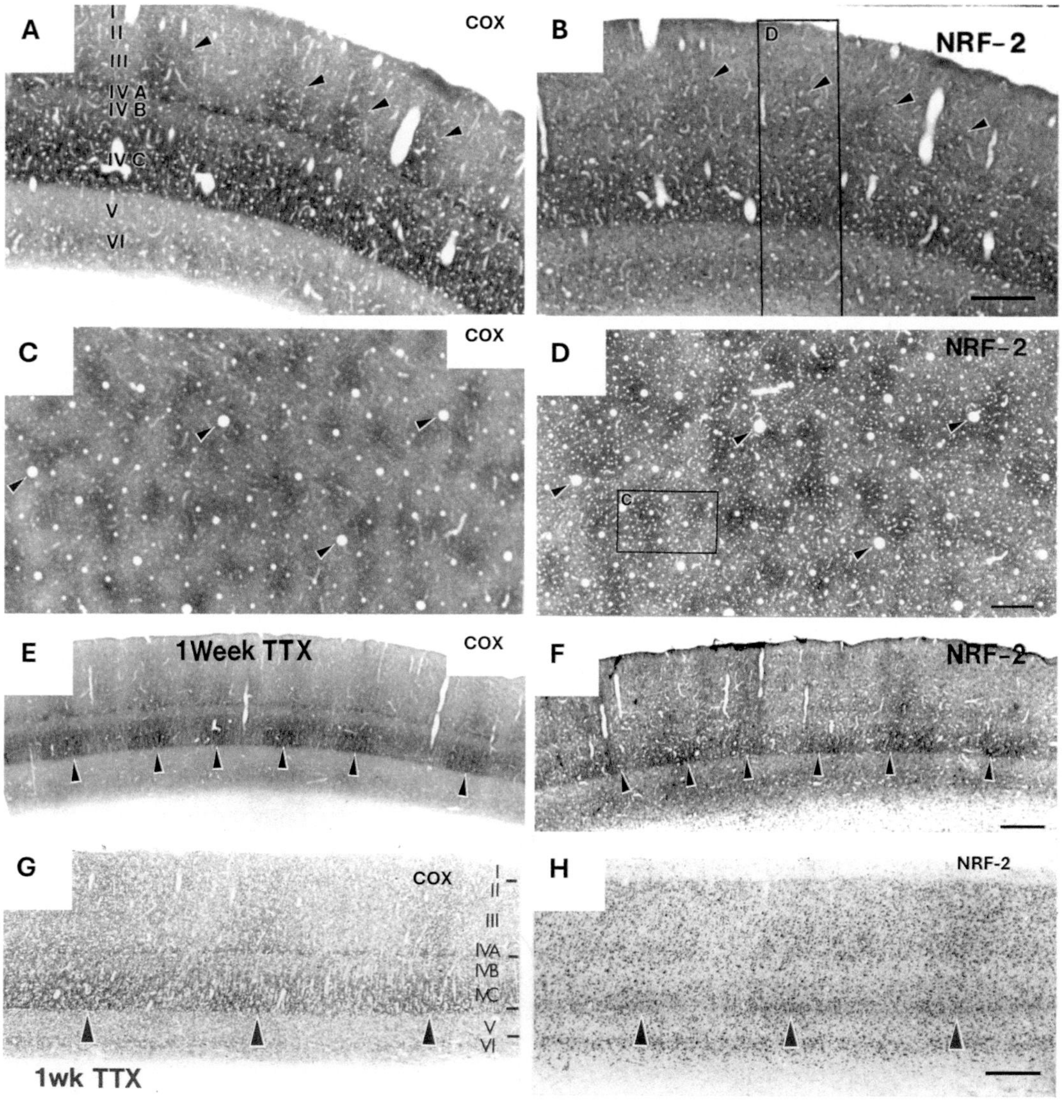

FIGURE 9.1 NRF-2 labeling in the macaque visual cortex is comparable to that of COX. (A and B) Two consecutive cross sections of macaque V1 show a similar labeling pattern with COX histochemistry or NRF-2α immunohistochemistry. Layer numbers are marked in A. Arrowheads in each point to layers II/III puffs that are in exact register. (Box in B labeled D should be ignored). (C and D) Two consecutive tangential sections highlight supragranular puffs that are also in exact register between COX and NRF-2α. Arrowheads in each indicate corresponding blood vessels for section alignment. (Box in D labeled C should be ignored). (E and F) After 1 week of intraocular TTX, alternating dark and light columns can be seen in both COX and NRF-2α-reacted sections, especially in layer IVC (arrowheads indicate respective dark columns). (G and H) Two consecutive fresh, unfixed cross sections of V1 reacted for COX histochemistry (G) and for *NRF-2α* mRNA (in situ hybridization of digoxigenin-labeled riboprobe) (H) after 1 week of monocular TTX. Alternating dark (arrowheads) and light columns are visible, especially in layer IVC, in both sections.

Scale bar in each set = 0.5 mm.

A–F are modified from Nie and Wong-Riley 1999; G–H are modified from Guo et al. 2000.

2, *4*, and *6a* mRNAs were analyzed with in situ hybridization of specific riboprobes as well as with RNase protection assays. Histochemical and immunohistochemical reaction products were measured semi-quantitatively with optical densitometry of individual neurons (Zhang and Wong-Riley 2000b).

In untreated cultures, GABPα subunit proteins are localized mainly in the nuclei of neurons, whereas those of GABPβ are in both the nuclei and cytoplasm in an approximate 1:1 ratio. After 2 h of depolarizing stimulation, the level of GABPβ is significantly increased in the nucleus ($P < 0.001$), with an eightfold increase in the nucleus-to-cytoplasmic ratio. This is consistent with the translocation of this subunit from the cytoplasm to the nucleus to complex with and transactivate the α subunit for target gene activation in response to increased neuronal activity. The level of GABPα is significantly upregulated in the nucleus after 5 h of stimulation, indicating additional de novo synthesis. Both subunits remain elevated in the nucleus at 10 h and are still above control levels after 40 h of KCl treatment ($P < 0.001$ for both) (Figure 9.2A and B). The control NeuN does not increase in labeling (Zhang and Wong-Riley 2000b).

By comparison, the mitochondrial (*COX2*) subunit mRNA in untreated neurons is localized to both the cell bodies and dendrites, whereas the mRNAs of the nuclear (*COX4* and *COX6a*) subunits

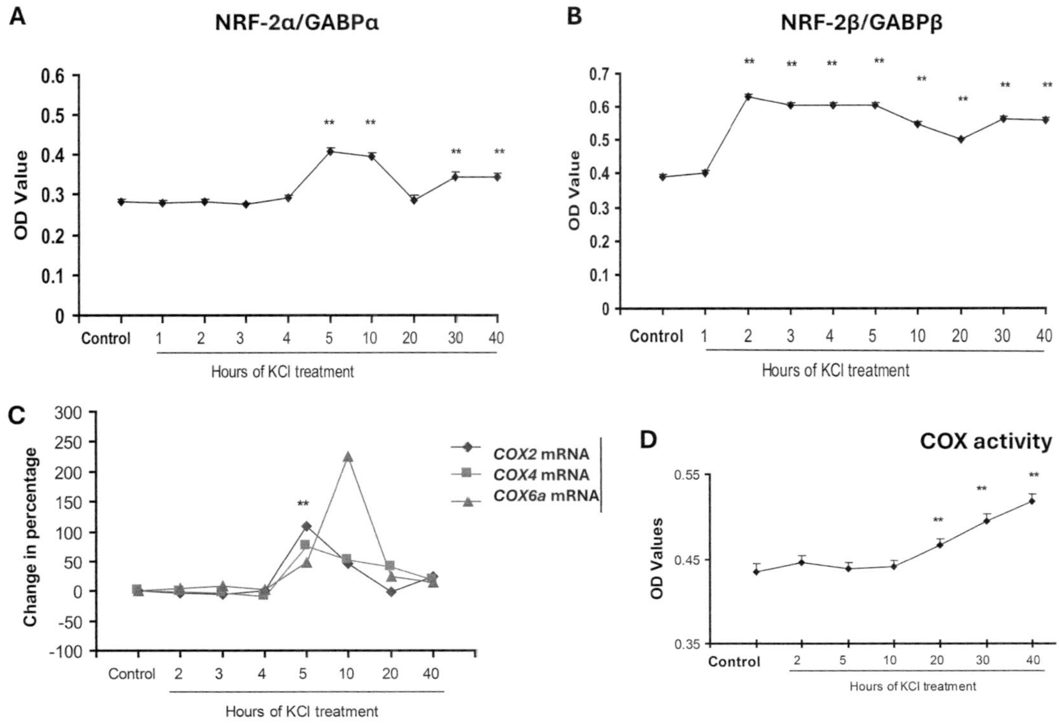

FIGURE 9.2 Upregulation of NRF-2 and COX by KCl in cultured primary neurons. (A) NRF-2α (GABPα) protein in neuronal nuclei is upregulated after 5 h of KCl stimulation and remains high at 10 h before leveling off. (B) NRF-2β (GABPβ) protein in the nucleus, on the other hand, is already elevated after 2 h of KCl and remains high throughout the course of stimulation. (C) Quantification of RNase protection assays shows that the message levels of *COX2*, *COX4*, and *COX6a* are all significantly upregulated after 5 h of KCl stimulation, but *COX6a* does not peak until after 10 h. All three transcripts level off after their peaks, but they all remain above control levels after 40 h of stimulation. (D) COX activity measured with single neuron optical densitometry shows a significant increase after 20 h of KCl and remains high after 40 h. All comparisons are with the respective controls.

$**P < 0.001$ for all comparisons with respective controls.

From Zhang and Wong-Riley 2000b.

are strictly within the cell somata. All three subunits are not upregulated until 5 h of depolarizing stimulation, and *COX6a* expression does not peak until 10 h (Figure 9.2C). COX activity is not significantly increased until 20 h of KCl stimulation and remains high at 30 h and 40 h (Figure 9.2D) (Zhang and Wong-Riley 2000b).

These studies indicate that both α and β subunits of NRF-2/GABP sense and respond to increased neuronal activity by upregulating their own levels, and their actions, especially that of the β subunit, precede those of *COX* subunit mRNAs from both genomes. They are consistent with NRF-2 being a bigenomic transcriptional coordinator of *COX* subunit genes from both the mitochondrial and nuclear genomes. All transcript upregulation precedes that of COX activity.

9.3.2 Ultrastructural Study of Depolarization-Induced Response of NRF-2 in Cultured Rat Visual Cortical Neurons

In cultured normal primary neurons, immunogold-silver-labeled particles marking NRF-2α and NRF-2β (antibodies from Nie and Wong-Riley 1999 and Vercauteren et al. 2008, respectively) are present ultrastructurally in both the nucleus and cytoplasm. In the nucleus, both subunits are associated mainly with euchromatin rather than heterochromatin, consistent with active transcription. In the cytoplasm, both subunits are associated primarily with free ribosomes and occasionally with the Golgi apparatus and the outer membrane of the nuclear envelope. The presence in the Golgi indicates either posttranslational processing or an association with a protein being processed there. Location at the outer membrane of the nuclear envelope suggests that the subunits are being translocated from the cytoplasm to the nucleus. However, other possibilities, such as interaction with other proteins or non-specific labeling, must be ruled out. Specificity is strengthened by the absence of labeling in the mitochondria (Yang et al. 2004).

Five hours of KCl-induced depolarizing stimulation increases the nucleus-to-cytoplasm ratio of both subunits six- to sevenfold ($P < 0.001$ for both) without increasing total labeling densities. This is consistent with a cytoplasm-to-nucleus translocation of both subunits of NRF-2 in response to heightened neuronal activity (Yang et al. 2004).

These results confirm the light microscopic findings (Section 9.3.1) that NRF-2α and β respond to increased activity by translocating from the cytoplasm to the nucleus to activate target genes (such as *COX* and other respiratory enzymes), thereby increasing energy capacity to fuel neuronal activity. The acute translocating response occurs before transcription and translation of both subunits for continued activation of target genes, if chronic upregulation should be necessary.

9.4 ULTRASTRUCTURAL DISTRIBUTION OF NRF-2 α AND β SUBUNITS IN RAT AND MACAQUE VISUAL CORTEX AND THEIR RESPONSE TO DECREASED NEURONAL ACTIVITY

As indicated in Chapter 5 (Section 5.6), there are three major metabolic neuronal types in the macaque monkey visual cortex, each with distinct structural, synaptic, and metabolic features. The level of COX is highest in the medium-sized type C cells and lowest in the smallest type A cells, with the large pyramidal type B cells in between (Wong-Riley et al. 1989a). Is there a differential distribution of NRF-2α and β in these three cell types, and which cell type exhibits the greatest response to retinal impulse blockade?

The above questions were addressed in a quantitative pre-embedding EM immunocytochemical study of rat and monkey visual cortex using the silver-intensified immunogold labeling (Wong-Riley et al. 2005b). Antibodies were the same as those described in Section 9.3.2.

In agreement with the cultured studies above, specific nuclear labeling of both NRF-2α and β subunits in both rats and monkeys is primarily over euchromatin rather than heterochromatin, and cytoplasmic labeling is over free ribosomes, Golgi apparatus, and occasionally the outer membrane of the nuclear envelope, but not in the mitochondria. A greater density of both subunits is observed in the nucleus, where they both operate, than in the cytoplasm. In both layer IVC and supragranular puffs of the macaque visual cortex, both NRF-2 subunits are densest in the COX-rich type C cells, lowest in the COX-poor type A cells, and intermediate in the moderately COX-rich pyramidal type B cells. Monocular impulse blockade with TTX induces a greater reduction of both NRF-2 subunits in type C cells, which exhibit the greatest loss of COX (Wong-Riley et al. 1989a, 2005b).

These in vivo experiments confirm the activity-dependent regulation of NRF-2 in neurons. Its distribution in the highly specialized primate visual cortex is virtually identical to that of COX, even at the level of cell types. This last point has not been reported thus far for any other transcription factor. Such a tight relationship places NRF-2 in a cardinal position to sense energy demand of neurons and to respond swiftly to intensified demand by translocating existing subunits from the cytoplasm to the nucleus, upregulating its own transcription and translation, and activating transcription of target respiratory genes, such as *COX*. When the energy demand of neurons lessens, NRF-2 subunits downregulate their own expression.

9.5 IS NRF-2 A MASTER REGULATOR OF ALL 13 SUBUNITS OF *COX* IN NEURONS?

The fact that NRF-2's expression is intimately related to that of COX, and that functional NRF-2 binding sites are found in a few *COX* subunit genes, begs the question: Is it possible that NRF-2, like NRF-1, regulates all 13 subunits of *COX* in neurons? As indicated at the beginning of this chapter, NRF-2 is already known to regulate transcription factor genes (*Tfam*, *Tfb1m*, and *Tfb2m*) responsible for the transcription and replication of mtDNA (Scarpulla 1997; Gleyzer et al. 2005), thereby indirectly regulating the three mitochondrial subunits of *COX*. What remains unknown is whether NRF-2 regulates all ten nuclear-encoded *COX* subunit genes in neurons. A series of experiments was conducted to document such regulation.

9.5.1 In Silico Analysis of *COX* Subunit Promoters for NRF-2 Binding Motifs

DNA sequences 1 kb upstream and 1 kb downstream (excluding the protein coding region) of the TSS of each of the 10 nuclear-encoded *COX* subunit genes, based on the rat genome database in GenBank™, were analyzed in silico. All promoters show putative NRF-2 binding sites as well as tandem repeats of the *GGAA* motif. Moreover, at least one putative binding site is found within 150 bp of the TSS of each *COX* subunit gene (Ongwijitwat and Wong-Riley 2005).

9.5.2 In Vitro EMSA and Supershift Assays

Electrophoretic mobility shift and supershift assays verified that NRF-2 binds to *COX4i1* (Virbasius and Scarpulla 1991; Carter et al. 1992; Virbasius JV et al. 1993; Wong-Riley et al. 2000), *COX5b* (Carter et al. 1992; Virbasius JV et al. 1993), *COX6a1* (Wong-Riley et al. 2000), *COX7a2* (Seelan et al. 1996), and *COX7c* (Seelan and Grossman 1997). In addition, *COX6b*, *COX7b*, *COX7c*, and *COX8a* form shift and supershift bands with NRF-2, with *COX4i1* serving as a positive control (Ongwijitwat and Wong-Riley 2005). For *COX4* gene, binding to tandem repeats of NRF-2/GABP sites has a 10- to 20-fold greater affinity than to a single site of NRF-2 (Carter and Avadhani 1994).

9.5.3 In Vivo ChIP Assays

In vivo ChIP assays on chromatin prepared from cultured primary neurons and anti-GABPα polyclonal antibodies (C20 from Santa Cruz Biotechnology, Santa Cruz, CA) yield immunoprecipitated *COX5a*, *COX5b*, *COX6c*, and *COX7a2* subunit promoters (Ongwijitwat and Wong-Riley 2005), comparable to the *COX4i1* and *COX6a1* positive controls, whereas the negative control *NRF-1* exon 8, without any NRF-2 binding sites, is not labeled, nor is the negative antibody control anti-nerve growth factor receptor p75 (Figure 9.3). Surprisingly, *COX6b*, *7b*, *7c*, and *8a* exhibit very little signal, raising the possibility that their binding sites might have been masked from the NRF-2 antibodies.

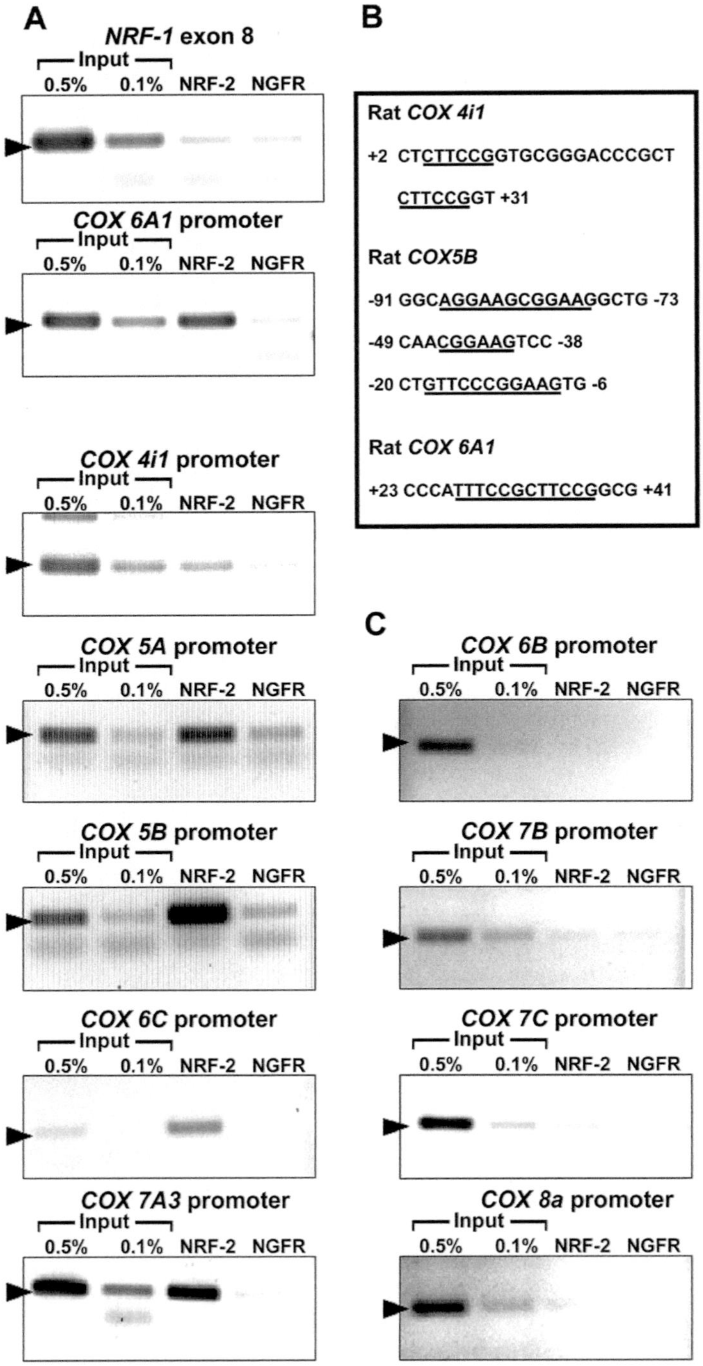

FIGURE 9.3 ChIP assays of NRF-2 with COX subunit promoters in primary neurons. (A) With *NRF-1* exon 8 as the negative control and *COX6a1* as the positive control, promoters of *COX4i1*, *5a, 5b*, *6c*, and *7a3* all show a positive band when neuronal chromatin is immunoprecipitated with NRF-2 antibodies (arrowheads), but not with antibodies against nerve growth factor receptor p75 (negative control, NGFR). Input controls are 0.5% and 0.1% chromatin. (B) DNA sequences and their positions with respect to the transcription start site (numbers flanking each sequence) and the NRF-2 binding sites underlined are shown for *COX4i1*, *5b*, and *6a1*. (C) Even though *COX6b*, *7b*, *7c*, and *8a* tested positive for NRF-2 binding, they do not show a positive signal with ChIP.

From Ongwijitwat and Wong-Riley 2005.

9.5.4 Promoter Site-Directed Mutagenesis Studies

To address the question of NRF-2's binding on the promoters of *COX6b*, *7b*, *7c*, and *8a* further, site-directed mutagenesis was used. The fact that each of these subunit genes has multiple putative NRF-2 binding sites necessitates point mutations in each site to determine the true functional ones (Figure 9.4). At least one functional site is identified for each of the four genes, and mutations of these cause significant reductions in promoter activity that range from 70–80% in *COX6b*, 60% in *COX7b*, 40–50% in *COX7c*, to 25% in *COX8a* (Ongwijitwat and Wong-Riley 2005). Site-directed mutagenesis and DNase-I footprinting of NRF-2 binding sites on bovine liver *COX7a2* also decrease its promoter activity (Seelan et al. 1996).

9.5.5 Dominant-Negative Constructs of NRF-2α and NRF-2β

Dominant negatives of NRF-2α and NRF-2β were tested for their effects on COX activity (Ongwijitwat and Wong-Riley 2005). The NRF-2α dominant-negative construct retains the Ets DNA binding domain but lacks the 56-amino-acid C-terminus that interacts with the β subunit.

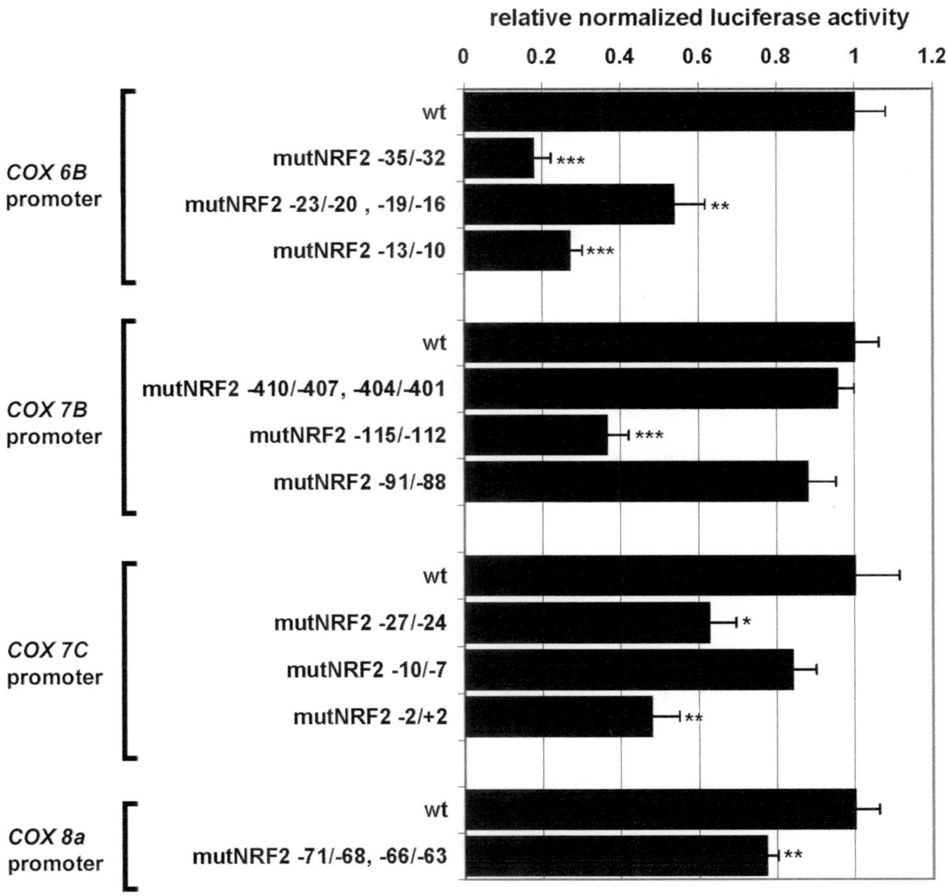

FIGURE 9.4 Site-directed mutagenesis of NRF-2 binding sites on COX promoters. Point mutations are made on putative NRF-2 binding sites on *COX6b*, *7b*, *7c*, and *8* promoters (positions with respect to TSS are indicated), and those that show significant reductions in promoter activity in comparison to wild-type (wt) controls are shown with asterisks.

*$P < 0.05$, **$P < 0.01$, ***$P < 0.001$.

From Ongwijitwat and Wong-Riley 2005.

The NRF-2β dominant-negative construct has the interaction domain with the α subunit but not the C-terminus transactivation domain. It binds to endogenous α subunit but fails to activate transcription, thereby suppressing NRF-2 activity. PC12 cells transiently transfected with a dominant-negative NRF-2α construct exhibit a 30% reduction in their COX activity, whereas those transfected with the dominant-negative NRF-2β construct suffer a milder but significant 14% decrease.

9.5.6 Effect of Silencing NRF-2 with RNA Interference on *COX* Subunit Gene Expression

Murine F9 cells were transfected with plasmid vectors expressing shRNA against two target sequences of murine *NRF-2α* mRNA (Ongwijitwat et al. 2006). Cells transfected with empty vectors served as controls. Three days after transfection, cells were harvested and assayed for protein and mRNA expressions with western blotting and RT-qPCR, respectively.

NRF-2α silencing leads to an 80–90% reduction in *NRF-2α* mRNA levels and more than 50% decrease in NRF-2α protein as compared to empty vector controls. All ten nuclear-encoded *COX* subunit gene transcripts are reduced 40–70%, but *NRF-1*, not known to be a target gene of *NRF-2*, does not change (Figure 9.5A). In addition, the expressions of *Tfam*, *Tfb1m*, *Tfb2m*, *SURF1*, *VDAC*, and *TOM20* are all significantly reduced by *NRF-2α* shRNA ($P < 0.04$–0.001) (Figure 9.5B) (Ongwijitwat et al. 2006). *Tfb2m* is not affected in these experiments, perhaps indicating a special case in F9 embryonic carcinoma cells, as *Tfb2m* is known to be regulated by NRF-2 in HeLa cells (Gleyzer et al. 2005).

Silencing NRF-2α also causes a 20% decrease in COX activity in F9 cells (Ongwijitwat et al. 2006). This value may appear modest. However, as COX is a crucial enzyme in oxidative metabolism, even a 20% reduction in activity induces a significant physiological impact. Another relevant point is that NRF-1 most likely contributes appreciably to maintaining some of the remaining COX activity.

9.5.7 NRF-2-Dependent Downregulation of *COX4i1* Promoter by TTX Blockade in Neurons

To determine if NRF-2 is partly responsible for the downregulation of *COX* transcripts by TTX, primary neurons were transfected with a vector (pGL3) containing the *COX4i1* promoter that drives the expression of the luciferase reporter gene. When exposed to 0.4 μM TTX for 6 days, transcript levels of *COX4i1* decrease by ~60%. However, point mutation of an NRF-2 binding site on the *COX4i1* promoter leads to ~50% reduction in promoter activity, and it no longer responds to TTX (Ongwijitwat et al. 2006).

9.5.8 Isolated NRF-2 Binding Sites Confer the Ability to Respond to TTX in Neurons

Is the interaction between NRF-2 and its *cis* element on the *COX* promoter alone necessary and sufficient for the cell's response to TTX? Primary neurons were transfected with a vector that contained an artificial construct of two sets of NRF-2 tandem pairs derived from the rat *COX4i1* promoter cloned upstream of an HIV basal promoter. The addition of multiple NRF-2 binding sites enhances the promoter activity by more than 100-fold compared to the unmodified promoter. TTX exposure for 6 days downregulates promoter activity to only ~30% of the untreated ones. The HIV basal promoter alone has a very low level of transcriptional activity with or without TTX.

The same experiment was performed with a *COX6a1* promoter construct that had different spacing between NRF-2 tandem repeats in each of the two pairs and a different surrounding sequence from that of *COX4i1*. Again, promoter activity increases by ~100-fold compared to basal HIV promoter alone, and TTX reduces activity by ~40% (Ongwijitwat et al. 2006).

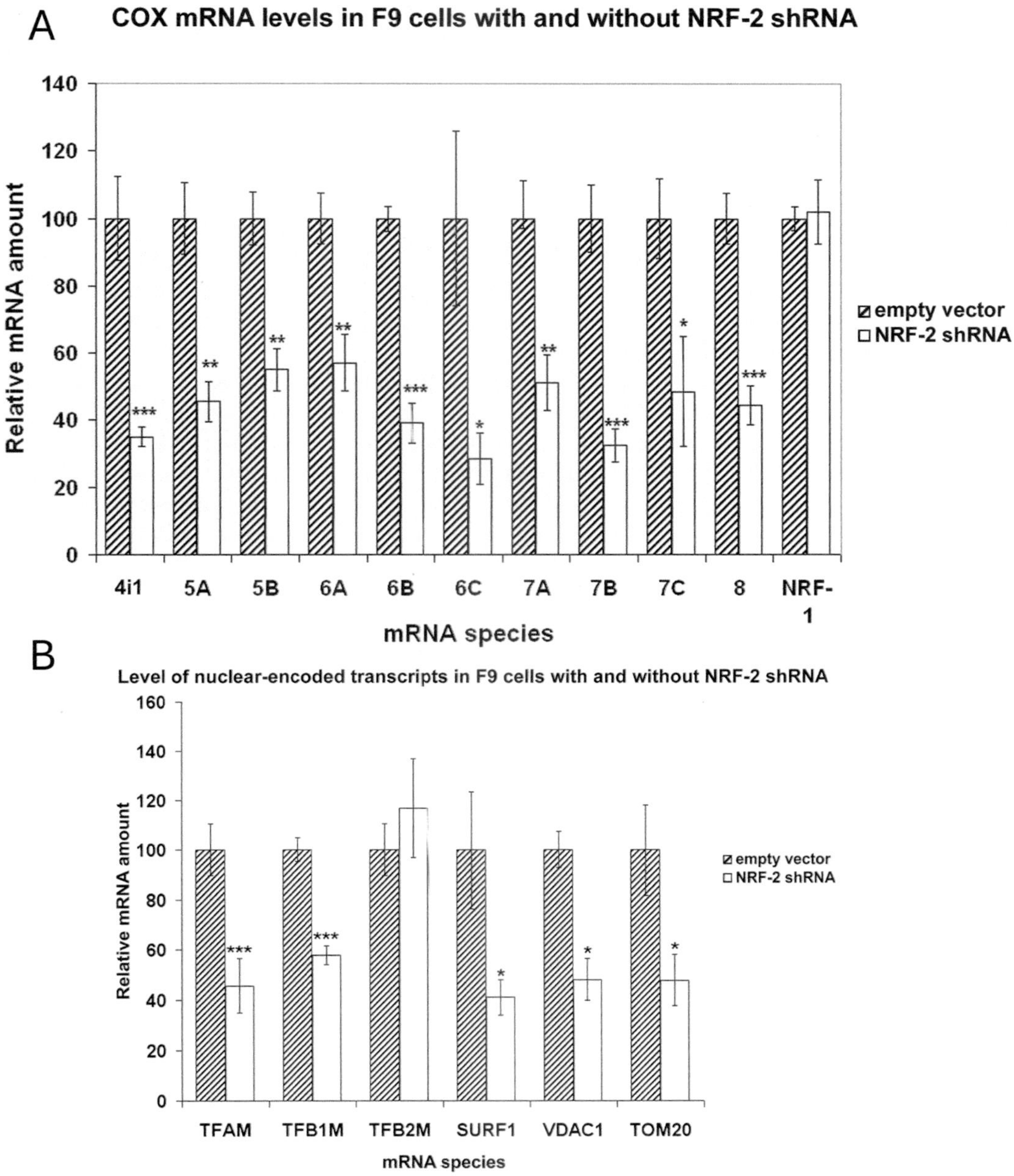

FIGURE 9.5 Effect of silencing NRF-2 on COX subunit gene expression. (A) All ten nuclear-encoded *COX* subunit mRNAs are significantly downregulated by RNA interference of *NRF-2* in F9 cells, but *NRF-1* does not show any change. (B) Transcript levels of *Tfam*, *Tfb1m*, *Tfb2m*, *SURF1*, *VDAC1*, and *TOM20* are all reduced significantly by *NRF-2* shRNA. All comparisons are made with respective controls transfected with empty vectors.

$*P < 0.05$, $**P < 0.01$, $***P < 0.001$.

From Ongwijitwat et al. 2006.

These experiments indicate that, at least for the two *COX* subunits tested, NRF-2 is necessary and sufficient for modulating *COX* subunit promoter activity in primary neurons. NRF-2 binding sites alone, and not the surrounding sequence, are sufficient for the response.

9.5.9 Conservation of NRF-2 Binding Sites Among Mouse, Rat, and Human *COX* Promoters

When the ten nuclear-encoded *COX* subunit gene promoters from mouse, rat, and human are aligned, a subset of NRF-2 binding sites (*GGAA* or *TTCC*) is found to be highly conserved among the three species, but the sequences surrounding these sites are variable, especially between rodents and humans. For every *COX* promoter, at least one NRF-2 binding site is conserved among the three species (Ongwijitwat and Wong-Riley 2005).

All the experiments discussed in this chapter point to the critical role of NRF-2 as a master regulator and coordinator of all 13 *COX* subunit genes from the two genomes. However, NRF-2 does not act alone, as NRF-1 is also found to regulate all mitochondria- and nuclear-encoded *COX* subunit genes (see Chapter 8). What, then, is the relationship between the two transcription factors? Silencing NRF-1 does not affect the transcript levels of *NRF-2* (Dhar et al. 2008), and silencing NRF-2 likewise does not alter the expression of *NRF-1* (Ongwijitwat et al. 2006). The two transcription factors, therefore, function independently but in parallel. For neurons, in which a ready supply and a speedy replenishment of energy is paramount to the maintenance of their activity, it is a distinct advantage to have redundant mechanisms to transcriptionally activate an essential energy-generating enzyme involving the coordination of the two genomes. As both NRF-1 and NRF-2 themselves respond to changes in neuronal activity, they both act as sensors to the upstream energy signals of neurons and swiftly adjust the transcriptional activity of their target genes, notably *COX*.

10 Specificity Proteins 1 (Sp1) and 4 (Sp4)

Transcriptional Regulators of All 13 COX Subunit and Na^+/K^+-ATPase Genes in Neurons

ABSTRACT

Do NRF-1 and NRF-2 suffice, or is COX regulated by yet another transcription factor? Binding motifs for specificity protein 1 (Sp1) have been noted but not characterized on several COX subunit genes. Does it also coordinate the regulation of all 13 COX subunits from the two genomes in neurons? By means of multiple approaches, Sp1 was indeed found to do so. However, the SP family has four members, and the less studied Sp4 is more brain-specific and more abundant in neurons than Sp1. We found that Sp4 directly regulates all ten nuclear COX genes and indirectly the three mitochondrial COX genes by controlling the expression of Tfam, Tfb1m, and Tfb2m in neurons. Sp4 also cooperates with NRF-1 in activating the expression of the Atp1b1 gene of Na^+/K^+-ATPase, but balances with NRF-1 in positively regulating Atp1a1. In addition, it stimulates Atp1a3, known to function in highly active neurons. Thus, Sp4 is another master regulator of all COX subunit genes and a coordinator of energy generation with energy consumption in neurons.

In the last two chapters, NRF-1 and NRF-2 have proven to be master regulators of all 13 *COX* subunit genes in neurons. Is it possible that COX is regulated by yet another transcription factor?

The binding motif of specificity protein 1 (Sp1) of the Sp/Krüppel-like factor (KLF) family of zinc finger transcription factors has been noted on the promoters of some *COX* subunit genes, including *COX4*, *5b*, *6a1*, *7a2*, and *7C* (Yamada et al. 1990; Basu et al. 1993; Grossman and Lomax 1997; Ongwijitwat and Wong-Riley 2004, 2005), as well as the *Tfam* gene (Larsson et al. 1997). Can this family also regulate all 13 *COX* genes in neurons?

Sp1 is the first C_2H_2-type zinc finger transcription factor isolated and cloned from mammalian cells after its presence was first discovered in the simian virus 40 (SV40) early promoter (Gidoni et al. 1984; Suske 1999). It is ubiquitously expressed in mammalian tissues and reportedly activates many genes, including housekeeping, tissue-specific, and cell cycle-regulated genes. It recognizes GC-rich sequences in CpG islands of housekeeping genes and other TATA-less genes. It has a high-affinity binding motif of "*GGGCGG*" or "*CCCGCC*" (the GC box), which often occurs in repeats (Gidoni et al. 1984; Lania et al. 1997; Wierstra 2008). However, its affinity for the GT box (*GGGGTGGGG*) or CT box (*CCCTCC*) (Letovsky and Dynan 1989; Tuthill et al. 2003) is much weaker. Murine embryos with Sp1 knockout are severely retarded in development and usually die during the 2nd week of gestation (around day 11) (Suske 1999).

Sp1 belongs to a family of four members: Sp1, Sp2, Sp3, and Sp4, all of which are found in humans, all contain three zinc finger DNA-binding domains close to their C-terminus, and all except Sp2 have two glutamine-rich transactivation domains in their N-terminus (Sp2 has only one such domain) (Lania et al. 1997; Suske 1999). Sp1 and Sp3 are ubiquitously present, Sp2 is found in various cell lines, and Sp4 is restricted to the brain and certain epithelia (Hagen et al. 1995; Suske 1999). Sp1, Sp3, and Sp4 can bind to the classical GC box with identical affinity, but Sp2 binds to a GT-rich

DOI: 10.1201/9781003507765-10

element and is less closely related to the other three. Whereas Sp members are activators, Sp3 can also be a repressor of Sp1- and Sp4-mediated transcription. Unlike Sp1, Sp4 is not able to act synergistically through adjacent binding sites (Hagen et al. 1995; Suske 1999). Knocking out Sp4 causes two-thirds of affected mice to die within a few days after birth, and the surviving ones are significantly smaller than their wild-type littermates (Supp et al. 1996).

10.1 DOES SP1 REGULATE ALL 13 *COX* SUBUNIT GENES?

Multiple approaches were used to determine if Sp1 regulates all 13 *COX* subunit genes from the two genomes (Dhar et al. 2013). In silico promoter analysis indicates that promoters of *COX4i1*, *5a*, *5b*, *6a1*, *6c*, *7a*, *7c*, and *8a* contain a typical Sp1 sequence motif "*GGGCGG*" or "*CCCGCC*", but *COX6b* and *7b* have an atypical "*GGGCGT*" or "*GGGCGA*" sequence. In vitro EMSA shows DNA–protein shift bands and Sp1-specific supershift bands (using polyclonal antibodies against Sp1, Santa Cruz Biotechnology) for all ten nuclear *COX* genes, and GM3 synthase promoter with a known Sp1-binding site is a positive control. Mutating the Sp1 binding site results in neither shift nor supershift complexes (Dhar et al. 2013).

In vivo ChIP assays of N2a cells immunoprecipitated with anti-Sp1 antibodies reveal that the proximal promoters of all ten nuclear *COX* genes, as well as those of *Tfam*, *Tfb1m*, and *Tfb2m*, coimmunoprecipitate with Sp1 antibodies, but not with NGFR antibodies. The positive control GM3 synthase shows a clear band, but the negative control β-actin has none (Dhar et al. 2013).

Knocking down *Sp1* with murine *Sp1* shRNA (Santa Cruz Biotechnology) in N2a cells results in an 83% reduction in *Sp1* mRNA and an 82% decrease in Sp1 protein ($P < 0.001$ for all). All 13 *COX* subunit mRNAs, as well as transcripts of *Tfam*, *Tfb1m*, and *Tfb2m*, are significantly downregulated compared to controls with scrambled shRNA vectors ($P < 0.001$ for all). COX I and COX IV-1 proteins are likewise reduced by 55% and 40%, respectively ($P < 0.001$ for both) (Dhar et al. 2013).

Overexpressing Sp1 using human *Sp1* cDNA (Open Bio-systems, Lafayette, CO), cloned into pcDNA Dest40 vector and transfected into N2a cells, leads to an 80-fold increase in *Sp1* mRNA and a 5-fold increase in Sp1 protein ($P < 0.001$ for both). All 13 *COX* subunit mRNAs and those of the three mitochondrial transcription factors are significantly upregulated as compared to empty vector controls ($P < 0.001$ for all). COX I and COX IV-1 proteins are increased by 55% ($P < 0.001$) and 40% ($P < 0.01$), respectively (Dhar et al. 2013).

Silencing *Sp1* with shRNA prevents the KCl-induced upregulation of *Sp1*, all 13 *COX* subunit genes, and the three mitochondrial transcription factor genes in primary neurons ($P < 0.001$ for all). On the other hand, overexpressing Sp1 rescues the TTX-induced downregulation of Sp1 itself and its 16 target genes in primary neurons ($P < 0.001$ for all) (Dhar et al. 2013).

Thus, Sp1 functionally regulates all ten *COX* subunit genes and the three mitochondrial-encoded ones indirectly by regulating *Tfam*, *Tfb1m*, and *Tfb2m* (Dhar et al. 2013). However, the fact that Sp1 is ubiquitously expressed but Sp4 is more brain-specific raises the question: How does Sp4 regulate *COX* subunit genes in neurons?

10.2 DOES SP4 REGULATE ALL 13 *COX* SUBUNIT GENES IN NEURONS?

10.2.1 Prevalence of Sp4 versus Sp1 in Brain, Primary Neurons, and N2a Cells

Western blotting of proteins from murine brain, cultured primary neurons of murine visual cortex, and N2a cells reveals Sp4 to be more abundant in brain and primary neurons than in N2a cells, whereas the opposite is true for Sp1 ($P < 0.001$ for all) (Figure 10.1A and B). When primary neurons are exposed to 20 mM KCl stimulation for 5 h or 0.4 μM TTX impulse blockade for 3 days, their Sp4 content is significantly up- and downregulated, respectively ($P < 0.001$ for both), indicating that Sp4 itself is regulated by neuronal activity (Johar et al. 2013) (Figure 10.1C and D).

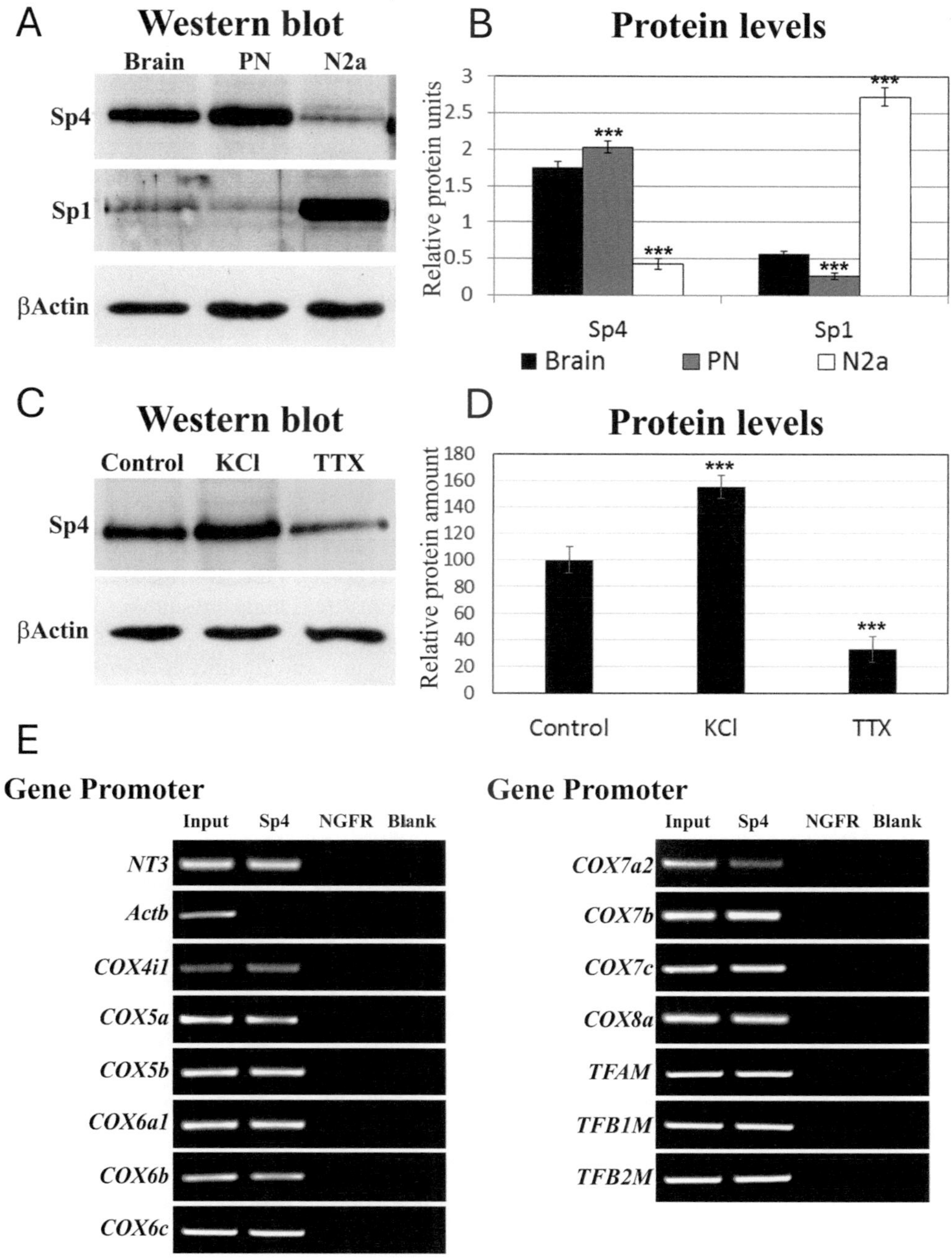

FIGURE 10.1 Sp4 in neurons and ChIP assays of its interaction with COX genes. (A and B) Sp4 protein is present in greater amounts in the brain and primary neurons (PN) than in N2a cells, whereas the opposite is true for Sp1. β-actin is the loading control. (C and D) Sp4 protein is upregulated by KCl stimulation and downregulated by TTX blockade. All comparisons are made with respective controls. $^{***}P < 0.001$. (E) ChIP assays indicate that Sp4 binds to promoters of all ten nuclear-encoded *COX*, as well as *Tfam*, *Tfb1m*, *Tfb2m*, and the positive control *NT3*, but not the negative β-actin (*Actb*) control in murine visual cortical chromatin. Neither antibodies against nerve growth factor receptor p75 (NGFR) nor the lack of antibodies (Blank) results in any band. Input lanes represent 0.5% chromatin without immunoprecipitation.

Modified from Johar et al. 2013.

10.2.2 In Vivo ChIP Assays of Sp4 Binding to *COX* Subunit Genes in Brain Tissues

When ChIP assays were performed with chromatin derived from murine visual cortex and anti-Sp4 antibodies (V20, sc645, Santa Cruz Biotechnology), positive bands were found for all ten nuclear *COX* subunit genes, *Tfam*, *Tfb1m*, *Tfb2m*, and the murine *NT3* promoter with a known Sp4 binding site, the positive control. The negative control *β-actin* exon 5 shows no band, nor does the negative control for immunoprecipitation using anti-NGFR p75 antibodies (Figure 10.1E) (Johar et al. 2013).

10.2.3 Effect of Sp4 Knockdown on *COX* Subunit Genes

When N2a cells are transfected with two Sp4 siRNA sequences cloned into the pLK0.1 TRC cloning vector (Plasmid 10878; Addgene, Cambridge, MA), their *Sp4* mRNA level is reduced by 70% and Sp4 protein by 67%. Their COX I and COX IV-1 proteins are decreased by 53% and 40%, respectively ($P < 0.001$ for all) (Figure 10.2A and B) (Johar et al. 2013). Transcripts of all 13 *COX* subunit genes, *Tfam*, *Tfb1m*, and *Tfb2m* are significantly downregulated ($P < 0.001$ for all). *Surf1* transcripts, with a known Sp binding site on the promoter of its gene (Cole and Gaston 1997), are also reduced ($P < 0.001$) (Johar et al. 2013).

Comparable experiments were performed on cultured primary neurons derived from murine visual cortex. Knocking down *Sp4* causes a 71% decrease in *Sp4* mRNA and a 65–75% reduction in all 13 *COX* subunit genes (except for a 47% decrease in *COX4i1*), *Tfam*, *Tfb1m*, *Tfb2m*, and *Surf1* ($P < 0.001$ for all) (Figure 10.2C). Surprisingly, *Sp1* transcripts are slightly but significantly increased ($P < 0.05$). Likewise, when primary neurons are transfected with *Sp1* shRNA, transcripts of all 13 *COX* subunit genes, *Tfam*, *Tfb1m*, *Tfb2m*, and *Surf1* do not undergo significant changes. However, *Sp4* message is slightly but significantly increased ($P < 0.05$) (Johar et al. 2013). These experiments demonstrate the dominance of Sp4 over that of Sp1 in primary neurons. However, when the action of either is suppressed, the other tends to compensate, at least partially, in neurons.

10.2.4 Effect of Sp4 Overexpression on *COX* Subunit Genes

N2a cells that were transfected with the pcDNA-Dest40 vector expressing human *Sp4* exhibit a 77-fold increase in their *Sp4* message and a tenfold increase in their Sp4 protein, as well as a 68% rise in COX I and a 78% increase in COX IV-1 proteins ($P < 0.001$ for all except $P < 0.01$ for *COX4i1*) (Figure 10.3A and B). Transcripts of all 13 *COX* subunit genes, *Tfam*, *Tfb1m*, *Tfb2m*, and *Surf1* are also significantly upregulated ($P < 0.001$ for all) (Johar et al. 2013).

When cultured primary neurons are transfected with *Sp4* expression vectors, a 28-fold increase in *Sp4* message results. Likewise, transcripts of all 13 *COX* subunit genes, *Tfam*, *Tfb1m*, *Tfb2m*, and *Surf1* are significantly elevated compared to empty vector controls ($P < 0.001$ for all) (Figure 10.3C) (Johar et al. 2013).

10.2.5 Regulation of *COX* Expression by Sp4 is Activity Dependent

Depolarizing stimulation of N2a cells with 5 h of 20 mM KCl yields a significant increase in *Sp4* transcripts and those of the 16 target genes described above compared to scrambled vector controls ($P < 0.001$ for all). Transfection with *Sp4* shRNA prevents such upregulation, whereas transfection with scrambled vector controls plus KCl does not (Figure 10.4A). The converse experiments of 0.4 μM TTX blockade for 3 days significantly reduce mRNAs of *Sp4* and its target genes ($P < 0.001$ for all except $P < 0.05$ for *COX6c*, *Tfb2m*, and *Surf1*), but transfection with *Sp4* expression vectors rescues all 16 target genes from being downregulated by TTX ($P < 0.001$ for all when compared with empty vector controls plus TTX) (Figure 10.4B) (Johar et al. 2013). The relative activity of COX in primary neurons and in N2a cells is significantly increased by KCl and decreased by TTX. Transfection with *Sp4* shRNA prevents the upregulation by KCl, whereas overexpressing Sp4 rescues the downregulation by TTX ($P < 0.001$ for all) (Figure 10.4C and D) (Johar et al. 2013).

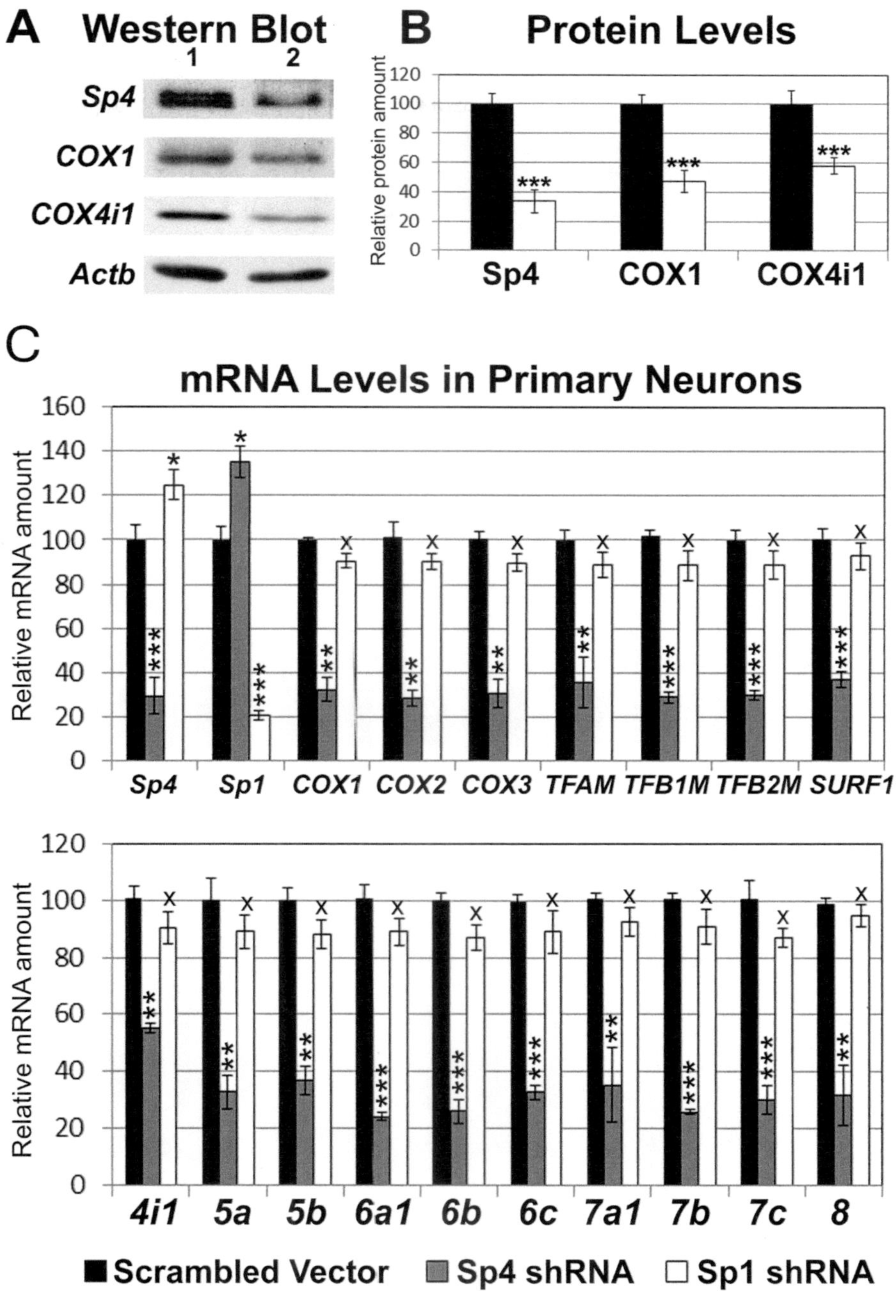

FIGURE 10.2 Effect of silencing Sp4 in neurons. (A and B) Western blots and quantitative analysis of Sp4, COX I, and COX4i1 proteins in control N2a cells transfected with scrambled vectors (lane 1) and in those after *Sp4* shRNA transfection (lane 2). $N = 6$ for each group. $^{***}P < 0.001$ compared to controls. (C) Transcript levels of *Sp4*, *Sp1*, all 13 *COX* subunits, *Tfam*, *Tfb1m*, *Tfb2m*, and *SURF1* in primary neurons transfected with scrambled vector controls, *Sp4* shRNA, or *Sp1* shRNA. Silencing *Sp4* induces a slight but significant increase in *Sp1* transcript, and vice versa. All target genes are downregulated by *Sp4* shRNA but not by *Sp1* shRNA. $N = 3$ for each data point. All comparisons are with scrambled vector controls. $^{**}P < 0.01$, $^{***}P < 0.001$, X = non-significant.

Modified from Johar et al. 2013.

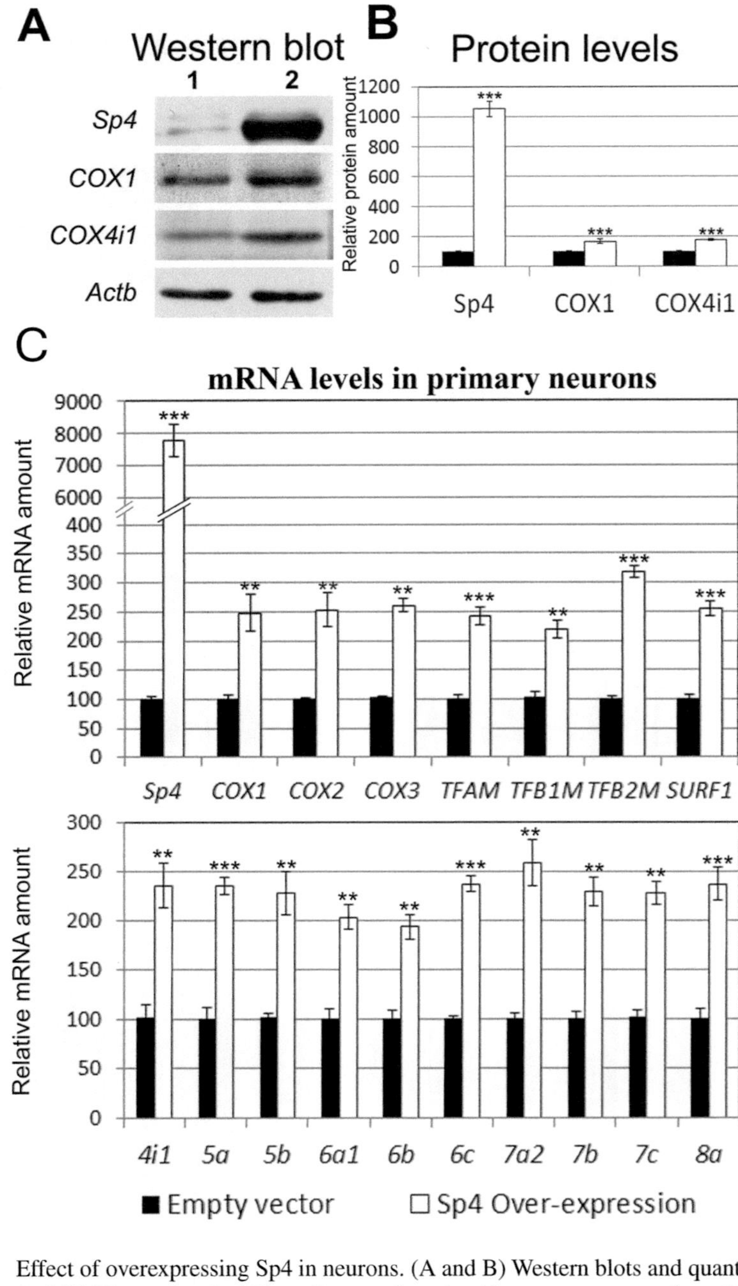

FIGURE 10.3 Effect of overexpressing Sp4 in neurons. (A and B) Western blots and quantitative analysis of Sp4, *COX I*, and *COX4iI* proteins in N2a cells transfected with empty vectors (lane 1) or with vectors overexpressing Sp4 (lane 2). $N = 6$ for each group. $^{***}P < 0.001$ compared to controls. (C) Transcript levels of *Sp4*, all 13 *COX* subunits, *Tfam*, *Tfb1m*, *Tfb2m*, and *SURF1* in primary neurons are all upregulated by Sp4 overexpression as compared to empty vector controls. $N = 3$ for each group. $^{**}P < 0.01$, $^{***}P < 0.001$.

Modified from Johar et al. 2013.

10.2.6 Conservation of Sp4 Binding Sites

Functional binding sites for Sp4 on 10 nuclear *COX* subunit genes, *Tfam*, *Tfb1m*, and *Tfb2m* are conserved among mice, rats, and humans with a 60–100% homology (Johar et al. 2013). As discussed above, Sp4 and Sp1 share the same *cis*-binding motif.

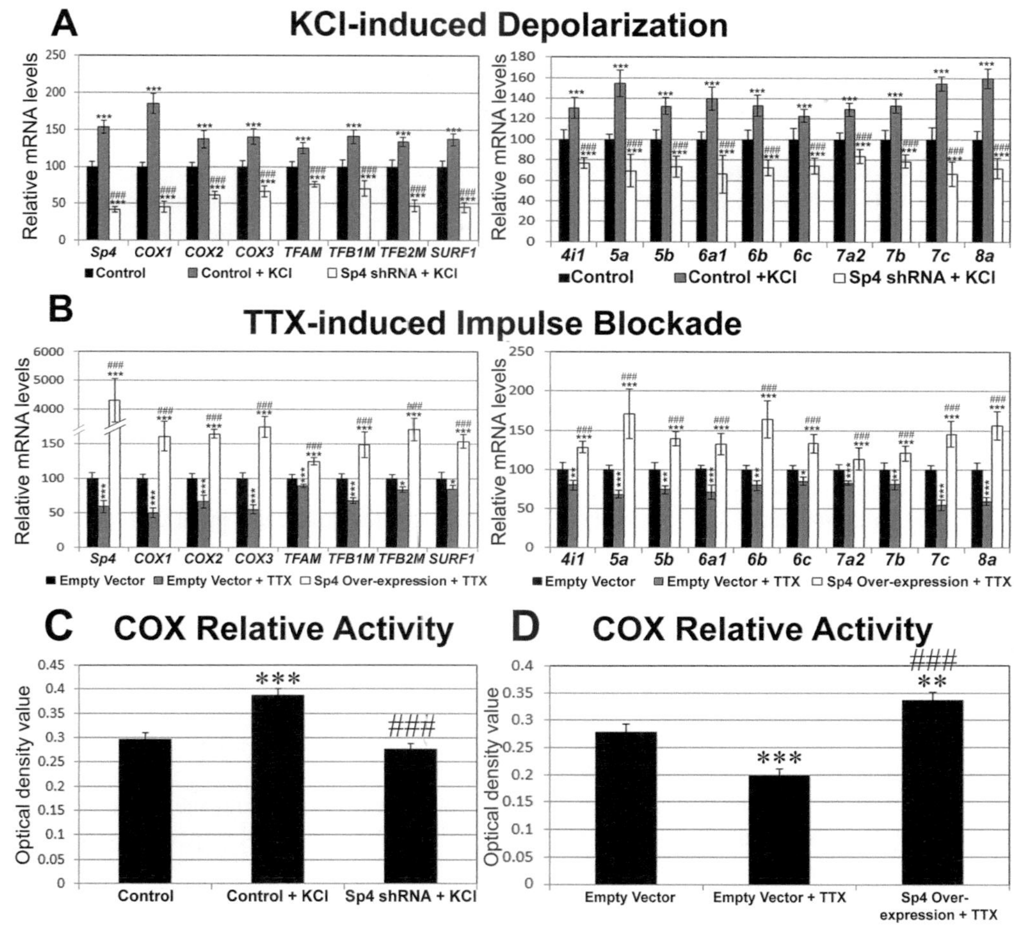

FIGURE 10.4 Activity-dependent regulation of Sp4 and the effect of altering Sp4 levels on COX in neurons. (A) Transcript levels of *Sp4* and 17 of its target genes are all upregulated by KCl stimulation in N2a cells, but *Sp4* shRNA prevents such an increase. (B) The same transcripts as in A are all downregulated by TTX, but overexpressing Sp4 rescues them from such a decrease. (C and D) Relative activity of COX measured with optical densitometry in N2a cells. KCl stimulation increases the activity, but *Sp4* shRNA prevents such an increase (C). On the other hand, TTX decreases the activity, but overexpression of Sp4 averts such a decrease (D). $N = 6$ for each data point in A and B. $N = 122$, 125, and 128 cells for each of the three groups in C, and 115, 162, and 207 cells for each of the three groups in D. All comparisons marked by asterisks are made with either scrambled vector (A and C) or empty vector (B and D) controls. Comparisons marked by # are made with either control + KCl or control + TTX.

$^{*}P < 0.05$, $^{**}P < 0.01$, $^{***}P < 0.001$, $^{\#\#\#}P < 0.001$.

Modified from Johar et al. 2013.

Clearly, in addition to NRF-1 and NRF-2, Sp1 and especially the dominant Sp4 in neurons directly regulate all ten nuclear subunits of *COX* and indirectly control the expression of the three mitochondrial *COX* by activating the three mitochondrial transcription factors *Tfam*, *Tfb1m*, and *Tfb2m*. Such bigenomic regulation is precise and is activity-dependent.

10.3 SP4 REGULATES THE EXPRESSION OF NA$^+$/K$^+$-ATPASE SUBUNIT GENES

What is the role of Sp4 in the transcriptional regulation of the major energy-consuming enzyme, Na^+/K^+-ATPase, in neurons?

In silico analysis of the proximal promoters of murine Na^+/K^+-ATPase subunit gene isoforms reveals several GC box motifs (*GGGCGG* or *CCCGCC*) on *Atp1a1* and *Atp1b1* promoters, whereas the *Atp1a3* promoter has a CT box motif (*CCTCCCC*). *Atp1b1* also has a slightly modified GC-box motif (*GCCGCCG* and *AGGCGGA*) close to its TSS (Johar et al. 2014).

In vitro EMSA studies indicate that Sp4 is more abundant in visual cortical extract than in HeLa extract, but the reverse is true for Sp1 ($P < 0.001$ for all). Sp3 has two isoforms in the HeLa extract. The longer 110 kb isoform exists in almost equal amounts in HeLa and visual cortical extracts, whereas the shorter isoform is virtually absent in the cortical extract. When incubated with the nuclear extract from murine visual cortex, shift bands with Sp1, Sp3, and Sp4 migrate together. A much more prominent supershift band appears, however, with Sp4 antibodies than with Sp1 or Sp3 antibodies. *Atp1a1*, *Atp1a3*, and *Atp1b1* each forms a shift band and a supershift complex quite prominently with Sp4 antibodies, more moderately with Sp3, and much weaker with Sp1. Of the two Sp sites tested for *Atp1b1*, one of them (site 2) shows a stronger signal (Johar et al. 2014).

In vivo ChIP assays with chromatin from murine visual cortex demonstrate that the promoters of *Atp1a1*, *Atp1a3*, *Atp1b1*, and the two positive controls, GM3 synthase and the neuron-specific neurotrophin 3 (*Ntf3*) with a known Sp4 binding site (Ishimaru et al. 2007) are all immunoprecipitated by Sp1, Sp3, and Sp4 antibodies. Furthermore, they all show the strongest signal with the Sp4 antibodies (Johar et al. 2014).

Mutating the Sp binding sites on the promoters of *Atp1a1*, *Atp1a3*, and *Atp1b1* causes significant decreases in their promoter activities compared to wild-type controls ($P < 0.001$ for *Atp1a1* and site 2 of *Atp1b1*; $P < 0.05$ for *Atp1a3* and site 1 of *Atp1b1*). Overexpressing Sp1, Sp3, or Sp4 leads to significant increases in the promoter activities of *Atp1a1*, *Atp1a3*, and *Atp1b1* ($P < 0.001$ for all). However, such upregulation is thwarted when the Sp binding sites on these promoters are mutated ($P < 0.05$ for all). Again, site 2 mutation in *Atp1b1* has a greater effect than altering site 1 (Johar et al. 2014).

Silencing *Sp1*, *Sp3*, or *Sp4* in either N2a cells or primary neurons from murine visual cortex with specific shRNA leads to a significant downregulation of *Sp1*, *Sp3*, and *Sp4*, respectively, their target genes *Atp1a1*, *Atp1a3*, *Atp1b1*, and the two positive controls *COX2* and *COX7c* ($P < 0.05$–0.001 compared to scrambled vector controls). Silencing Sp4 has a greater effect on target transcripts in primary neurons than in N2a cells. *Ntf3* message, however, shows an increase with each of the three Sp knockdowns, confirming that *Ntf3* is negatively regulated by Sp factors (Ishimaru et al. 2007). Silencing *Sp1*, *Sp3*, or *Sp4* in N2a cells also causes a significant downregulation of gene products of these three transcription factors and those of Atp1a1, Atp1a3, and Atp1b1 ($P < 0.001$ for all) (Johar et al. 2014).

The complementary experiments of overexpressing *Sp1*, *Sp3*, or *Sp4* result in protein and transcript expression patterns exactly opposite to those of silencing, that is, they are all significantly upregulated in both N2a cells and primary neurons ($P < 0.05$–0.001). Again, Sp4 has a greater effect on *Atp* and *COX* target genes in primary neurons than in N2a cells. *Ntf3*, on the other hand, is downregulated by an overexpression of each of the three *Sp* factors ($P < 0.05$–0.001) (Johar et al. 2014).

Both Sp1 and Sp4 respond to changes in neuronal activity by upregulating their transcripts with KCl stimulation and downregulating them with TTX inhibition ($P < 0.05$ for Sp1 and 0.001 for Sp4). However, Sp3 is unaffected in either case (Figure 10.5A). Mutating Sp binding sites prevents the target *Atp* promoters (*Atp1a1*, *Atp1a3*, and *Atp1b1* site 2) from being upregulated by KCl (Figure 10.5B). Transfecting N2a cells with *Sp1* or *Sp4* shRNA also prevents the increase in *Atp* and *COX* transcripts by KCl, but silencing *Sp3* still allows some upregulation of target transcripts, indicating that Sp3 is not regulated by neuronal activity (Figure 10.5C). The converse experiment of overexpressing Sp factors prevents the downregulation of target transcripts by TTX ($P < 0.05$–0.001) (Figure 10.5D) (Johar et al. 2014).

Functional Sp binding sites on the promoters of *Atp1a1*, *Atp1a3*, and *Atp1b1* are highly conserved among humans, mice, and rats (Johar et al. 2014).

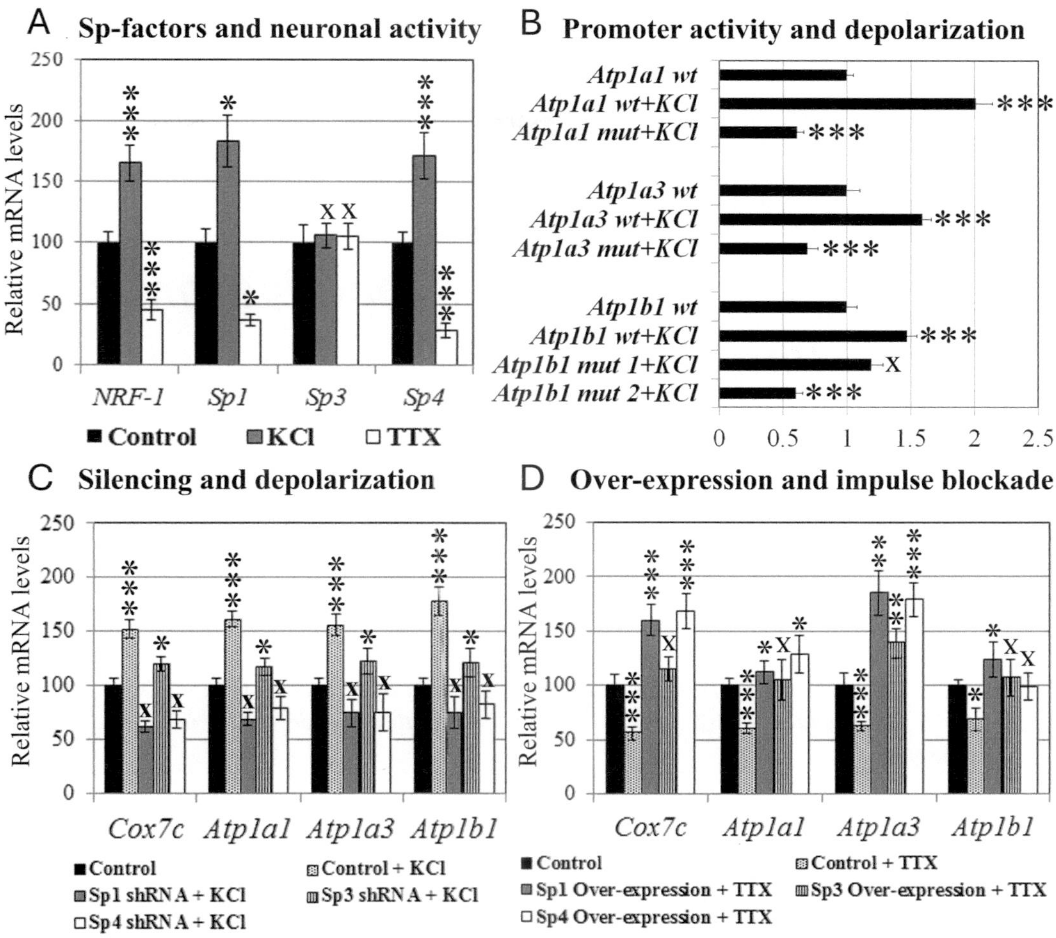

FIGURE 10.5 Activity-dependent regulation of Na^+/K^+-ATPase subunit genes by Sp factors. (A) With NRF-1 serving as a positive control, transcript levels of *Sp1* and *Sp4* are upregulated by KCl and downregulated by TTX, but *Sp3* is not affected. (B) KCl depolarization significantly increases activity in wild-type (wt) promoters of *Atp1a1*, *Atp1a3*, and *Atp1b1*, but mutating the Sp site prevents such an increase in *Atp1a1*, *Atp1a3*, and site 2 (mut 2) but not site 1 (mut 1) of *Atp1b1*. (C) Silencing Sp1 or Sp4, but not Sp3, with shRNA prevents the upregulation of transcripts of *COX7C*, *Atp1a1*, *Atp1a3*, and *Atp1b1* by KCl. (D) Overexpressing Sp1, Sp3, or Sp4 rescues *COX7C*, *Atp1a1*, *Atp1a3*, and *Atp1b1* transcripts from being downregulated by TTX. $N = 5$ for each data point. All comparisons are made with respective controls.

***$P < 0.05$, **$P < 0.01$, ***$P < 0.001$, X = non-significant.**

From Johar et al. 2014.

Hence, in comparison to NRF-1, Sp4 (and to a lesser extent Sp1) is a much stronger regulator of neuronal Na^+/K^+-ATPase in that it positively regulates all three functional subunits (*Atp1a1*, *Atp1a3*, and *Atp1b1*). It synergizes with NRF-1 in stimulating the transcription of *Atp1b1* but opposes NRF-1 in activating *Atp1a1* to achieve a more balanced regulation. It also directly activates the expression of *Atp1a3*, whose gene product is known to be present exclusively in neurons, especially tonically active neurons, and is capable of repolarizing membrane potential faster after depolarization than the α1 subunit (Edwards et al. 2013). At the transcriptional level, then, Sp4 effectively coordinates the coupling between energy consumption and energy generation in neurons.

11 Peroxisome Proliferator-Activated Receptor-γ Coactivator 1α (PGC-1α)

A Transcriptional Coactivator in Regulating COX in Neurons

ABSTRACT

A strong coactivator of NRF-1 and NRF-2 in non-neuronal tissues is PGC-1α (peroxisome proliferator-activated receptor gamma coactivator-1α). It was originally isolated and cloned from a brown fat cDNA library by Spiegelman's group. Among its many functions, including thermogenesis, response to environmental stressors, and heme biogenesis, PGC-1α also induces the expression of NRF-1 and NRF-2 and is regarded as a master regulator of mitochondrial biogenesis. We found that PGC-1α responds to changes in neuronal activity both in culture and in vivo earlier than NRF-1 or NRF-2, consistent with its role in activating these transcription factors. The signaling mechanism for depolarization-induced activation of PGC-1α is mediated by p38 MAP kinase and calcium channels in neurons.

In 1999, Wu et al. from Bruce Spiegelman's laboratory published a paper indicating that a cold-inducible coactivator of nuclear receptors, PGC-1, "stimulates a powerful induction of NRF-1 and NRF-2 gene expression". PGC-1, or peroxisome proliferator-activated receptor gamma (PPARγ) coactivator-1, was isolated and cloned from a brown fat cDNA library the year before by the same laboratory (Puigserver et al. 1998). When induced by cold, PGC-1 greatly enhances the transcriptional activity of PPARγ and thyroid hormone receptors on promoters of the uncoupling protein UCP-1 in brown fat. Ectopic expression of PGC-1 in white fat precursor cells, normally containing low levels of PGC-1 and UCP-1, not only upregulates both but also transcripts of *COX2*, *COX4*, and *β* subunit of ATP synthase, as well as cellular content of mtDNA (Puigserver et al. 1998). Ectopic expression of PGC-1 in myoblasts or myotubes increases the mRNAs of *NRF-1*, *NRF-2α*, and their target gene *Tfam* (Wu et al. 1999). Ectopic expression of PGC-1 in myotubes enhances oxygen consumption 1.9-fold in the presence of a chemical uncoupler, FCCP, and 3.2-fold with the addition of an ATP synthase inhibitor, oligomycin, indicating that PGC-1 can augment mitochondrial uncoupling and the total respiratory capacity of mitochondria. Co-transfecting 3T3 fibroblasts with PGC-1 and NRF-1 plus a luciferase reporter construct with four NRF-1 binding sites in its minimal promoter results in a sizable increase in transcriptional activity, suggesting that PGC-1 can activate NRF-1. Transfecting C2C12 myoblast cells with a PGC-1 expression vector raises the density of their mitochondria, but co-transfecting these cells with a dominant-negative allele of NRF-1 blocks such an effect, indicating that PGC-1 mediates mitochondrial biogenesis via its activation of NRF-1. Binding sites on PGC-1 for NRF-1 and vice versa have been identified, strongly supporting their interactions (Wu et al. 1999). These experiments led Wu et al. (1999) to propose that environmental signals (e.g., cold, high-caloric diet) activate PGC-1, which mediates both adaptive thermogenesis and mitochondrial biogenesis by activating UCP, NRF-1, and NRF-2 in muscle cells. An orphan nuclear receptor, estrogen-related receptor α (ERRα or ERR1), is also an effector of PGC-1 in

DOI: 10.1201/9781003507765-11

regulating genes involved in mitochondrial energy metabolism (Schreiber et al. 2004). PGC-1 and ERRα also jointly induce SIRT3, a mitochondrial sirtuin implicated in ROS detoxification (reviewed in Scarpulla et al. 2012). Besides its general role in regulating metabolic homeostasis and mitochondrial biogenesis, PGC-1 also initiates tissue-specific functions, such as fiber type switching in skeletal muscles in response to exercise, and gluconeogenesis, heme biogenesis, and fatty acid oxidation in the liver during fasting (Yoon et al. 2001; Baar et al. 2002). In response to energy deprivation, PGC-1 is activated by deacetylation via SIRT1 and by phosphorylation via AMPK (AMP-activated protein kinase) to enhance mitochondrial biogenesis and ATP production (reviewed in Scarpulla et al. 2012). PGC-1, then, coordinates energy generation with the metabolic demands of cells. The level of PGC-1 protein is affected by Ca^{2+} signaling mechanisms, AMPKα activity, thyroid hormone, and post-translational phosphorylation (Irrcher et al. 2003).

After the discovery of a homologue of PGC-1, which Spiegelman's laboratory named PGC-1β, they renamed the original PGC-1, PGC-1α (Lin et al. 2002). Even though both coactivators share similar tissue distribution and both have the highest expression in brown adipose tissue and heart, their mRNA levels are regulated differently in response to cold and during adipocyte differentiation. Both coactivators share a conserved amino acid motif for the docking of host cell factor, which is implicated in cell cycle regulation and viral infection. The level of PGC-1β is extremely low in the brain (Lin et al. 2002).

Even though PGC-1α is found in the brain (Puigserver et al. 1998), very little was known about its function in neurons prior to 2006. Below are a few experiments done on neurons.

11.1 ACTIVITY-DEPENDENT REGULATION OF PGC-1α IN NEURONS AND ITS RELATIONSHIP WITH NRF-1 AND NRF-2

Does PGC-1α respond to functional inactivation in neurons in vitro and in vivo? If so, is the time course of response different from that of NRF-1 and NRF-2?

11.1.1 TTX Blockade in Cultured Primary Neurons

Cultured primary neurons were exposed to 0.4 μM TTX for 2, 4, or 6 days, and untreated sister cultures served as controls (Liang and Wong-Riley 2006). PGC-1α is the first to downregulate its mRNA by 41% after 2 days of inactivation ($P < 0.05$), when no change is detected in *NRF-1* and *NRF-2* (both α and β subunits). After 4 days of TTX, PGC-1α's mRNA is reduced by 62.6% ($P < 0.0$) and that of *NRF-2β* by 31.8% ($P < 0.05$), but neither *NRF-2α* nor *NRF-1* exhibits any change. By 6 days post-TTX, transcripts of all four factors are significantly reduced (PGC-1α by 82.7%, *NRF-1* by 41.8%, *NRF-2α* by 54%, and *NRF-2β* by 60.6% ($P < 0.001$ for PGC-1α's mRNA and $P < 0.01$ for the other three) (Figure 11.1A). Clearly, PGC-1α responds to impulse blockade earlier than either NRF-1 or NRF-2α, and NRF-2β reacts sooner than NRF-2α.

Western blots reveal that PGC-1α protein is reduced after 4 days of TTX and remains low after 6 days ($P < 0.05$ and $P < 0.01$, respectively). Levels of NRF-1, NRF-2α, and NRF-2β proteins are unchanged until 6 days post-TTX ($P < 0.05$ for all) (Liang and Wong-Riley 2006) (Figure 11.1B). Again, changes in PGC protein precede those of NRF-1 and NRF-2 in neurons.

11.1.2 Monocular Enucleation in Rats

In vivo monocular enucleation in rats leads to a 29% decrease in PGC-1α's transcripts in the contralateral, deprived visual cortex 4 days later ($P < 0.05$), while those of *NRF-1* and *NRF-2β* are not detected until 1 week post-enucleation, when PGC-1α's mRNA exhibits a further decrease (48.8% for PGC-1α, 38.5% for *NRF-1* and 30.5% for *NRF-2β*, $P < 0.01 - 0.001$). By 2 weeks

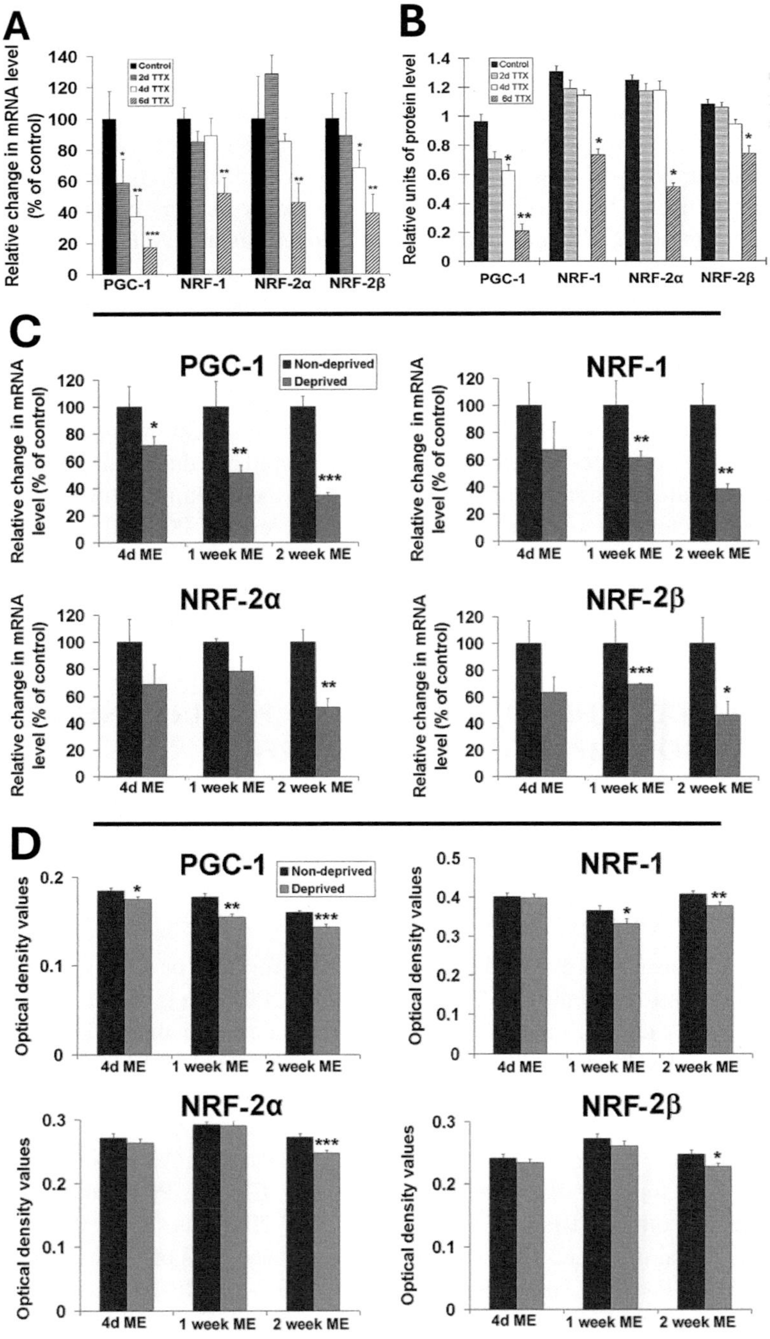

FIGURE 11.1 Activity-dependent regulation of PGC-1α by decreased neuronal activity. (A and B) In cultured primary neurons, PGC-1α mRNAs are significantly reduced by 2 days of TTX (A) and its protein by 4 days (B). However, *NRF-2β* message is not reduced until after 4 days, and its protein, as well as both the message and proteins of NRF-1 and NRF-2α, are not decreased until 6 days after inactivation. (C) With monocular enucleation (ME), PGC-1α message in the deprived visual cortex is downregulated 4 days later, whereas those of NRF-1 and NRF-2β are not reduced until after 1 week, and that of NRF-2α after 2 weeks. (D) The same time course of response is true for the levels of PGC-1α and NRF-1 proteins, but those of NRF-2α and NRF-2β are not decreased until 2 weeks post-enucleation. All comparisons are made with controls.

$*P < 0.05$, $**P < 0.01$, $***P < 0.001$.

Modified from Liang and Wong-Riley 2006.

post-enucleation, PGC-1α's mRNAs are reduced by 65.5%, *NRF-1* by 62.0%, *NRF-2β* by 53.5%, and *NRF-2α* by 48.4% ($P < 0.05 - 0.001$) (Figure 11.1C) (Liang and Wong-Riley 2006).

Monocular enucleation induces comparable changes in protein expressions of the four factors. Again, PGC-1α downregulates its proteins after 4 days, ahead of the 1-week change in NRF-1 and the 2-week alteration in NRF-2α and β ($P < 0.05 - 0.001$) (Figure 11.1D) (Liang and Wong-Riley 2006).

In short, PGC-1α adjusts its transcripts and proteins swiftly to changes in the functional activity and energy demand of neurons. The fast response of PGC-1α is consistent with its proposed roles in sensing the environment, inducing NRF-1 and NRF-2, and being a master regulator of mitochondrial biogenesis and metabolic homeostasis (Wu et al. 1999; Puigserver and Spiegelman 2003).

11.2 QUANTITATIVE IMMUNOCYTOCHEMISTRY OF DEPOLARIZATION-INDUCED EXPRESSION OF PGC-1α IN NEURONS

Does PGC-1α also respond to increasing neuronal activity? To address this, cultured primary neurons were exposed to 20 mM KCl and harvested 0.5, 1, 3, 5, 7, and 10 h later. They were then processed for EM and post-embedding immunogold labeling of anti-PGC-1 antibodies (sc-13067, Santa Cruz Biotechnology) (Meng et al. 2007). Like NRF-1 and NRF-2, PGC-1α-immunogold particles are localized to the nucleus, where they associate primarily with euchromatin, and in the cytoplasm, mainly over free ribosomes and less so the Golgi apparatus and the outer membrane of the nuclear envelope, but not in mitochondria. KCl depolarizing stimulation sharply increases the labeling density of PGC-1α at 0.5 h, peaking at 1 h, remaining high at 3 h, gradually declining at 5 and 7 h, and returning to baseline at 10 h. At all times, the density of immunogold particles is higher in the nucleus than in the cytoplasm (Meng et al. 2007). The time-dependent increase in both compartments indicates that PGC-1α responds swiftly (within 0.5 h) to increased neuronal activity by synthesizing new proteins (in the cytoplasm) and translocating them into the nucleus. PGC-1α presumably recruits transcription factors such as NRF-1 and NRF-2 to activate target genes (including *COX*) for the generation of ATP. The earlier adjustment of PGC-1α's message and protein than those of NRF-1 and NRF-2 is consistent with this mechanism (Zhang and Wong-Riley 2000b; Yang et al. 2004, 2006; Liang and Wong-Riley 2006; Liang et al. 2006b; Meng et al. 2007; Liang et al. 2010).

11.3 p38 MAP KINASE AND CALCIUM CHANNELS MEDIATE SIGNALING IN DEPOLARIZATION-INDUCED ACTIVATION OF PGC-1α IN NEURONS

What is the signaling mechanism by which PGC-1α is activated by depolarizing stimulation in neurons? In non-neuronal tissues, PGC-1α is activated by p38 mitogen-activated protein kinase (p38 MAPK) either directly through phosphorylation (Knutti et al. 2001; Puigserver and Spiegelman 2003) or indirectly by binding cAMP-response element, which are activated by p38 MAPK (Cao et al. 2004; Akimoto et al. 2005). The signaling pathway for PGC-1α in neurons was unclear prior to 2010.

If p38 MAPK is the signaling mechanism, then its level should be altered by neuronal activity. When primary neurons were stimulated by 20 mM KCl depolarization, the levels of both p38 MAPK and its phosphorylated form p-p38 MAPK were significantly elevated at 15 min and continued rising at 30 and 60 min (Liang et al. 2010) (Figure 11.2A–D). This prompt response is earlier than that of PGC-1α (30 min for both message and protein). Pretreatment with a p38 MAPK inhibitor, SB203580 (2.5 and 5 μM) for 30 min suppresses the upregulation of both p38 and p-p38 by KCl ($P < 0.05 - 0.01$) (Figure 11.2E and F). The specificity of p38 MAPK was tested by the addition of an inhibitor of JNK (c-Jun N-terminal kinase, a member of the MAPK family), JNKI-1 (1 μM), in the pretreatment. Although this inhibitor drastically reduces JNK protein, it has only a slight, though significant, effect on the expression of p-p38 (Liang et al. 2010) (Figure 11.2G and H). This indicates that p38 MAPK is an early sensor of increased neuronal activity.

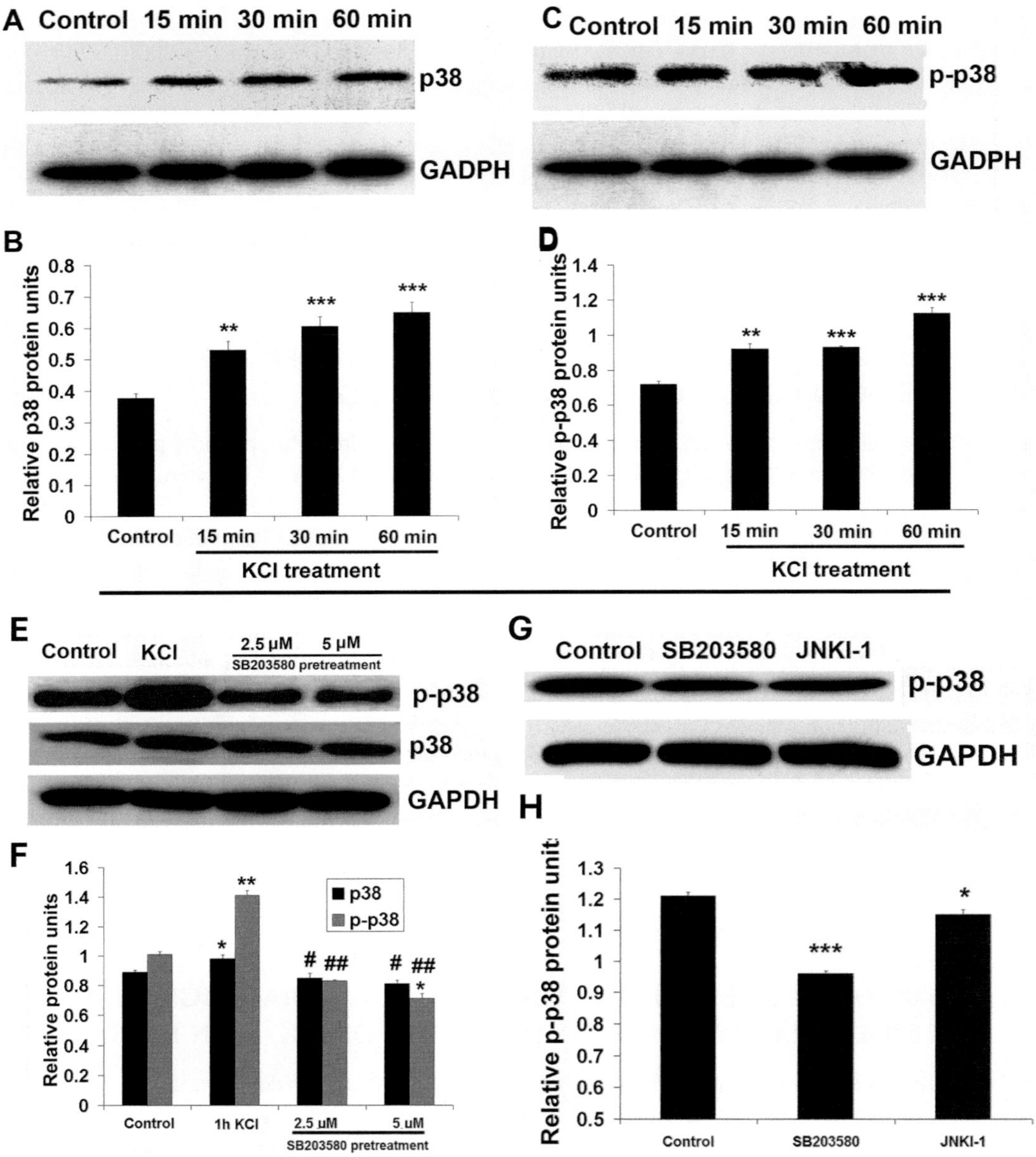

FIGURE 11.2 Activity-dependent regulation of PGC-1 by increased neuronal activity and the effect of p38 MAPK inhibitors. (A–D) A time-dependent increase in the protein levels of both p38 MAPK (p38; A and B) and phosphor-p38 MAPK (p-p38; C and D) by 15, 30, and 60 min of KCl stimulation in primary neurons. (E and F) After 30 min of pretreatment with 2.5 μM or 5 μM of SB203580 (a p38 MAPK inhibitor), primary neurons are no longer able to upregulate their p38 and p-p38. (G and H) Pretreatment with 5 μM SB203580 is much more effective than 1 μM JNKI-1 (an inhibitor of JNK) in reducing p-p38 MAPK level in N2a cells.

All * comparisons are with controls. $^*P < 0.05$, $^{**}P < 0.01$, $^{***}P < 0.001$. All # comparisons are with KCl alone. $^{\#}P < 0.05$, $^{\#\#}P < 0.01$.

Modified from Liang et al. 2010.

Pretreatment with SB203580 significantly suppresses the upregulation of PGC-1α protein at 1 h of KCl stimulation to below control levels in both primary neurons (Figure 11.3A and B) and N2a cells. By comparison, pretreatment with JNKI-1 is much less effective (Figure 11.3C) (Liang et al. 2010). These findings are consistent with PGC-1α being activated by p38 MAPK in neurons.

To determine the role of calcium channels in the activation of PGC-1α, primary neurons were pretreated with a calcium channel blocker, nifedipine, for 30 min before exposure to KCl for 5 h (at which time the PGC-1α mRNA has reached its peak level). At all three concentrations (10, 50, and 100 μM), nifedipine effectively suppresses the upregulation of PGC-1α's message and protein ($P < 0.05$ for all) (Figure 11.3D–F) (Liang et al. 2010).

To further verify the importance of p38 MAPK in signaling PGC-1α in neurons, primary neurons and N2a cells were transfected with p38 shRNA (Santa Cruz Biotechnology). Control shRNA plasmid-A (Santa Cruz Biotechnology) or scrambled shRNA serves as a negative control. Knocking down p38 MAPK significantly decreases transcript levels of both p38 MAPK and PGC-1α ($P < 0.001$). The reductions are 73% and 82%, respectively, in N2a cells, and 76% and 84%, respectively, in primary neurons compared with empty vector controls. Scrambled shRNA yields comparable results to empty vectors. Likewise, protein expressions of total p38 MAPK, p-p38 MAPK, and PGC-1α are significantly decreased by the silencing of p38 MAPK ($P < 0.001$ for all) (Figure 11.3G and H) (Liang et al. 2010).

The above experiments document the importance of p38 MAPK and calcium channels in mediating the signaling mechanism that activates PGC-1α in response to depolarizing stimulation in neurons.

11.4 PGC-1α AND ROS

PGC-1α plays an important role in regulating reactive oxygen species. On the one hand, PGC-1α and PGC-1β are strongly induced by ROS, and on the other, they regulate a complex ROS defense system (St-Pierre et al. 2006). *PGC-1α* null mice are extremely sensitive to oxidative stress caused by agents such as MPTP and kainic acid. MPTP induces greater dopaminergic cell death in the substantia nigra pars compacta, and kainic acid causes more apoptotic cell death in CA1 hippocampal neurons in *PGC-1α* ko mice than in controls (St-Pierre et al. 2006). Clearly, PGC-1α is at the center of a network of homeostatic regulatory mechanisms that control energy metabolism and ROS detoxification.

11.5 PGC-1α IN THE RETINA

PGC-1α (and PGC-1β) are highly expressed in the murine retina, especially in photoreceptor cells (Egger et al. 2012). Surprisingly, knocking out *PGC-1α* does not adversely affect either the morphology or the function of dark-adapted retina. However, exposure to 15,000 lux of white light for 2 h followed by 24 h recovery results in photoreceptor cell damage in *PGC-1α*-null but not the wildtype mice (Egger et al. 2012). Ten days after light exposure, scotopic ERG is still blunted in the knockout but not the WT mice. Moreover, PGC-1α (and PGC-1β) expressions are reduced in mouse models of retinitis pigmentosa (Egger et al. 2012).

11.6 PGC-1-RELATED COACTIVATOR (PRC)

Another coactivator structurally and functionally related to PGC-1 was discovered by Andersson and Scarpulla (2001), who named it PGC-1-related coactivator or PRC. Like PGC-1, PRC interacts with transcription factors, such as NRF-1, and coactivates their target genes. Unlike PGC-1, however, PRC is not upregulated during thermogenesis in brown fat, but rather, it is rapidly induced by serum growth factors in quiescent fibroblasts that reenter the cell cycle. PRC orchestrates a protective response to metabolic stress by co-activating genes involved in inflammation, proliferation, and

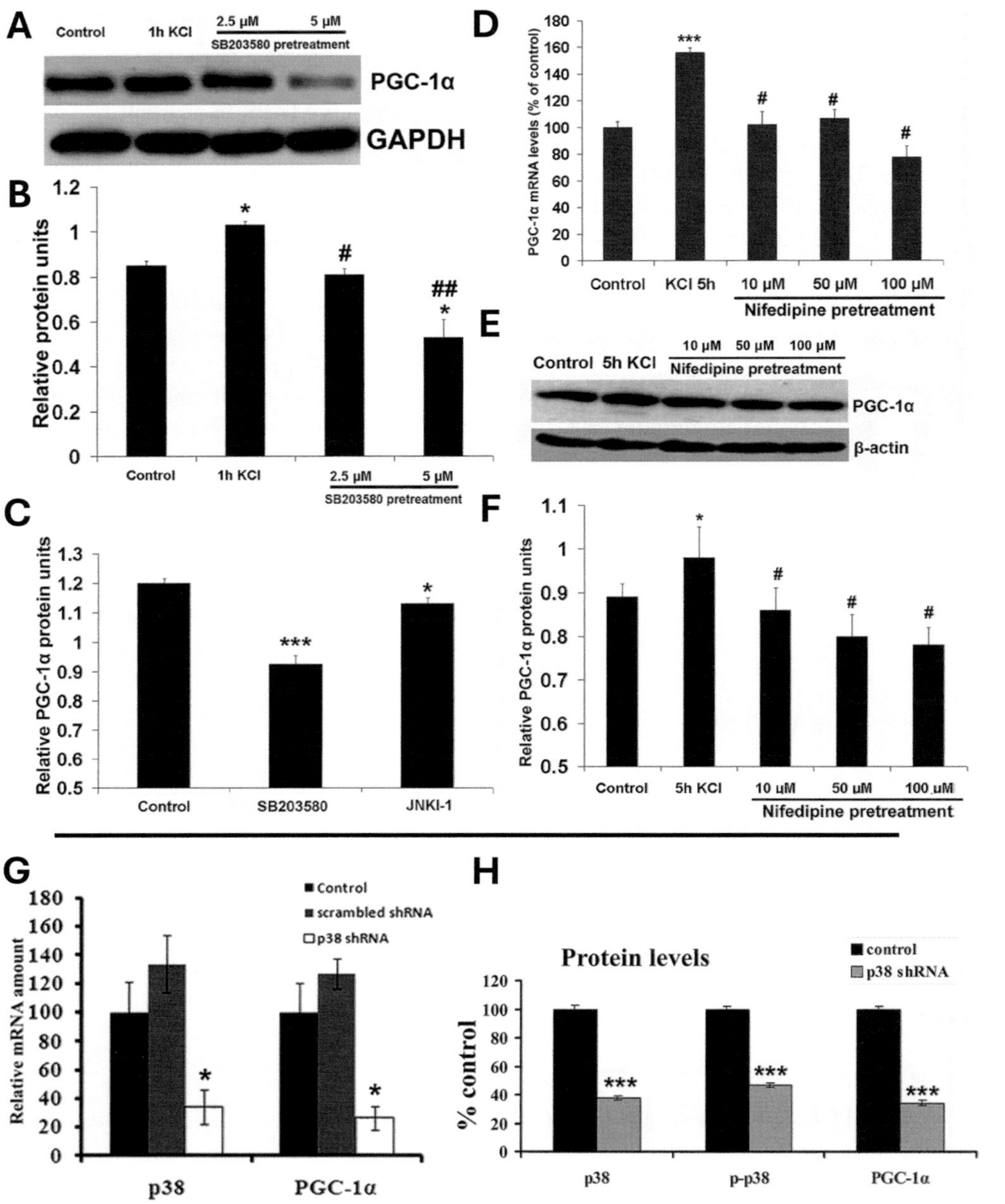

FIGURE 11.3 Effect of p38 MAPK and calcium channel inhibitors on PGC-1α. (A and B) The level of PGC-1α protein in primary neurons is upregulated by 1 h of KCl stimulation. However, 30 min of 2.5 μM or 5 μM SB203580 pretreatment prior to KCl prevents such an increase, and the level returns to that of control (at 2.5 μM) or below that of control (at 5 μM). (C) Pretreatment with 1 μM JNKI-1 is less effective than SB203580 in preventing the increase, though it is significantly lower than that of the control. PGC-1α mRNA (D) and protein (E–F) levels are increased by 5 h KCl simulation in primary neurons. However, pretreatments with 10, 50, or 100 μM of the calcium channel blocker nifedipine prevent such an increase, and the level returns to that of controls. (G) Transcript levels of p38 MAPK and PGC-1α in N2a cells are significantly reduced by silencing p38 with shRNA, but not with scrambled shRNA. (H) Levels of p38, p-p38, and PGC-1α proteins in N2a cells are all significantly decreased by *p38* shRNA.

All * comparisons are with controls. $^{*}P < 0.05$, $^{***}P < 0.001$. All # comparisons are with KCl alone. $^{\#}P < 0.05$, $^{\#\#}P < 0.01$, $^{\#\#\#}P < 0.001$.

Modified from Liang et al. 2010.

metabolic reprogramming. The PRC stress program is induced as part of an adaptive response to apoptosis and premature senescence (Gleyzer and Scarpulla 2013). In a thyroid oncogene cell line, PRC is found to interact with multiple transcription factors, including NRF-1, NRF-2, and especially ERR1 in regulating energy metabolism, cell cycle, RNA processing, as well as the expression of microRNA (Dumont et al. 2018). Relatively little is known about PRC in neurons.

Much remains to be uncovered about the vital roles that the PGC-1 family (PGC-1α, PGC-1β, and PRC) plays in sensing the environment and coordinating a network of transcription factors to maintain the metabolic health of neurons and how their dysfunction may be linked to various neurological diseases. The finding that PGC-1 family regulates mitochondrial density prompted the suggestion that they be used as therapeutic agents in neurodegenerative diseases such as Parkinson's and Alzheimer's, in which mitochondrial loss is a common phenotype (Wareski et al. 2009).

12 Tight Coupling of Neuronal Activity and Energy Metabolism

NRF-1 Transcriptionally Coregulates Mediators of Glutamatergic Neurotransmission, Energy Consumption, and Energy Generation

ABSTRACT

The possibility that NRF-1 mediates the tight coupling between neuronal activity and energy metabolism at the transcriptional level was tested on several genes involved in glutamatergic neurotransmission. Multiple molecular biological approaches reveal that NRF-1 positively regulates the gene expressions of N-methyl-D-aspartate (NMDA) receptor subunits 1 and 2B (Grin1 and Grin 2b), α-amino-3-hydroxy-5-methyl-4-isoxazole propionic acid (AMPA) receptor subunit 2 (Gria2), neuronal nitric oxide synthase (Nos1), and the motor that transports the NMDA receptor 2B from the cell body exclusively to dendrites, KIF17 (Kif17). Thus, NRF-1 is strategically positioned to coordinate the expressions of mediators of energy generation (COX), energy expenditure (Na^+/K^+-ATPase), and glutamatergic neurotransmission.

As discussed in Chapter 2, the brain, weighing merely 3 lb, consumes as much as 20% of the total energy of the body (Clark and Sokoloff 1999). Much of this energy is utilized for repolarizing membrane potentials after depolarization, that is, postsynaptic events of excitatory synapses (Wong-Riley 1989) mediated by excitatory neurotransmitters, typically glutamate and its receptors. If neuronal activity and energy metabolism are tightly coupled, is it possible that this coupling also exists at the transcriptional level? If COX, the mediator of energy metabolism, is transcriptionally regulated by a set of transcription factors (NRF-1, NRF-2, and Sp4), the question is whether these same factors regulate mediators of excitatory transmission, that is, glutamatergic receptors? In this and the next two chapters, experiments documenting a tight coupling between neuronal activity and energy metabolism at the transcriptional level are discussed.

Glutamate is the most abundant amino acid and the major excitatory neurotransmitter in the mammalian brain (McEntee and Crook 1993). Its action depends on three families of multimeric, ligand-gated, ionotropic receptors: *N*-methyl-D-aspartate (NMDA), α-amino-3-hydroxy-5-methyl-4-isoxazole propionic acid (AMPA), and kainate (KA) receptors, as well as three groups of metabotropic, G-protein-coupled (mGluR) receptors. Each receptor type comprises multiple subunits (Sugiyama et al. 1987; Nakanishi 1992; Seeburg 1993; Hollmann and Heinemann 1994; Traynelis et al. 2010).

DOI: 10.1201/9781003507765-12

12.1 NRF-1 COREGULATES NMDA GLUTAMATE RECEPTORS AND *COX*

NMDA receptor channels are highly permeable to Ca^{2+} when the voltage-dependent Mg^{2+} block is removed (Mori and Mishina 1995). These receptors, widely distributed across brain regions, serve important roles in neuronal development, plasticity, learning, and memory by mediating most of the excitatory synapses in the CNS (Choi 1988; Mori and Mishina 1995). However, dysfunctional NMDA receptors lead to excitotoxicity, the basis of many brain pathologies (Rothman and Olney 1995). There are seven subunits in NMDA receptors: NR1 (or GluN1), NR2A-D (GluN2A-D), and NR3A,B (GluN3A,B). The receptor is tetrameric, with two obligatory GluN1 and either two GluN2 or a combination of GluN2 and GluN3, the exact composition dictates receptor function (Monyer et al. 1994). GluN2 binds glutamate, and GluN1 and GluN3 bind glycine, a necessary ligand for receptor function (Traynelis et al. 2010). GluN2 has 4 isoforms: GluN2A, 2B, 2C, and 2D. GluN2A and 2B induce long-term potentiation (Schotanus and Chergui 2008). The temporal and spatial distributions vary among the four GluN2 subunits (Monyer et al. 1994). GluN2B is the most prominent one in the adult cortex, GluN2A is relatively abundant in the adult hippocampus, whereas GluN2C and 2D are, in general, relatively sparse, and GluN3A and 3B are primarily embryonic (Goebel and Poosch 1999).

In silico analysis of proximal promoters of murine NMDA (*Grin*) subunit genes indicates that *Grin1* and *Grin2b* promoters bear atypical NRF-1 binding motifs but have the invariant GCA core present in *COX* subunit genes (Dhar et al. 2008). *Grin2a*, *2c*, *2d*, and *3b* promoters possess atypical GC-rich motif but without the GCA core. *Grin3a* lacks both typical and atypical NRF-1 binding sites and, thus, serves as a negative control (Dhar and Wong-Riley 2009).

In vitro EMSA assays reveal shift and specific supershift bands for *Grin1* and *Grin2b* when reacted with HeLa nuclear extract and NRF-1 antibodies, but other *Grin* subunits do not. In vivo ChIP assays demonstrate that *Grin1* and *Grin2b* promoters coimmunoprecipitate with NRF-1 antibodies, but *Grin2a* and *Grin3a* do not (Figure 12.1A). Site-directed mutagenesis of NRF-1 binding sites significantly reduces the promoter activity of *Grin1* and *Grin2b*, and transfecting N2a cells or primary neurons with *NRF-1* shRNA leads to a significant reduction in *Grin1* and *Grin2b* transcripts, but *Grin2a* and *Grin3a* are unaffected (Figure 12.1B). Likewise, NRF-1 silencing downregulates the proteins of the former two but not the latter two subunits. Depolarizing stimulation of N2a cells or primary neurons significantly upregulates mRNAs of all four *Grin* subunits. Transfection with *NRF-1* shRNA, however, prevents the upregulation of *Grin1* and *Grin2b* by KCl (20 mM for 5 h) but has no effect on *Grin2a* and *Grin3a*. Conversely, all four Grin subunits are downregulated by TTX (0.4 μM for 3 days), and overexpressing NRF-1 in primary neurons prevents such reductions in *Grin1* and *Grin2b* transcripts, but not those of *Grin2a* and *Grin3a* (Figure 12.1C) (Dhar and Wong-Riley 2009).

NRF-1, therefore, links glutamatergic synaptic transmission and energy metabolism at the transcriptional level. The high homology among mice, rats, and humans for NRF-1 binding on promoters of *Grin1*, *Grin2b*, and *COX* underscores the evolutionary importance of this conservation.

12.2 NRF-1 COREGULATES AMPA GLUTAMATE RECEPTORS AND *COX*

The other major type of ligand-gated, ionotropic glutamatergic receptors mediating fast, excitatory synaptic transmission are AMPA receptors. They are also tetrameric, composed of homo- and heteromeric forms of four subunits: GluR1-4 (GluR A-D, or GluA1-4) (Traynelis et al. 2010). GluA1-3 exist in most neurons, whereas GluA4 is regionally distributed and is expressed mainly in early postnatal life. AMPA receptors are critical in regulating neuronal functions, including synaptogenesis, circuit formation, homeostatic scaling, and synaptic plasticity involving long-term potentiation and long-term depression via activity-dependent insertion and

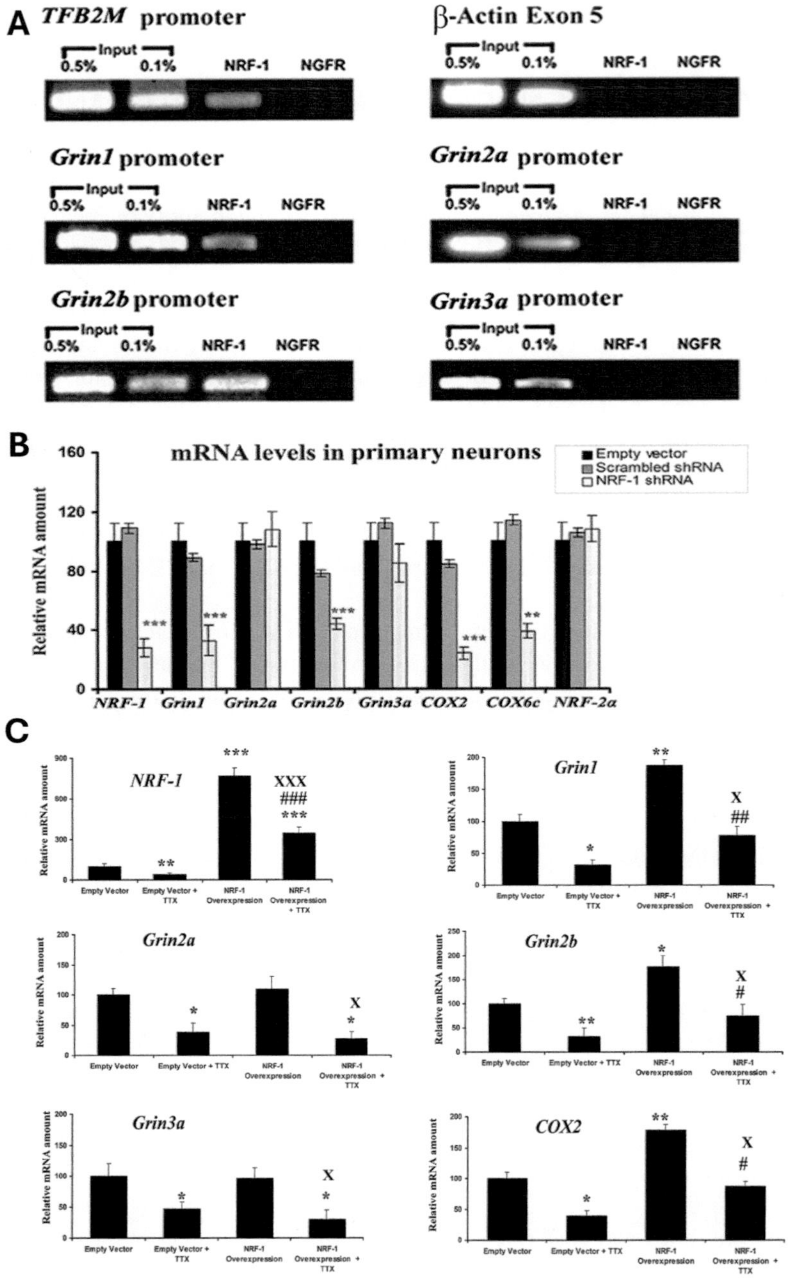

FIGURE 12.1 NRF-1 coregulates NMDA receptor subunit genes and *COX*. (A) ChIP assays show that *Grin1* and *Grin2b* promoters are coimmunoprecipitated by anti-NRF-1 antibodies, but not those of *Grin2a* and *Grin3a*. *Tfb2m* serves as the positive control, and β-*actin* exon 5 as the negative control. Anti-nerve growth factor receptor p75 (NGFR) is a negative control for co-IP. Input lanes are chromatin without immunoprecipitation. (B) Primary neurons transfected with *NRF-1* shRNA exhibit significant decreases in mRNA levels of *NRF-1* itself, *Grin1*, *Grin2b*, and positive controls *COX2* and *COX6c* as compared to those transfected with empty vectors or with scrambled shRNA. However, *Grin2a*, *Grin2b*, and *NRF-2α* remain unchanged. (C) NRF-1 overexpression in primary neurons significantly increases transcript levels of *NRF-1* itself, *Grin1*, *Grin2b*, and *COX2* as compared to empty vector controls, but *Grin2a* and *Grin3a* are not affected. Whereas TTX induces a downregulation of all transcript levels, only those that respond to NRF-1 overexpression (i.e., *NRF-1*, *Grin1*, *Grin2b*, and *COX2*) are rescued from being suppressed by TTX. $N = 6$ for each data point or group.

All * comparisons are with controls. *$P < 0.05$, **$P < 0.01$, ***$P < 0.001$. All # comparisons are with empty vector + TTX. $^{\#}P < 0.05$, $^{\#\#}P < 0.01$, $^{\#\#\#}P < 0.001$. All X comparisons are with NRF-1 overexpression. $^{X}P < 0.05$, $^{XXX}P < 0.001$.

Modified from Dhar and Wong-Riley 2009.

removal of receptor subunits at synaptic sites (Nakanishi 1992; Seeburg 1993; Diering and Huganir 2018). Among the four subunits, GluA2 is unique in that it undergoes mRNA editing by converting a conserved neutral glutamine 607 in the channel pore to a positively charged arginine (Sommer et al. 1991). This Q/R switch renders GluA2-containing receptors generally Ca^{2+}-impermeable, whereas GluA2-lacking receptors are highly permeable to Ca^{2+}. The latter are inwardly rectifying due to voltage-dependent blockage of an outward current flow by intracellular polyamines (Burnashev et al. 1992; Tanaka et al. 2000; Isaac et al. 2007; Diering and Huganir 2018).

In silico analysis of DNA sequences 1 kb 5' upstream and 100 bp 3' beyond the TSS of the four AMPA receptor subunit genes (*Gria1-4*) in the murine genome database indicates that *Gria2* promoter has a typical NRF-1 binding site, whereas the other three subunit genes possess the atypical NRF-1 binding motif and lacking the GCA core (Dhar et al. 2009b). An earlier study by Myers et al. (1998) noted that the promoter of *Gria2* gene has an NRF-1 binding site, but the binding was not rigorously tested.

Both in vitro EMSA/supershift and in vivo ChIP assays confirm direct interactions between *Gria2* promoter and NRF-1, whereas *Gria1*, *Gria3*, and *Gria4* do not (Dhar et al. 2009b). Site-directed mutagenesis of the NRF-1 binding site reduces Gria2 promoter activity by 42% ($P < 0.05$). Silencing NRF-1 with shRNA downregulates GluA2 protein in N2a cells by 60% ($P < 0.05$) and *Gria2* message in N2a cells and primary neurons by 70–80% ($P < 0.01$) (Figure 12.2A) (Dhar et al. 2009b). On the other hand, suppressing NRF-1 expression has no effect on the protein and transcript levels of the other three AMPA receptor subunits, indicating that the effect is specific. Transfecting primary neurons with *NRF-1* shRNA suppresses the depolarization-induced upregulation of *Gria2* mRNA but has no effect on those of *Gria1*, *Gria3*, and *Gria4*. Overexpressing NRF-1 in primary neurons prevents the downregulation of *Gria2* transcripts and protein induced by TTX blockade, but does not alter the other three *Gria* subunits, confirming the specificity of NRF-1 (Figure 12.2B and C). Mutating the NRF-1 binding site on the *Gria2* promoter also prevents the KCl-induced increase in promoter activity (Dhar et al. 2009b). These experiments strongly indicate that the activity-dependent expression of GluA2 involves, at least in part, the transcriptional control by NRF-1 in neurons. The NRF-1 binding motif with a GCA core on the *Gria2* promoter is highly conserved among humans, mice, and rats (Dhar et al. 2009b).

The distribution of GluA2 in the monkey visual cortex is similar, but not identical to that of COX (Wong-Riley and Jacobs 2002). In layers II/III, GluA2-immunoreactive patches are in perfect register with COX-rich puffs. Layer IVA is marked by a honeycomb-like pattern of GluA2, strongly resembling that of COX. However, this pattern exists only prenatally (fetal day 139) and is absent in the adult. The significance of the transient expression of GluA2 is unclear, but it suggests that AMPA receptors undergo a developmental switch from GluA2-containing to GluA2-lacking, facilitating fast excitatory transmission in layer IVA. Layer IVC, having the highest density of COX, is virtually devoid of GluA2 except at the base, the IVCβdark (see Chapter 4). The absence of GluA2 in IVC is evidently necessary to enhance Ca^{2+} influx and promote processing of geniculocortical input. NMDA receptors are also plentiful in IVC (Wong-Riley et al. 1998a), the site of the highest level of spontaneous activity among all cortical layers (DeYoe et al. 1995). Layer VI is only moderately reactive for COX but is rich in GluA2 that labels many pyramidal neurons. With monocular TTX blockade, light and dark bands of GluA2 appear in layer VI, in line with those of COX in IVC, indicating that GluA2 expression is activity-dependent (Wong-Riley and Jacobs 2002). Likewise, GluA2 responds to TTX inactivation and KCl stimulation in cultured primary neurons by downregulating and upregulating, respectively, the expressions of its protein and message (Bai and Wong-Riley 2003).

GluA2 knockout mice exhibit normal gross brain morphology but abnormal behavior, including altered grooming, locomotion, motor coordination, spatial learning, and non-spatial learning performance in the water maze (Gerlai et al. 1998). Long-term potentiation in these mutants is markedly enhanced, and the kainite-induced Ca^{2+} permeability is increased 9-fold, supporting widespread

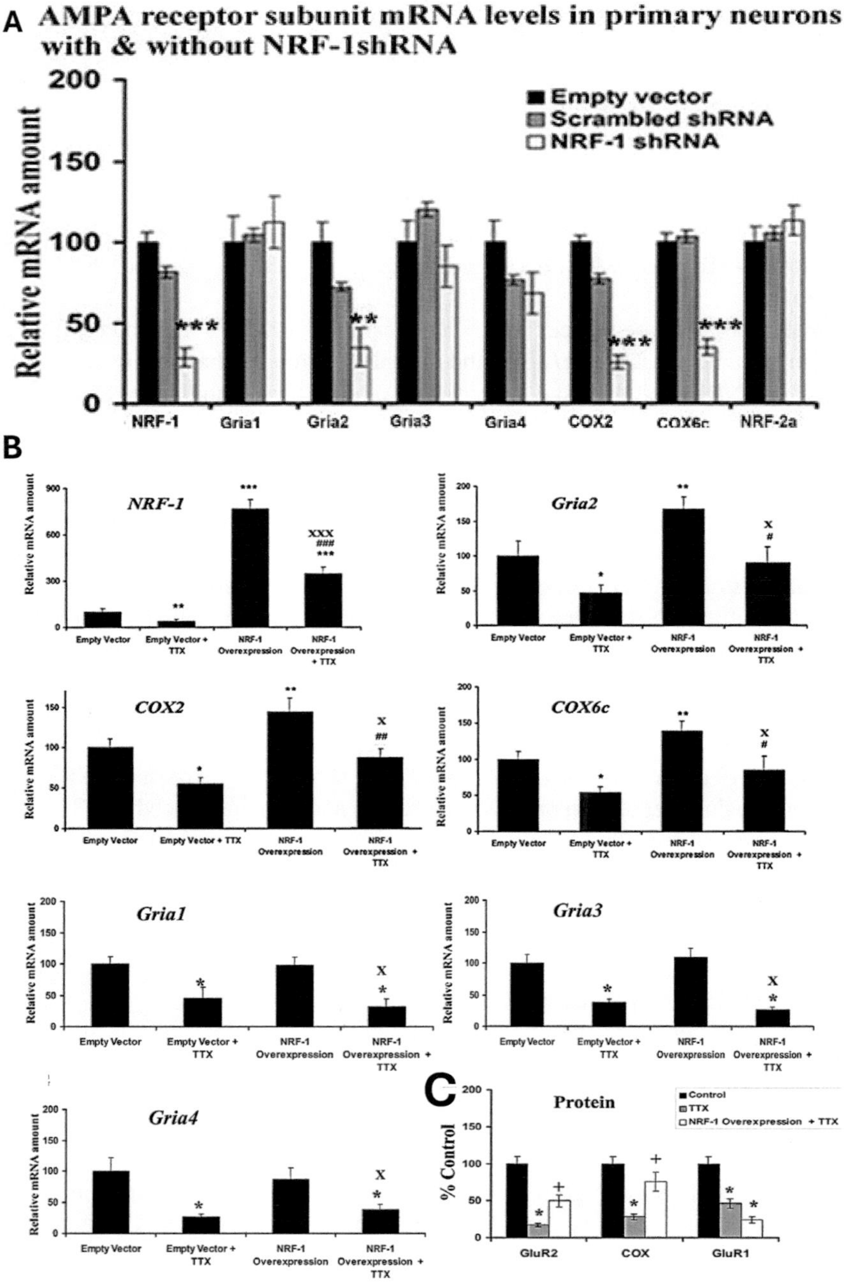

FIGURE 12.2 NRF-1 coregulates the AMPA receptor subunit 2 gene and *COX*. (A) Primary neurons transfected with *NRF-1* shRNA show significant reductions in transcript levels of *NRF-1*, *Gria2*, *COX2*, and *COX6c*, but not those of *Gria1*, *Gria3*, *Gria4*, and *NRF-2α*, as compared to those transfected with empty vectors or scrambled shRNA. (B) Overexpressing NRF-1 in primary neurons upregulates the transcript levels of *NRF-1*, *Gria2*, *COX2*, and *COX6c*, but not those of *Gria1*, *Gria3*, and *Gria4*. Overexpressing NRF-1 also protects itself and its target genes (*Gria2*, *COX2*, and *COX6c*) from being downregulated by TTX, but has no effect on *Gria1*, *Gria3*, and *Gria4*. (C) Levels of GluR2, COX, and GluR1 proteins are all reduced by TTX, but only those of GluR2 and COX are rescued from the effects of TTX by NRF-1 overexpression. $N = 6$ for each data point or group.

All * comparisons are with controls. *$P < 0.05$, **$P < 0.01$, ***$P < 0.001$. All # comparisons are with empty vector + TTX. $^{\#}P < 0.05$, $^{\#\#}P < 0.01$, $^{\#\#\#}P < 0.001$. All X comparisons are with NRF-1 overexpression. $^{X}P < 0.05$, $^{XXX}P < 0.001$. The + values in C are compared to TTX.

Modified from Dhar et al. 2009b.

increases in excitability (Jia et al. 1996). Thus, in dictating Ca^{2+} permeability in AMPA receptors, GluA2 plays an important role in controlling synaptic plasticity and behavior. Its constitutive presence or dynamic insertion during AMPA receptor trafficking is also critical in curbing excitotoxicity and neuronal death (Liu and Zukin 2007).

12.3 NRF-1 COREGULATES NEURONAL NITRIC OXIDE SYNTHASE AND *COX*

A downstream signaling molecule of NMDA receptors is the short-lived gaseous neurotransmitter, nitric oxide (NO). Ca^{2+} entry via NMDA receptors triggers calmodulin binding onto nitric oxide synthase (NOS), which converts L-arginine to L-citrulline with the release of NO. NO, in turn, diffuses in postsynaptic cells and across synapses to presynaptic terminals (and astrocytic processes), where it activates guanylate cyclase to form the second messenger cyclic GMP, activating cGMP-dependent protein kinase to alter the gating of ion channels and facilitate a host of synaptic functions (Garthwaite 1991, 2008).

Three distinct isoforms of NOS are prominent in three different cell types: neuronal (NOS1 or nNOS), inducible (NOS2 or iNOS; mainly in macrophages but inducible in many cell types), and endothelial (NOS3 or eNOS). NO mediates synaptic transmission, macrophage activation, and vascular relaxation, respectively, in these cell types (Nathan and Xie 1994; Dawson & Dawson 1996; Garthwaite 2008). NOS1 exists at low concentrations in many cells but at high concentrations in select groups of neurons, especially GABAergic ones (Seress et al. 2005).

A putative NRF-1 binding site was reported but not functionally characterized for the human *NOS1* promoter (Hall et al. 1994). When the promoters of murine *Nos1*, *2*, and *3* genes were analyzed in silico, no homology was found for the reported human NRF-1 site, which lacks a GCA core deemed invariant previously (Dhar et al. 2008). Further probing revealed a single NRF-1 binding motif with a GCA core in exon 1 of the murine *Nos1* gene (Dhar et al. 2009c) designated by Genomatix as a promoter region. It has 85–95% homology with both the rat and human *Nos1* gene. On the other hand, *Nos2* and *Nos3* promoters lack both typical and atypical NRF-1 binding motifs.

In vitro EMSA using HeLa nuclear extract demonstrates that *Nos1* promoter forms a DNA-protein shift band and a supershift complex with NRF-1 antibodies. Neither *Nos2* nor *Nos3* proximal promoters yield any shift band, nor does mutating the NRF-1 binding site in *Nos1* (Dhar et al. 2009c). In vivo ChIP assays with N2a cell chromatin reveal that the *Nos1* promoter is immunoprecipitated by NRF-1 antibodies, but those of *Nos2* and *Nos3* are not. Mutating the NRF-1 site significantly reduces *Nos1* promoter activity ($P < 0.05$), and transfecting N2a cells or primary neurons with *NRF-1* shRNA downregulates NOS 1 protein and mRNAs ($P < 0.05 - 0.01$), but not those of NOS2 and NOS3 (Dhar et al. 2009c) (Figure 12.3A). Silencing NRF-1 prevents the upregulation of *Nos1* transcripts by KCl stimulation, but *Nos2* and *Nos3* are unaffected by either KCl or *NRF-1* shRNA (Figure 12.3B). Conversely, TTX significantly reduces the transcripts of all three *Nos*, but only *Nos1* is rescued by the overexpression of NRF-1 (Figure 12.3C) (Dhar et al. 2009c). Silencing NRF-1 causes an 80% decrease in the expression of guanylyl cyclase (*Gucy1a2*) (Figure 12.3D), a downstream target of the NO pathway, and its own promoter has a putative NRF-1 binding site (Dhar et al. 2009c).

Hence, NRF-1 directly regulates the expression of critical subunit genes of NMDA receptors and their downstream effectors, NOS1, as well as possibly guanylyl cyclase in neurons.

12.4 NRF-1 REGULATES KINESIN SUPERFAMILY PROTEIN KIF17

The wide range influence of NRF-1 on the transcriptional regulation of multiple key players of glutamatergic synaptic transmission raises the question of whether it also controls the expression of a kinesin gene, KIF17, the transporter of glutamatergic receptors?

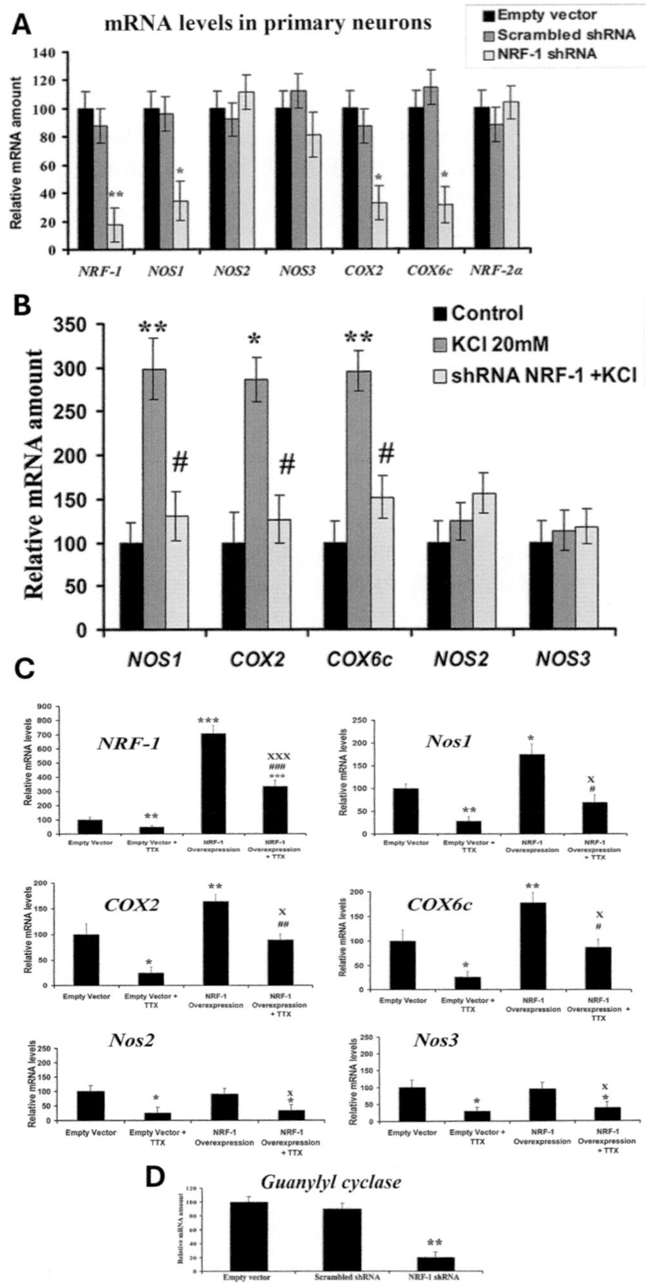

FIGURE 12.3 NRF-1 coregulates neuronal nitric oxide synthase (*Nos1*) and *COX*. (A) Silencing NRF-1 in primary neurons induces a downregulation of mRNA levels of *NRF-1*, *Nos1*, *COX2*, and *COX6c*, but not those of *Nos2*, *Nos3*, or *NRF-2α*. (B) Depolarizing stimulation with KCl for 5 h upregulates message levels of *Nos1*, *COX2*, and *COX6c*, but not those of *Nos2* and *Nos3* in N2a cells. In the presence of *NRF-1* shRNA, however, KCl is no longer able to increase the transcript levels of *Nos1*, *COX2*, and *COX6c*, but *Nos2* and *Nos3* are not affected. (C) Overexpressing NRF-1 in primary neurons upregulates mRNA levels of *NRF-1*, *Nos1*, *COX2*, and *COX6c*, but not those of *Nos2* and *Nos3*. NRF-1 overexpression also prevents itself and its target genes (*Nos1*, *COX2*, and *COX6c*) from being suppressed by TTX, but it has no effect on *Nos2* and *Nos3*. (D) Transfecting N2a cells with *NRF-1* shRNA significantly reduces the message levels of a major subunit of guanylyl cyclase (*Gucy1a2*) as compared to those transfected with empty vector or scrambled shRNA controls. $N = 6$ for each data point or group in A, C, and D. Experiments were done in triplicate in B.

All * comparisons are with controls. *$P < 0.05$, **$P < 0.01$, ***$P < 0.001$. All # comparisons are with empty vector + TTX. $^{\#}P < 0.05$, $^{\#\#}P < 0.01$, $^{\#\#\#}P < 0.001$. All X comparisons are with NRF-1 overexpression. $^{X}P < 0.05$, $^{XXX}P < 0.001$.

Modified from Dhar et al. 2009c.

The kinesin superfamily proteins (KIFs) are ATP-dependent molecular motors that carry protein complexes, membranous organelles, and mRNAs along microtubules to targeted, intracellular destinations (Hirokawa 1998; Hirokawa et al. 2009). They comprise 14 distinct families varying in structure and functions. Kinesins play important roles in neuronal development, synaptic transmission, intraflagellar transport, and photoreceptor outer segment renewal (Hirokawa and Takemura 2004; Hirokawa et al. 2009; Insinna et al. 2009; Silverman and Leroux 2009). KIF17 is a member of the kinesin-2 family that is plus-end directed. It is a homodimer with a pair of head motor domains that bind microtubules, a coiled-coil stalk, and a tail domain that binds cargos. Its functions vary widely in different cell types (Hirokawa and Takemura 2004; Wong-Riley and Besharse 2012). In neurons, KIF17 binds the postsynaptic density-95/disc large/zona occludens-1 (PDZ) domain of Mint 1 (mLin10) within a scaffolding protein complex and transports vesicles containing the GluN2B subunit of NMDA receptors from the cell body exclusively to dendrites (Setou et al. 2000). It also transports kainate receptor GluK5 and potassium Kv4.2 channels in a somatodendritic direction (Kayadjanian et al. 2007; Chu et al. 2006). Overexpressing KIF17 enhances GluN2B-dependent learning and memory (Wong et al. 2002). Functional blockade or knockdown of KIF17 disrupts GluN2B expression and its transport to synaptic sites, whereas overexpressing GluN2B upregulates KIF17 expression (Guillaud et al. 2003). Parallel changes of motors and their cargo raise the intriguing question of whether they share the same regulatory mechanism, even at the transcriptional level?

The promoter region of the murine *Kif17* gene spanning 1 kb upstream and up to 200 bp downstream of the TSS (excluding protein-coding sequence) was analyzed in silico for NRF-1 and NRF-2α binding sites (Dhar and Wong-Riley 2011). Two atypical NRF-1 motifs (A and B) are revealed, both with the invariant GCA core and bearing 85–95% homology with rat and human *Kif17* promoters. The two putative NRF-2α binding sites are not in tandem repeat. Analysis of neuronal *Kif3a* and *Kif5b* promoters uncovers neither typical nor atypical NRF-1 binding motifs.

Both A and B NRF-1 binding sites form shift and supershift bands with EMSA, and mutating the NRF-1 site obliterates the bands. Putative NRF-2α sites, however, do not generate shift or supershift bands (Dhar and Wong-Riley 2011). ChIP assays with chromatin from primary neurons demonstrate that *Kif17* promoter immunoprecipitates with NRF-1 antibodies, but *Kif1a* promoter does not (Figure 12.4A). NRF-2α antibodies do not immunoprecipitate any *Kif17* promoter. Mutating the A and B NRF-1 binding sites on *Kif17* reduces its promoter activity by 60% and 35%, respectively ($P < 0.01$ and 0.05, respectively). Silencing NRF-1 with shRNA in primary neurons downregulates both KIF17 protein and *Kif17* message by 70% ($P < 0.05$ for both) but has no effect on the expressions of *Kif1a* or *NRF-2α*. Silencing NRF-2α likewise has no effect on the expression of *Kif17* (Dhar and Wong-Riley 2011). *Kif17* responds to changes in neuronal activity, in that silencing NRF-1 prevents its message and protein upregulation by KCl stimulation (Figure 12.4B and C), and overexpressing NRF-1 rescues them from being downregulated by TTX (Figure 12.4D and E). The negative control *Kif1a* is unaffected by these manipulations (Dhar and Wong-Riley 2011).

In conclusion, NRF-1 coordinates the expression of genes responsible for the production and transportation of key mediators of glutamatergic neurotransmission, GluN2B and KIF17. NRF-1 definitively facilitates tight coupling between neuronal activity and energy metabolism by co-regulating all 13 subunits of *COX* (see Chapter 8) and mediators of glutamatergic transmission via NMDA and AMPA receptors, through downstream nitric oxide synthase and possibly guanylyl cyclase, to the transporter (KIF17) of a critical glutamate receptor subunit. Moreover, NRF-1 regulates important subunits of a major energy consumer, Na^+/K^+-ATPase (Johar et al. 2012) and ensures that energy production matches energy utilization at the molecular and cellular levels (Figure 12.5). To date, no other transcription factor has been proven to accomplish as much. However, NRF-1 does not function alone. The following two chapters will reveal its partners in transcriptionally coordinating energy metabolism and neuronal activity.

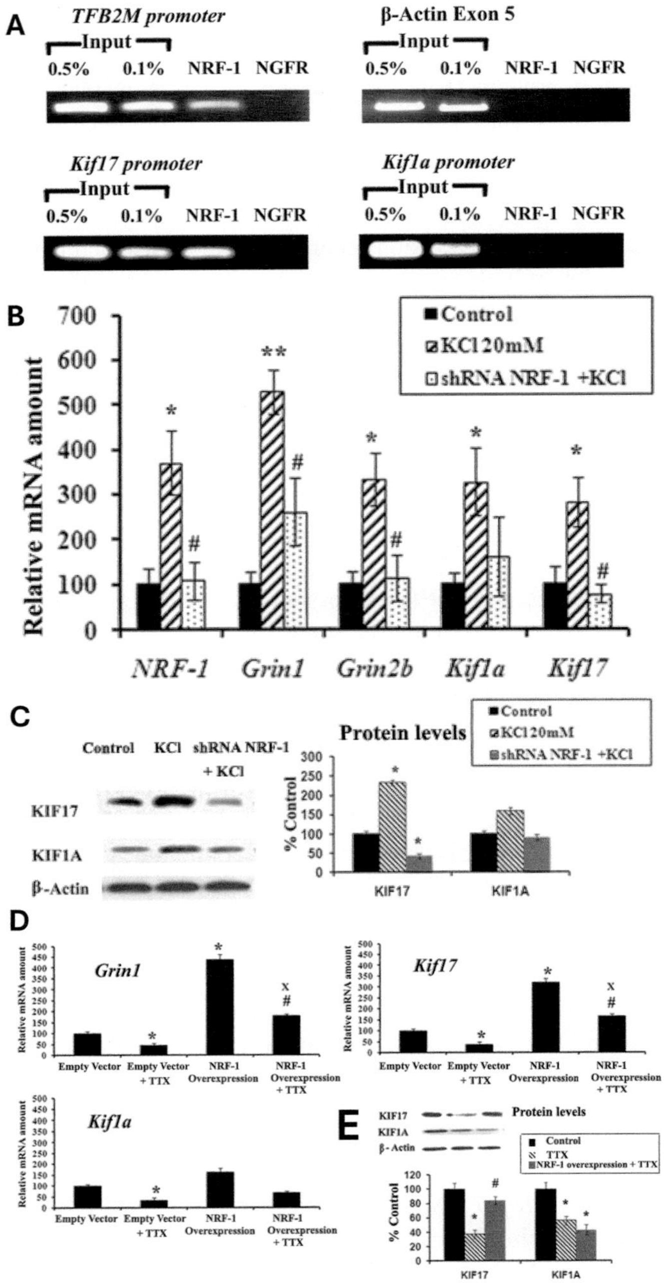

FIGURE 12.4 NRF-1 regulates *Kif17* in neurons. (A) ChIP assays indicate that *Kif17*, but not the *Kif1a* promoter, is immunoprecipitated by antibodies against NRF-1. *Tfb2m* serves as the positive control, and *β-actin* exon 5 as the negative control. Anti-NGFR does not yield any positive band. Input chromatins are taken before immunoprecipitation. (B) Transcripts for *NRF-1*, *Grin1*, *Grin2b*, *Kif1a*, and *Kif17* are all upregulated by 20 mM KCl for 5 h in primary neurons. However, silencing NRF-1 with shRNA in these neurons prevents the upregulation of *NRF-1*, *Grin1*, *Grin2b*, and *Kif17*, but not of *Kif1a*. (C) Western blots and quantitation indicate that KCl stimulation increases KIF17 and KIF1A proteins (though only those of KIF17 reach significance). Silencing NRF-1 with shRNA prevents such an increase for KIF17, but not for KIF1A. (D) Overexpressing NRF-1 in primary neurons upregulates transcript levels of *Grin1* and *Kif17*, but not *Kif1a*. It also prevents *Grin1* and *Kif17* mRNAs from being downregulated by TTX, but not that of *Kif1a*. (E) TTX significantly reduces the levels of KIF17 and KIF1A proteins, and overexpressing NRF-1 rescues the former but not the latter from the effect of TTX.

All * comparisons are with controls. *$P < 0.05$, **$P < 0.01$. All # comparisons are with KCl (B), with empty vector + TTX (D), or with TTX (E). $^{\#}P < 0.05$. All X comparisons are with NRF-1 overexpression. $^{X}P < 0.05$.

Modified from Dhar et al. 2009c.

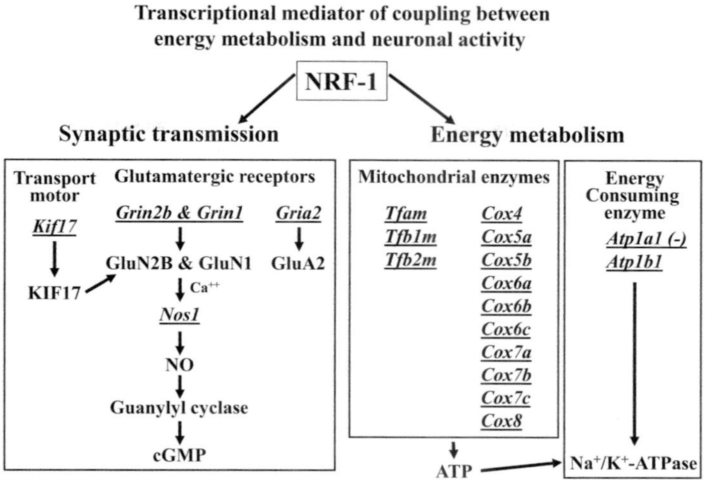

FIGURE 12.5 A schematic diagram of NRF-1's role as a transcriptional mediator of the tight coupling between energy metabolism and neuronal activity. On the synaptic transmission side, NRF-1 regulates the expression of *Grin1* and *Grin2b* of NMDA and *Gria2* of AMPA glutamatergic receptor subunit genes, the downstream *Nos1* gene, the *Kif17* gene (whose product transports GluN2B in neurons), and possibly the gene for guanylyl cyclase. On the energy metabolism side, NRF-1 regulates the expression of all ten nuclear-encoded *COX* subunit genes and mitochondrial transcription factor genes (*Tfam*, *Tfb1m*, and *Tfb2m*) that direct the transcription of the three mitochondrial-encoded *COX* subunit genes. NRF-1 also negatively regulates *Atp1a1* but positively controls *Atp1b1* gene of Na^+/K^+-ATPase, the major energy-consuming enzyme in neurons. All target genes of NRF-1 are underlined, and all are positively regulated except for *Atp1a1*, which is indicated by a (-) sign.

13 NRF-2 Transcriptionally Coregulates Genes for Glutamatergic Neurotransmitter Receptors, COX, and BDNF

ABSTRACT

NRF-2 regulates the same glutamatergic receptor subunit genes as NRF-1, that is, Grin 1 and Grin2b of NMDA receptors and Gria2 of AMPA receptors. The independent yet parallel mechanisms of these two transcription factors underscore the critical nature of their target genes in neurons. NRF-2 also regulates the expression of the only coding exon, exon IX, of the gene for brain-derived neurotrophic factor (BDNF), a crucial player in both developing and mature neurons. Thus, NRF-2 not only mediates the transcriptional coupling between energy metabolism (COX) and neuronal activity (glutamatergic receptors), but also the expression of BDNF, which functions at the crossroad of mitochondrial bioenergetics and glutamatergic neural transmission.

If NRF-1 coregulates *COX* and key subunits of glutamatergic receptors, what is the possibility that NRF-2 does the same? If so, does NRF-2 regulate identical or different subunit genes? Three mechanisms are possible for NRF-2's function with NRF-1: complementary (each regulates a different set of genes), concurrent and parallel (both regulate the same genes), or a combination of complementary and concurrent/parallel. To address these issues, multi-pronged approaches were utilized to probe NRF-2's role in the coupling process of neuronal activity and energy metabolism.

13.1 NRF-2 REGULATES THE SAME NMDA RECEPTOR SUBUNIT GENES AS NRF-1

When the proximal promoters of all murine NMDA receptor subunit genes (*Grin1*, *Grin2a-d*, and *Grin3a-b*) are analyzed in silico for putative NRF-2 binding sites, "*GGAA*" or its complement "*TTCC*," in tandem repeat (up to 24 bp separation), all promoters except for *Grin3b* possess this motif (Priya et al. 2013a).

However, in vitro EMSA studies with N2a nuclear extract indicate that only *Grin1, Grin2b*, and *Grin3a* form DNA-protein shift bands and DNA-protein-NRF-2 antibodies supershift complexes, whereas *Grin2a*, *Grin2c*, *Grin2d*, and *Grin3b* do not. Moreover, in vivo ChIP assays with chromatin from either N2a cells or primary neurons reveal that *Grin1* and *Grin2b* promoters alone are immunoprecipitated by NRF-2 (Priya et al. 2013a).

DOI: 10.1201/9781003507765-13

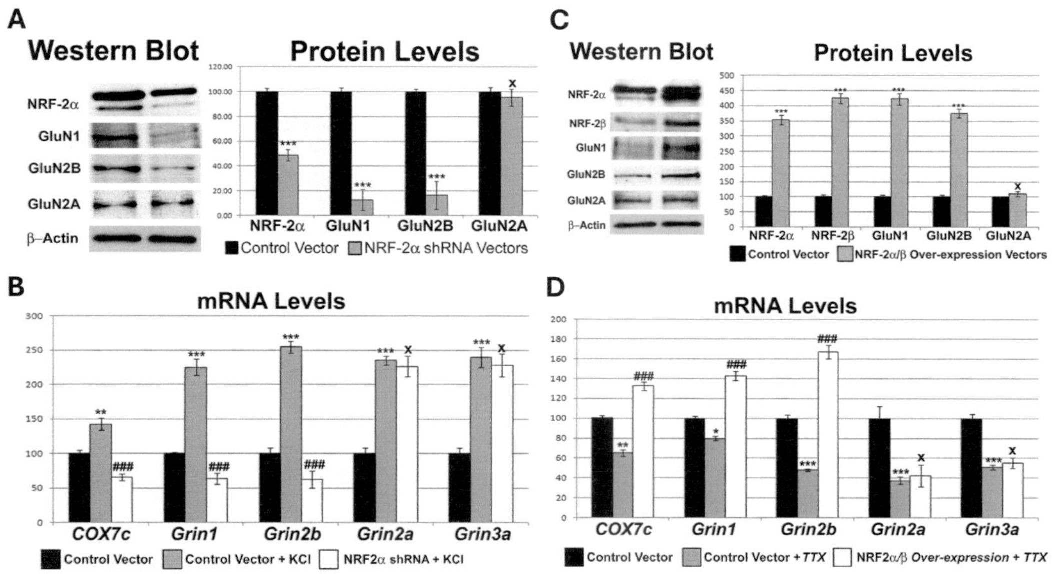

FIGURE 13.1 NRF-2 regulates the same NMDA receptor subunit genes as NRF-1. (A) Silencing NRF-2α with shRNA in N2a cells induces a significant reduction in protein levels of NRF-2α itself, GluN1, and GluN2B, but not GluN2A. (B) Depolarizing stimulation with 20 mM KCl for 5 h upregulates transcript levels of *COX7c*, *Grin1*, *Grin2b*, *Grin2a*, and *Grin3a* as compared to controls. However, such an increase is prevented by the presence of *NRF-2α* shRNA for the first three mRNAs, but not for *Grin2a* and *Grin3a*. (C) Overexpressing both NRF-2α and NRF-2β in N2a cells significantly upregulates NRF-2α, NRF-2β, GluN1, and GluN2B proteins, but not GluN2A. (D) Impulse blockade with 0.4 μM TTX for 3 days significantly reduces transcript levels of *COX7c*, *Grin1*, *Grin2b*, *Grin2a*, and *Grin3a*. However, NRF-2α/β overexpression rescues the first three transcripts from being downregulated by TTX, but not those of *Grin2a* and *Grin3a*.

All * comparisons are with respective controls. *P < 0.05, **P < 0.01, ***P < 0.001. ###P < 0.001 and X = non-significant, both when compared to controls in C, to KCl in B, and to TTX in D.

Modified from Priya et al. 2013a.

Site-directed mutagenesis of NRF-2 binding sites reduces promoter activity of *Grin1* by 26% and *Grin2b* by 41% ($P < 0.001$ for both) (Priya et al. 2013a). It also prevents the upregulation of *Grin1* and *Grin2b* promoter activity by KCl (20 mM for 5 h), indicating that NRF-2 is necessary for the activity-dependent increase. Silencing NRF-2α with shRNA significantly reduces GluN1 and GluN2B proteins by 87% and 83%, respectively ($P < 0.001$ for both) (Figure 13.1A) and *Grin1* and *Grin2b* transcripts by 38.5% and 45.0%, respectively ($P < 0.01 - 0.001$). It also prevents the upregulation of both transcripts by KCl (Figure 13.1B). On the other hand, overexpressing both NRF-2α and NRF-2β (enabling full transcriptional activity) in N2a cells increases *Grin1* and *Grin2b* proteins by 422% and 376%, respectively (Figure 13.1C) and mRNAs by 158.6% and 221.7%, respectively ($P < 0.001$ for all). It also rescues *Grin1* and *Grin2b* transcripts from being downregulated by TTX (0.4 μM for 3 days) (Figure 13.1D). Neither silencing nor overexpressing NRF-2 affects the expression of the other NMDA receptor subunit genes (*Grin2a*, *2c*, *2d*, *3a*, and *3b*). The NRF-2 binding sites are conserved between rats and mice for *Grin1* and among rats, mice, and humans for *Grin2b* (Priya et al. 2013a).

Hence, NRF-2 positively regulates the same NMDA receptor subunit genes as NRF-1 (see Chapter 12, Section 12.1) in a concurrent and parallel manner. Both also regulate all 13 *COX* subunit genes in neurons (see Chapters 8 and 9). As silencing either factor does not affect the other (Ongwijitwat et al. 2006; Dhar et al. 2008), NRF-1 and NRF-2 each mediates the coupling of neuronal activity and energy metabolism independently of each other. The advantage of this seemingly redundant mechanism is likely to ensure safety and precision, such that genes critical for neuronal functioning are dually regulated. When one factor is disabled or malfunctions, the other is there to sustain the vital task of supplying sufficient energy to at least partially support neuronal activity. The close matching of NRF-2 and COX distributions in the primate visual cortex (discussed in Chapter 9) further indicates that NRF-2's regulation of COX is even more precise than that of NRF-1.

13.2 NRF-2 REGULATES THE SAME AMPA RECEPTOR SUBUNIT GENES AS NRF-1

Does NRF-2 also regulate AMPA receptor genes? In silico analysis of proximal promoters of all four subunit genes (*Gria1*, *Gria2*, *Gria3*, and *Gria4*) suggests putative tandem binding motifs (separated by up to 24 bp) on all four *Gria* genes (Priya et al. 2014b). However, EMSA yields shift and supershift bands only for *Gria1* and *Gria2*, and the binding of NRF-2 to *Gria2* is stronger than that to *Gria1*. ChIP assays with chromatin from either N2a cells or primary neurons reveal that only the *Gria2* promoter is immunoprecipitated by NRF-2 antibodies, but not the other three (*Gria1*, *Gria3*, and *Gria4*).

Mutating the NRF-2 binding site significantly reduces the promoter activity of *Gria2* and prevents its upregulation by KCl. Transfecting N2a cells with two shRNA plasmid vectors targeting *NRF-2α* mRNA reduces GluA2 protein by 25% (Figure 13.2A) and *Gria2* mRNA by 28% ($P < 0.05$ and <0.01, respectively) but has no effect on *Gria1*, *Gria3*, and *Gria4* transcripts. Overexpressing NRF-2α and NRF-2β in N2 cells increases GluA2 protein by 175% (Figure 13.2C) and *Gria2* message by 163% ($P < 0.001$ for both) but fails to do so for the other three AMPA receptor subunits. Silencing or overexpressing NRF-2 has a stronger effect on primary rat cortical neurons, as *Gria2* message is reduced by 48% ($P < 0.05$) and increased by 174% ($P < 0.001$), respectively (Figure 13.2B and D). Suppressing NRF-2 in either N2a cells or primary neurons prevents the upregulation of their *Gria2* message by KCl, and overexpressing NRF-2 rescues these transcripts from being downregulated by TTX (Figure 13.2B and D). The NRF-2 binding site on the *Gria2* promoter is highly conserved among mice, rats, and humans (Priya et al. 2014b).

Again, NRF-2 regulates the same AMPA receptor subunit gene, *Gria2*, as NRF-1 (see Chapter 12, Section 12.2) in a concurrent and parallel fashion. As discussed above, this mitigates the possible breakdown of neuronal communication when the malfunctioning of one factor is partially rescued by the other. No doubt, additional factors are also involved (see Chapter 14) in the vital task of coupling neuronal activity and energy metabolism.

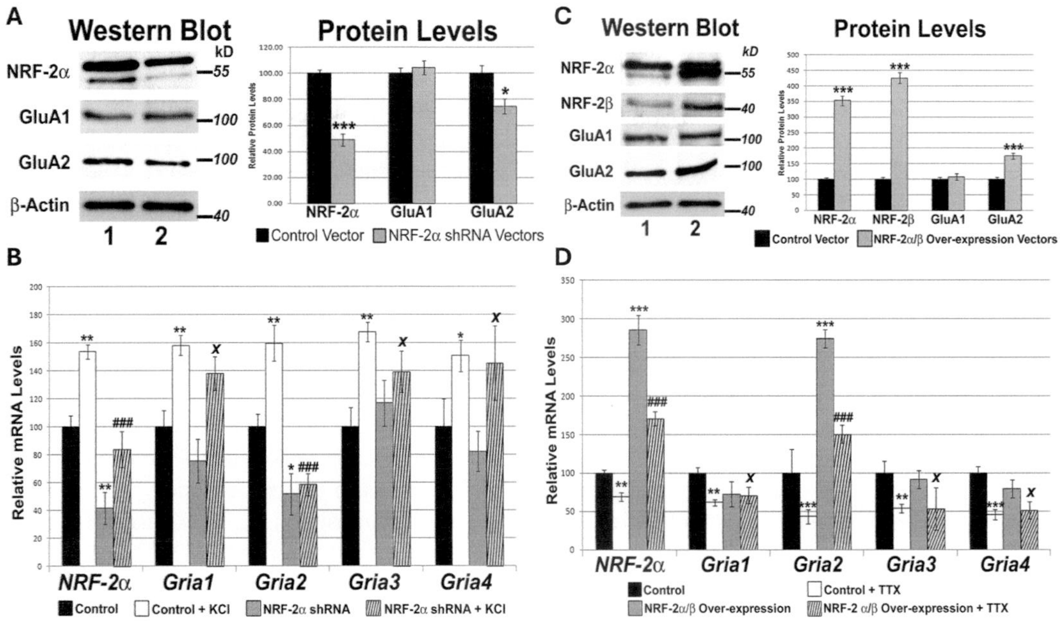

FIGURE 13.2 NRF-2 regulates the same AMPA receptor subunit gene as NRF-1. (A) Silencing NRF-2α with shRNA in N2a cells significantly reduces levels of NRF-2α and GluA2 proteins, but not of GluA1. (B) KCl stimulation for 5 h in primary neurons significantly upregulates *NRF-2α*, *Gria1*, *Gria2*, *Gria3*, and *Gria4* transcripts. *NRF-2α* shRNA significantly reduces *NRF-2α* and *Gria2* mRNAs, but not those of others. Such silencing also prevents *NRF-2α* and *Gria2* transcripts from being upregulated by KCl, but it has no effect on those of *Gria1*, *Gria3*, and *Gria4*. (C) Overexpressing both NRF-2α and NRF-2β in N2a cells significantly increases NRF-2α, NRF-2β, and GluA2 proteins, but not those of GluA1. (D) TTX blockade for 3 days in primary neurons significantly reduces *NRF-2α*, *Gria1*, *Gria2*, *Gria3*, and *Gria4* transcripts. Overexpressing NRF-2α/β rescues *NRF-2α* and *Gria2* message from being downregulated by TTX, but it has no effect on those of *Gria1*, *Gria3*, and *Gria4*.

All * comparisons are with respective controls. *$P < 0.05$, **$P < 0.01$, ***$P < 0.001$. ###$P < 0.001$ and X = non-significant, both when compared to KCl in B and to TTX in D.

Modified from Priya et al. 2014b.

13.3 NRF-2 REGULATES BRAIN-DERIVED NEUROTROPHIC FACTOR (BDNF) GENE

13.3.1 Brain-Derived Neurotrophic Factor (BDNF)

Brain-derived neurotrophic factor (BDNF) is the second neurotrophin discovered (Barde et al. 1982) after nerve growth factor, and it plays an important role in neurogenesis, differentiation, survival, synaptic transmission, synaptic plasticity, and neural degeneration (McAllister et al. 1995; Patterson et al. 1996; Lessmann 1998; Poo 2001). It is widely expressed in both the developing and adult mammalian CNS (Conner et al. 1997; Katoh-Semba et al. 1997). BDNF is of special interest because it functions at the interface of metabolism and synaptic plasticity. On the one hand, it stimulates energy metabolism, enhances the respiratory control index, and increases glucose utilization in neurons (Burkhalter et al. 2003; Markham et al. 2004). Concurrently, it augments glutamate release and facilitates the delivery of NMDA and AMPA receptors to their synaptic sites (Carvalho et al. 2008; Martin and Finsterwald 2011). This significant dual role raises an intriguing question: Is the *Bdnf* gene regulated by a transcription factor that coregulates energy metabolism and neuronal activity?

13.3.2 The *Bdnf* Gene

The *Bdnf* gene is unique in that in rodents it contains nine exons, eight of which are non-coding (exons I to VIII) and only exon IX is coding, yielding the BDNF protein (Aid et al. 2007). Each of the non-coding exons has its own promoter and may be spliced to the coding exon, resulting in tissue- and age-specific expressions. The human *BDNF* gene, on the other hand, was originally reported to have seven non-coding and one coding exons (Liu et al. 2005) but later discovered to have 11 exons, including nine functional promoters and a single exon (also named exon IX) coding for the mature BDNF. The splicing of various non-coding exons to the coding one yields various transcripts expressed in a tissue- and region-specific manner (Pruunsild et al. 2007). BDNF is synthesized as a precursor protein (pro-BDNF) and secreted as a mixture of pro-BDNF and mature BDNF (Aid et al., 2007).

Despite the functional significance of BDNF and the detailed structural analysis of its gene, relatively little is known about its transcriptional regulation. Much attention has focused on the non-coding exon IV (previously known as exon III), whose promoter has a binding site for cAMP response element binding protein (CREB) and another for the calcium-responsive factor (CaRF) (Tao et al. 2002; Zheng et al. 2012). Very little was known prior to 2016 about the *cis* element in the promoter region of the coding exon IX that may regulate its expression.

13.3.3 Transcriptional Regulation of *Bdnf* Coding Exon IX by NRF-2

As described above, the rat *Bdnf* gene has eight non-coding exons and a single coding exon IX, with IXA providing the precursor portion of the pro-BDNF (Aid et al. 2007). To determine the relative expression of each of the nine exons in the rat visual cortex, total RNA was extracted, cDNA synthesized, and mRNA levels determined by RT-qPCR (Nair and Wong-Riley 2016). Clearly, the expression of exon IX is at least 10-fold higher than any of the other exons, including exon IXA. Exons II, V, and VII are almost non-detectable (Figure 13.3A).

In silico analysis of rat and mouse *Bdnf* exon IX 1 kb 5′ upstream and 1 kb 3′ downstream of the TSS uncovers four putative tandem NRF-2 binding sites upstream of the coding portion of exon IX with high homology between rats and mice, and partial homology with humans (Nair and Wong-Riley 2016) (Figure 13.3B). EMSA demonstrates that only the two sites closer to the translation start codon (ATG) and within exon IXA yield shift and supershift bands for NRF-2. In vivo ChIP with chromatin from rat visual cortex confirms that both site I and site II are immunoprecipitated by NRF-2 antibodies. Mutating these two sites reduces their promoter activity to 62.4% and 52% of controls, respectively ($P < 0.05$ for both) (Nair and Wong-Riley 2016). Silencing NRF-2α with shRNA in primary neurons decreases BDNF protein to 53% of controls and its mRNA for exon IX to 43% of controls ($P < 0.001$ for both) (Figure 13.3C and D). Overexpressing NRF-2α/β in primary neurons increases levels of BDNF protein to 167% of controls and its message to 248.5% of controls ($P < 0.001$ for both) (Figure 13.3E and F). The expression of *Bdnf* exon IX is activity-dependent, as its transcripts are increased by KCl stimulation to 142% of controls in cultured visual cortical neurons but are reduced by TTX to 42% of controls ($P < 0.001$ for both). However, in the presence of *NRF-2* shRNA, KCl can no longer upregulate BDNF, and overexpressing NRF-2α/β rescues BDNF transcripts from being downregulated by TTX (Figure 13.3D and F).

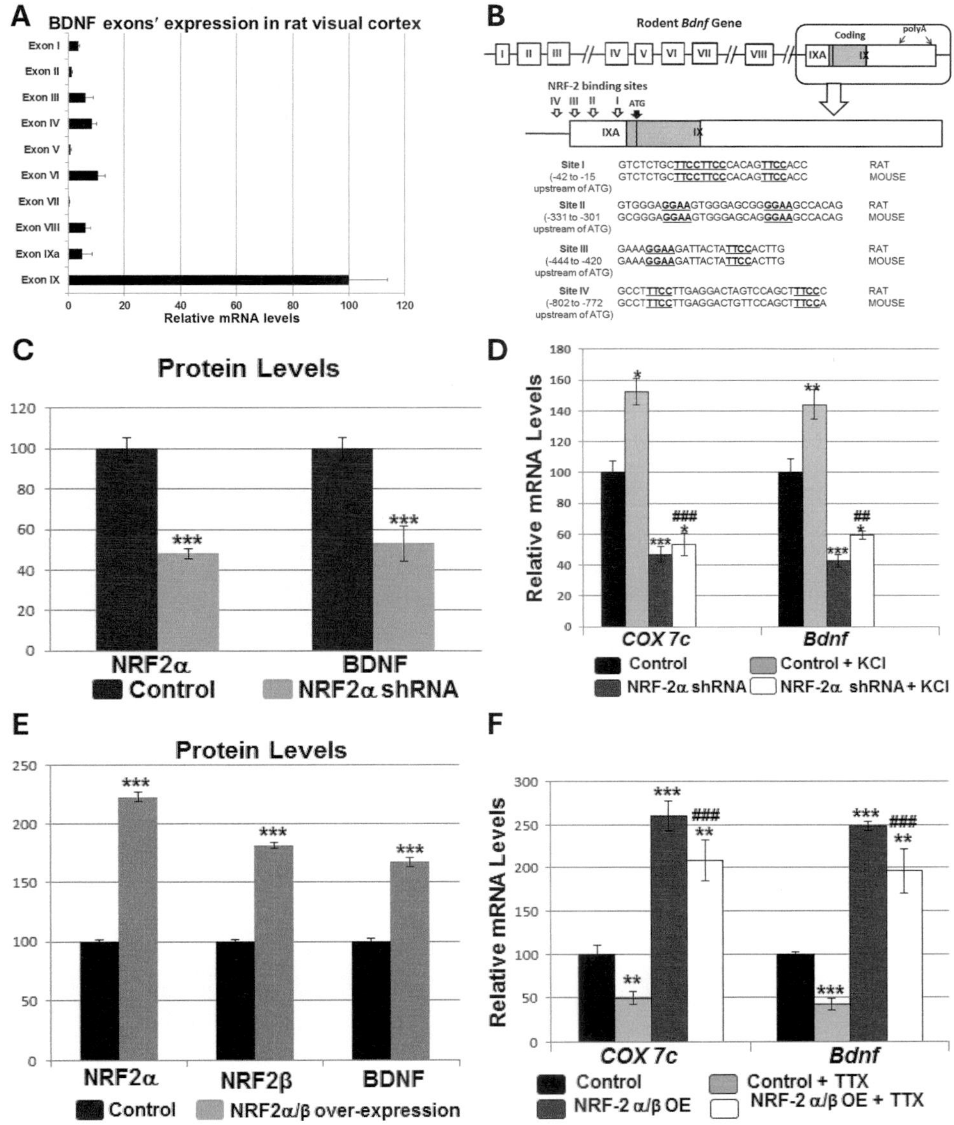

FIGURE 13.3 NRF-2 regulates BDNF gene. (A) Expression of each of the nine BDNF exons in the rat visual cortex. In comparison to the only coding exon, exon IX (expressed as 100%), all of the other exons, including exon IXA, are expressed at a 10-fold or lower level. (B) A schematic diagram of the rat *Bdnf* gene, with the eight non-coding exons and the single, coding exon IX, including exon IXA (not drawn to scale). The locations of the four putative tandem repeat NRF-2 binding sites (sites I-IV) are indicated for both rat and mouse. ATG indicates the translation start site, which is 21 nucleotides downstream of the transcription start site of exon IX. Sites I and II proved to be functional, but not sites III and IV. (C) Transfecting primary neurons with *NRF-2α* shRNA induces a significant reduction in levels of NRF-2α and BDNF proteins. (D) *COX7c* and *Bdnf* transcripts in primary neurons are significantly upregulated by KCl and downregulated by *NRF-2α* shRNA. Silencing NRF-2α also prevents these transcripts from being augmented by KCl. (E) Levels of NRF-2α, NRF-2β, and BDNF proteins are significantly upregulated by NRF-2α/β overexpression in primary neurons. (F) Overexpressing NRF-2α/β significantly upregulates *COX7c* and *Bdnf* transcripts in primary neurons and rescues them from being downregulated by TTX.

All * comparisons are with respective controls. *$P < 0.05$, **$P < 0.01$, ***$P < 0.001$. All # comparisons are with KCl (D) or TTX (F). ##$P < 0.01$, ###$P < 0.001$.

Modified from Nair and Wong-Riley 2016.

Clearly, NRF-2 not only regulates all 13 *COX* subunit genes (see Chapter 9) but also critical NMDA and AMPA receptor subunit genes, thereby mediating the coupling between energy metabolism and neuronal activity at the transcriptional level. It also directly controls the expression of the single coding exon of BDNF, whose gene product plays a pivotal role in both mitochondrial bioenergetics and glutamatergic neural transmission (Figure 13.4).

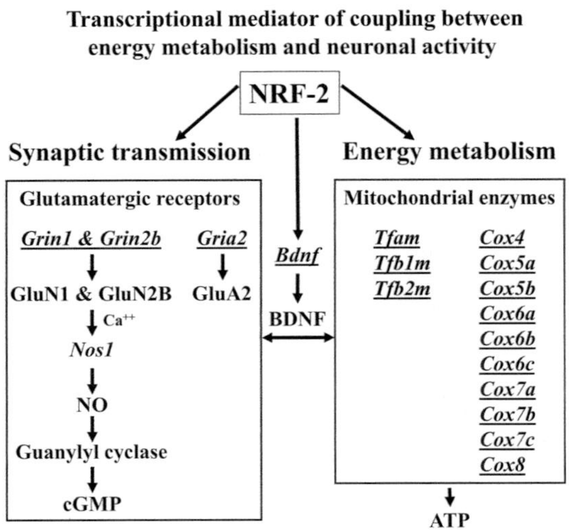

FIGURE 13.4 A schematic diagram of NRF-2's role as a transcriptional mediator of the tight coupling between energy metabolism and neuronal activity. NRF-2 regulates the same glutamatergic receptor subunit genes (*Grin1*, *Grin2b*, and *Gria2*) for synaptic transmission and the same *Tfam*, *Tfb1m*, *Tfb2m*, and the 10 nuclear-encoded *COX* genes for energy generation as NRF-1. In addition, NRF-2 also regulates the only coding exon of *Bdnf* gene, whose gene product is involved in both energy metabolism and synaptic transmission.

14 Sp4 Mediates Tight Coupling of Neuronal Activity and Energy Metabolism by Transcriptionally Coregulating Glutamatergic, GABAergic, Na^+/K^+-ATPase, and COX Genes in Neurons

ABSTRACT

Within the Sp family of transcription factors, it is Sp4, and not Sp1 or Sp3, that regulates the expression of glutamatergic receptor subunit genes, specifically Grin1 and Grin2b of NMDA receptors and Gria2 of AMPA receptors, in a concurrent and parallel fashion with NRF-1 and NRF-2. In addition, Sp4 complements NRF-1 and NRF-2 in controlling the expression of the Grin2a gene. Moreover, Sp4 specifies the transcription of the inhibitory GABAA receptor α1 and α2 subunit genes (Gabra1 and Gabra2), thereby contributing to the excitation-inhibition balance in neurons. Finally, chromosome conformation capture reveals that glutamatergic subunit genes (Grin1, Grin2b, Gria2, and Nos1) interact with COX subunit genes in the same transcription factory. This strengthens the tight coupling between neuronal activity and energy metabolism at the molecular level.

As mentioned in Chapter 10, Sp1 is among the first and the most well-studied transcription factors. However, in neurons, it is Sp4 that plays the dominant role. Besides regulating all 13 subunits of *COX* from the two genomes, does Sp4 also control the expression of neurotransmitter receptor genes, thereby coupling neuronal activity and energy metabolism at the transcriptional level?

14.1 SP4 COREGULATES NMDA GLUTAMATE RECEPTORS AND *COX*

In silico analysis indicates that, among the promoters of the murine NMDA receptor subunit genes, *Grin1, Grin2a*, and *Grin2c* possess the GC box motif (GGGCGG), and *Grin2b* has the GC box with a tandem CT box (CCCTCC) (Priya et al. 2013b).

The prevalence of Sp1, Sp3, and Sp4 was compared in nuclear extracts of murine visual cortical tissue, N2a cells, and HeLa cells, using TATA box binding protein as the loading control. Sp4 is most abundant in cortical tissue, less so in N2a cells, and least in HeLa cells. The opposite is true for Sp3 and Sp1, which are most prevalent in HeLa, moderate in N2a, and least in cortical tissue. Both Sp1 and its phosphorylated form are present in all three cell types, but only HeLa and N2a cells contain the smaller isoforms of Sp1 (Priya et al. 2013b).

In vitro EMSA shows specific Sp4 shift and supershift bands for *Grin1, Grin2a*, and *Grin2b*, but not *Grin2c*. On the other hand, Sp1 and Sp3 bands are absent for *Grin1* and *Grin2a* and meager for *Grin2b*. In vivo ChIP assays with chromatin from the visual cortex confirm the EMSA results in that

DOI: 10.1201/9781003507765-14

Grin1, Grin2a, and *Grin2b* (but not *Grin2c*) promoters are immunoprecipitated by Sp4 antibodies. On the other hand, neither Sp1 nor Sp3 antibodies can immunoprecipitate any of the *Grin* promoters (Priya et al. 2013b).

Mutating the Sp4 binding sites on *Grin1, Grin2a*, and *Grin2b* promoters reduces their activity by 34%, 38%, and 31%, respectively ($P < 0.001$ for all). All three promoters respond to KCl stimulation (20 mM for 5 h) by increasing their activity by 44%, 50%, and 43%, respectively ($P < 0.001$ for all). However, such activity-dependent increases are abolished by mutating Sp4 binding sites. Likewise, transfecting N2a cells or primary neurons with plasmid vectors expressing *Sp4* shRNA reduces the expression of *Grin1, Grin2a*, and *Grin2b* by 34 -78% ($P < 0.01 - 0.001$) and their proteins by 33 -55% ($P < 0.05 - 0.001$) (Figure 14.1A and B). Silencing *Sp1* or *Sp3* has no effect on the

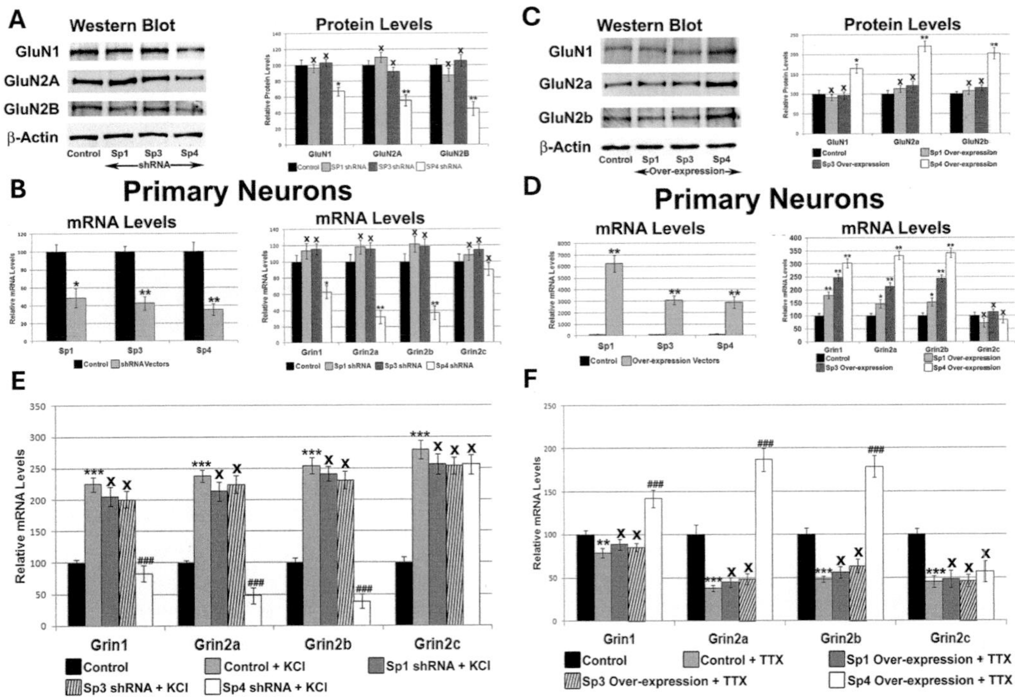

FIGURE 14.1 Sp4 regulates NMDA receptor subunit genes. (A) Western blots and quantitation of GluN1, GluN2A, and GluN2B proteins downregulated by *Sp4* shRNA, but not by silencing Sp1 or Sp3 in N2a cells. (B) The same is true for the detrimental effect of *Sp4* shRNA on transcript levels of *Grin1, Grin2a*, and *Grin2b*, but not *Grin2c*, in primary neurons. However, *Sp1* and *Sp3* shRNAs yield no effect. (C) Overexpressing Sp4, but not Sp1 or Sp3, significantly increases GluN1, GluN2A, and GluN2B proteins in N2a cells. (D) Message levels of *Grin1, Grin2a*, and *Grin2b*, but not *Grin2c*, are significantly upregulated by Sp4 overexpression in primary neurons. Surprisingly, the first three transcripts are also increased by Sp1 and Sp3 overexpression, most likely due to competitive binding to the same Sp site normally occupied by Sp4. (E) KCl stimulation for 5 h upregulates *Grin1, Grin2a, Grin2b*, and *Grin2c* transcripts in N2a cells. However, *Sp4* shRNA, but not *Sp1* or *Sp3* shRNA, prevents such a rise by KCl in the first three sets of transcripts, though it has no effect on *Grin2c*. (F) TTX induces a significant downregulation of *Grin1, Grin2a, Grin2b*, and *Grin2c* mRNAs in N2a cells. However, Sp4 overexpression, but not that of Sp1 and Sp3, rescues the first three transcripts from being suppressed by TTX. *Grin2c* shows no change. $N = 3$ for each data point in A–D. $N = 6$ for each data point in E and F.

All * comparisons are with respective controls. *$P < 0.05$, **$P < 0.01$, ***$P < 0.001$. All # comparisons are with KCl (E) or TTX (F). ###$P < 0.001$. All X = non-significant as compared to controls (A–D), to KCl (E), or to TTX (F).

Modified from Priya et al. 2013b.

expression of any of the four *Grin* genes (*Grin1* and *Grin2a-c*) in N2a cells or primary neurons (Priya et al. 2013b).

Overexpressing Sp4 in N2a cells significantly elevates transcript levels of *Grin1, Grin2a*, and *Grin2b* by 53%, 152%, and 224%, respectively, and protein levels by 65%, 121%, and 102%, respectively ($P < 0.001$ for all) (Figure 14.1C). *Grin2c* is unaffected. Overexpressing Sp1 or Sp3 does not alter the transcript or protein amounts of the *Grin* genes. Overexpressing Sp4 in primary neurons strikingly increases *Grin1, Grin2a*, and *Grin2b* mRNAs by 203%, 231%, and 241%, respectively ($P < 0.001$ for all), but the expression of *Grin2c* is unchanged (Figure 14.1D). Surprisingly, overexpressing Sp1 and Sp3 in primary neurons also upregulates *Grin1, Grin2a*, and *Grin2b* (but not *Grin2c*) transcripts, though less pronounced than with Sp4 overexpression (Priya et al. 2013b). Artificially multiplying Sp1 and Sp3 may facilitate their binding to the same motif on the *cis* element as Sp4 in primary neurons, even though they normally don't do so.

Grin1, Grin2a, and *Grin2b* transcripts respond well to changes in neuronal activity, with KCl-induced increases of 125%, 138%, and 154%, respectively, and TTX-triggered (0.4 μM for 3 days) decreases of 21%, 63%, and 52%, respectively ($P < 0.001$ for all). *Sp4* shRNA prevents the upregulation of these transcripts by KCl (Figure 14.1E), and overexpressing Sp4 blocks their downregulation by TTX (Figure 14.1F). *Grin2c* also responds well to both KCl and TTX, but its transcripts are unaffected by either shRNA or overexpression of Sp4. Silencing or overexpressing either Sp1 or Sp3 has no effect on the activity-dependent expression of the four *Grin* subunit genes in N2a cells (Priya et al. 2013b).

These experiments clearly indicate that Sp4, not Sp1 or Sp3, is instrumental in the activity-dependent responses of *Grin1, Grin2a*, and *Grin2b* genes in neurons. The fact that Sp4 regulates not only the same NMDA receptor subunit genes (*Grin1* and *Grin2b*) as NRF-1 and NRF-2, but also the *Grin2a* subunit gene, not controlled by the other two, signifies that it operates in a concurrent/parallel as well as a complementary fashion with NRF-1 and NRF-2. It also contributes to an optimal proportion of GluN2A and GluN2B expressed in different neurons. As all three transcription factors control the expression of the 13 subunits of *COX*, together they execute the tight coupling between energy metabolism and NMDA receptor-mediated neuronal activity.

14.2 SP4 COREGULATES AMPA GLUTAMATE RECEPTORS AND *COX*

What about AMPA receptor subunit genes? Does Sp4 participate in their regulation? Myers et al. (1998) reported a Sp1 binding site for *Gria2* using promoter deletion analysis and EMSA/supershift assays. A more in-depth study was subsequently conducted by Priya et al. (2014a).

Computer-assisted search for putative Sp binding sites on the proximal promoters of *Gria1–4* indicates that *Gria1* has GC-rich regions resembling the Sp binding motif, *Gria2* has the atypical, GT box, whereas *Gria3* and *Gria4* have the typical GC box motif. EMSA studies with murine visual cortical nuclear extract reveal shift and supershift bands for *Gria2* with Sp4 antibodies, but not with Sp1 or Sp3 antibodies. In contrast, *Gria1, Gria3*, and *Gria4* do not yield shift or supershift bands with Sp4 antibodies. In vivo ChIP assays with chromatin from murine visual cortex confirm that the *Gria2* promoter is immunoprecipitated by Sp4 antibodies, but not promoters for *Gria1, Gria3*, and *Gria4* with putative Sp binding motifs (Priya et al. 2014a).

Site-directed mutagenesis of the Sp4 binding site on the *Gria2* promoter severely reduces the promoter activity by 56% ($P < 0.001$) and prevents its upregulation by KCl, indicating that the binding of Sp4 is necessary for activity-dependent regulation. Knocking down Sp4 expression with shRNA in N2a cells decreases *Gria2* mRNAs by 62% and proteins by 41% ($P < 0.001$ and $P < 0.05$, respectively) (Figure 14.2A), but *Gria1, Gria3*, and *Gria4* are unaffected. Silencing Sp1 or Sp3 is also ineffective in perturbing any of the four *Gria* subunit genes. Primary neurons yield comparable results (Figure 14.2B). Overexpressing Sp4 in N2a cells upregulates *Gria2* mRNAs by 135% and GluA2 proteins by 47% ($P < 0.001$ for both) (Figure 14.2C). The same is true in primary neurons (Figure 14.2D). Overexpressing Sp4 does not alter either transcript or protein levels of

Gria1, Gria3, or *Gria4* in both cell types. By the same token, overexpressing Sp1 or Sp3 has no effect on mRNA or protein concentrations of *Gria1–4* in N2a cells. However, they do upregulate those of *Gria2* in primary neurons, though to a lesser extent than with Sp4 (Priya et al. 2014a). Again, artificially increasing Sp1 and Sp3 levels may facilitate their binding to Sp4 sites that they normally do not occupy.

KCl-induced depolarization upregulates *Gria2* transcripts by 110% ($P < 0.001$), but knocking down Sp4 blocks the activity-dependent increase (Figure 14.2E). TTX downregulates *Gria2* transcripts by 44% ($P < 0.01$), but overexpressing Sp4 rescues the activity-dependent decrease (Figure 14.2F). On the other hand, neither Sp1 nor Sp3 exerts any influence on activity-dependent

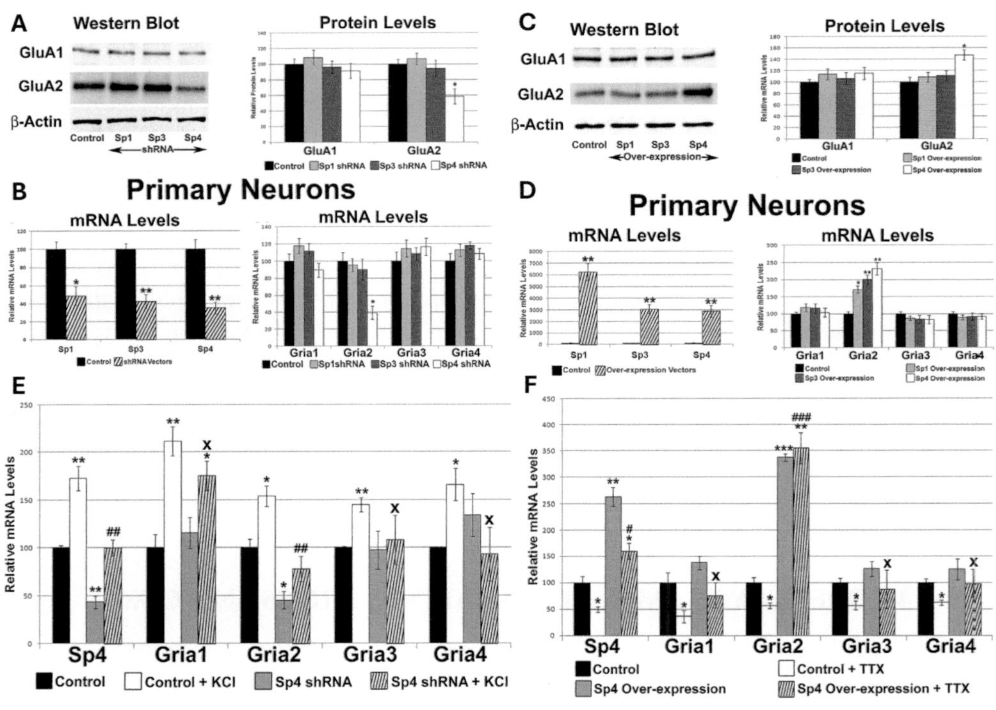

FIGURE 14.2 Sp4 regulates AMPA receptor subunit genes. (A) Western blots and quantitation of GluA1 and GluA2 protein levels. The latter is downregulated by *Sp4* shRNA, but not by silencing Sp1 or Sp3 in N2a cells. (B) The same is true for the detrimental effect of *Sp4* shRNA on transcripts of *Gria2*, but not of *Gria1, Gria3*, and *Gria4* in primary neurons. However, *Sp1* and *Sp3* shRNAs yield no effect. (C) Overexpressing Sp4, but not Sp1 or Sp3, significantly increases proteins of GluA2, but not of GluA1, in N2a cells. (D) Transcripts of *Gria2*, but not of *Gria1, Gria3*, and *Gria4*, are significantly upregulated by Sp4 overexpression in primary neurons. Surprisingly, *Gria2* transcripts are also increased by Sp1 and Sp3 overexpression, though not to the same extent as Sp4. (E) KCl stimulation for 5 h upregulates transcripts of *Sp4* itself, *Gria1, Gria2, Gria3*, and *Gria4* in primary neurons. However, *Sp4* shRNA, which downregulates *Gria2* transcripts, also prevents them from being increased by KCl. Silencing Sp4 has no effect on the expressions of *Gria1, Gria3*, and *Gria4*. (F) TTX induces a significant downregulation of *Sp4, Gria1, Gria2, Gria3*, and *Gria4* in primary neurons. However, Sp4 overexpression rescues *Sp4* and *Gria2*, but not the other transcripts, from being suppressed by TTX. $N = 3$ for each data point.

All * comparisons are with respective controls. *$P < 0.05$, **$P < 0.01$, ***$P < 0.001$. All # comparisons are with KCl (E) or TTX (F). #$P < 0.05$, ##$P < 0.01$, ###$P < 0.001$. All X = non-significant as compared to KCl (E) or to TTX (F).

Modified from Priya et al. 2014a.

changes in *Gria2* expression, establishing Sp4 as the major driving force. All three Sp factors (Sp1, 3, and 4) are ineffective in altering activity-induced changes in the expressions of *Gria1, Gria3*, and *Gria4* (Priya et al. 2014a). The functional Sp4 binding site on the *Gria2* promoter is highly conserved among mice, rats, and humans.

Thus, Sp4 is the only transcription factor of the Sp family that mediates the tight coupling between energy metabolism and neuronal activity at the molecular level, and it does so not only in a concurrent and parallel manner with NRF-1 and NRF-2 (for *Grin1, Grin2b*, and *Gria2*) but also is complementary to them (for *Grin2a*).

14.3 SP4 REGULATES INHIBITORY GABAERGIC RECEPTORS IN NEURONS

What about inhibitory neural transmission? It is as much a part of neural activity as excitation. How are inhibitory neurotransmitter receptors regulated transcriptionally? Relatively little is known, but a few of the GABAergic receptor genes have been examined.

14.3.1 GABA and $GABA_A$ Receptors

Gamma-aminobutyric acid (GABA) is the major inhibitory neurotransmitter in the CNS (Krnjević 1970; Olsen et al. 1988). There are two major receptor types: the fast-acting ionotropic $GABA_A$ type and the slower-acting metabotropic $GABA_B$ type, which are G-protein-coupled receptors (Shaye et al. 2021). The ligand-gated type A receptors ($GABA_AR$) are prevalent in neurons, and their fast action in response to GABA binding opens the chloride channels to hyperpolarize the postsynaptic neurons. $GABA_AR$ are pentameric proteins composed of distinct polypeptide types (α, β, γ, δ, and others). Each type is represented by a family of genes with high amino acid sequence identities but diverse functions (Olsen and Tobin 1990; Möhler 2006). Thus far, 19 $GABA_AR$ subunit genes have been identified in humans (α1-α6, β1-β3, γ1-γ3, δ, ε, π, θ, and ρ1-ρ3), but the main adult $GABA_AR$ is made up of two α, two β, and one γ subunit (Ghit et al. 2021). Various subunits are more prevalent in a regional and laminar fashion in the CNS and in non-neural tissues (Huntsman et al. 1994; Ghit et al. 2021). Dysfunctions of $GABA_AR$ are implicated in several disorders, including anxiety, epilepsy, memory deficits, and schizophrenia (Möhler 2006).

Nair et al. (2016) tested the possibility that the same transcription factor, Sp4, which regulates some of the key excitatory neurotransmitter receptor subunit genes, also regulates three major $GABA_AR$ subunit genes, *Gabra1* ($GABA_AR\alpha1$), *Gabra2* ($GABA_AR\alpha2$), and *Gabra3* ($GABA_AR\alpha3$).

The expression of $GABA_AR\alpha1$ increases with age and plateaus in the adult and the aging brain (Laurie et al. 1992; Golshani et al. 1997; Yu et al. 2006). As $GABA_AR\alpha1$ gains dominance in the receptor complex with development, the kinetics of the $GABA_AR$ also accelerate threefold (Bosman et al. 2002; Murphy et al. 2005). The α1 subunit has a high affinity for binding GABA and benzodiazepine (Ghit et al. 2021), and it is the primary mediator of benzodiazepine's sedative, amnesic, and anticonvulsant actions (Rudolph et al. 1999). During the critical period of visual cortical development, the α1-containing $GABA_ARs$ exclusively drive ocular dominance plasticity (Fagiolini et al. 2004).

$GABA_ARs$ containing the α2 subunits have slower kinetics than those with α1 (Möhler 2006), and their distributions vary from those of α1 (Huntsman et al. 1994). Whereas α1 is found in most GABAergic synapses at all postsynaptic sites, $GABA_AR\alpha2$ is located only in certain synapses and often preferentially positioned on axon initial segments of pyramidal neurons in the cortex and hippocampus (Nusser et al. 1996). Whereas neurons that express $GABA_AR\alpha1$ mediate sedation, those expressing $GABA_AR\alpha2$ facilitate anxiolysis (Möhler 2006) and are involved in modulating neuronal firing (Fagiolini et al. 2004).

Unlike $GABA_ARα1$, the expression of $GABA_ARα3$ is highest during early postnatal life but decreases with age in both the cortex and the brain stem (Liu and Wong-Riley 2004; Yu et al. 2006). The pattern of distribution can be similar to but also different from $GABA_ARα1$ (Yu et al. 2006). $GABA_ARs$ with α3 subunits have lower affinity for GABA and slower kinetics than those with α1 (Verdoorn 1994). The developmental switch from α3 to α1 is the basis for faster kinetics and shorter decay time of the mature $GABA_AR$ (Okada et al. 2000). Neurons that express $GABA_ARα3$ are thought to process sensory motor function associated with the schizophrenic phenotype (Möhler 2006).

14.3.2 Sp4's Role in Regulating $GABA_Aα1$ and α2 Receptor Subunit Genes

In silico analysis indicates that the ubiquitously expressed *Gabra1*gene promoter has many typical Sp binding motifs (the GC box) and a TATA box upstream of the TSS. The *Gabra2* promoter also has multiple typical as well as atypical Sp binding motifs. The *Gabra3* promoter, on the other hand, shows only atypical GT and CT boxes that share homology between rats and mice, but not humans (Nair et al. 2016).

EMSA studies with visual cortical nuclear extract and *Gabra1, 2*, and *3* subunit probes containing Sp4 putative sites reveal DNA-protein shift bands and DNA-protein-Sp4 antibodies supershift bands for *Gabra1* and *Gabra2*, but not for *Gabra3*. Likewise, ChIP assays reveal *Gabra1* and *Gabra2* immunoprecipitation by Sp4 antibodies, but not *Gabra3* (Nair et al. 2016). Site-directed mutagenesis of the Sp4 binding sites reduces promoter activity of *Gabra1* by 53% and *Gabra2* by 61% ($P < 0.001$ for both). Knocking down Sp4 with shRNA in cultured primary neurons significantly decreases *Gabra1* and *Gabra2* mRNAs by 56% and 54%, respectively, and their proteins by 36.8% and 26.7%, respectively ($P < 0.01$ for all) (Figure 14.3A and B).

Overexpressing Sp4 in primary neurons significantly increases *Gabra1* and *Gabra2* transcripts to 210% and 293%, respectively, and their proteins to 182% and 144%, respectively ($P < 0.001$ for all) (Figure 14.3C and D). It also returns *Sp4, Gabra1*, and *Gabra2* messages downregulated by *Sp4* shRNA the day before to control levels in primary neurons. *Gabra3* message and protein, on the other hand, are unaffected by changes in Sp4 (Nair et al. 2016).

Both *Gabra1* and *Gabra2* subunits respond to changes in neuronal activity (Figure 14.3B and D). KCl depolarization upregulates *Gabra1* and *Gabra2* transcripts by 164% and 152% ($P < 0.05$ for both), whereas TTX impulse blockade downregulates them by 63% and 62%, respectively ($P < 0.001$ and $P < 0.05$, respectively). In the presence of *Sp4* shRNA, however, KCl no longer increases the two transcripts, and when Sp4 is overexpressed, TTX is unable to suppress them. Moreover, Sp4 overexpression rekindles KCl-induced upregulation of *Gabra1* and *Gabra2* transcripts previously inhibited by *Sp4* shRNA in primary neurons. Such treatments have no effect on the *Gabra3* message or protein. The functional Sp4 binding sites for *Gabra1* and *Gabra2* are conserved among mice, rats, and humans (Nair et al. 2016).

The above studies highlight the importance of Sp4 in transcriptionally regulating two major subunit genes of $GABA_AR$, *Gabra1* and *Gabra2*, but not *Gabra3*. Such regulation is activity-dependent. Whether Sp4 also controls the other $GABA_AR$ subunit genes (such as β and γ) is yet unknown. The fact that Sp4 also regulates many excitatory glutamatergic receptor subunit genes (see above) indicates that it is positioned to balance excitation with inhibition to achieve a more homeostatic state of neuronal functioning. Such co-regulation is made even more remarkable as Sp4 also mediates the tight coupling between neuronal activity and energy metabolism by controlling the expression of a key energy generator, COX, and a major energy consumer, Na^+/K^+-ATPase (see Chapter 10, Sections 10.2 and 10.3). Sp4 is justifiably a master transcriptional regulator that coordinates excitatory and

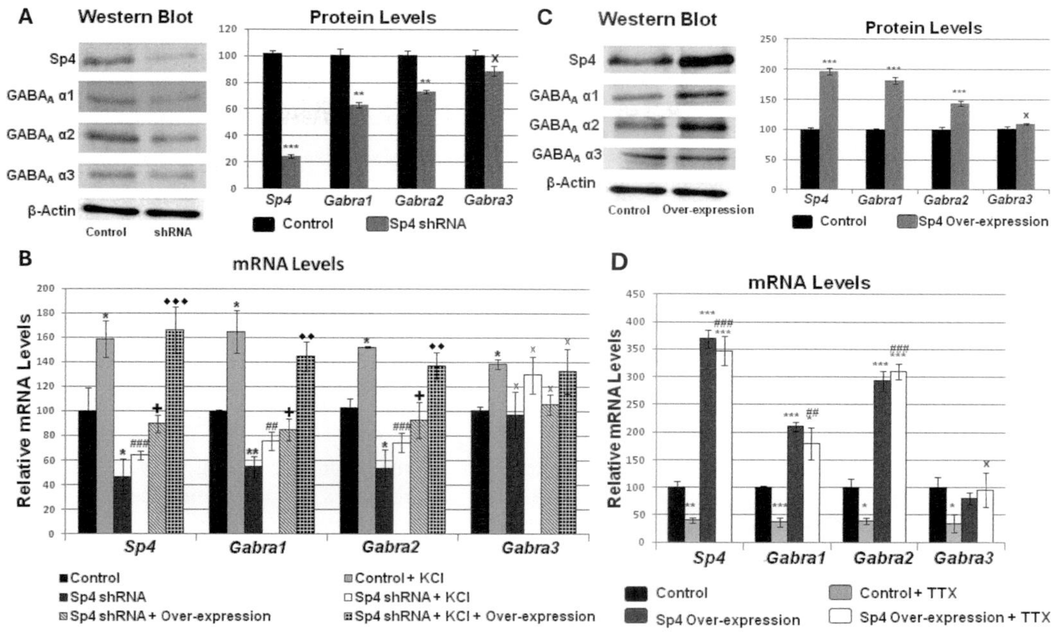

FIGURE 14.3 Sp4 regulates GABA$_A$ receptor subunit genes in primary neurons. (A) Western blots and quantitation indicate that silencing Sp4 with shRNA suppresses protein levels of Sp4 itself, GABA$_A$Rα1 (labeled Gabra1), and GABA$_A$Rα2 (Gabra2), but not GABA$_A$Rα3 (Gabra3). (B) *Sp4, Gabra1, Gabra2*, and *Gabra3* transcripts are all upregulated by KCl, but only the first three are significantly downregulated by *Sp4* shRNA, which also inhibits these three transcripts from being augmented by KCl. Transfecting primary neurons first with *Sp4* shRNA, then with Sp4 overexpression rescues the transcripts from being downregulated, whereas such treatments plus KCl significantly upregulate them. *Gabra3* is not affected by either silencing or overexpressing Sp4. (C) Overexpressing Sp4 increases Sp4, Gabra1, and Gabra2 proteins, but has no effect on Gabra3. (D) Overexpressing Sp4 not only increases *Sp4, Gabra1*, and *Gabra2* transcripts but also prevents their downregulation by TTX. *Gabra3*, on the other hand, responds to TTX inactivation but not to Sp4 overexpression. $N = 3$ for all data points.

All * comparisons are with respective controls. *$P < 0.05$, **$P < 0.01$, ***$P < 0.001$. All # comparisons are with KCl (B) or TTX (D). ##$P < 0.01$, ###$P < 0.001$. All ♦comparisons are with *Sp4* shRNA + KCl in B. ♦♦$P < 0.01$, ♦♦♦$P < 0.001$. All X = non-significant, when compared to *Sp4* shRNA + KCl in B, to controls in C, and to TTX alone in D.

Modified from Nair et al. 2016.

inhibitory neuronal functioning with the consumption and production of energy required for sustaining neuronal communications.

The role of Sp4 as a transcriptional mediator of coupling between energy metabolism and neuronal activity is schematically represented in Figure 14.4. Sp4 regulates the expression of all *COX* subunit genes and those of *Grin1, Grin2b*, and *Gria2* in a concurrent and parallel manner with NRF-1 and NRF2. It also coregulates *Atp1b1* with NRF-1. Moreover, Sp4 complements NRF-1 and NRF-2 by controlling the expressions of *Grin2a, Gabra1, Gabra2, Atp1a1*, and *Atp1a3*. No doubt, the spheres of influence are likely to extend far beyond these genes.

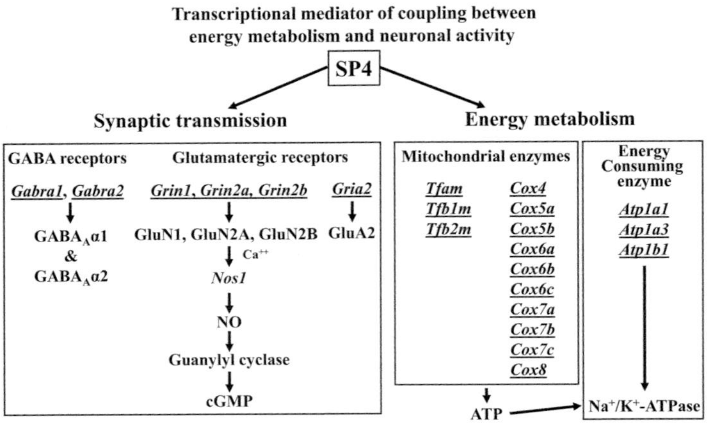

FIGURE 14.4 Schematic diagram of Sp4 as a transcriptional mediator of coupling between neuronal activity and energy metabolism. For synaptic transmission, Sp4 regulates glutamatergic receptor (*Grin1, Grin2a, Grin2b*, and *Gria2*) and GABAergic receptor (*Gabra1* and *Gabra2*) subunit genes. For energy metabolism, Sp4 controls the gene expression of energy-generating enzymes and transcription factors (all 10 nuclear *COX* subunits, *Tfam, Tfb1m*, and *Tfb2m*) as well as energy-consuming enzymes (*Atp1a1, Atp1a3*, and *Atp1b1*).

14.4 3C OF GLUTAMATERGIC RECEPTOR AND *COX* GENES

We have seen in Chapter 7 (Section 7.3) that the ten nuclear-encoded *COX* subunit genes and the three transcription factor genes responsible for mitochondrial DNA transcription all reside in the same "transcription factory". If there is true tight coupling between neuronal activity and energy metabolism, and if the same set of transcription factors (NRF-1, NRF-2, and Sp4) regulates all 13 *COX* subunit genes and critical genes of glutamatergic receptors (Chapters 12–14), is it possible that all these genes are transcribed in the same transcription factory?

To answer this question, the chromosome conformation capture (3C) method was again employed (Dhar and Wong-Riley 2010). As the 10 nuclear-encoded *COX* subunit genes are known to interact with each other (Dhar et al. 2009a), only two were chosen in testing interactions with glutamatergic receptor and *Nos* subunit genes. Control templates were generated by artificially pairing *Cox4i1* or *Cox8a* subunit genes (representing the largest and the smallest nuclear-encoded *COX* subunits) with *Grin1, Grin2b, Gria2*, and *Nos1*, as well as with *Grin3a, Gria4*, and *Nos3* genes (Figure 14.5A). In the murine genome, *Cox4i1* and *Cox8a* are on chromosomes 8 and 19, respectively, *Grin1, Grin2b*, and *Grin3a* are on chromosomes 2, 6, and 4, respectively, *Gria2* and *Gria4* are on chromosomes 3 and 9, respectively, and both *Nos1* and *Nos3* are on the same chromosome 5 but are 94.42 Mb apart, not forming a gene cluster. A positive 3C control was an internal cross-linking product between two *Bgl*II sites located 1.6 kb apart within the coding region of the calreticulin (*CalR*) gene. Three sets of negative controls were also used: a) possible cross-linking between the *Bgl*II site of *CalR* and the *Bgl*II site of *Cox4i1*, b) no cross-linking agent (formaldehyde), and c) no ligase.

3C assays show that the genomic loci for *Grin1, Grin2b, Gria2*, and *Nos1* all interact with those of *Cox4i1* and *Cox8a* at the transcriptional level in primary neurons, but not in C2C12 muscle cells, indicating that they are neuron-specific (Figure 14.5B). No interactions exist between *COX* subunit genes and *Grin3a, Gria4*, or *Nos3*, or with the gene for calreticulin, a non-mitochondrial protein. The fact that *Nos1* and *Nos3* are on the same chromosome, yet *Cox4i1* and *Cox8a* interacted only with *Nos1* and not with *Nos3*, indicates that such interactions are highly specific. All negative 3C controls show no interactions (Dhar and Wong-Riley 2010).

To determine if such genomic interactions are activity-dependent, primary neurons were stimulated with 20 mM KCl for 5 h or impulse blocked with TTX for 6 days before being subjected to the 3C analysis. KCl induces a 224–540% increase in cross-linking frequency compared to non-treated controls ($P < 0.05$), whereas TTX causes a 68–83% decrease in such frequency among the genes ($P < 0.05$) (Figure 14.5C). To determine if specific inhibitors of COX disrupt these interactions, N2a cells were exposed to 300 µM KCN for 5 h (a dose and time that inhibit COX but do not induce cell death). This results in a 60–80% reduction in cross-linking frequency among the *COX* and neurochemical genes ($P < 0.05$) (Dhar and Wong-Riley 2010) (Figure 14.5D).

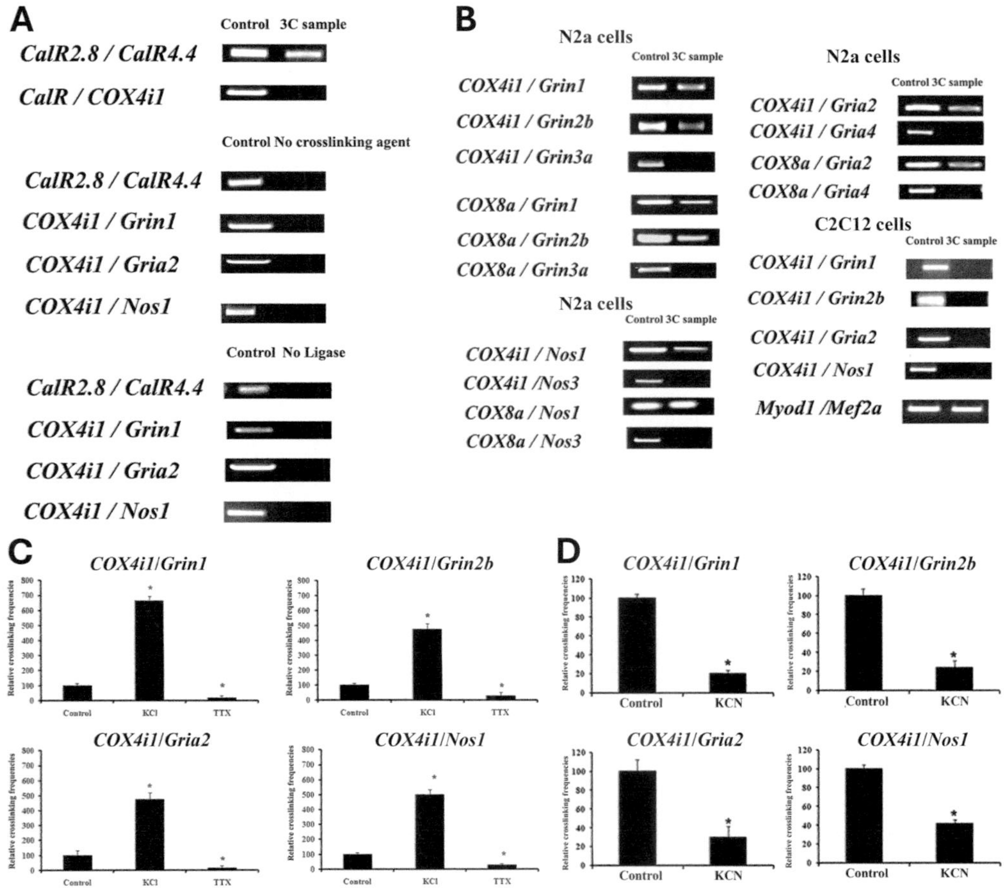

FIGURE 14.5 Chromosome conformation capture (3C) of subunit genes of glutamatergic neurochemicals and *COX*. (A and B) Control templates are generated from artificially linking restriction fragments of various genes after *Bgl*II digestion. Calreticulin (*CalR*) *2.8* and *4.4* form the positive 3C control, and *CalR* and *COX4i1* form the negative control. Other negative controls are no crosslinking agent and no ligase. Interactive pairs are seen between *COX4i1* or *COX8a* and *Grin1, Grin2b, Gria2*, and *Nos1*, but not with *Grin3a, Gria4*, or *Nos3* in N2a cells. In C2C12 muscle cells, interactions are evident between muscle-specific transcription factors *Myod1* and *Mef2a*, but not between *COX4i1* and any of the neurochemical genes, indicating that they are neuron-specific. (C) Interactions between *COX4i1* and *Grin1, Grin2b, Gria2*, and *Nos1* in primary neurons are upregulated by KCl and downregulated by TTX. (D) However, in the presence of 300 µM KCN, an inhibitor of COX, for 5 h, the frequencies of interactions are significantly reduced.

All * comparisons are with respective controls. *$P < 0.05$.

Modified from Dhar and Wong-Riley 2010.

Thus, neurons have devised an efficient mechanism for coordinating the transcriptional coupling of energy metabolism and glutamatergic neurotransmission at the molecular level. When neuronal activity is increased, critical transcription factors such as NRF-1, NRF-2, and Sp4 are recruited, presumably by PGC-1α (and possibly related factors), to assemble all relevant genes in the same transcription factory. With the assistance of RNA polymerase II, genes for glutamatergic transmission and *COX* will be transcribed (and later translated), so that energy production matches the energy requirement of neuronal activity. When neuronal activity is reduced or blocked, molecular interactions are downregulated, intermolecular interactions are reduced, and transcription rates are decreased. While neuronal activity dictates the orchestration of the whole ensemble, it cannot be sustained without a ready supply of energy derived mainly from the mitochondria. Homeostatic control is exquisitely executed at the molecular level (Wong-Riley 2012).

In conclusion, the last three chapters document the tight coupling between neuronal activity and energy metabolism at the transcriptional level. This is accomplished by a set of at least three transcription factors: NRF-1, NRF-2, and Sp4 that coregulates genes mediating synaptic transmission and energy metabolism in both concurrent/parallel and complementary fashions (Figure 14.6). Future research will, no doubt, uncover many more genes, such as metabotropic glutamatergic and GABAergic receptor subunit genes as well as transcription factors, such as CREB or YY1, involved in this vital process.

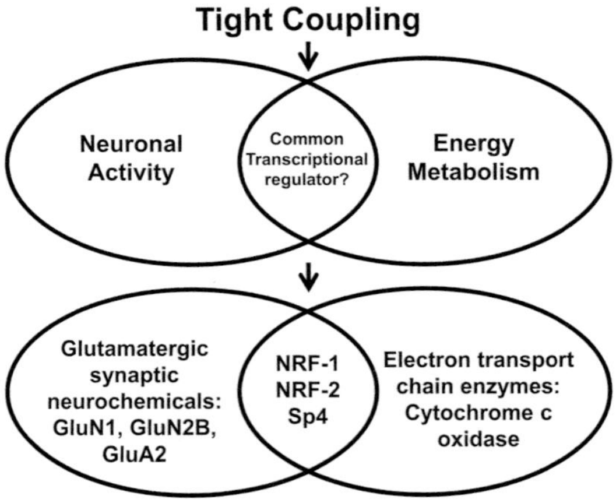

FIGURE 14.6 Schematic rendition of the tight coupling between neuronal activity and energy metabolism mediated by a common set of transcription factors: NRF-1, NRF-2, and Sp4. Shown here are genes positively regulated by all three factors. Additional genes regulated by each are presented in this and the previous two chapters.

15 Critical Period of Postnatal Brain Stem Respiratory System Development in Rats

ABSTRACT

A striking revelation by COX is the mechanism underlying a "critical period" of brain stem respiratory development in rats. Against a general increase with postnatal age, the level of COX activity unexpectedly plummets at postnatal day P12 in multiple brain stem respiratory nuclei. This is paralleled by a similar trend in excitatory neurochemicals. However, inhibitory neurochemicals exhibit a significant upregulation at P12. Many other neurochemical changes also occur within this brief developmental time window. The synaptic imbalance induced by suppressed excitation and enhanced inhibition at P12 is confirmed with patch-clamp recordings of hypoglossal motoneurons at P12–13. The ventilatory response to acute hypoxia is the weakest at P13. BDNF and its TrkB receptors are markedly reduced at P12–13 in multiple respiratory nuclei. An agonist of TrkB partially reverses the synaptic imbalance, whereas a TrkB antagonist accentuates it. The critical period is deemed necessary to allow a transient dominance of inhibition to fine-tune the synaptic network for optimal respiratory functioning.

The "critical period" is a term originally used to describe a time of imprinting for graylag birds (Jaynes 1957). It was subsequently used for the period of song acquisition in birds and language development in humans (Doupe and Kuhl 1999). It also applies to an age when visual and auditory maps in the barn owl align for accurate sound localization with respect to visual space (Knudsen and Knudsen 1990). The critical period in the visual cortex has been extensively studied and will be discussed in the next chapter. The focus of this chapter is on the brain stem respiratory development in the rat, where a critical period was described but not well characterized before 2002.

Can cytochrome *c* oxidase tell us anything about the critical period of respiratory system development? Such a relationship seems far-fetched. How can a metabolic enzyme reveal anything about an important period of postnatal development? Before delving into this subject, a background review is in order.

Respiration is clearly vital to life. It is mediated by a network of nuclei in the brain stem and spinal cord that regulates the diaphragm, intercostal, and other airway respiratory muscles via the phrenic, intercostal, and specific cranial nerves. The orchestration of such coordinated activities is complex and yet almost autonomous and without any conscious control by the individual. This control is virtually instantaneous at birth, when the infant mammal expands its lungs and adapts to atmospheric breathing. However, not all components of this network are mature at birth. A period of refinement is necessary during postnatal development.

A stark indication that the postnatal period may be a vulnerable time comes from human cases of Sudden Infant Death Syndrome (SIDS). What is SIDS? According to the Panel of the National Institute of Child Health and Human Development (NICHD) in 1989, SIDS is "The sudden death of an infant under one year of age which remains unexplained after a thorough case investigation, including performance of a complete autopsy, examination of the death scene, and a review of the clinical history". The peak incidence of SIDS is surprisingly not at birth, but rather, between the 2nd and 4th months of life, that is, a possible critical period. While the cause remains elusive, Filiano and Kinney (1994) proposed three risk factors for SIDS: a vulnerable infant, an exogenous stressor or stressors, and a critical developmental period in homeostatic control. When and only when all three exist simultaneously does SIDS occur.

DOI: 10.1201/9781003507765-15

Is there a critical period in respiratory development? It has been defined as a time window "devoted to structural and/or functional shaping of the neural system subserving respiratory control" (Carroll 2003). Perturbations during this period can lead to respiratory plasticity that may persist with lifelong consequences. Such perturbations include hypoxia, hypercapnia, intermittent hypoxia, injury, stress, and pharmacological agents (Mitchell and Johnson 2003). Similar disturbances outside this period have little or no long-term effects.

What is the exact timing of the critical period? It obviously varies among animal species. Rats are a popular animal model due to their immaturity at birth, their relatively large litter size (about a dozen), and their ease of handling. Many studies indicate that the first two postnatal weeks are the most critical in this species. Nuclear magnetic resonance spectroscopy of rat brain stem slices reveals a metabolic maturation between P11 and P17 (Schweitzer et al. 1992). Exposing rat pups between P10 and P25 to intermittent hypoxia impedes spatial learning during the juvenile stage and causes hyperactivity in juvenile male (but not female) rats (Row et al. 2002). Hyperoxia during the first two weeks attenuates the phrenic nerve and ventilatory responses to acute hypoxia in the adult (Bavis et al. 2002; Bisgard et al. 2003). This period coincides with the maturation of carotid chemoreceptors, which also occurs between the first and second postnatal weeks in the rat (Kholwadwala and Donnelly 1992). The carotid sinus nerve response to hypoxia is mature by P8 in the rat (Bamford et al. 1999), and carotid body denervation at P7–8 causes greater mortality and respiratory abnormality than before or after that time, suggesting that this period is critical for the development of chemosensitivity (Serra et al. 2001). The hypercapnic response, on the other hand, is well developed at birth (Belegu et al. 1999). However, the response falls throughout the first week to a minimum at P8 before rising to higher levels at P10 through adulthood (Stunden et al. 2001). These authors concluded that P8 is the critical window of low sensitivity to hypercapnia. A subsequent study, which focused only on P1, 3, 7, and 10, also shows a low point of response to hypercapnia at P7 (Wickström et al. 2002). However, a later study of three rat strains did not find a decrease in hypercapnic response at P6 and P7, but rather, CO_2 sensitivity is variable from P0 to P14 and does not increase significantly until P15 in all three strains (Davis et al. 2006). Nonetheless, in the 1990s and early 2000s, many investigators in the respiratory community tended to lean toward P7–P8 as a critical period of respiratory development.

15.1 RESPIRATORY CONTROL CENTERS IN THE BRAIN STEM

Respiratory control centers form a network located in three main regions of the brain stem: pons, dorsal medulla, and ventral medulla. They are also known as the pontine respiratory group (PRG), the dorsal respiratory group (DRG), and the ventral respiratory group (VRG), respectively (Bianchi et al. 1995; Feldman et al. 2003; Wong-Riley and Liu 2005) (Figure 15.1). The PRG is made up of the lateral and medial parabrachial nuclei (PB_L and PB_M) and the Kölliker–Fuse (KF) nucleus. It receives input from the medulla and takes part in respiratory rhythm modulation, diaphragmatic motor control, vocalization, and the control of airway muscles during exercise and sleep (Alheid et al. 2004). The DRG is composed mainly of the nucleus tractus solitarius (NTS), with the ventrolateral subnucleus as a major center (NTS_{VL}). It receives inputs from many sources, notably a direct one from chemoreceptors in the carotid body. Its neurons are mainly inspiratory, and they fire in synchrony with those of the phrenic nerves (Hilaire et al. 1990; Duffin et al. 1995). The VRG is subdivided functionally into a rostral (rVRG) and a caudal (cVRG) group. The rVRG includes the Bötzinger complex (BötC), the pre-Bötzinger complex (pre-BötC or PBC), and parafacial respiratory group (pFRG), all of which are in the rostroventrolateral medulla (RVLM). The BötC contains a major group of inhibitory expiratory neurons (Duffin et al. 1995; Ezure et al. 2003), whereas the PBC and pFRG are centers for respiratory rhythmogenesis (Onimaru et al. 1988; Smith et al. 1991; Rekling and Feldman 1998; Onimaru and Homma 2003). The cVRG is made up primarily of the nucleus retroambiguus (NRA), whose neurons are mainly expiratory premotor ones (Duffin et al. 1995; Shen and Duffin 2002).

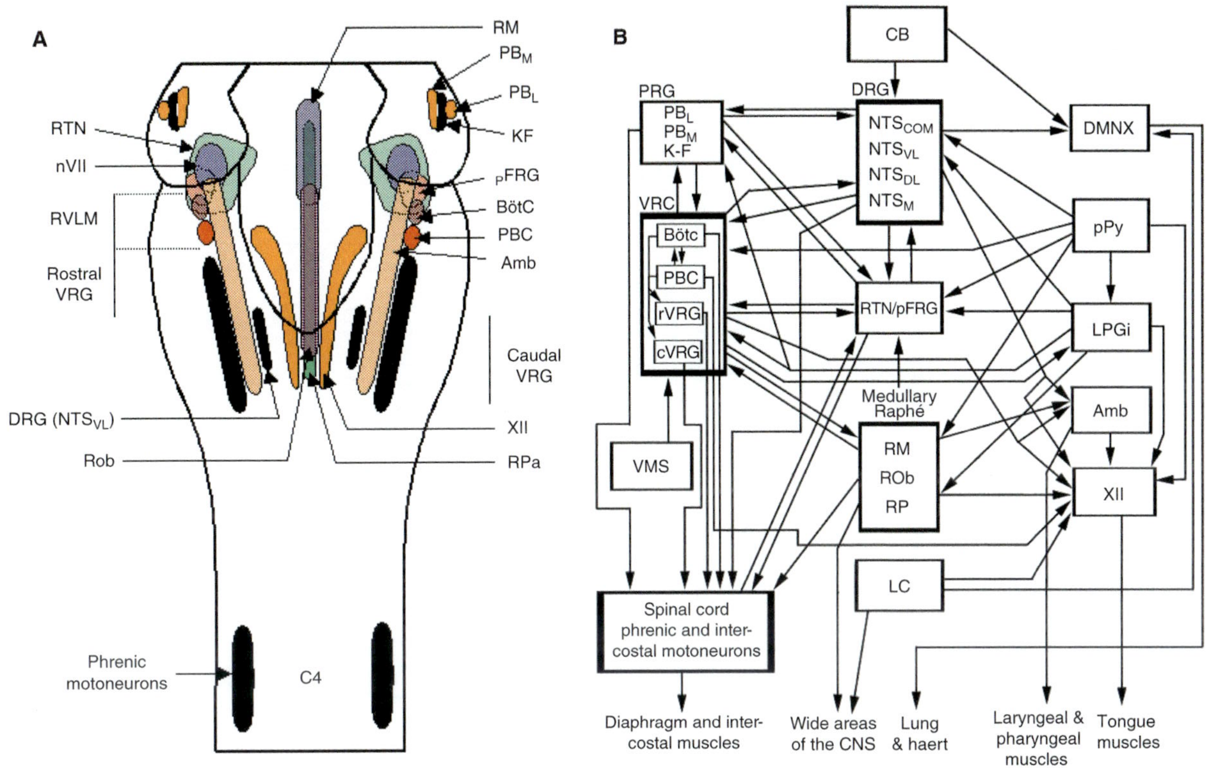

FIGURE 15.1 (A) Schematic longitudinal diagram of brain stem nuclei associated with respiratory control. The upper portion is pons, the middle portion is the medulla, and the lower portion is the spinal cord. (B) Schematic representation of the connections among the respiratory-associated nuclei. Amb, nucleus ambiguus; BötC, Bötzinger complex; CB, carotid body; DMNX, dorsal motor nucleus of the vagus; DRG, dorsal respiratory group; KF, Kölliker–Fuse nucleus; LC, locus coeruleus; LPGi, lateral paragigantocellular nucleus; NTS, nucleus tractus solitarius (subnuclei: com, commissural; DL, dorsolateral; M, medial; VL, ventrolateral); PBC, pre-Bötzinger complex; PB_L, lateral parabrachial nucleus; PB_M, medial parabrachial nucleus; pFRG, parafacial respiratory group; pPy, parapyramidal region; PRG, pontine respiratory group; RM, raphé magnus; ROb, raphé obscurus; RP, raphé pallidus; RTN, retrotrapezoid nucleus; RVLM, rostroventrolateral medulla; VMS, ventral medullary surface; VRC, ventral respiratory column; VRG, ventral respiratory group (c, caudal; r, rostral); XII, hypoglossal nucleus. For landmark reference: C4, cervical 4 level; nVII, facial nucleus.

A. From Wong-Riley and Liu 2005; B. From Wong-Riley et al. 2013.

Besides the above, other nuclei involved in airway control include the upper airway motoneurons of the nucleus ambiguus (NA or Amb), the dorsal motor nucleus of the vagus (DMNX), and the hypoglossal nucleus (XII), which innervates the tongue and pharyngeal muscles to maintain upper airway patency during respiration (Núñez-Abades et al. 1992; St-John 1998; Jordan 2001). The inferior olivary nucleus (IO), which is intimately associated with the cerebellum, may also be involved in coordinating ventilatory skeletal muscles with upper airway muscles (Ruigrok 1997; Harper et al. 1998).

An intricate network of central chemoreceptors is dispersed diffusely in the CNS. They include the cerebellum, posterior hypothalamus, and various brain stem regions such as the raphé nuclei (raphé magnus RM, raphé obscurus ROb, and raphé pallidus RP forming a rostro-caudal midline column), locus coeruleus, NTS, retrotrapezoid nucleus (RTN), and an area near the ventral surface of the medulla. These chemoreceptors also contribute to the ventilatory output in response to intracellular pH and normal or abnormal mixture of gases and hydrogen ions in the blood (Nattie 2001; Okada et al. 2001; Putnam 2001; Messier et al. 2004).

15.2 SURPRISING REVELATION BY CYTOCHROME *c* OXIDASE

When cytochrome *c* oxidase histochemistry was first used to study the development of the rat's respiratory system (Sprague-Dawley or SD rats), the focus was on every day from the day of birth (P0) to P7, then skipped to P14 and P21. A transient decrease or plateau in COX activity was noted at P3–4 in a number of brain stem respiratory nuclei, but otherwise the trend is a gradual increase in activity with age, which is not surprising (Liu and Wong-Riley 2001) (Figure 15.2A and B). However, when COX activity was evaluated every day from P0 to P21 (when brain stem respiratory nuclei have reached their maturity) in the PBC, a totally unexpected pattern appeared: against a gradual increase in COX activity with age, a sudden and significant drop occurs at P12 not correlated with a steady growth in cell size (Liu and Wong-Riley 2002; Liu et al. 2003) (Figure 15.2C). This pattern is also present in multiple respiratory-related nuclei in the brain stem (including the NTS_{VL}, Amb, XII, Rob, DMNX, and the medial accessory nucleus of the inferior olivary nucleus IO_{MA}), but not in the non-respiratory somatosensory cuneus nucleus (CN) (Liu and Wong-Riley 2003). What cause(s) such a precipitous fall in metabolic activity?

15.3 NEUROCHEMICAL CHANGES DURING THE CRITICAL PERIOD

As discussed in Chapters 2–4, much of the energy demand, hence COX activity, in neurons is for repolarizing membrane potentials after excitatory, depolarizing stimulation. If COX activity suddenly drops at P12, it implies a sudden fall in excitation and possibly a rise in inhibition. This requires further investigation.

15.3.1 A Time of Synaptic Imbalance

To test for the possibility of a synaptic imbalance, various neurochemicals were assessed immunohistochemically and semi-quantitatively analyzed with optical densitometry within individual neurons (Liu and Wong-Riley 2002, 2005) (Figure 15.3). Consistently, developmental trends of the excitatory neurochemical glutamate and the obligatory NMDA receptor subunit 1, GluN1, follow that of COX, that is, a general increase with age, with a gentle dip or plateau at P3–P4 but a significant fall at P12. On the other hand, trends for the inhibitory neurochemicals GABA, $GABA_B$ receptors, and glycine receptors exhibit a slight elevation at P3–4 but a distinct and significant rise at P12. These opposing trends exist in multiple respiratory-related nuclei (PBC, NTS_{VL}, Amb, XII, IO_{MA}, and DMNX), but not in the non-respiratory cuneate nucleus (CN). Surprisingly, the level of AMPA receptor subunit GluA2 also increases at P12. As GluA2 reduces the permeability of Ca^{2+}, a rise in its expression may signal suppression of excitability within respiratory-related nuclei at P12.

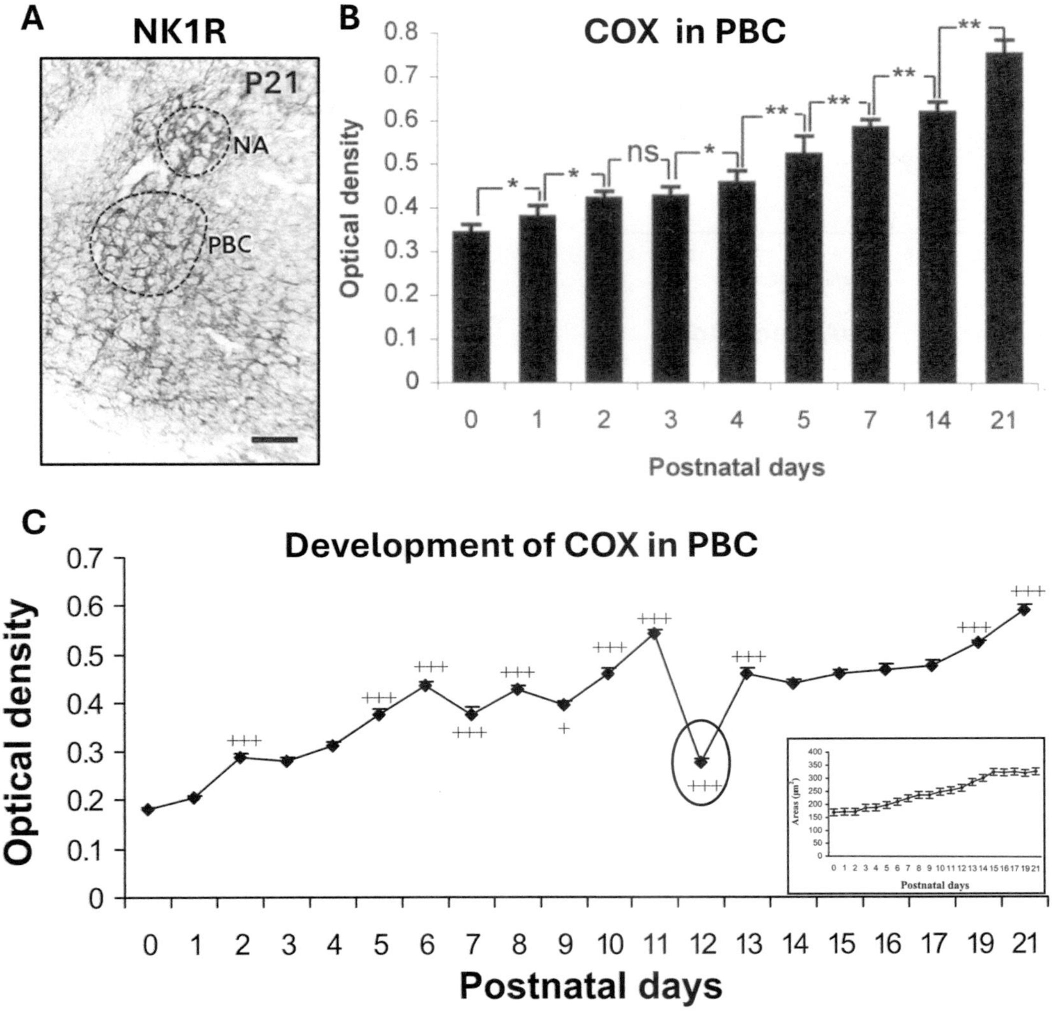

FIGURE 15.2 Development of COX in the PBC. (A) Immunohistochemical labeling of neurokinin-1 receptor (NK1R), a marker of the pre-Bötzinger complex (PBC; within the lower dotted circle) and the nucleus ambiguus (NA; within the upper dotted circle) in the brain stem of a P21 rat. (B) Optical densitometric analyses of COX labeling in the PBC at P0 (day of birth), P1, P2, P3, P4, P5, P7, P14, and P21. There is an apparent, gradual increase in the intensity of labeling with age. (C) When COX labeling is studied every day from P0 to P21, a sharp decrease in the intensity is noted at P12 (data point within the oval) against a gradual increase with age. This abrupt change is not correlated with a smooth increase in cell size with age (inset). Each data point in C represents 150–350 neurons measured. Statistical comparisons between successive age groups are based on Tukey's Studentized range test. A and inset in C are from Liu and Wong-Riley 2002, B is from Liu and Wong-Riley 2001; and C is modified from Liu et al. 2003.

$^{*}P < 0.01$, $^{**}P < 0.001$ (in B); $^{***}P < 0.001$ (in C).

Clearly, a synaptic imbalance of suppressed excitation and heightened inhibition is prevalent among respiratory-related nuclei but not in the non-respiratory CN at P12 in the rat. Reduced excitation lessens energy demand, and the level of COX naturally falls. Enhanced inhibition requires relatively little energy because, as stated in Chapter 2 (Section 2.3.3.1), repolarization after hyperpolarization is mainly passive. The state of enhanced inhibition may render the rat pups more vulnerable to respiratory failure. Indeed, carotid body denervation close to and within this period

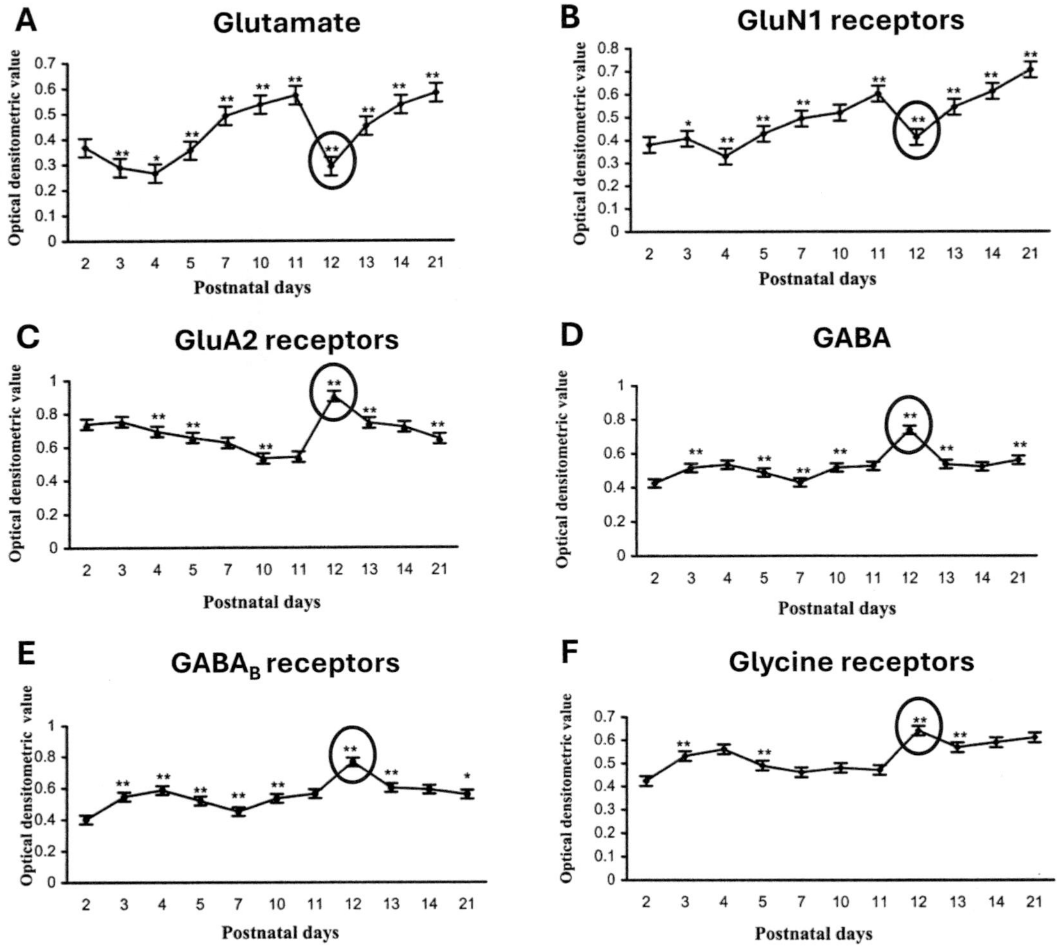

FIGURE 15.3 Neurochemical development in the PBC. Optical densitometric analyses of immunoreaction products of glutamate (A), GluN1 receptors (B), GluA2 receptors (C), GABA (D), GABA$_B$ receptors (E), and glycine receptors (F) in the PBC during the first three postnatal weeks. Note the abrupt fall in the expressions of glutamate and GluN1 and the sharp rise in those of GluA2, GABA, GABA$_B$, and glycine receptors, all at P12 (data points are within ovals). Each data point represents 150–350 neurons measured. Tukey's Studentized range test for comparing successive age groups.

$*P < 0.05$, $**P < 0.01$.

Modified from Liu and Wong-Riley 2002.

delays and prolongs the metabolic development of the PBC (Liu et al. 2003). Moreover, when rat pups are exposed to 5 days of sustained hypoxia (11% O_2), those exposed from P11 to P15 have absent hypoxic ventilatory response (HVR), decreased hypercapnic response, and a high rate of unexplained mortality, whereas those exposed before (P1–5) or after (P21–25) that period do not experience the same effects (Mayer et al. 2014).

15.3.2 Developmental Switches in Chloride Transporter Dominance and in GABAergic and Glycinergic Receptor Subunit Dominance

The time of synaptic imbalance is accentuated by several switching events that take place postnatally: a) GABAergic transmission transitions from depolarization to hyperpolarization (Ben-Ari et al. 2007). This occurs at different times in different systems and different species. b) The functional switch is caused by a change in intracellular chloride concentration with the maturation of the chloride-transporting mechanism (Delpire 2000; Ben-Ari et al. 2007). Earlier in development, the Cl^--importer Na^+-K^+-$2Cl^-$ co-transporter 1 (NKCC1) dominates, and the increased intracellular Cl^- concentration results in a depolarizing current through the GABA (and glycine) receptor chloride channels (Plotkin et al. 1997). With development, the neuron-specific Cl^-- exporter K^+-Cl^- co-transporter 2 (KCC2) predominates, and the reduced intracellular Cl^- concentration drives the inward flow of Cl^- (through inhibitory receptors) and hyperpolarizes the membrane (Payne et al. 1996). In the brain stem, NKCC1 immunoreactivity decreases from P0 to P21, with an abrupt and statistically significant fall at P12 in the PBC, Amb, and XII, whereas KCC2 immunoreactivity increases from P0 to P21, and the two trends switch dominance at around P12 (Liu and Wong-Riley 2012) (Figure 15.4A). c) The kinetics of GABAergic transmission transitions from slow to fast with development (Ben-Ari et al. 2007). This may coincide with the switch in dominance of specific $GABA_A$ receptor subunits. As discussed in Chapter 14 (Section 14.3.1), $GABA_AR\alpha3$ has slow kinetics, and $GABA_AR\alpha1$ is much faster. In both the PBC and NTS_{VL}, the expression of $GABA_AR\alpha3$ is relatively high at P0 but declines with age, and that of $GABA_AR\alpha1$ is low at birth and increases with age; these two events intersect around P12 (Figure 15.4B). The expression of $GABA_AR\alpha2$, on the other hand, remains relatively low and constant during development (Liu and Wong-Riley 2004, 2006). d) Like GABA, glycine is a major inhibitory transmitter that undergoes a developmental transition from depolarization to hyperpolarization due to changes in intracellular Cl^- concentration (Kirsch 2006). Glycine receptors (GlyR) also undergo developmental switches in subunit dominance. These receptors are also heteropentameric ligand-gated Cl^- channels made up of α subunits ($\alpha1$-$\alpha4$) with or without β subunits (Dutertre et al. 2012). $GlyR\alpha1$ has the fastest decay kinetics, $GlyR\alpha2$ the slowest, and $GlyR\alpha3$ is intermediate (Singer et al. 1998; Dutertre et al. 2012). $\alpha4$ is found mainly in the retina (Heinze et al. 2007). In the PBC, Amb, XII, and NTS_{VL}, the expressions of $GlyR\alpha2$ and $GlyR\alpha3$ with slow decay kinetics are relatively high at P2, but decline after 1–1½ weeks, reaching the lowest levels at P21. On the other hand, $GlyR\alpha1$ level increases with age and significantly so by P12. The expression of $GlyR\beta$ rises from P2 to P12, then declines slightly until P21. Thus, the switch in dominance from the neonatal $GlyR\alpha2$ and $GlyR\alpha3$ to the adult $GlyR\alpha1$ occurs around P12 (Liu and Wong-Riley 2013a) (Figure 15.4C and D). e) A developmental switch from predominantly long-duration, slower decay constant GABAergic to short-duration, faster decay constant glycinergic synaptic currents has been reported in various parts of the brain (Ben-Ari et al. 2007). In the developing hypoglossal motoneurons (XII), the percent contribution of GABA to miniature inhibitory postsynaptic current (mIPSC) decreases from P0 to P16 (from 60.12% to 20.48%), and that of glycine increases over the same period (from 30.05% to 78.91%). The two trends cross at P7, but the greatest difference starts at P13 (Figure 15.4E and F). The GABA and glycine co-release component of mIPSCs decreases from 8–10% at P0-P3 to 0.6–1.8% between P4 and P16 (Gao et al. 2011).

These developmental switches underpin a transitional period around P12 in the rat, when GABA and glycine have attained their potent inhibitory influence on brain stem respiratory-related neurons.

15.3.3 Developmental Changes in Other Neurochemicals

15.3.3.1 Developmental Changes in NMDA Receptor Subunits

Besides the ubiquitous and obligatory GluN1, other NMDA receptor subunits undergo developmental changes. Specifically, GluN2A expression increases gradually from P2 to P11, but falls significantly at P12 before rising to a plateau until P21 in the PBC, Amb, XII, and NTS_{VL}, but not in the non-respiratory CN (Liu & Wong-Riley 2010c). The expressions of GluN2B and GluN2C remain relatively constant from P2 to P21, except for a significant fall of GluN2C in the PBC at P12 and a more gradual but non-significant decline in this subunit from P10 to P12 in the NTS_{VL}. An apparent switch in dominance from the age-dependent decrease in GluN2D to a developmental increase in GluN3B occurs in these same nuclei, with their trends intersecting around P13 (Liu & Wong-Riley 2010c). This implies a transition from the neonatal, slowest decay kinetics of GluN2D (Cull-Candy et al. 2001) to the more mature, faster decay of GluN3B during respiratory functional maturation.

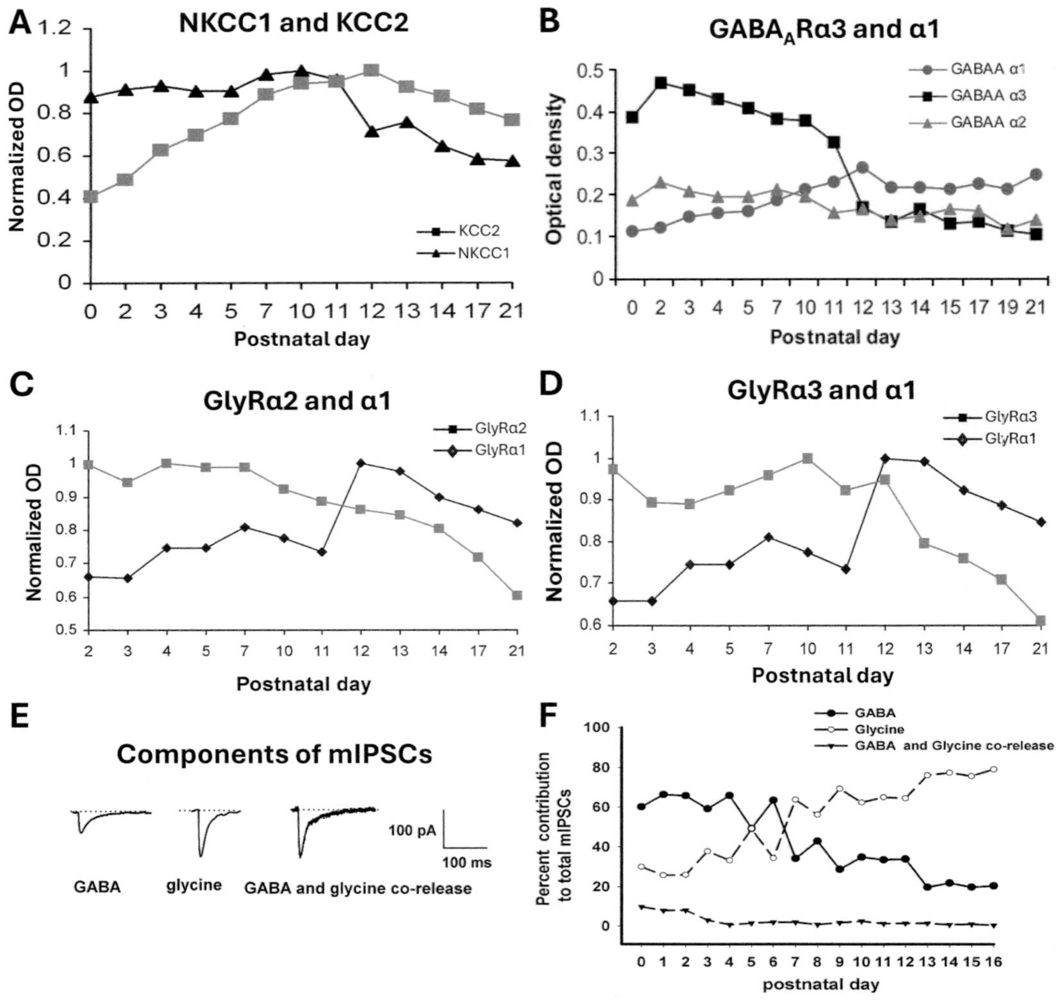

FIGURE 15.4 Developmental switching of Cl⁻ transporters and inhibitory receptor subunits. (A) NKCC1 immunoreactivity in PBC neurons decreases with age, whereas that of KCC2 increases with age. The two trends intersect around P11. (B) $GABA_A$ receptor α3 immunoreactivity in PBC neurons decreases with age, whereas that of $GABA_A$Rα1 increases with age. The two trends intersect at P12. The expression of $GABA_A$Rα2 remains at a relatively steady, low level during the first three postnatal weeks. Immunoreactivity of both glycine receptor α2 (C) and α3 (D) in PBC neurons decreases with age, whereas that of GlyRα1 increases with age. The trends of both GlyRα2 and GlyRα3 intersect with that of GlyRα1 at P12. Each data point represents 700 neurons for NKCC1 and 1200 neurons for KCC2 (A), 150–500 for each GABA receptor subunit (B), and 800 for each glycine receptor subunit (C and D) measured. (E) Representative mIPSCs mediated by GABA, glycine, and their co-release in hypoglossal neurons. (F) The percentage contributions of GABA and glycine to total mIPSCs. GABA has an age-dependent decrease but glycine has an age-dependent increase in their respective contribution from P0 to P16. The greatest difference between their contributions is at P13 and thereafter.

A: from Liu and Wong-Riley 2012, B: from Liu and Wong-Riley 2004, C and D: Modified from Liu and Wong-Riley 2013a; E and F: from Gao et al. 2011.

When present, GluN3B renders the NMDA receptors less permeable to Ca^{2+} and less sensitive to Mg^{2+} block (Cavara et al. 2010). The fall in GluN2A rather than GluN2B at P12 is significant, as NMDA receptors with GluN1/GluN2A normally mediate peak current densities four times larger and peak channel open probability 2- to 4-fold higher than those with GluN1/GluN2B (Chen et al. 1999). The abrupt fall in GluN2A indicates a more intense suppression of excitation around P12.

15.3.3.2 Developmental Changes in Serotonin Receptor Subunits, Its Synthesizing Enzyme, and Its Transporter

Serotonin or 5-hydroxytryptamine (5-HT) is an indolaminergic neurotransmitter involved in multiple physiological functions, including cardiorespiratory, motor, sensory, and thermoregulatory activities, as well as affective, cognitive, aggressive, sexual, and arousal behaviors (Zifa and Fillion 1992; Jacobs and Fornal 1997; Okaty et al. 2019). In the brain stem, these functions also include respiratory modulation, eupneic ventilation, and CO_2 chemoreception (Hilaire and Duron 1999; Hodges and Richerson 2010). Serotonergic neurons in the brain stem are distributed mainly in the midline medullary raphé nuclei and the ventral medulla (parapyramidal area) (Dahlstroem and Fuxe 1964; Connelly et al. 1989). They innervate almost all brain areas, including all brain stem respiratory-related nuclei as well as phrenic motor nuclei in the spinal cord (Holtman et al. 1984; Connelly et al. 1989; Voss et al. 1990). Transgenic mice lacking serotonin neurons exhibit severe apnea and high perinatal mortality (Hodges et al. 2009).

Serotonin receptors comprise seven gene families with at least 14 subtypes divisible into metabotropic G-protein-coupled ones with seven transmembrane domain (5-$HT_{1A\text{-}F}$, 5-$HT_{2A\text{-}C}$, 5-$HT_{4/5A/5b/6/7}$, with 5-HT_{1C} later re-designated as 5-HT_{2C}) and an ionotropic one (5-HT_3) (Zifa and Fillion 1992; Hoyer et al. 1994; Sharp and Barnes 2020). Whereas the metabotropic receptors mediate slow signaling, the ionotropic 5-HT_3 receptors facilitate rapid, excitatory synaptic transmission. All 5-HT receptors are postsynaptic with subtype-specific distribution, but 5-HT_{1A} and 5-HT_{1B} are also expressed as autoreceptors in 5-HT neurons to inhibit the release of 5-HT (Sharp and Barnes 2020). Activation of 5-HT_{1A} increases phrenic nerve activity (Valic et al. 2008) and exerts respiratory rhythm generation (Hilaire et al. 1997). However, activation of 5-HT_{1B} inhibits phrenic motoneurons and XII (Singer et al. 1996; Di Pasquale et al. 1997). Thus, the actions of these two receptor types counterbalance each other. When 5-HT_{1A} and 5-HT_{2A} are co-activated, the receptors are involved in the depolarization of PBC neurons (Schwarzacher et al. 2002). Among the receptors, 5-HT_{2A} plays an important role in respiratory regulation, including respiratory rhythmogenesis, respiratory long-term facilitation, and gasping (Baker-Herman and Mitchell 2002; Peña and Ramirez 2002; Tryba et al. 2006). Exogenous 5-HT exerts a biphasic phrenic nerve activity, with initial transient inhibition followed by prolonged facilitation known as phrenic long-term facilitation mediated primarily by 5-HT_{2A} (Liu et al. 2011). In a subset of SIDS victims, the binding of 5-HT_{1A} and 5-$HT_{2A/C}$ is significantly reduced in medullary nuclei that support cardiorespiratory homeostasis (Kinney 2005; Cummings et al. 2024).

A detailed analysis of the developmental trends of 5-HT_{1A}, 5-HT_{1B}, and 5-HT_{2A} in the rat brain stem indicates that the level of 5-HT_{2A} is maintained at relatively high levels from P2 to P21, except for the precipitous fall at P12 in the PBC, Amb, and XII ($P < 0.05$ for all), but not in the CN (Liu and Wong-Riley 2008) (Figure 15.5A–C and F). The drop from P10 to P12 is more gradual in the three raphé nuclei (RM, ROb, and RP) and RTN/pFRG, so the day-to-day change is not statistically significant (Liu and Wong-Riley 2010a). The expressions of 5-HT_{1A} and 5-HT_{1B} are quite high from P2 to P11 in the three raphé nuclei, PBC, Amb, XII, and RTN/pFRG, but they all fall abruptly at P12 and remain relatively low thereafter (Liu & Wong-Riley 2010a).

In short, all three serotonergic receptor subtypes (1A, 1B, and 2A) in multiple nuclei of the brain stem respiratory system are expressed at high levels from birth until the middle of the second postnatal week, when they fall either precipitously at P12 or gradually toward P12, before they recover or reach a plateau until P21. Such widespread reduction in 5-HT receptor expression around P12 should attenuate the reported net stimulatory effect of 5-HT on respiratory output (Lindsay and Feldman 1993; Hodges and Richerson 2010) with potential adverse effects on respiratory control.

In addition to 5-HT receptors, the levels of 5-HT synthesizing enzyme tryptophan hydroxylase (TPH) and serotonin transporter (SERT) in the raphé nuclei, PBC, Amb, and RTN are high between P2 and P11, fall distinctly at P12, and plateau thereafter until P21 (Liu and Wong-Riley 2010b) (Figure 15.5D and E).

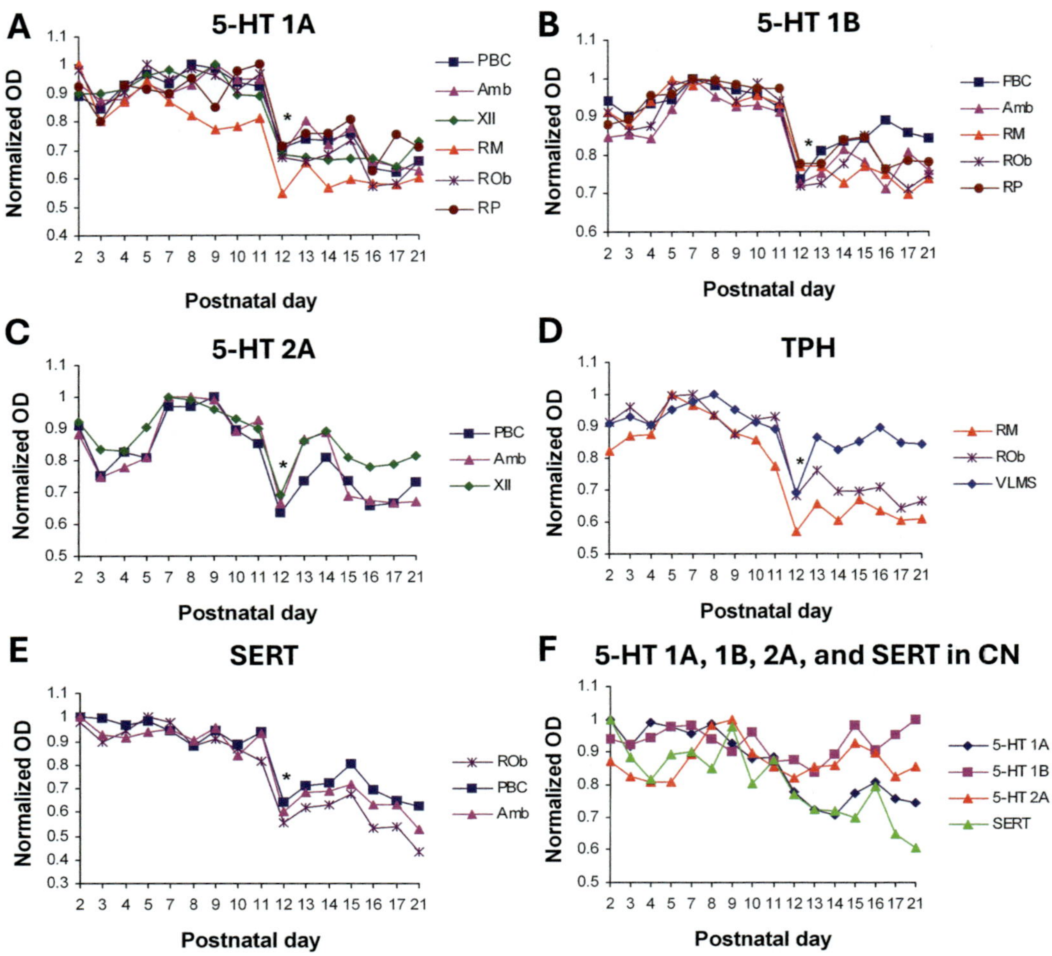

FIGURE 15.5 Development of serotonergic neurochemicals. Optical densitometric analyses of immunoreaction product of 5-HT_{1A} (A), 5-HT_{1B} (B), 5-HT_{2A} (C), TPH (D), and SERT (E) in various respiratory-related brain stem nuclei from P2 to P21. Note that the level of labeling is the highest from P2 to P11, but they all have a sudden fall at P12, followed by a slight rise or a plateau until P21. This trend does not occur in the non-respiratory cuneate nucleus (CN) (F). Approximately 600–1000 neurons were measured for each nucleus at each age for each neurochemical. For abbreviations, see legend in Figure 15.1.

Modified from Wong-Riley et al. 2013.

Together with synaptic imbalance marked by heightened inhibition and suppressed excitation, reduced expressions of serotonin, its major receptors, and its transporters around P12 severely compromise respiratory regulation in the brain stem respiratory network and render the animal less capable of responding adequately to respiratory insults, such as hypoxia or hypercapnia. The insult would be much more severe during sleep, when the firing of serotonergic neurons is decreased (McGinty and Harper 1976), threatening airway patency (Horner 2007), and when respiratory activities are normally depressed (Olson and Simon 1996).

15.3.3.3 Developmental Changes in BDNF and TrkB

In Chapter 13 (Section 13.3), the importance of BDNF is briefly discussed. As a neurotrophic factor, BDNF and its high-affinity tyrosine protein kinase B (TrkB) receptor play important roles in neuronal survival, differentiation, cell migration, neurite outgrowth, synapse formation, stabilization, and plasticity during development (Hofer and Barde 1988; Yoshii and Constantine-Paton 2010). BDNF enhances excitatory synapses by promoting glutamate release from glutamatergic neurons and increases the amplitude of their excitatory postsynaptic current (EPSCs). BDNF also increases the amplitude of inhibitory postsynaptic current (IPSCs) onto GABAergic neurons, hence reducing the efficacy of inhibitory transmission (Wardle and Poo 2003). BDNF depresses evoked GABA and glycine IPSCs (eIPSCs) in the pain pathway of the spinal cord (Bardoni et al. 2007). Importantly, BDNF-TrkB signaling is essential for the development of brain stem respiratory system (Erickson et al. 1996; Katz 2005; Ogier et al. 2007), and dysregulation of BDNF leads to severe respiratory dysfunction in Rett's syndrome (Katz 2014). If BDNF-TrkB is pro-excitation and suppresses inhibition, is it possible that their expressions are altered during the critical period of respiratory development?

A systematic analysis of the developmental expressions of BDNF and TrkB from P0 to P21 demonstrates that they are highly expressed during that period except for a precipitous fall at P12 or P12-P13 in the PBC, Amb, NTS_{VL}, NTS_{COM} (commissural nucleus of NTS, another recipient of peripheral chemoafferents), RTN/pFRG, and XII (Liu and Wong-Riley 2013b; Gao et al. 2014) (Figure 15.6A–E and G–K). Both BDNF and TrkB transcripts in XII are high at the end of the first postnatal week but are significantly reduced at P12 (Gao et al. 2014) (Figure 15.6F and L).

Neonatal hyperoxia (90% O_2 from P0 to P10) delays the fall in the expressions of COX, BDNF, TrkB, 5-HT_{1A}, and 5-HT_{2A} from P12 to P14 in PBC, XII, and NTS_{VL}, whereas the non-respiratory CN is not affected (Mu et al. 2018). It also delays the expressions of TPH and SERT from P12 to P14 in RM and RP. However, hyperoxia in the neonate can only delay but not eliminate the critical period, when these neurochemicals are significantly downregulated.

15.3.3.4 Developmental Changes in PACAP

Another neurochemical that affects respiration is pituitary adenylate cyclase-activating polypeptide (PACAP), a neuropeptide originally isolated from an extract of bovine hypothalamus (Miyata et al. 1989). It shares 68% homology at its N-terminus with vasoactive intestinal polypeptide (VIP), but it is 1000-fold more potent than VIP in stimulating adenylate cyclase in the pituitary (hence its name). PACAP is widely found in the brain and peripheral organs (Arimura 1992; Vaudry et al. 2009). It is implicated in neural development, neuromodulation, neurotransmission, metabolic homeostasis, and transcription at central and peripheral targets (Vaudry et al. 2009; Rudecki and Gray 2016; Rivnyak et al. 2018). It also modulates glutamatergic, catecholaminergic, and cholinergic neurotransmission with neuroprotective and homeostatic functions (Hannibal et al. 2000; Liu et al. 2000; MacDonald et al. 2005).

PACAP is present in several respiratory nuclei and tracts in the brain stem (Uddman et al. 1991; Légrádi et al. 1994; Kausz et al. 1999). It powerfully stimulates breathing through its activation of arterial chemoreceptors and maintains normal pulmonary vascular tone (Runcie et al. 1995; Vaudry et al. 2009). The most intriguing finding is that mice deficient in the PACAP gene die suddenly after birth due to decreased respiratory chemoresponse and apnea, but the mortality rate does not peak until the second postnatal week (Gray et al. 2001; Wilson and Cummings 2008), when sudden changes in COX and neurochemicals implicate a critical period in respiratory development.

What is the expression of PACAP in the brain stem during normal postnatal development? A detailed temporal analysis of PACAP38 immunoreactivity (the C-terminus 38 residues, the major one expressed in the nervous system) from P2 to P21 in the PBC, Amb, XII, and NTS_{VL} implicates PACAP's involvement in the critical period. Its expression rises from P2 to peak around P10, followed by a significant fall at P12 and a slight rise thereafter in the four nuclei examined ($P < 0.05$ for all), but the level within the CN remains relatively constant (Liu and Wong-Riley 2019) (Figure 15.7). High expressions of PACAP in the respiratory nuclei throughout most of the first three postnatal weeks underscore its importance in the normal development and functioning of these nuclei. However, its sudden fall at P12 in normal animals renders them more vulnerable to respiratory insults. Indeed, piglets exposed to acute intermittent hypercapnic hypoxia exhibit reduced levels of PACAP and its

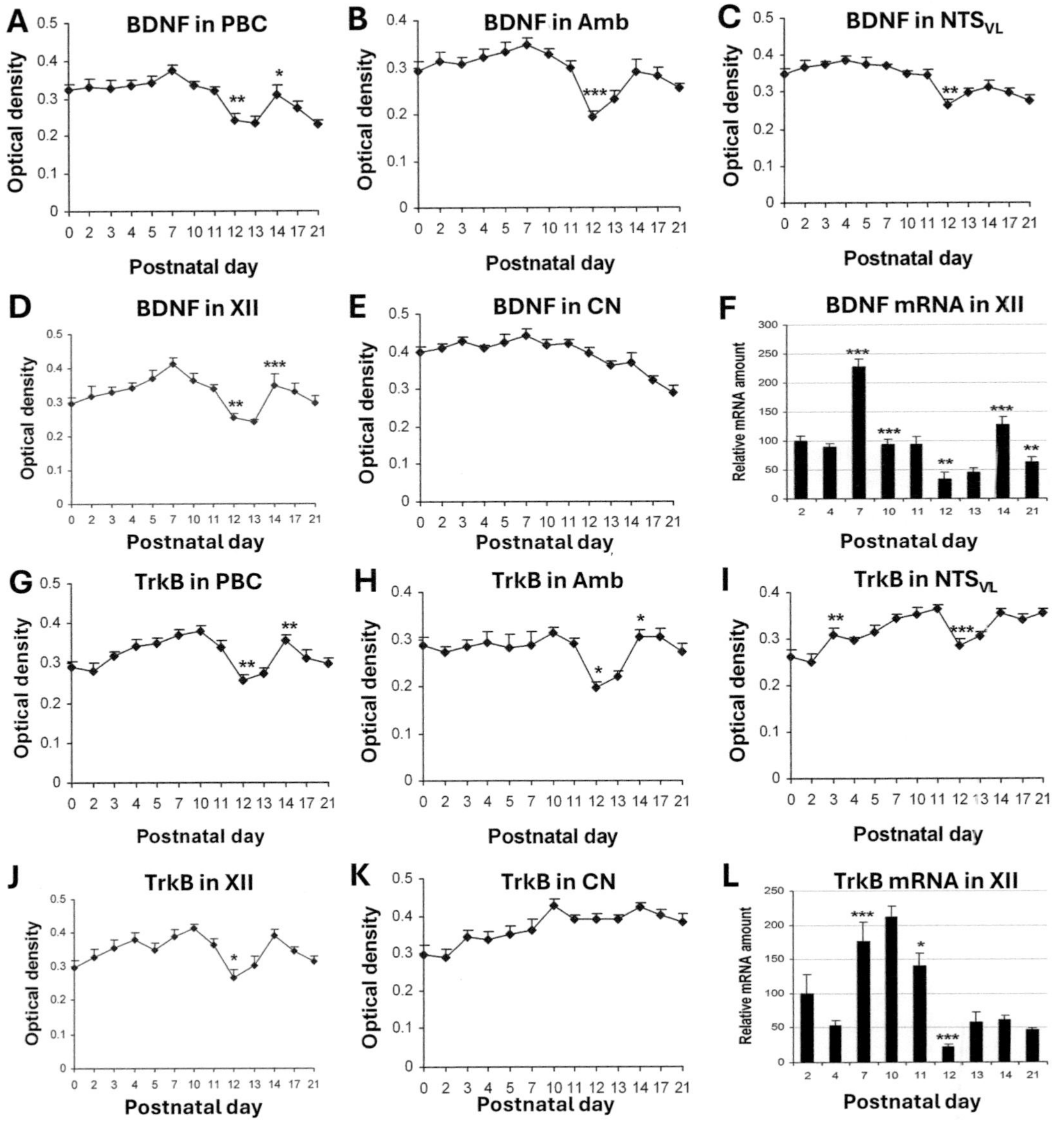

FIGURE 15.6 Development of BDNF and TrkB. Optical densitometric analysis of immunoreaction product of BDNF (A–E) and TrkB (G–K) in different respiratory-related nuclei and the non-respiratory CN from P0 to P21. Note the sudden fall in BDNF and TrkB expressions at P12 that often remains low at P13 before rising again at P14 in PBC, Amb, NTS$_{VL}$, and XII, but not in CN. Each data point represents about 700 (for BDNF) and 1,100 (for TrkB) neurons measured. mRNA levels of BDNF (F) and TrkB (L) in XII at various age groups from P2 to P21. The levels rise sharply for both at P7 but fall to their lowest point at P12. $N = 3$ for each time point. ANOVA indicates a significant difference among the age groups, and Tukey's test indicates significant differences between one age group and the next.

$*P < 0.05$, $**P < 0.01$, $***P < 0.001$.

A–C, G–I, E, and K are from Liu and Wong-Riley 2013b; D, F, J, and L are from Gao et al. 2014.

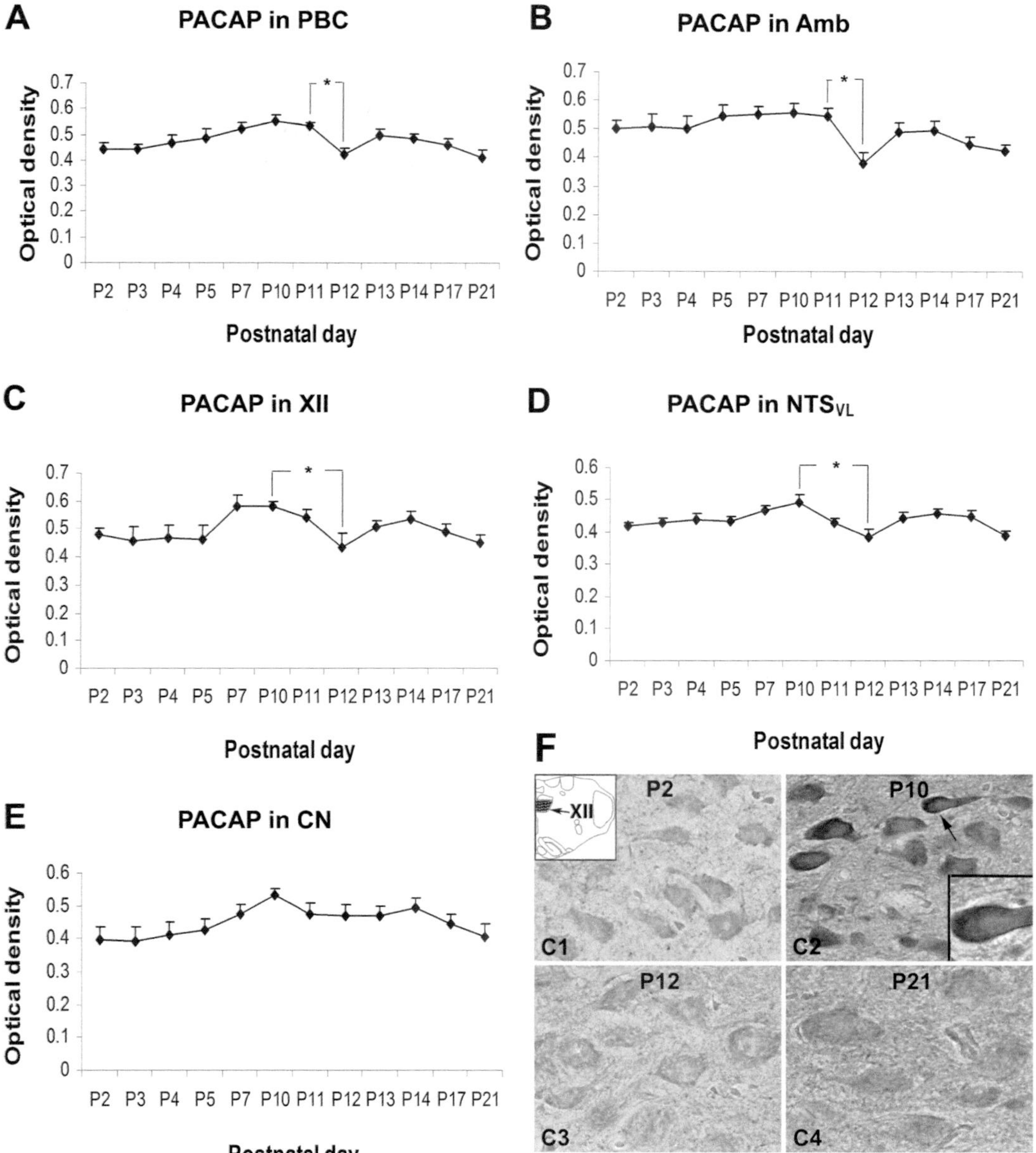

FIGURE 15.7 Development of PACAP. Optical densitometric analysis of PACAP immunoreactivity in neurons of PBC (A), Amb (B), XII (C), NTS_{VL} (D), and CN (E) from P2 to P21. Note that the level peaks around P10 but falls significantly at P12 in the first four nuclei (**P* < 0.05) but not in the CN. Each data point represents 500–800 neurons measured from 8 rats. (F) Representative examples of PACAP immunoreactivity in XII at P2, P10, P12, and P21. Inset in P2 shows the location of XII in the rat brain stem.

Modified from Liu and Wong-Riley 2019.

cognate receptor PAC1 in their brain stem nuclei (Huang et al. 2017a), and abnormalities in the PACAP system are present in SIDS brains (Huang et al. 2017b; Tóth et al. 2023).

Hence, PACAP's downregulation during the critical period of normal brain stem development (Liu and Wong-Riley 2019) deserves to be seriously considered as a contributing factor to SIDS.

15.4 PHYSIOLOGICAL RESPONSES TO HYPOXIA DURING THE CRITICAL PERIOD

Metabolic and neurochemical changes in multiple brain stem respiratory-related nuclei described above around P12 point to a critical period of heightened inhibition, suppressed excitation, and reduced expressions of important trophic factors, including BDNF and PACAP, during normal development. Are ventilatory responses abnormal during this period?

15.4.1 Ventilatory Responses in Normoxia and Acute Hypoxia

The ventilatory development was investigated in normal Sprague-Dawley rats aged P0 to P21 with whole body plethysmography under normoxia (5 min) followed by acute hypoxia (5 min of 10% O_2 + 90% N_2). Each pup was studied every fifth day to avoid the cumulative effect of hypoxia. By using multiple litters and staggering the days, every day from P0 to P21 was covered (Liu et al. 2006). Results show that A) under normoxia (N), there is an age-dependent increase in the breathing frequency (*f*) from birth, peaking at P13, followed by a decline. The tidal volume (V_T) increases with age, but when normalized to body weight, it decreases with age, except for a significant increase at P13. The minute ventilation ($\dot{V}_E$, product of V_T and *f*) normalized to body weight shows minor fluctuations from P0 to P12, a significant rise at P13, followed by a gradual decrease until P21. B) During 5 min of hypoxia (H), there is a biphasic $\dot{V}_E$ response (initial increase followed by a decrease) from P3 onward. However, whereas the ratio of $\dot{V}_{E(H)}$ to $\dot{V}_{E(N)}$ is generally >1 during development, it is <1 after the first 1–2 min from P13 to P16, with the lowest value at P13. C) The H:N ratio during the first 30 s and the last min for *f*, V_T, and $\dot{V}_E$ all show a distinct fall at P13, before the ratios for V_T and $\dot{V}_E$ rise again and that for *f* declines further until P21. D) The H:N ratios averaged over 5 min of hypoxia for *f*, V_T, and $\dot{V}_E$ all drop precipitously at P13. The ratios for V_T and $\dot{V}_E$ rise again until P21, whereas that for *f* remains low thereafter (Figure 15.8A–C). The hypoxic ventilatory response, therefore, is influenced by both *f* and V_T before P13, but is dominated by V_T after P13 in the rat. E) Although the body weight increases steadily with age, the body temperature shows a significant rise between P3 and P4 and a distinct peak at P13.

The above study clearly indicates that, during normal development, a window exists (around P12–13 in the rat) when the animal's response to hypoxia is the weakest, concurrent with many changes in neurochemicals and in COX within brain stem respiratory nuclei. Toward the end of the second postnatal week, then, is rightfully a critical period of respiratory development in the rat.

15.4.2 Metabolic Rate during Normoxia and Acute Hypoxia

As minute ventilation is influenced by the metabolic rate, that is, by the rate of oxygen consumption ($\dot{V}o_2$) and carbon dioxide production ($\dot{V}co_2$), is the weak response to hypoxia at P13 accompanied by an inadequate metabolic rate? When rat pups are monitored daily throughout the first three postnatal weeks (Liu et al. 2009), their ventilation and metabolic rates are surprisingly not correlated at P0–P1, when $\dot{V}o_2$ and $\dot{V}co_2$ in normoxia are at their lowest, but $\dot{V}_E/\dot{V}o_2$ and $\dot{V}_E/\dot{V}co_2$ ratios are the highest. This means that newborn pups have a high rate of ventilation but a relatively low rate of metabolism. The hyperventilation may compensate for the limited alveolar surface area (Burri et al. 1974) and poor efficiency in gas exchange, whereas the low metabolic rate may be due to a weak or absent thermoregulatory response and moderate tissue growth at this time (Mortola 1987). Acute hypoxic challenge (7 min) during this period induces a relatively low ventilatory response due to low chemosensitivity but relatively high metabolic rate as compared to normoxia.

The rest of the first postnatal week is the opposite of P0–P1 (Liu et al. 2009). Under normoxia, the $\dot{V}_E/\dot{V}o_2$ and $\dot{V}_E/\dot{V}co_2$ ratios are reduced but stable with improved gas exchange (Burri et al. 1974; Blanco 1995), and the metabolic rate is relatively high for tissue growth and to compensate for heat loss over a greater surface area-to-body mass ratio (Mortola 1987; Sant'Anna and Mortola 2002).

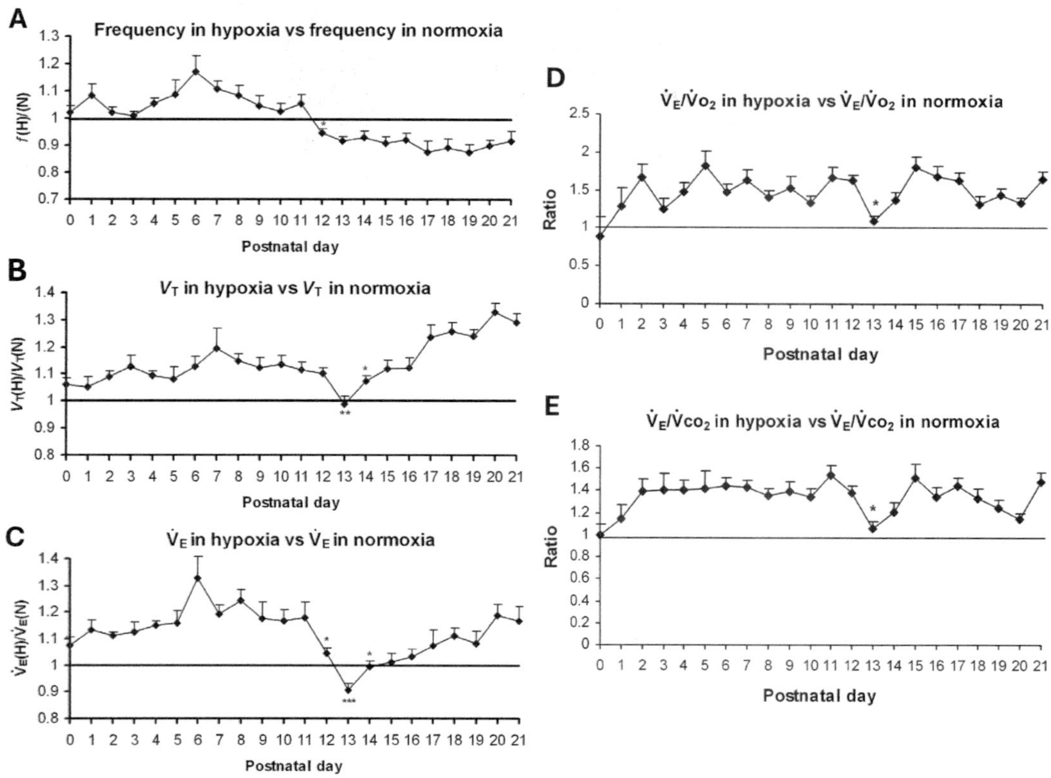

FIGURE 15.8 Development of ventilation and metabolic rate in normoxia and hypoxia. Comparisons of frequency (*f*, A), tidal volume (V_T, B), and minute ventilation ($\dot{V}_E$, C) in 5 min of acute hypoxia versus those in normoxia in rat pups, daily, from P0 to P21. The values are all above 1 from P0 to P12, with a peak at P6 (for *f* and $\dot{V}_E$) or P7 (for V_T). At P12, the ratios for *f* and $\dot{V}_E$ fall precipitously, and at P13, the ratios for V_T and $\dot{V}_E$ are at their lowest, falling just below 1 for V_T and significantly below 1 for $\dot{V}_E$. Both ratios rise significantly at P14 and continue to rise until P21, whereas that for *f* remains below 1 through P21. The ratio between $\dot{V}_E/\dot{V}_{O2}$ in 7 min of hypoxia and $\dot{V}_E/\dot{V}_{O2}$ in normoxia (D), and that between $\dot{V}_E/\dot{V}_{CO2}$ in hypoxia and $\dot{V}_E/\dot{V}_{CO2}$ in normoxia (E), are plotted for each day from P0 to P21. Note that after P0, the lowest ratio for both is at P13, the only time point that shows a statistically significant difference between any two adjacent age groups. A–C: approximately 12–17 animals from seven litters were studied for each day from P0 to P21. Another six animals from a single litter were studied daily from P0 to P21 under normoxia to serve as additional reference. D, E: A total of 18–30 rats from 11 to 16 litters were tested for each day from P1 to P21, and 13 rats from seven litters were tested for P0. All normoxic data are normalized as 1. Tukey's test.

$^*P < 0.05$, $^{**}P < 0.01$, $^{***}P < 0.001$.

A–C: modified from Liu et al. 2006; D and E: modified from Liu et al. 2009.

When challenged with acute hypoxia, the system responds with increased ventilation but with a protective hypometabolism against hypoxia, consistent with the maxim that "the higher the normoxic $\dot{V}o_2$, the greater its decrease during hypoxia" (Mortola 2004). Increase in ventilation, on the other hand, promotes the delivery of O_2 to the mitochondria (Frappell et al. 1992) for the generation of ATP.

After an initial adjustment from P8 to P10, the system is more stable during the second and third postnatal weeks (except for P13), with steady metabolic rates and ventilation-to-metabolic ratios both in normoxia and acute hypoxia. However, a major change occurs at P13, marked by sudden enhancements of the metabolic rate and $\dot{V}_E/\dot{V}o_2$ and $\dot{V}_E/\dot{V}co_2$ ratios in normoxia, as well as a significant increase in f, V_T, V_E, and an abrupt rise in body temperature. Causes of these changes are likely multifaceted (Liu et al. 2009). When challenged with acute hypoxia at P13, both the metabolic rate and $\dot{V}_E/\dot{V}o_2$ and $\dot{V}_E/\dot{V}co_2$ ratios fail to keep pace with the amplified values under normoxia, resulting in an inadequate response. At P13, then, the animal is operating at a relatively low, compromised metabolic rate under hypoxia (Figure 15.8D and E). If hypoxia had been more severe or prolonged during this time, the animal's comparatively low metabolic and ventilatory response and low energy production would be insufficient to meet the increased metabolic demands against hypoxia, and the survival of the animal would be jeopardized. By P14, the ventilatory response to hypoxia remains low, but the metabolic response returns to normal (Liu et al. 2006). The respiratory quotient ($\dot{V}co_2/\dot{V}o_2$) fluctuates during the first three postnatal weeks, with a small but statistically insignificant peak during hypoxia at P13.

These results reinforce the conclusion that toward the end of the second postnatal week is a critical period of respiratory development in the normal rat. When confronted with acute hypoxia, both the ventilatory and metabolic responses are weakest at this time (Liu et al. 2006, 2009). This vulnerability applies to both genders (Liu and Wong-Riley 2013c). Sustained hypoxia between P11 and P15 induces a high rate of unexplained mortality (Mayer et al. 2014). In a rat model of bronchopulmonary dysplasia induced by perinatal hyperoxia, acute ventilatory sensitivity to hypoxia is reduced at P12 to P14 (Mouradian et al. 2019).

15.5 ELECTROPHYSIOLOGICAL CHANGES DURING THE CRITICAL PERIOD

15.5.1 Evidence of Synaptic Imbalance

Thus far, metabolic, neurochemical, and physiological data collectively point toward the end of the second postnatal week as a critical period of respiratory development. What can be gleaned from synaptic development at a single neuronal level? Are there abrupt changes?

Hypoglossal motoneurons (XII) are an appropriate model for daily (P0-P16) testing with whole-cell patch-clamp recordings (Gao et al. 2011). These neurons maintain upper airway patency (see Section 15.1), and their abnormalities are implicated in SIDS and obstructive sleep apnea (Remmers et al. 1978; Lavezzi et al. 2010).

Remarkably, the amplitude and charge transfer of miniature excitatory postsynaptic current (mEPSCs) in XII are significantly reduced at P12–13, whereas those of miniature inhibitory ones (mIPSCs) and their mean frequency are significantly increased ($P < 0.05 - 0.001$). The rise time and decay time kinetics of mEPSCs and mIPSCs accelerate with age. The amplitude and frequency of spontaneous EPSCs (sEPSCs) decrease significantly at P12–13, whereas those of sIPSCs are significantly enhanced ($P < 0.05$ for all) (Figure 15.9A–D). Thus, at the single cell level, a state of synaptic imbalance exists with suppressed excitation and enhanced inhibition at P12–13 in a major brain stem respiratory nucleus in rats.

Several factors likely contribute to the synaptic imbalance in hypoglossal neurons during the critical period: (1) Adjustments in the rate and amount of neurotransmitters released would impact the frequency and amplitude of postsynaptic currents. (2) A change in the properties of postsynaptic receptors would alter their amplitude and kinetics. Examples include a switch in the dominance of specific receptor subunits for $GABA_A$ and glycine receptors from neonatal to mature and from slow to fast kinetics around P12 (see Section 15.3.2). (3) Alterations in the rate and magnitude of energy metabolism, such as a fall in COX at P12 in many respiratory nuclei (Section 15.2), would play a role. Suppressed excitation during the critical period reduces the energy demand at this time. (4) A downregulation of neuromodulators, such as serotonin and its receptors (Section 15.3.3.2), would modify synaptic transmission. For example, serotonin modulates evoked glycinergic IPSCs in XII (Engelhardt et al. 2010). (5) Reduced expressions of PACAP (Section 15.3.3.4) would further disrupt synaptic transmission during the critical period.

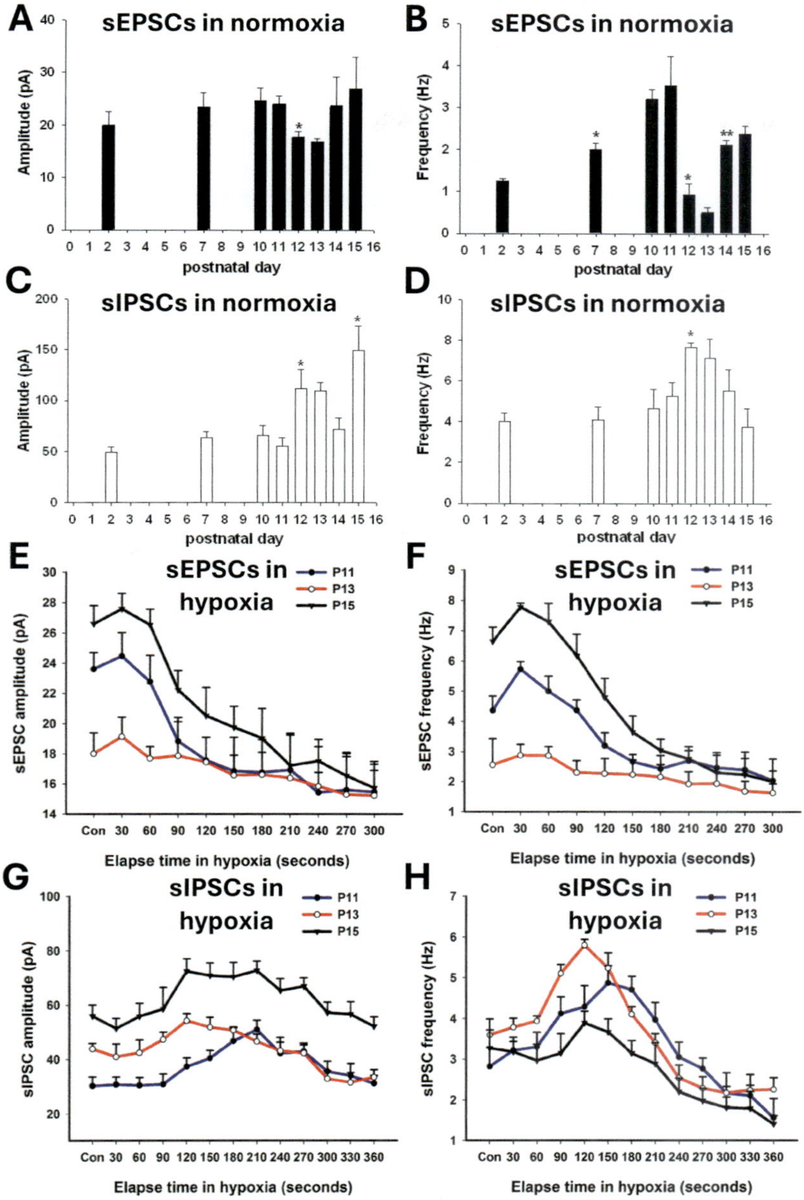

FIGURE 15.9 sEPSCs and sIPSCs in normoxia and hypoxia. Whole-cell patch-clamp recordings of sEPSCs (amplitude and frequency, A and B, respectively) and sIPSCs (amplitude and frequency, C and D, respectively) in hypoglossal (XII) neurons under normoxia at P2, P7, and daily from P10 to P15. Note that both the amplitude and frequency of sEPSCs are significantly reduced at P12–P13, whereas those of sIPSCs are significantly increased at that time. $N = 7–11$ for each data point. Tukey's test for successive age comparisons. $*P < 0.05$, $**P < 0.01$. (E-H) sEPSCs and sIPSCs at baseline (Control or Con) and at 30 sec intervals in acute hypoxia (95% N_2 and 5% CO_2) for 5 min at P11 (blue), P13 (red), and P15 (black). Note that both the amplitude and frequency of sEPSCs (E and F) are at their lowest at P13 at baseline, and their response to hypoxia is also at the lowest compared to the other two age groups. Hypoxia induces a biphasic response at P11 and P15, but not at P13. On the other hand, the amplitude of sIPSCs (G) at P13 is between the other two age groups, and the frequency of sIPSCs (H) at that age is the highest among the three groups. Hypoxia induces an initial decrease, then a rise, followed by a fall in both amplitude and frequency of sIPSCs for all three age groups, but the greatest rise is at P13.

A–D: modified from Gao et al. 2011. E–H: modified from Gao et al. 2015.

15.5.2 Effect of Acute Hypoxia at Different Ages

When NTS_{VL} neurons, recipients of carotid body input, in brain stem slices were exposed to acute hypoxia (95% N_2 and 5% CO_2) for 5–6 min, P13 neurons showed the weakest response in both the amplitude and frequency of sEPSCs but the frequency of their sIPSCs was increased as compared to those aged P11 or P15 (Gao et al. 2015) (Figure 15.9E–H).

Clearly, at the synaptic level, P12–13 represents a critical period of synaptic imbalance, during which the response to hypoxia is the weakest.

15.5.3 Role of BDNF in Synaptic Imbalance

Does BDNF play any role in synaptic imbalance during the critical period? As discussed in Section 15.3.3.3 above, BDNF normally potentiates excitatory and attenuates inhibitory synaptic transmission. A significant reduction in its expression around P12 in multiple respiratory nuclei would lead to suppressed excitation and enhanced inhibition. If so, then exogenous BDNF should at least partially reverse the synaptic imbalance. This was tested with whole-cell patch-clamp recordings on XII, daily, from P0 to P16 (Gao et al. 2014) (Figure 15.10). Indeed, exogenous BDNF significantly increases the normally reduced frequency of sEPSCs but decreases the normally heightened amplitude and frequency of sIPSCs during the critical period (P12–13). Moreover, exogenous BDNF reduces the normally enhanced frequency of mIPSCs at P12–13. These effects are partially blocked by K252a, a TrkB receptor antagonist. Hence, reduced levels of BDNF and TrkB toward the end of the second postnatal week contribute significantly to synaptic imbalance in brain stem respiratory nuclei at that time.

The role of BDNF was further tested in a central respiratory chemosensor, NTS_{VL}, which also exhibits suppressed sEPSC and enhanced sIPSC amplitudes during the critical period (Gao et al. 2015). These neurons were monitored with patch-clamp recordings after in vivo applications of either a TrkB agonist, 7,8-DHF (7,8-dihydroxyflavone, at 5 mg kg^{-1}), or a TrkB antagonist, ANA-12 (at 2.46 mmole kg^{-1}) (Gao et al. 2015). Strikingly, intraperitoneal injections of 7,8-DHF once a day for 2 days during the critical period (P12–13) prevent the synaptic imbalance at that time (Figure 15.11A–D), whereas application of ANA-12 once a day for only one day during the critical period accentuates the imbalance (Figure 15.11E–H). Treatment of either agent before the critical period (P10–11) has no effect, whereas 7,8-DHF after the critical period (P14–15) increases the frequency of sEPSCs and decreases the amplitude of sIPSCs, though to a much lesser degree than during the critical period. ANA-12 treatment after the critical period reduces the amplitude and frequency of sEPSC and increases those of sIPSCs in XII, an effect opposite to that of 7,8-DHF. In vivo ANA-12 significantly attenuates the amplitude of mEPSCs and potentiates the amplitude and frequency of mIPSCs during the critical period.

These results further strengthen the contributory role of BDNF downregulation in the synaptic imbalance during the critical period of respiratory development in rats.

15.6 IS A CRITICAL PERIOD NECESSARY FOR RESPIRATORY DEVELOPMENT?

What are the reasons underpinning the synaptic imbalance during the critical period? Why would normal development permit or even necessitate such an imbalance?

Both electrophysiological studies in XII (Gao et al. 2015) and ultrastructural studies in the cerebellum (Mjaatvedt and Wong-Riley 1988) indicate that excitatory synapses develop and mature earlier than inhibitory ones. During the first week and a half of brain stem respiratory nuclei development, expressions of glutamate, NMDA receptors, and COX increase with age, and levels of the trophic factors BDNF, neuromodulators serotonin, and PACAP are all relatively high, promoting the establishment of an excitatory network (see above). However, once the excitatory innervation of the network is launched, refinement of respiratory circuits requires enhanced inhibitory modulation to

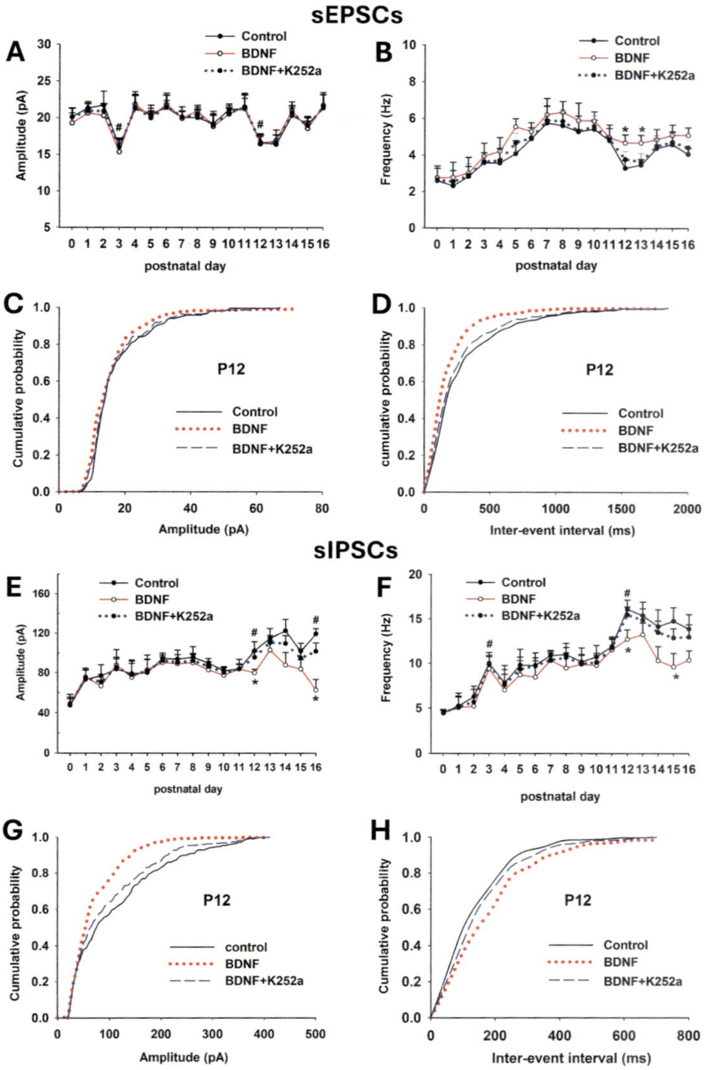

FIGURE 15.10 Effect of BDNF on sEPSCs and sIPSCs in hypoglossal (XII) neurons. Whole-cell patch-clamp recordings of amplitude (A) and frequency (B) of sEPSCs and amplitude (E) and frequency (F) of sIPSCs in XII neurons from P0 to P16, daily. Each neuron was recorded before (control), after 2 min of 100 pM BDNF (red), then after BDNF-plus-K252a, a TrkB antagonist (dotted line). The amplitude of sEPSCs (A) is significantly low at P3 ($N = 11$) and at P12–13 ($N = 15$) (*$P < 0.05$ for both), but the applications of BDNF and BDNF-plus-K252a do not result in any change, as verified by the lack of change in the cumulative probability of amplitude at P12 (C). The frequency of sEPSCs is significantly low at P12–13 (B), but BDNF increases it, and this increase is significant only at P12–13 (*$P < 0.05$ for both. $N = 15$ at P12 and 17 at P13) (D). The cumulative probability plot for the inter-event interval at P12 shows a shift to the left after BDNF at P12, indicating an increase in frequency. BDNF-plus-K252a returns frequency to control levels at every age (B). On the other hand, the amplitude of sIPSCs (E) increases significantly at P12 ($N = 18$) and P16 ($N = 10$) (*$P < 0.05$ for both), and the frequency (F) also increases at P3 and P12 (*$P < 0.05$ for both). BDNF induces significant decreases in the amplitude at P12 and P16 (E) and significant decreases in the frequency at P12 and P15 (F) (*$P < 0.05$ for both). The cumulative probability plot of sIPSC amplitude at P12 shows a shift to the left after BDNF (G), indicating a reduction of amplitude, and K252a reverses it. The cumulative probability of inter-event interval at P12 shows a shift to the right after BDNF (H), verifying a decrease in sIPSC frequency, and K252a again reverses that. Day-to-day comparisons are made with Tukey's studentized test.

Modified from Gao et al. 2014.

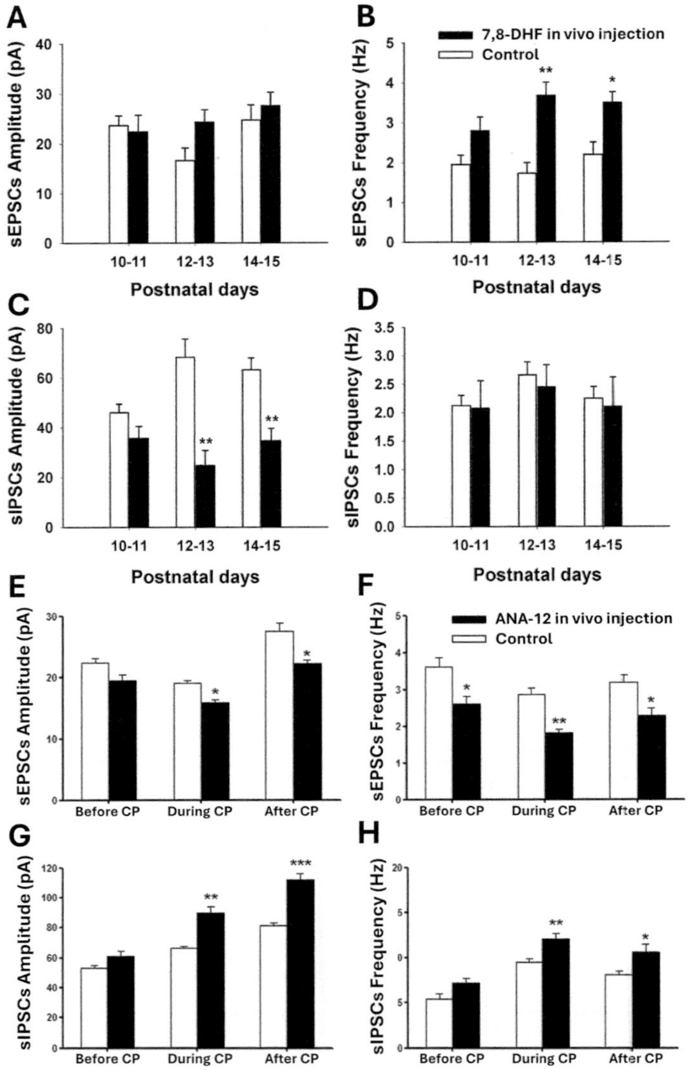

FIGURE 15.11 Effect of *in vivo* 7,8-DHF or ANA-12 on sEPSCs and sIPSCs at different ages. (A-D) Amplitude and frequency of sEPSCs and sIPSCs in NTS_{VL} neurons after *in vivo* injections of 7,8-DHF (5 mg/kg, i.p.), once a day for 2 days before (P10–11), during (P12–13), and after (P14–15) the critical period (CP), and recorded on the 3rd day. Control animals received a comparable volume of vehicle (10% DMSO). Under normal (control) conditions, these neurons, like those of XII, exhibit reduced sEPSCs and heightened sIPSCs during the CP. 7,8-DHF prevents such an imbalance by enhancing sEPSCs and suppressing sIPSCs, reaching significance for sEPSC frequency and sIPSC amplitude during (and after) the CP. (E–H) Amplitude and frequency of sEPSCs and sIPSCs in XII neurons after *in vivo* injections of ANA-12 (2.46 mM/kg, i.p.), once a day for 1 day only, and recorded the next day before (P8), during (P12–13), and after (P16) the CP. Control animals received a comparable volume of vehicle (10% DMSO). Contrary to the effect of 7,8-DHF, ANA-12 accentuates the synaptic imbalance during the CP by suppressing sEPSCs and enhancing sIPSCs during that time. The values are significant for the amplitude and frequency of both sEPSCs and sIPSCs during and after the CP, as well as for the frequency of sEPSCs before the CP. $N = 9–11$ for each data point in A-D, and $N = 10–23$ for each data point in E–H. Tukey's test for successive age comparisons.

$^*P < 0.05$, $^{**}P < 0.01$.

Modified from Gao et al. 2015.

fine-tune the respiratory functions and connections. This developmental transition necessitates a transient downregulation (but not elimination) of the trophic factors and neuromodulators so that inhibitory synapses can flourish. Indeed, GABA and glycine are dramatically upregulated, and their receptors undergo switches in dominance from the neonatal to the mature forms, and the increased influence of KCC2 over NKCC1 enables the transition and strengthening of hyperpolarizing inhibitory synapses over depolarizing excitatory ones for GABA and glycine. Reduced excitation during this critical period downregulates the energy demand, and the level of COX temporarily plummets. This creates a window of vulnerability. When challenged with respiratory stressors such as acute hypoxia, the animals exhibit the weakest response. With more severe stressors such as sustained hypoxia, the response may be woefully insufficient, and death may ensue (Sections 15.3.1 and 15.4.2). The vulnerability appears to extend to the immune system. When given a sublethal dose of endotoxin 2 days after a nonlethal dose of influenza A virus, 12-day-old rat pups unexpectedly die, whereas younger or older ones do not die from the same treatment (Blood-Siegfried et al. 2002). These findings have significant implications for SIDS, as seemingly healthy infants die unexpectedly during a narrow window of postnatal development if the three risk factors collide at the same time. In comparing brain stem developmental time scales between rodents and humans, Ballanyi (2004) considers P11–14 in rodents (a critical period of respiratory development discussed in this chapter) to be comparable to human postnatal 2 to 4 months of age, the peak incidence of SIDS.

In conclusion, the critical period of respiratory development is a transitional time between immaturity and maturity. It is a natural, *necessary* phase of *normal* development and should not be eliminated (Wong-Riley and Liu 2008; Wong-Riley et al. 2013, 2019). Fortunately, it is only transitory, and, in time, a balanced state of excitation and inhibition is achieved, likely driven by both genetic and environmental forces.

16 Critical Period of Postnatal Primary Visual Cortical Development in Rats

ABSTRACT

The pioneering work of Hubel and Wiesel in the 1960s and 1970s opened a floodgate of research on the critical period of visual cortical development in mammals. In rats, cytochrome c oxidase shows an age-dependent increase in activity from the day of eye opening (P14-P15) to maturity (P35-P36), except for a sudden and significant fall from the end of the 4th to the end of the 5th postnatal week. Patch-clamp recordings reveal a precipitous decline in spontaneous EPSCs and a marked rise in sIPSCs during this period. Concurrently, expressions of excitatory neurochemicals, BDNF, and TrkB receptors are transiently downregulated, whereas those of inhibitory $GABA_A$ receptors are continually upregulated. Such an imbalance of suppressed excitation and enhanced inhibition explains the reduced demand for energy supply, hence reduced COX activity. Application of an agonist of TrkB, 7,8-DHF, in vivo partially reversed the synaptic imbalance, whereas a TrkB antagonist, ANA-12, accentuated the imbalance during this transient period. Monocular deprivation for 4 days is most detrimental during this period of synaptic imbalance, justifiably considered a critical period. Combined results from the visual cortex and brain stem respiratory system point to a set of basic principles regarding critical periods of postnatal development. The existence of a master regulator or a set of regulators deserves serious consideration in future studies.

Studies on the critical period of brain stem respiratory development (Chapter 15) prompt the question: What can COX reveal about the critical period of visual cortical development in rats?

The mammalian visual cortex is not fully mature at birth. A great deal of structural and functional growth and adjustments take place postnatally (Fagiolini et al. 1994). In the rat, even shortly after eye opening (P14 in Sprague–Dawley or P15 in Long-Evans rats), visual acuity is only half the adult value, mean receptive field size is 5 to 6 times that of the adult, and selectivity for orientation and direction of movement is "almost absent" (Fagiolini et al. 1994).

The concept of a "critical period" of postnatal visual development was introduced more than half a century ago by Hubel and Wiesel. They found that monocular lid suture during the first few months of a kitten's life significantly reduced the size of deprived LGN neurons and drastically decreased the number of visual cortical cells driven by the closed eye (Wiesel and Hubel 1963a, 1963b; Hubel and Wiesel 1970). The period of susceptibility appears suddenly around the start of the 4th postnatal week, peaks between the 4th and 5th week, and disappears around the end of the 3rd postnatal month. Three months of monocular deprivation from birth resulted in limited recovery in cortical physiology, even when the closed lid was opened for 5 years. Monocular deprivation in the adult for a year had no detectable effect. Their later study in macaque monkeys monocularly enucleated or lid sutured from the 2nd or 3rd postnatal week for varying periods of time yielded similar and even more drastic results, in that normal eye columns in the primary visual cortex expanded greatly at the expense of markedly shrunken deprived eye columns (Hubel et al. 1977). The open eye was at a distinct competitive advantage by either sprouting into the neighboring column or maintaining its wide territory, while the deprived eye afferents retracted more than they normally would (Hubel et al. 1977). The period of susceptibility was recognized as the critical

 DOI: 10.1201/9781003507765-16

period of visual cortical development, and changes in the ocular dominance were known as ocular dominance plasticity. In macaque monkeys, monocular lid suture at 1 week of age led to the greatest shrinkage in columns, whereas suturing at 12 weeks had no effect, indicating that the first few weeks are most critical in the development of binocular vision in the primate (Horton and Hocking 1997).

Following these groundbreaking studies, many laboratories discovered that perturbations of visual experiences during the critical period result in nearly permanent disruption of the structural and/or functional organization of the visual cortex. Whereas retinotopic maps and orientation selectivity are generally innately established and sustained by spontaneous neural activity, visual experience refines the binocular matching of stimuli, including orientation preference, in cortical neurons (reviewed in Espinosa and Stryker 2012). Monocular deprivation in rodents decreases excitability, visual acuity, contrast sensitivity, and binocular matching in cortical neurons, and dark rearing prolongs the critical period and reduces the selectivity of visual cortical neurons for orientation and direction of movement (Cynader 1983; Stafford 1984; Fagiolini et al. 1994; Gianfranceschi et al. 2003; Nataraj et al. 2010). In the cat, monocular deprivation during the critical period until 35 days or older results in a permanent loss of binocular depth perception, whereas similar treatment at 4 months (after the critical period) has no adverse effect (Timney 1983). Alternating occlusion for 10 or 21 days during the critical period also reduces the number of binocular neurons in the visual cortex (Presson and Gordon 1979).

Ocular misalignment induced surgically (resulting in esotropia or exotropia) or optically (with a high-powered minus lens, causing anisometropia) during the critical period also leads to a permanent loss of binocular neurons and stereopsis in the visual cortex of cats and monkeys (Berman and Murphy 1981; Crawford et al. 1993). Surprisingly, as little as 3 days of optically imposed strabismus at the height of the critical period in monkeys enhances binocular suppression, but interocular spatial phase disparity is not affected (Zhang et al. 2005).

Blakemore and Cooper (1970) reported that kittens raised in the dark except for a few hours in the day viewing only vertical or horizontal stripes had their visual cortical neurons preferring stimuli orientation that they were raised in and not stimuli within 20° of the inappropriate orientation (anisotropy). However, a later systematic, quantitative study failed to substantiate such a bias (Stryker and Sherk 1975). Further, animals were responsive to stimulus orientations that matched their innate selectivity but unresponsive to those that were not, indicating that orientation selectivity is innate, and that visual experience serves to maintain rather than develop such selectivity (reviewed in Espinosa and Stryker 2012). Still another study of stripe rearing in juvenile mice using two-photon calcium imaging indicates that deeper (but not superficial) layer II/III neurons do show an increased bias toward the experienced orientation, suggesting an instructive mechanism, at least for those neurons (Kreile et al. 2011). Deeper neurons were not examined.

Regarding direction-selectivity, it seems to differ among species. Whereas visual experience is needed for the development of such selectivity in ferrets and cats, it is not necessary for mice (reviewed in Espinosa and Stryker 2012). Striate neurons in young infant macaque monkeys also have a much lower overall directional sensitivity than those in adults, indicating that visual experience contributes to the development of this attribute (Hatta et al. 1998). Mice, on the other hand, possess both orientation- and direction-selective neurons at eye opening and retain them even when reared in darkness (Rochefort et al. 2011).

The critical period, then, is when important receptive field properties, such as ocular dominance, orientation, and directional selectivity, are becoming fully developed and fine-tuned (Hubel and Wiesel 1970; Fagiolini et al. 1994; Tan et al. 2022).

The timing of the critical period varies among species, but the exact timing for each species is not always clear. Typically, only a few time points were examined by the investigators, and the extent of deprivation varies widely among studies, even for the same species (Rothblat and

Schwartz 1979; Guire et al. 1999). In monkeys, sensitivity to monocular form deprivation lasts throughout the first 2 years of life (Crawford et al. 1993). However, photopic threshold spectral sensitivity is complete by 6 months of age, while scotopic functions are mature by 3 months (Crawford et al. 1993). Horton and Hocking (1997) stated that the "primate visual cortex is most vulnerable to deprivation during the first weeks of life" without any apparent delay in onset. Such vulnerability pertains to ocular dominance column shrinkage, whereas vulnerability to amblyopia may extend up to 2 years (Horton, personal communication). In cats, the height of susceptibility is between the 4th and 5th postnatal weeks and disappears by the end of the 3rd month (Hubel and Wiesel 1970). In hooded rats, the critical period reportedly begins at the end of the 3rd postnatal week, peaks between the 4th and 5th week, before declines to reach the adult state after the 5th week (Fagiolini et al. 1994). However, a later study claimed the period to be at least 2 weeks longer (Guire et al. 1999). In mice, the critical period for experience-dependent plasticity of binocular response has been identified as between P19 and P32 (Gordon and Stryker 1996). More recent studies using visual evoked potentials claim that the critical period ends at about 5 weeks of age (Frenkel and Bear 2004).

The timing of the critical period is alterable experimentally. The roles of GABA and BDNF in the opening and closure of this period will be discussed below. Of special significance are the extracellular chondroitin sulphate proteoglycans that form perineuronal nets around neurons (especially GABAergic, fast-spiking, parvalbumin-positive basket cells) and act as brakes to close the critical period (Pizzorusso et al. 2002; Bavelier et al. 2010). Closure can be delayed by means of chondroitinase or dark rearing (Pizzorusso et al. 2002; Bavelier et al. 2010). Many other brake agents have also been found, such as myelin, Nogo receptors, and Lynx protein (reviewed in Bavelier et al. 2010).

16.1 CYTOCHROME *c* OXIDASE REVEALS A PERIOD OF LOW METABOLIC ACTIVITY DURING POSTNATAL DEVELOPMENT OF THE RODENT VISUAL CORTEX

Insights gained from probing the critical period of the brain stem respiratory system (Chapter 15) point to COX as a sensitive indicator of the precise period when synaptic imbalance is prominent and when the animal is most vulnerable to external stressors directed at that specific system. Will COX reveal similar changes in the visual cortex? The brain stem studies also highlight the importance of (a) understanding the *normal* course of development (the controls) before one can determine the abnormal; and (b) close temporal spacing in monitoring the development of COX (*daily* in the rat) before, during, and after the suspected critical period so that one can capture unexpected changes in metabolic demands of developing neurons.

When the expression of COX was analyzed with single neuron optical densitometry in each of the visual cortical layers II to VI for each postnatal day from the day of eye opening (P14 in SD rats) to P36 (when cortical neurons presumably have matured), a common trend emerges for each of the five cellular layers. The metabolic activity of neurons does not follow a smooth path of increases with age. It rises from P14 and peaks at P27 before a sudden precipitous fall at P28, followed by a further decrease to the lowest point between P30 and P32, before rising gradually to match the P27 level at P35 (Zhang et al. 2018) (Figure 16.1A).

The same trend exists in the hooded Long-Evans (LE) rats, whose eyes do not open until a day later (P15). This shifts the trend by one day, that is, the peak is reached at P28, falls at P29, remains low until P33, before rising to P28 level at P36 (Zhang et al. 2018) (Figure 16.1B).

COX, therefore, strikingly reveals a transient period between the end of the 4th to the end of the 5th postnatal week when a pronounced decline in the metabolic activity of visual cortical neurons exists in rats. What is the cause of such a drastic change?

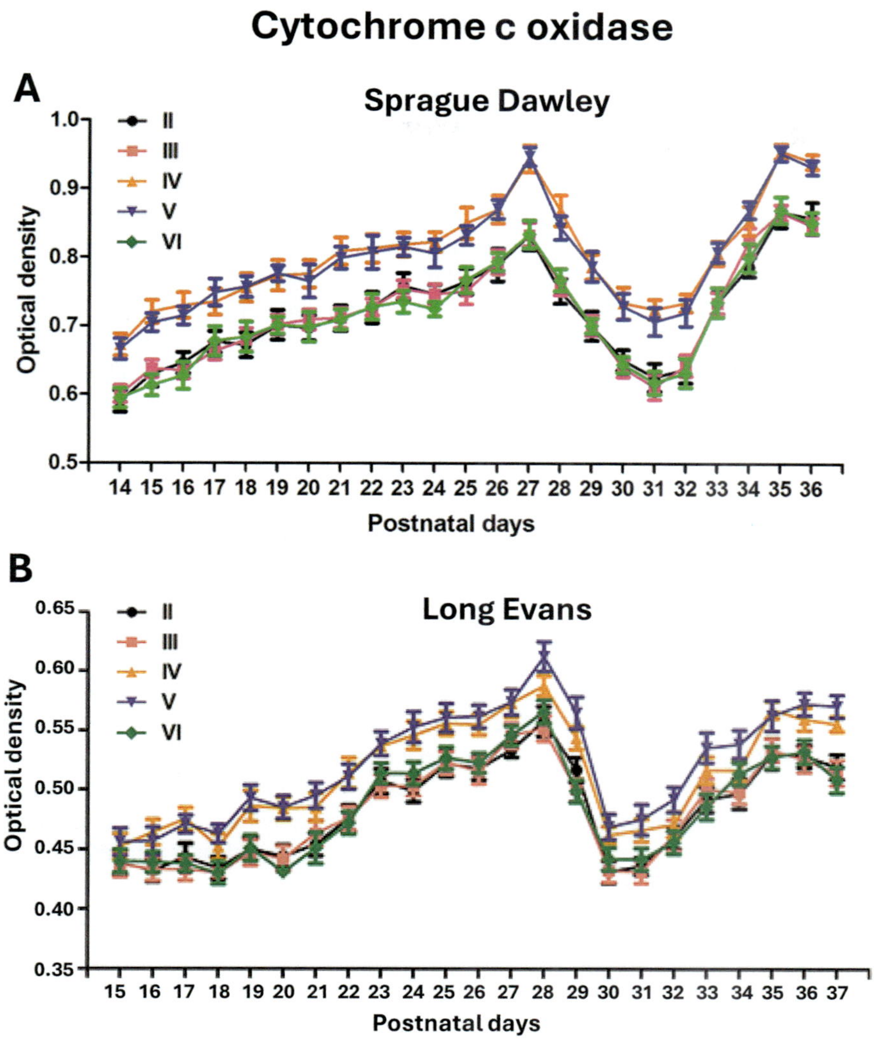

FIGURE 16.1 Optical densitometric analyses of reaction product of COX histochemistry in single neurons of each of five cellular layers (II to VI) in the visual cortex of Sprague–Dawley (A) and Long-Evans (B) rats for each of the postnatal days, from P14 to P36 in A and from P15 to P37 in B. The developmental trends are similar among the five layers. Note that after reaching a peak at P27 (in A) and P28 (in B), there is a sudden fall in the expression starting at P28 (in A) or at P29 (in B), reaching the lowest point at P30-P32, before returning to peak levels at P35 (in A) or at P36 (in B). N = 100–150 (for A) or 50-100 (B) neurons from at least 2 litters for each layer at each time point tested.

Modified from Zhang et al. 2018.

16.2 ELECTROPHYSIOLOGICAL CHANGES DURING VISUAL CORTICAL DEVELOPMENT: EVIDENCE OF A SYNAPTIC IMBALANCE

16.2.1 Spontaneous Excitatory and Inhibitory Synaptic Currents (sEPSCs and sIPSCs)

Whole-cell patch-clamp recordings of layer V pyramidal neurons of the SD rat visual cortex were monitored daily, from P14 to P36. The frequency of excitatory sEPSCs increases to a greater extent than the amplitude from P14 to P27, indicating an increase in excitatory drive. Both parameters, however, fall precipitously from P28 to P33 before rising to P27 levels at P34 and remain so to P36 (Figure 16.2A–D). The fall in the amplitude of sEPSCs is likely due to a downregulation of postsynaptic excitatory receptors. The decline in the frequency probably results from a decrease in excitatory drive and an increase in inhibitory input. The overall 10-90% rise time of sEPSCs does not alter significantly with age, but the decay time increases markedly with age, indicating longer periods of sustained depolarization with maturity (Zhang et al. 2018).

Like sEPSCs, the frequency of inhibitory sIPSCs rises more steeply than that of amplitude from P14 to P27. However, in contrast to sEPSCs, both parameters of sIPSCs increase significantly at P28 and remain high until P33, before a precipitous fall at P34 to the P27 level, followed by a plateau until P36 (Figure 16.2E–H). The marked upsurge in the amplitude of sIPSCs between P28 and P33 indicates an increase in inhibitory (GABAergic) receptors as well as a more powerful surge of inhibitory Cl^- current induced by a longer and/or more frequent opening of Cl^- channels by KCC2. The prominent rise in the frequency of sIPSCs during that time signifies an increase in presynaptic inhibitory spiking activity. Both the 10-90% rise time and decay kinetics of sIPSCs accelerate with age (Zhang et al. 2018).

Thus, a state of synaptic imbalance exists from the end of the 4th to the end of the 5th postnatal week, with suppressed excitation and enhanced inhibition, coinciding temporally with the period of reduced COX activity in visual cortical neurons.

16.2.2 Miniature Excitatory and Inhibitory Synaptic Currents (mEPSCs and mIPSCs)

The miniature excitatory and inhibitory synaptic currents were also monitored in layer V pyramidal neurons at ages P23, 25, and daily from P27 to P36 (Zhang et al. 2018). A significant decrease in the amplitude of mEPSCs occurs at P28 and remains low until P33 before a significant increase ensues at P34. This trend mimics that of sEPSCs and indicates a postsynaptic adjustment of excitatory neurotransmitter receptors (see below). The reduction in the frequency of mEPSCs between P28 and P33 is evident, but it does not reach statistical significance. Nonetheless, such reductions reflect a decrease in excitatory drive to visual cortical neurons.

On the other hand, the amplitude of mIPSCs rises steadily from P23 and peaks at P32, before returning to P31 level at P34. This is likely caused by a steady increase in the density of postsynaptic inhibitory (GABAergic) receptors as well as a longer and/or more frequent opening of Cl^- channels during that time. The frequency of mIPSCs steadily increases from P23 to P29 followed by a sizable increase between P30 and P33, before a marked fall at P34 and a slight rise until P36. The prominent rise in both sIPSCs and mIPSCs during the 5th postnatal week signifies a strong inhibitory drive to cortical neurons at that time.

16.2.3 The Excitatory/Inhibitory (E/I) Ratio and AMPA/NMDA Receptor-Mediated Evoked EPSC (eEPSC) Ratio

When evoked EPSCs (eEPSCs) and IPSCs (eIPSCs) were recorded from the *same* neurons, the excitatory/inhibitory (E/I) ratio is found to be significantly lower at P28, the period of synaptic imbalance, compared to ratios before (P18 and P25) and after (P36) that period ($P < 0.001$ for all) (Zhang et al. 2018) (Figure 16.3A and B). No significant difference is found when P18, P25, and P36 are compared with each other. This is notable because P25 is considered part of the critical period by several investigators (Perkins and Teyler 1988; Fagiolini et al. 1994; Guire et al. 1999). The fall in the E/I ratio during the period of synaptic imbalance underscores the concomitant plunge in excitation and the prominent upsurge in inhibition at that time.

When the ratio of AMPA- versus NMDA receptor-mediated eEPSCs was tested and compared before, during, and after the period of synaptic imbalance (i.e., P23-24, P28, and P35-36, respectively), no difference was found among the three age groups (Figure 16.3C and D). This indicates that there is no obvious dominance of either AMPA or NMDA receptor types during the ages tested.

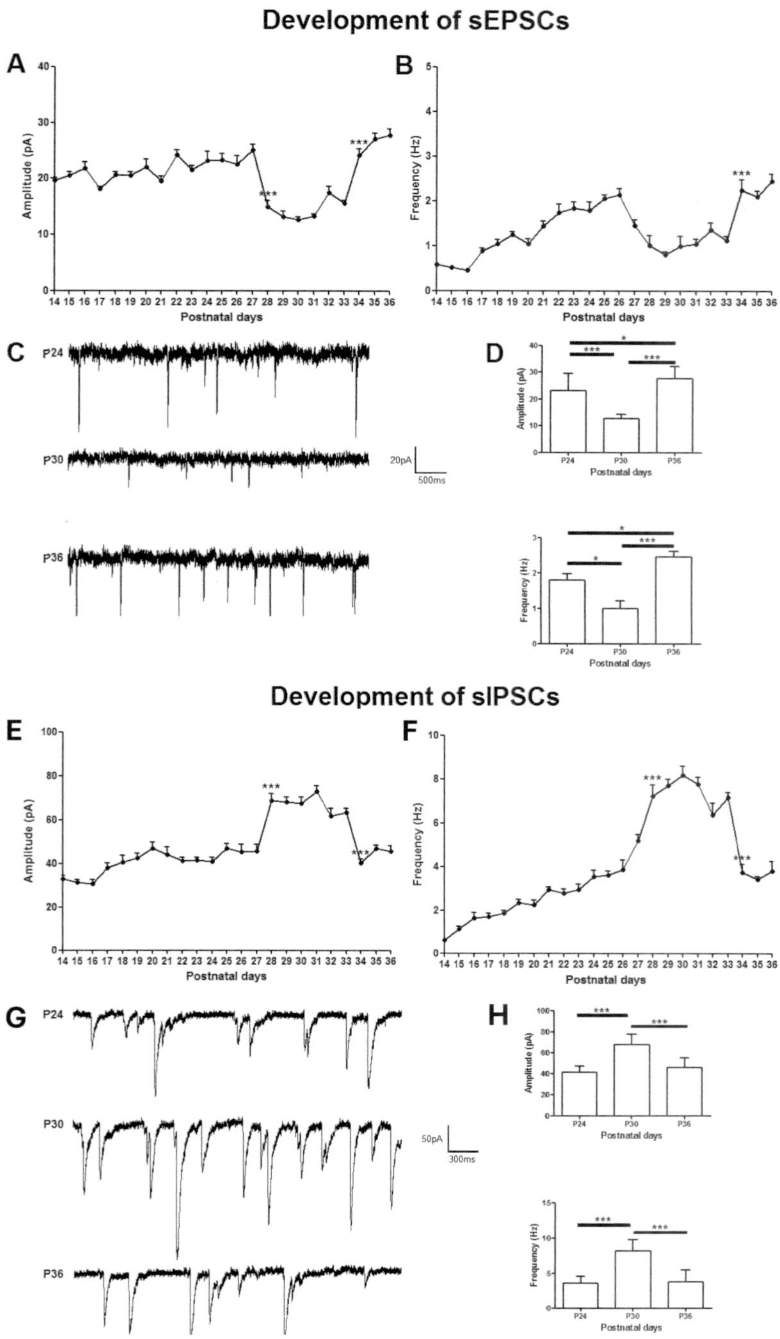

FIGURE 16.2 Development of sEPSCs and sIPSCs in the rat visual cortex. (A and B) Mean amplitudes and frequencies of sEPSCs in layer V pyramidal neurons were recorded daily from P14 to P36. Note that the amplitudes plunge sharply at P28 ($P < 0.001$) and remain low until rising significantly at P34 ($P < 0.001$). The frequencies decrease more gradually from P27 to P29 and remain low until P32, before rising sharply at P34 ($P < 0.001$). $N \geq 15$ from at least 2 litters for each time point tested. (C) Sample traces of sEPSCs at P24, P30, and P36. (D) All pairwise comparisons are significant. $*P < 0.05$, $***P < 0.001$. (E and F) Mean amplitudes and frequencies of sIPSCs in layer V pyramidal neurons were recorded daily from P14 to P36. Note that both parameters rise significantly at P28 ($P < 0.001$ for both) and remain high before falling abruptly at P34 ($P < 0.001$ for both). $N \geq 15$ from at least 2 litters for each time point tested. (G) Sample traces of sIPSCs at P24, P30, and P36. (H) Pairwise comparisons are significant between P24 and P30, and between P30 and P36 for both parameters. ($P < 0.001$ for all).

Modified from Zhang et al. 2018.

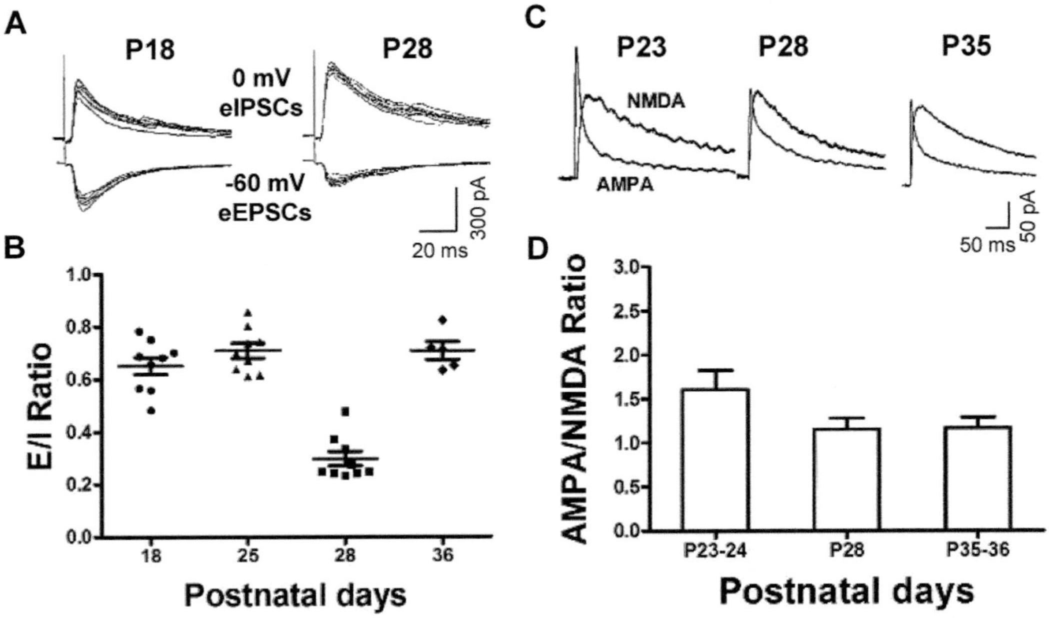

FIGURE 16.3 E/I ratio and AMPA/NMDA receptor-mediated eEPSC ratio. (A) Sample tracings of evoked (e) IPSCs (voltage clamped at 0 mV) and eEPSCs (voltage clamped at -60 mV) recorded at P18 and P28. Compared to those at P18, the amplitude of eIPSC is greater, and that of eEPSC is less at P28. (B) In comparing the E/I ratio, the one at P28 is significantly lower than those at P18, P25, and P36. $N = 9$ at P18, 9 at P25, 9 at P28, and 5 at P36. Six stimulations for each cell. One-way ANOVA and Tukey's test. $P < 0.0001$ when P28 is compared to each of the other three time points, but no difference when those three time points are compared with each other. (C) Representative tracings of AMPA and NMDA receptor-mediated eEPSCs at P23-24, P28, and P35. (D) The AMPA/NMDA synaptic ratio at the three time points shows no significant difference among them. $N = 8$ at P23-24, 6 at P28, and 6 at P35-36.

From Zhang et al. 2018.

16.3 NEUROCHEMICAL CHANGES DURING THE CRITICAL PERIOD

16.3.1 Developmental Changes in NMDA and AMPA Receptors

Expressions of the obligatory NMDA receptor subunit 1 (GluN1) and AMPA receptor subunit 1 (GluA1) were analyzed with single-cell optical densitometry in neurons of cortical layers II to VI, daily, from P14 to P36, in SD rats (Zhang et al. 2018). Developmental trends of both are similar among the layers and are virtually identical to those of COX. There is an age-dependent increase in the expression of GluN1 and GluA1 from P14 to P27, peaking at P27. However, both fall abruptly at P28, reaching the lowest point at P30-31 before rising gradually until attaining P27 levels at P35 (Figure 16.4A and C). The intensity of immuno-labeling centered on the cytoplasm is comparable to that on the cell membrane (Figure 16.4B). The decrease in excitatory glutamatergic receptors from P28 to P33 is concurrent with reductions in sEPSC and mEPSC amplitudes (Sections 16.2.1 and 16.2.2).

The expressions of GluN1 and GluA1 were also tested in the hooded LE rats, and the trends are equivalent to those of SD rats except for a one-day shift, as in the case of COX (Zhang et al. 2018). These results indicate that electrophysiological changes in EPSCs are paralleled by depressed glutamatergic receptor expressions during the period of synaptic imbalance in the rat's visual cortex.

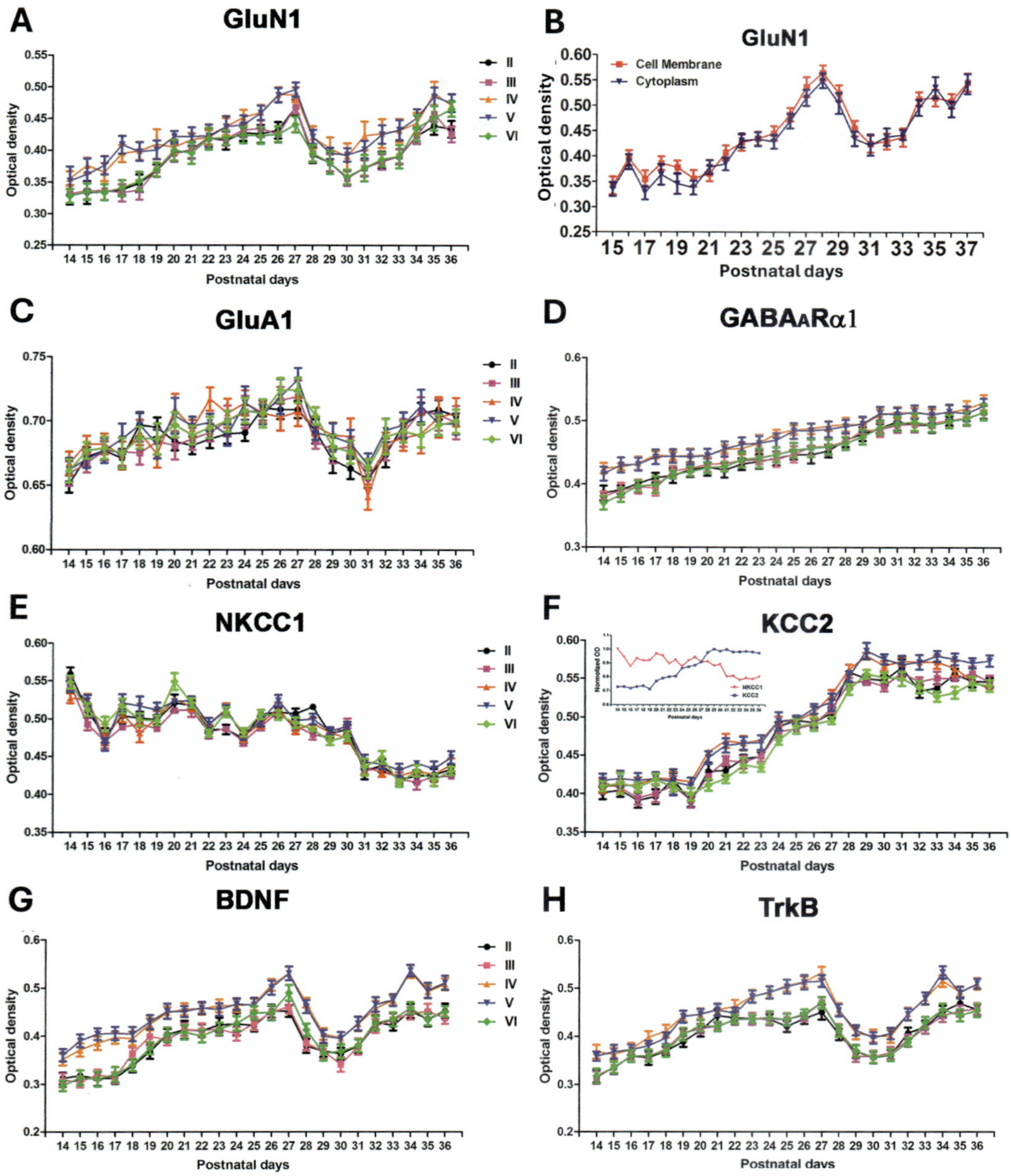

FIGURE 16.4 Optical densitometric analyses of reaction product of immunohistochemistry against GluN1 (A and B, with B indicating that the intensities of labeling in the cell membrane and cytoplasm are virtually identical), GluA1 (C), GABA$_A$Rα1 (D), NKCC1 (E), KCC2 (F), BDNF (G), and TrkB (H) in single neurons of each of five cellular layers in the visual cortex. Note the fall in the expressions of GluN1, GluA1, BDNF, and TrkB between P28 and P33/P34. On the other hand, GABA$_A$Rα1 expression continues to rise with age. The expression of NKCC1 decreases with age, whereas that of KCC2 increases with age. The two trends intersect at around P27 (inset in F). N = 100-150 neurons from at least 2 litters for each time point tested.

Modified from Zhang et al. 2018.

16.3.2 Role of GABA

The first hint that intracortical inhibition may be involved in deprivation amblyopia came from studies by Duffy and his colleagues (Duffy et al. 1976; Burchfiel and Duffy 1981). They administered bicuculline, a $GABA_A$ receptor antagonist, either intravenously or microiontophoretically into the visual cortex of monocularly deprived cats. Bicuculline restored binocularity of almost half of the neurons studied, as they responded to deprived eye input. Exogenous glutamate was not effective. However, a similar experiment done in Blakemore's laboratory failed to identify any neurons dominated by the deprived eye (Sillito et al. 1981). These authors postulated that bicuculline simply induced a nonspecific increase in responsiveness and that intracortical GABAergic inhibition served a passive rather than an active role in the effect of monocular deprivation. The issue became more complex when continuous intracortical infusion of a $GABA_A$ receptor agonist, muscimol, into the visual cortex of monocularly deprived kittens shifted ocular dominance to the deprived eye (Reiter and Stryker 1988). This is contrary to the proposal that GABA is necessary for the dominance shift to the non-deprived eye. A follow-up study showed that thalamocortical afferents from the closed eye expanded in cortical regions silenced by muscimol, whereas the non-silenced portions were shrunken (Hata and Stryker 1994). These investigators concluded that muscimol-induced suppression of postsynaptic spike activity somehow favored the less active deprived eye. However, it remains unclear why neurons in non-deprived columns should be more disadvantaged? One possibility is that their intact geniculocortical afferents activate more GABAergic interneurons to, in turn, exert greater suppression on them.

Is GABA necessary and sufficient for neural plasticity? When a glutamate receptor antagonist, APV (D-2-amino-5-phosphovaleric acid), was applied subdurally over the rodent somatosensory barrel cortex during the critical period, it partially blocked plastic changes in somatotopic patterning of thalamocortical afferents induced by contralateral whisker cauterization (Schlaggar et al. 1993). Again, selective disruption of postsynaptic activation prevents critical period plasticity, as does $GABA_A$ receptor antagonist (see above). One postulate is that the integrity of glutamatergic receptors is necessary for neural plasticity. Another is that suppressing glutamatergic receptors indirectly facilitates GABAergic inhibition. However, if GABA plays the key role, it should enhance plasticity rather than blocking it.

Nevertheless, GABA and a major parvalbumin subtype of GABAergic neurons remain a major focus in studies of OD plasticity (reviewed in Levelt and Hübener 2012; Reh et al. 2020; Quast et al. 2023). In mice with genetic disruption of an isoform of GAD (GAD_{65}), the onset of their critical period is delayed until a $GABA_A$ receptor agonist, benzodiazepine (diazepam), is given. The same agent can also trigger a precocious onset of the critical period in young, normal mice (Hensch et al. 1998; Fagiolini and Hensch 2000). The maturation of GABAergic inhibition is proposed to be necessary in closing the critical period, as reducing GABAergic transmission with 3-mercaptopropionic acid or picrotoxin partially reactivates OD plasticity in adult visual cortex (Harauzov et al. 2010). Transplanting inhibitory neurons from the embryonic medial ganglionic eminence to the visual cortex reactivates OD plasticity past the critical period into adulthood (Southwell et al. 2010; Davis et al. 2015). Such induction is dependent on intact vesicular GABA transporters, the absence of which fails to initiate OD plasticity (Priya et al. 2019).

To determine the developmental pattern of $GABA_A$ receptors in the visual cortex, immunohistochemical analyses of the α1 subunit ($GABA_AR\alpha1$) in cortical neurons of normal SD rats aged P14 to P35 and in normal LE rats, P15 to P36, daily, were undertaken. Results reveal a steady, age-dependent increase in $GABA_AR\alpha1$ expression in all five cellular layers in both rat strains (Zhang et al. 2018) (Figure 16.4D). Moreover, the increase is sharper and more significant ($P < 0.001$) between P27 and P31 in SD rats, before tapering off at P32. These findings are consistent with an increase in GABAergic inhibition during the critical period, and that $GABA_AR\alpha1$ sets "the time of onset of a critical period of plasticity" (Möhler 2006).

16.3.3 A Switch in Dominance from NKCC1 to KCC2

As discussed in Chapter 15, GABA is initially depolarizing and excitatory due to the higher expression of the chloride importer NKCC1 during early postnatal development. The transition to hyperpolarization and inhibition is thought to be coincidental with the progressive dominance of the chloride exporter KCC2 over that of NKCC1.

Developmental trends of these two chloride transporters were examined in the rat visual cortex (Zhang et al. 2018). Single neuron optical densitometric analysis of NKCC1- and KCC2-immunoreactive neurons from P14 to P35, daily, reveals an age-dependent decrease in the expression of NKCC1 and an age-dependent increase in that of KCC2, with a plateau reached at about P30-31 for both in all five cellular layers (Figure 16.4E and F). The two trends intersect at around P27 (Figure 16.4F inset). This switch in dominance appears to occur one day earlier than the sharp rise in sIPSCs at P28 that is sustained until P33 (Section 16.2.1). Together, these changes point to the time between the end of the 4th to the end of the 5th postnatal weeks as the period of heightened inhibition in the rat visual cortex, consistent with the assumption that GABAergic inhibition undergoes a maturational process during the critical period (Section 16.3.2).

16.3.4 Role of BDNF During the Critical Period

Despite the inordinate focus on the inhibitory GABAergic transmission as critical in the opening and closure of the critical period, the fact that excitatory synaptic transmission and the expressions of glutamatergic GluN1 and GluA1 all plunge during the same period indicates that inhibitory maturation is not the only determining factor. An active suppression of excitation also plays an important role, leading to a state of marked synaptic *imbalance*. This concept is contrary to the prevailing view that the critical period is a time of synaptic *balance* between excitation and inhibition (Hensch and Fagiolini 2005; Takesian and Hensch 2013).

Brain-derived neurotrophic factor has been proposed as necessary for the maturation of GABAergic inhibition in the visual cortex during the critical period (Hanover et al. 1999; Huang et al. 1999). Transcripts and immunoreactivity of BDNF rise after eye opening (Bozzi et al. 1995; Rossi et al. 1999), seemingly in parallel with the maturation of GABAergic transmission. Genetically overexpressing BDNF in mice reportedly hastens the maturation of cortical inhibition and opens as well as closes the critical period earlier than in the wild type (Hanover et al. 1999; Huang et al. 1999). Reducing BDNF levels by dark rearing reportedly retards the maturation of GABAergic inhibition (Gianfranceschi et al. 2003).

The developmental trends of BDNF and its high-affinity TrkB receptors in the visual cortex of SD rats were analyzed in layers II to VI neurons from P14 to P35, daily (Zhang et al. 2018). Results indicate that immunoreactivity of both BDNF and TrkB rises from P14 to P27 and peaks at P27 in all five cellular layers. However, they both suddenly plummet at P28, just like COX, GluN1, and GluA1, and remain low until reaching the lowest point at P30, before returning gradually to the P27 level by P34 (Figure 16.4G and H). A similar trajectory holds true in LE rats, except it is shifted a day later. Clearly, the developmental trends of BDNF and TrkB parallel closely those of COX and excitatory GluN1 and GluA1, as well as the early phase of $GABA_A R\alpha 1$ development, but is opposite that of $GABA_A R\alpha 1$ during the period of synaptic imbalance.

BDNF likely enhances the growth and development of excitatory neurons and excitatory synapses for the establishment of a genetically programmed neuronal network. It also likely facilitates the growth of GABAergic neurons and the initial growth of excitatory, depolarizing GABAergic synapses. However, the complete maturation of these synapses into inhibitory, hyperpolarizing ones requires the transient *downregulation* of BDNF/TrkB and, directly or indirectly, the suppression of excitatory synapses, culminating in a state of synaptic imbalance.

The relationship between BDNF and excitation versus inhibition was further tested using agonist and antagonist of TrkB receptors (7,8-DHF at 5 mg kg^{-1} and ANA-12 at 1 mg kg^{-1}, respectively) given in vivo once a day for 2 days before, during, and after the period of synaptic imbalance and their effects tested in visual cortical neurons on the 3rd day, that is, at P25, P31, and P37 (Zhang et al. 2018). During the period of synaptic imbalance (P31), 7,8-DHF significantly enhances the amplitudes and frequencies of sEPSCs but suppresses both parameters of sIPSCs. On the other hand, ANA-12 reduces both the amplitudes and frequencies of sEPSCs while enhancing both parameters of sIPSCs during the same period. These findings affirm that BDNF augments excitation and suppresses inhibition. TrkB agonist partially reverses the synaptic imbalance, whereas its antagonist accentuates the imbalance.

Several reasons may account for the existing dogma that BDNF enhances GABAergic inhibition during the critical period of visual cortical development. First, no studies other than Zhang et al. (2018) did a day-to-day analysis of BDNF expression in the rodent visual cortex between P14 and P35, so the sudden fall between the end of the 4th to the end of the 5th postnatal week was neither anticipated nor investigated. Second, even when a fall was present, it was ignored. For example, a distinct drop in *BDNF* mRNA level at P30 was evident in a published figure but not mentioned by the authors (Bozzi et al. 1995; their Figure 2), possibly because of the large error bar. Third, the period between P27 and P35 was not always examined. For example, the immunohistochemical description of BDNF shows the P20 and adult data points, but nothing in between (Rossi et al. 1999; their Figure 2). Thus, a continuous rise in BDNF with age for the first 5 postnatal weeks was assumed but not verified. Fourth, a fall in the *BDNF* message level is shown for P28 in transgenic mice with BDNF overexpression (Huang et al. 1999; their Figure 1D), but the error bar is sizable. With larger sample sizes, perhaps changes missed in the above studies would have reached significance.

The hypothesis that BDNF has a positive relationship with GABA inhibition is also contrary to the documented actions of BDNF in regions outside the visual cortex. In those areas, BDNF enhances excitation and attenuates GABA-mediated inhibition (Levine et al. 1995; Tanaka et al. 1997; Poo 2001; Henneberger et al. 2002; Cheng and Yeh 2003; Martin and Finsterwald 2011). In the superior colliculus of BDNF-deficient mice, the frequencies of spontaneous action potentials are significantly reduced, whereas the amplitudes of GABAergic evoked, spontaneous, and miniature IPSCs are elevated (Henneberger et al. 2002). These same authors also found that exposing normal collicular neurons to the TrkB inhibitor K252a mimicked BDNF deprivation. In the hippocampus, BDNF augments evoked glutamate release from presynaptic neurons when the postsynaptic neuron is glutamatergic but not when it is GABAergic (Schinder et al. 2000). BDNF decreases the efficacy of inhibitory transmission by downregulating KCC2, the Cl^- exporter (Rivera et al. 2002; Wardle and Poo 2003). These reports are consistent with Zhang et al.'s (2018) findings that the expressions of BDNF and its TrkB receptors during the period of synaptic imbalance parallel those of excitatory glutamate receptors (i.e., all are suppressed) and opposite to the enhancement of inhibitory GABAergic receptors. Further evidence supporting this relationship is discussed in the next section.

16.4 EFFECT OF ACUTE MONOCULAR DEPRIVATION

Zhang et al. (2018) studied the effects of acute, 4-day monocular lid suture before (P19-P23), during (P28-P32), and after (P35-P39) the period of synaptic imbalance on layer V pyramidal neurons in the visual cortex of SD rats. The choice of this rat strain has interrelated disadvantages and advantages. The main disadvantage is that each visual cortical hemisphere is dominated more by contralateral input and has less ipsilateral input than the hooded rats (Lund 1965). It is not the best strain to study binocular rivalry. However, this confers its main advantage in having a largely non-deprived ipsilateral hemisphere as an internal control for the effect of visual deprivation alone, apart from binocular rivalry. The developmental trends of neurochemicals from the day of eye opening to the end of the 5th postnatal week are virtually the same between SD and LE rats, except for a one-day delay in the latter, and the retinal layer thickness measured with optical coherence tomography and Nissl stain is also equivalent between the two strains (Zhang et al. 2018).

Monocular deprivation is most detrimental during the period of synaptic imbalance, when the amplitudes and frequencies of both sEPSCs and IPSCs are significantly reduced (Figure 16.5), as are the expression levels of all neurochemicals examined (COX, GluN1, $GABA_ARα1$, BDNF, and TrkB) in the contralateral, deprived neurons compared to the non-deprived, ipsilateral controls (see Figure 16.7). In vivo administration of 7,8-DHF during that time partially or completely reverses the effects of deprivation on synaptic activity (by increasing the amplitudes and frequencies of sEPSCs and decreasing those of sIPSCs) (Figure 16.5) and on all neurochemicals (i.e., augmenting the expressions of COX, GluN1, BDNF, and TrkB, but reducing that of $GABA_ARα1$) (see Figure 16.7).

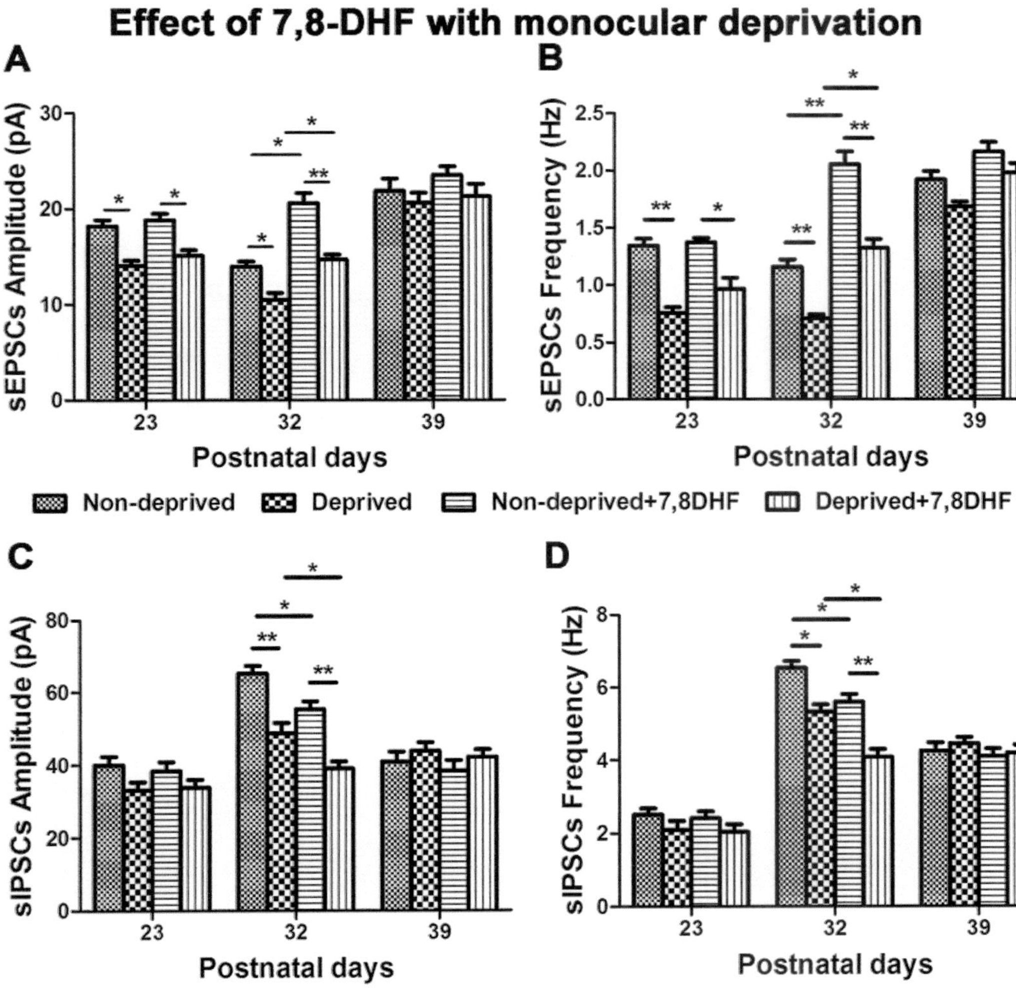

FIGURE 16.5 Effect of monocular lid suture (MD) with and without in vivo 7,8-DHF on sEPSCs and sIPSCs at different time points. Amplitudes and frequencies of sEPSCs (A and B) and sIPSCs (C and D) in layer V pyramidal neurons of non-deprived (ipsilateral) and deprived (contralateral) visual cortices were recorded 4 days after MD at three time points (P23, P32, or P39). Note that MD significantly reduces both amplitudes and frequencies of both sEPSCs and sIPSCs at P32 but not at earlier or later times for sIPScs, though also at earlier times for sEPSCs. 7,8-DHF has a marked effect on both sEPSCs and sIPSCs only at P32. It significantly increases both amplitudes and frequencies of sEPSCs in both non-deprived and deprived neurons but significantly decreases both parameters in both non-deprived and deprived neurons at this time. $N = 6–8$ at each time point tested.

$^*P < 0.5$, $^{**}P < 0.01–0.001$.

Modified from Zhang et al. 2018.

In contrast, ANA-12 reduces the already suppressed excitation but enhances the inhibitory synaptic transmission and $GABA_ARα1$ expressions (Figures 16.6 and 16.7). Administering these agents at P19-P23 or P35-P39 has minimal or no effect. These studies strengthen the significant role BDNF-downregulation plays in the transient synaptic imbalance during the critical period.

In conclusion, the above experiments document the period of synaptic imbalance from the end of the 4th to the end of the 5th postnatal week as the *critical period* of visual cortical development in the rat, when monocular deprivation is most damaging. This period coincides with the "peak" critical period reported for rats (Fagiolini et al. 1994) and mice (Gordon and Stryker 1996). Deprivation

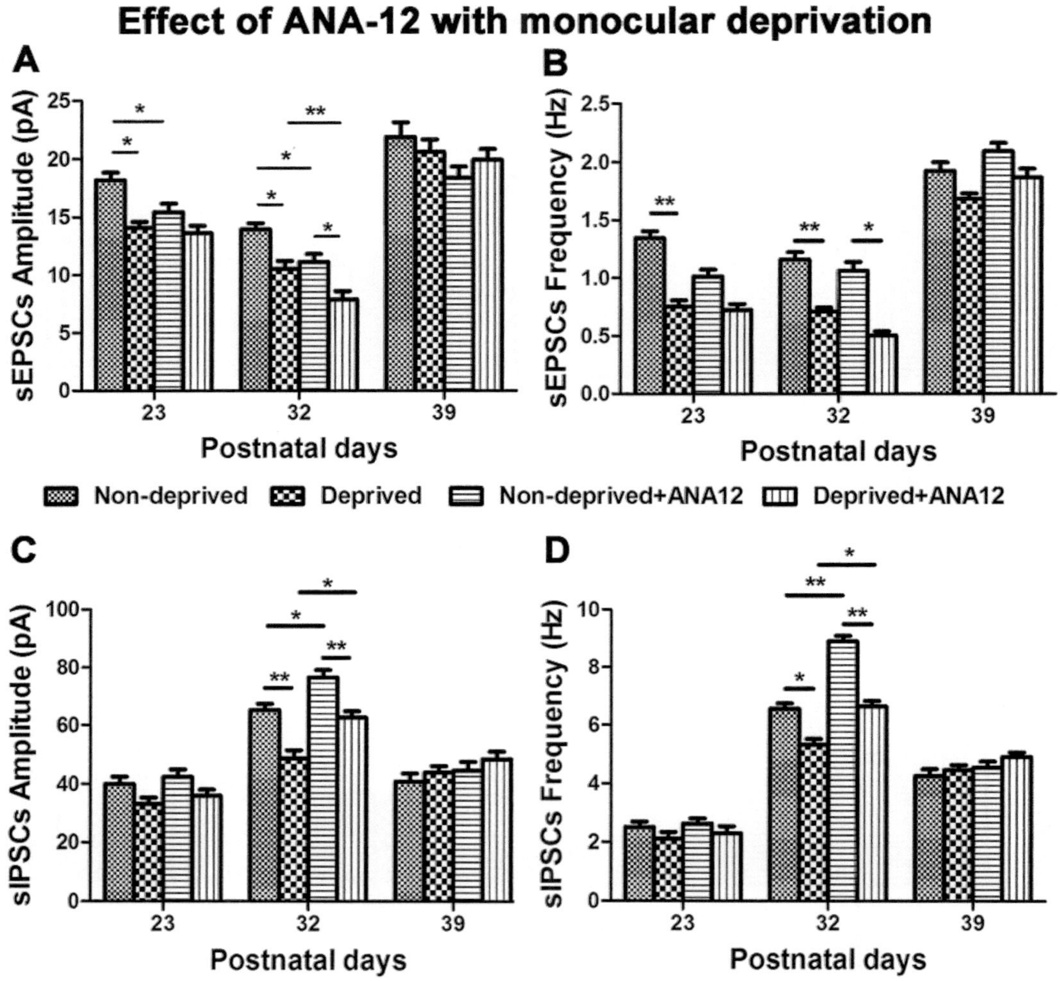

FIGURE 16.6 Effect of MD on sEPSCs and sIPSCs with and without in vivo ANA-12 at different time points. Amplitudes and frequencies of sEPSCs (A and B) and sIPSCs (C and D) in layer V pyramidal neurons of non-deprived (ipsilateral) and deprived (contralateral) visual cortices recorded 4 days after MD at three time points (P23, P32, or P39). The effect of MD alone is the same as in Figure 16.4. ANA-12 significantly reduces the amplitudes of sEPSCs at P32 and at P23 in non-deprived neurons, but it reduces both amplitudes and frequencies of deprived neurons only at P32. On the other hand, it significantly increases both amplitudes and frequencies of sIPSCs for both non-deprived and deprived neurons only at P32, the period of synaptic imbalance. $N = 6–8$ at each time point tested.

$^*P < 0.5$, $^{**}P < 0.01–0.001$.

Modified from Zhang et al. 2018.

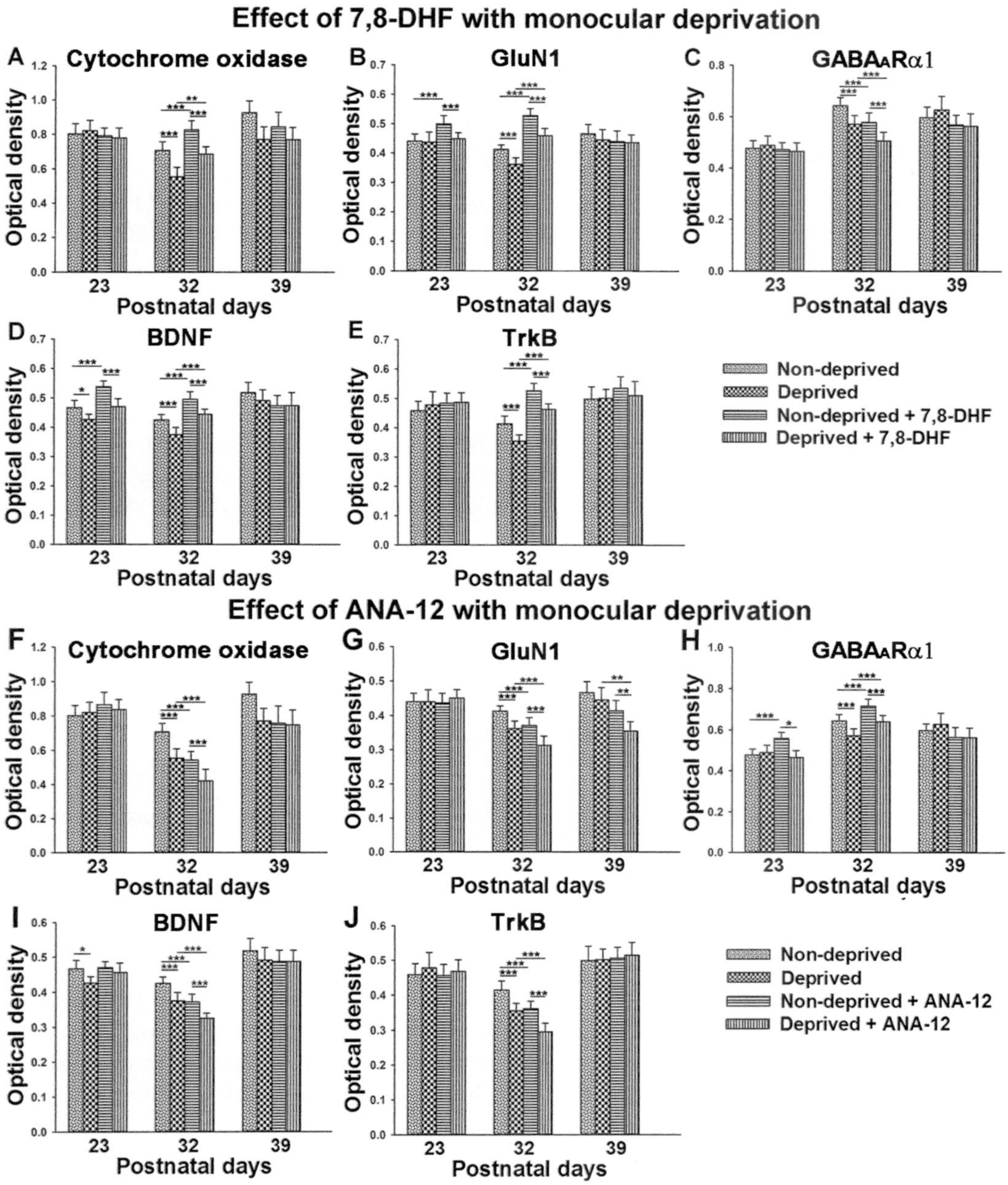

FIGURE 16.7 Effect of MD on neurochemical expression without or with in vivo 7,8-DHF or ANA-12. Optical densitometric analyses of reaction product of COX (A and F) and immunoreaction product of GluN1 (B and G), GABA$_A$Rα1 (C and H), BDNF (D and I), and TrkB (E and J) in single, non-deprived and deprived layer V neurons at three time points (P23, P32, and P39) after 4 days of MD with and without in vivo 7,8-DHF or ANA-12. The most prominent effects are at P32 for all neurochemicals examined. Whereas 7,8-DHF upregulates the expressions of COX, GluN1, BDNF, and TrkB, but downregulates that of GABA$_A$Rα1 in both non-deprived and deprived neurons at P32, ANA-12 has the opposite effect of downregulating COX, GluN1, BDNF, and TrkB, but upregulates that of GABA$_A$Rα1 in both non-deprived and deprived neurons at P32. 7,8-DHF exerts some effect on GluN1 and BDNF at P23, and ANA-12 on GluN1 at P39 and GABA$_A$Rα1 at P23. $N = 150$ neurons from at least 2 litters for each time point tested.

$^{**}P < 0.01$, $^{***}P < 0.001$.

Modified from Zhang et al. 2018.

after this period has virtually no effect, and deprivation before this period (P19-P23) causes only minor changes, such as decreased amplitudes and frequencies of sEPSCs (but not of sIPSCs) and reduced BDNF (but not TrkB) expression (Zhang et al. 2018).

16.5 IS THE CRITICAL PERIOD NECESSARY?

If the critical period is a time of synaptic imbalance and exceptional vulnerability, is it necessary for normal development? In the rat cortex, excitatory synapses develop and mature before inhibitory ones (Sutor and Luhmann 1995). This is crucial to establishing the genetically determined network of extrinsic and intrinsic connections. BDNF is vital during this time by promoting synapse formation and synaptic efficacy, and neuronal activity, in turn, enhances BDNF expression (Thoenen 1995; Yoshii and Constantine-Paton 2010). However, the refinement of neuronal receptive field properties and the strengthening of appropriate synaptic connections require inhibitory modulations. GABA-mediated inhibition is essential in the development of receptive field substructure, ocular dominance, orientation sensitivity, and directional selectivity in visual cortical neurons (Wolf et al. 1986; Hensch and Stryker 2004); much of these are environmentally sharpened. A transient downregulation of BDNF and excitatory neurochemicals enables inhibitory synapses to mature and flourish, but it inevitably induces synaptic imbalance and a state of heightened vulnerability. Nevertheless, the critical period is a *necessary* rite of passage to facilitate experience-dependent synaptic refinement, plasticity, and maturation. This period involves many players (reviewed in Espinosa and Stryker 2012; Hensch and Quinlan 2018; Jenks et al. 2021), and more undoubtedly will be uncovered in the coming years.

16.6 BASIC PRINCIPLES GUIDING CRITICAL PERIODS OF DIFFERENT NEURAL SYSTEMS

A few basic principles are gleaned from studies of the critical periods in the brain stem respiratory system and the visual cortex of rats (Wong-Riley 2021):

1. Postnatal development does not proceed in a straight, smooth path. Rather, a brief period of abrupt physiological, neurochemical, and metabolic changes exists, lasting about two days in the respiratory system and about a week in the visual cortex of rats.
2. The timing differs between the two systems: toward the end of the 2nd week for the respiratory system, and between the end of the 4th to the end of the 5th postnatal week for the visual cortex in rats. Timing can be experimentally delayed, for example, by carotid body denervation in the former and dark rearing in the latter, but not completely eliminated.
3. Before this period, excitatory synapses develop earlier than inhibitory ones to establish a genetically determined neuronal network. BDNF/TrkB plays a significant role in enhancing the growth and development of excitatory synapses, including the growth of depolarizing GABAergic ones.
4. Once excitatory synapses are established, BDNF/TrkB must be transiently downregulated to enable inhibitory GABAergic synapses to fully develop, mature, and flourish. This is essential to fine-tune neuronal properties and strengthen experience-dependent synaptic modifications and connections. It permits the systems, hence the organisms, to modulate their physiological properties and communication network to optimally respond to their surroundings. It is a critical period of plasticity molded by the interplay between nature and the environment.
5. During this critical period, cytochrome *c* oxidase levels are significantly reduced.
6. The reasons for a metabolic decline include not only a sudden surge of inhibitory synapses and inhibitory neurochemicals but also an abrupt suppression of excitatory synapses and excitatory neurochemicals, resulting in a state of *synaptic imbalance*.

7. In concert with and possibly initiating changes in excitatory neurochemicals and COX, BDNF/TrkB levels are *depressed*, not elevated, during the critical period.
8. The state of synaptic imbalance renders the system more vulnerable to external insults, such as hypoxia in the respiratory system or visual perturbations in the visual system. Abnormal disturbances are most detrimental during this time compared to before or after this period. Unchecked disruptions may lead to long-term maladjustments.
9. The critical period is a *necessary* phase of postnatal development, without which the system cannot fully transition from immaturity to maturity.
10. Once the inhibitory synapses are well-established, the excitatory ones return to full functioning, and a mature state of *synaptic balance* is achieved with more stabilized levels of BDNF/TrkB, excitatory and inhibitory neurochemicals, as well as metabolic enzymes such as COX to optimize neuronal functioning.

16.7 IS THERE A MASTER REGULATOR OF THE CRITICAL PERIOD?

The multifactorial complexities of the critical period uncovered in the past five decades are daunting, yet the more primitive brain stem respiratory system and the higher-order visual cortex share many fundamental principles as summarized above. The inevitable and tempting question is, does a master regulator or a set of regulators of the critical period exist for both, if not all, systems?

The important role that BDNF plays in both systems is discussed above. Two additional candidates, PACAP and thyroid hormone, deserve serious consideration: (A) The hypothalamic PACAP, introduced in Chapter 15 (Section 15.3.3.4), is involved in neural development, protection, transmission, modulation, transcription, and metabolic homeostasis (Vaudry et al. 2009; Johnson et al. 2019). Its expression, as that of BDNF, is reduced in multiple brain stem respiratory nuclei during the critical period (Liu and Wong-Riley 2019), and knockout of its gene induces sudden deaths in mice during the second postnatal week (Wilson and Cummings 2008), around the time of the critical period for respiratory system development. In mouse cortical neurons, PACAP stimulates the expression of *BDNF* message indirectly by potentiating glutamate release, enhancing NMDA receptor activation, and modulating synaptic plasticity (Pellegri et al. 1998). The distribution of PACAP mRNA is widespread in the developing and adult rat brain (Skoglösa et al. 1999), but whether it undergoes any distinct change during the critical period of visual cortical development is not yet known. (B) Thyroid hormone (TH) is essential for neurogenesis, neuronal differentiation, myelination, and synaptogenesis, and its receptors are widely distributed in the CNS (Ahmed et al. 2008; Bernal 2017). TH is involved in energy metabolism, mitochondrial function, BDNF regulation, and GABAergic synaptic development (Shulga et al. 2009; Westerholz et al. 2013; Mullur et al. 2014; Oefele et al. 2024). More recently, TH has been suggested as a regulator of the critical period for brain development (Batista and Hensch 2019), but whether its level is altered during the critical period of visual cortical or respiratory system development is not yet known. Future studies will shed more light on the definitive role of thyroid hormone in regulating (or not) the critical period of postnatal development.

Other candidates will likely surface in the coming years. No doubt, studies will reveal a complex interplay among multiple regulators in orchestrating the critical period in each system.

16.8 OPEN ISSUES

The examples given for synaptic imbalance during the critical period in the brain stem respiratory system and the visual cortex came from a single lab. Replications by other labs are paramount to confirm or refute these results. Additional systems and species should be examined to determine the universality (or lack thereof) of the principles outlined above. Inquiries should also extend to other players (neurochemicals, intracellular and extracellular components, genetic and epigenetic determinants, microRNA, and glia, to name just a few).

17 Role of Cytochrome *c* Oxidase in Human Diseases

Inherited Mitochondrial DNA Mutations, Nuclear DNA Mutations, and Neurodegenerative Diseases

ABSTRACT

Mitochondria are at the heart of health and disease, and COX deficiency is a common phenotype in many pathologies. Being a bigenomic enzyme, mutations in either the mtDNA or the nuclear genome that perturb COX genes or genes for its dozens of assembly factors cause metabolic dysfunction. Oxidative stress in normal aging and neurodegenerative diseases such as Alzheimer's, Parkinson's, and Huntington's diseases, as well as amyotrophic lateral sclerosis, whether genetic or idiopathic, all exhibit metabolic distress and often deficits in COX function. However, currently, there is no direct proof that mitochondrial or COX dysfunction is the primary cause rather than secondary in neurodegenerative diseases.

The central role of cytochrome *c* oxidase in aerobic energy metabolism among virtually all living organisms underpins its critical significance in life. Disruption of its structure or function can lead to dire consequences. Being the terminal enzyme of the mitochondrial electron transport chain, without which oxidative metabolism cannot be carried to completion, COX activity or content measured biochemically, immunohistochemically, or histochemically often indicates the integrity of the energy capacity in a cell, tissue, or organ, whether the source of pathology is COX itself, other members of the respiratory chain, or agents involved in energy metabolism. Mitochondria are ubiquitous and are at the center of health and disease, and mitochondrial pathologies often perturb multiple systems, especially highly oxidative ones, such as the heart, skeletal muscles, kidney, liver, and the brain. Not surprisingly, COX deficiency is common in most mitochondrial diseases. COX activity is reduced, oxidative phosphorylation is perturbed, and varying degrees of lactic acidosis exist, together with isolated or combinations of myopathy, cardiomyopathy, encephalopathy, and pathologies of other organs.

Whereas the clinical symptoms of mitochondrial diseases frequently involve multiple organs, initial biochemical and histochemical data are often derived from readily available blood samples and muscle biopsies. Among the earliest clinical reports of familial mitochondrial myopathies was a case of multiple family members, all with varying degrees of muscle pathologies but a common maternal linkage (Hudgson et al. 1972). Another early account of a patient with COX deficiency in skeletal and heart muscles also noted a subacute necrotizing encephalomyelopathy (known as Leigh's disease, first reported by Leigh in 1951) involving the CNS (Willems et al. 1977). DiMauro's group highlighted COX deficiencies that are fatal during infancy or reversible spontaneously (DiMauro et al. 1980, 1983). However, the precise genetic causes of these early patients were not determined.

As discussed in Chapter 1, COX is one of only four proteins in mammalian cells that have bigenomic sources: mitochondrial and nuclear. Mutations in either genome can result in multisystem pathologies. This chapter will briefly highlight key findings involving mutations in the three

DOI: 10.1201/9781003507765-17

mitochondrial subunit genes (*COX1*, *2*, and *3*) and mutations in the nuclear *COX* subunit genes, as well as nuclear genes encoding assembly factors for COX. Finally, some neurodegenerative diseases that result in dysfunction of energy metabolism, exemplified by COX deficiency, will be included. Space limitation necessitates restricting discussions to only a few major findings. However, a wealth of literature awaits interested readers and researchers.

17.1 COX DEFECTS ARISING FROM MITOCHONDRIAL DNA MUTATIONS

17.1.1 mtDNA Mutations

Before delving into specific mitochondrial *COX* gene mutations, it is helpful to recall four main characteristics of mtDNA mutations: (a) Maternal inheritance: in most species, including humans, paternal mitochondria contribute little, if any, to the mitochondrial population of zygotes; hence, virtually all mtDNAs are maternally inherited. (b) Cytoplasmic inheritance: the mitochondrial genome is separate from and independent of the nuclear one, thus the pattern of inheritance is non-Mendelian. (c) Homoplasmy and heteroplasmy: each cell can have all copies of its mtDNA be identical (homoplasmic) or, more likely, different (heteroplasmic). The severity of a mutation is dependent on the relative proportion of the two populations. (d) Point mutation or large-scale deletions: a point mutation of a single nucleotide attacks a single gene among the 13 protein-coding genes or the 22 tRNA genes. The mutation can be a transition (replace purine with purine, and pyrimidine with pyrimidine) or a transversion (replace purine with pyrimidine, or vice versa). It can be a silent/synonymous mutation (a base change that does not alter the amino acid sequence), a missense/nonsynonymous mutation (a single base change that results in coding for a different amino acid), or a nonsense/stop mutation (a point mutation that results in a premature stop codon). Large-scale deletions often affect several genes in the mtDNA.

The first disease with a clearly identified mtDNA mutation is Leber's hereditary optic neuropathy (Wallace et al. 1988). It is a transversion mutation of subunit 4 of complex I (*MT-ND4* or mitochondrial NADH dehydrogenase subunit 4) gene. Other ND genes have since been implicated in this disease as well (*MT-ND1*, *MT-ND6*, and *MT-ND4L*; Achilli et al. 2012). Many point mutations of mtDNA have been uncovered; notably, a single point mutation in a transfer RNA gene in the mtDNA often perturbs the translation of several mitochondrial genes and causes multisystem defects. Examples include a point mutation in the tRNA(Lys) gene that results in myoclonic epilepsy with ragged-red fibers (MERRF) (Sparaco et al. 1995), and a point mutation in the tRNA(Leu)(UUR) gene that causes mitochondrial myopathy, encephalopathy, lactic acidosis, and stroke-like episodes (MELAS) (Goto et al. 1990). Large-scale deletions of mtDNA (from 2.0 to 7.0 kb) generate a myriad of symptoms known as Kearns–Sayre syndrome (Zeviani et al. 1988b) and chronic progressive external ophthalmoplegia (Schon et al. 1989). Clinically, single-point mutations can cause different symptoms in different patients, and the same set of symptoms can result from different mutations. For the studies described below, it is important to bear in mind that single case reports require verification by many additional studies to attain statistical significance, and point mutations can be benign polymorphisms.

17.1.2 Mutations in the Mitochondrial *COX* Subunit Genes

As *COX1*, *2*, and *3* encode the catalytic core of the enzyme, their mutations, if severe, are likely to result in lethality either *in utero* or in early infancy. The severity of the deficit in energy metabolism and clinical pathologies is dependent on the relative abundance of mutated mtDNA in each tissue. Those embryos and infants that survive point mutations or microdeletions of *COX* genes presumably have a higher proportion of non-mutated than mutated mtDNA, or the prevalence of mutated mtDNA has not reached the threshold. Tissues and organs differ in their energy demands and likely manifest varying thresholds of mutated mtDNA before pathological symptoms appear. In cybrids generated by fusing ρ^0 (without mtDNA) osteosarcoma cells with platelets from patients with stop-codon mutations in *COX1*, the threshold level for preserving cellular respiration and ATP synthesis is about 40% mutated mtDNA (D'Aurelio et al. 2001). Other tissues likely have higher or lower thresholds.

17.1.2.1 *COX1* Mutations

Although Leigh's disease is most frequently associated with mutations in COX assembly factor genes (see Section 17.3), an adult onset of this disease has been reported in a patient with a novel stop codon mutation in the *COX1* gene (m.6570G>A; p.G26X) (m denotes mitochondria; p is protein, and X represents a non-specific stop for proteins) (Poole et al. 2019). A heteroplasmic, missense mutation m.6597C>A (p.Q232K) in the *COX1* gene was also found in a MELAS-like encephalomyopathic patient (Lamperti et al. 2012). In a unique upper motor neuron diseased patient, a heteroplasmic 5-bp microdeletion in the 5' end of *COX1* caused a premature stop codon and a truncated COX I subunit, COX deficiency, and reduced levels of COX II, III, and IV subunits, indicating a defective assembly of the holoenzyme (Comi et al. 1998). Likewise, a stop codon mutation in *COX1* led to a reduction in COX II protein, implying that COX assembly was affected (D'Aurelio et al. 2001). A heteroplasmic m.6489C>A missense mutation in *COX1* converted a highly conserved Leu196 to Ileu196 in a young patient with epilepsia partialis continua and decreased COX activity in muscle fibers (Varlamov et al. 2002). Missense mutations in *ATP6* and in *COX1* (m.6268C>T; p.A122V) were described in a 75-year-old Caucasian female with myalgic encephalomyelitis and chronic fatigue syndrome, possibly triggered by an Epstein–Barr virus 24 years earlier (Tang-Siegel et al. 2024).

17.1.2.2 *COX2* Mutations

Among the first reports of *COX2* mutations was a patient with early-onset multisystem disease and COX deficiency in muscles harboring a nonsense mutation (m.7896G>A) that was heteroplasmic in muscles, blood, and fibroblasts (Campos et al. 2001). A heteroplasmic, missense mutation (m.7706G>A; p.A41T) in the *COX2* gene was present in a child with Alpers–Huttenlocher-like disease (Uusimaa et al. 2003). A heteroplasmic variant in the *COX2* gene (m.7887G>A; p.G101D) was uncovered in an adult male patient with progressive cerebellar ataxia, neuropathy, and exercise intolerance. His muscles displayed COX deficiency and ragged red fibers (Baty et al. 2021).

17.1.2.3 *COX3* Mutations

As mtDNA is transcribed as a polycistronic transcript, a 9205delTA microdeletion in the *ATP6* gene not only removes the stop codon of *ATP6*, but also affects the cleavage sites in the *ATP8/ATP6/CO3* transcripts, thereby interfering with mRNA processing for *ATP6* and *COX3*. (Hejzlarová et al. 2015). The lack of functional ATP synthase and reduced COX content severely limits the production of ATP, but the threshold effect appears to be ~90% heteroplasmic mutation. A heteroplasmic missense mutation in the *COX3* gene (m.9861T>C; p.F219L) was detected in high frequency in Alzheimer's disease brains, and the prevalence is higher in parietal and temporal cortices and the caudate nucleus than in the hippocampus (Castora et al. 2019). A nonsense variant of *COX3* (m.9553G>A; p.W116*) (*represents a stop codon) was reported in a 47-year-old female with MELAS syndrome whose COX activity and subunit proteins were both reduced (Wang et al. 2021). The frequency of *COX3* mutations with MELAS-like syndrome awaits additional case studies.

In recent years, an increasing number of reports point to the prevalence of *COX1*, *2*, *3*, and nuclear structural gene mutations in various types of cancers, including colon, prostate, lung, and breast cancers (Petros et al. 2005; Gallardo et al. 2006; Namslauer and Brzezinski 2009; Arnold et al. 2013; Afkhami et al. 2020; de Oliveira et al. 2023). These mutations reduce COX activity, impair oxidative phosphorylation, and increase ROS levels. Caution is warranted, however, when interpreting disease causation and pathology with the mere presence of a mutation. Simple polymorphism must be ruled out, and a statistically significant sample size is essential.

17.2 MUTATIONS IN THE NUCLEAR-ENCODED *COX* SUBUNIT GENES

Many, but not all, of the nuclear-encoded *COX* subunit genes, including the tissue-specific isoforms (see Chapter 1, Table 1.1), have been reported to have mutations, and some of them have been associated with a disease phenotype (Sinkler et al. 2017). Most reports, however, are single cases, and prevalence as well as relevance to disease states require verifications with a larger sample size and documented consistency. With that caveat in mind, a few representative cases are described below.

A biallelic variant in *COX4i1* in two siblings born to consanguineous parents was associated with a Leigh syndrome-like phenotype of encephalopathy, developmental regression, hypotonia, intellectual disability, and seizure (Pillai et al. 2019). A homozygous missense mutation in the *COX4i2* gene caused exocrine pancreatic insufficiency, dyserythropoietic anemia, and calvarial hyperostosis (Shteyer et al. 2009). A homozygous mutation in *COX5A* in two siblings led to early-onset pulmonary arterial hypertension, isolated COX deficiency, lactic acidemia, and failure to thrive (Baertling et al. 2017). A 5 bp deletion in a splicing element of intron 2 of the *COX6A1* gene caused a phenotype of Charcot–Marie–Tooth disease (Tamiya et al. 2014), an inherited neuropathy with multiple mutation loci. A *COX6a2* deletion in mice disrupted perineuronal nets (discussed in Chapter 16), decreased the ATP-to-ADP ratio, and increased oxidative stress in fast-spiking, parvalbumin-positive interneurons. A mutation in this gene was evident in a human patient with mental and neurological abnormalities (Sanz-Morello et al. 2020). A homozygous missense mutation in exon 2 of the *COX6B1* gene (c.221G>A; p.R19H) (c denotes cDNA) in two siblings led to severe infantile encephalomyopathy (Massa et al. 2008). Another *COX6B1* missense mutation occurred in a patient with encephalomyopathy, hydrocephalus, and hypertrophic cardiomyopathy (Abdulhag et al. 2015). Deleting the heart-specific *COX7a1* gene in mice caused dilated cardiomyopathy early in development, but the condition improved and was stabilized by 6 months of age, apparently by incorporating the liver isoform of *COX7a2* into the cardiac COX holoenzyme (Hüttemann et al. 2012). As *COX7B* is the only *COX* gene encoded in the X chromosome, a *de novo* mutation in this gene in 2 simplex cases and a nonsense mutation in a familial case all exhibited X-linked microphthalmia with linear skin lesions (Indrieri et al. 2012). However, these phenotypes were also present in patients with a nonsense mutation in the *NDUFB11* gene (also on the X chromosome), which encodes a component of complex I (van Rahden et al. 2015). A homozygous splice site mutation in the *COX8A* gene that causes a frameshift in exon 2, reduced *COX8A* transcripts, and instability of the holoenzyme led to Leigh-like syndrome with leukodystrophy and severe epilepsy (Hallmann et al. 2016). Finally, a homozygous splice site mutation in the *NDUFA4* gene in two siblings caused COX deficiency and Leigh-like syndrome (Pitceathly et al. 2013). Likewise, a biallelic *NDFUA4* deletion was found in a patient with Leigh syndrome (Misceo et al. 2024).

Various nuclear-encoded COX subunit genes are implicated in cancers, such as breast cancer (Zhang et al. 2016; de Oliveira et al. 2023). Again, definitive proof is lacking, and more data are needed to establish statistical significance.

17.3 MUTATIONS IN GENES ASSOCIATED WITH COX ASSEMBLY

Mutations in COX assembly factor genes cause isolated or general COX deficiency and reduced COX activity in specific tissues. The most common pathology is Leigh syndrome, originally attributed, as stated above, to mutations in the *SURF1* gene (Tiranti et al. 1998; Zhu et al. 1998; Robinson 2000), which remains the most prevalent cause of this disease. However, an increasing number of cases implicate the involvement of COX subunit structural genes (see Sections 17.1 and 17.2) as well as other COX assembly factor genes, such as *COX10* (Antonicka et al. 2003), *COX15* (Halperin et al. 2020), *PET100* (Lim et al. 2014), and *TACO1* (translational activator of *COX1*; Weraarpachai et al. 2009), indicating that this syndrome is genetically heterogeneous. Indeed, the Leigh syndrome spectrum (Leigh syndrome and Leigh-like syndrome) is associated with numerous genes, including all five complexes of the mitochondrial inner membrane as well as many other proteins (reviewed in Rahman and Thorburn 2020).

Mutations in the heme chaperone protein COA7 led to spinocerebellar ataxia with axonal neuropathy (Higuchi et al. 2018; Ouchi et al. 2023). A homozygous missense mutation in the *COX10* gene caused an isolated COX defect in a consanguineous family (Valnot et al. 2000). A newborn male with two heterozygous variants (c.232G>A; p.G78R and c.452C>G; p.S151Ter) in the *COX15* gene (for heme a synthase) reportedly had hypertrophic cardiomyopathy and hyperlacticaemia in his serum and cerebrospinal fluid (Galvão de Oliveira et al. 2021). Among the copper chaperones, a mutation in the *SCO1* gene was implicated in three patients with developmental and epileptic encephalopathy and hypopituitarism (Barbato et al. 2024). Various mutations in the *SCO2* gene caused a variety of phenotypes, such as spinomuscular atrophy that mimics Werdnig-Hoffmann disease (Salviati et al. 2002) and early-onset axonal Charcot-Marie-Tooth disease (Rebelo et al. 2018; Gangfuβ et al. 2022). Two heterozygous mutations in *COX11* (a missense mutation c.739C>A; p.P247T inherited from the father and a frameshift mutation c766dupA; p.T256Nfs*8 inherited from the mother) were found in an 11-year-old girl exhibiting Leigh-like symptoms, including hypotonia, global developmental delay, cerebellar signs, ankle clonus, and dystonia (Caron-Godon et al. 2023). Two unrelated patients had the homozygous nonsense variant (c.244C>T; p.R82*) in their *COX16* gene, and they exhibited hypertrophic cardiomyopathy, encephalopathy, severe lactic acidosis, and isolated COX deficiency (Wintjes et al. 2021). Menkes kinky hair syndrome (trichopoliodystrophy) is an X-linked recessive neurodegenerative disease caused by a mutation not in *COX*, but in the copper-transporting ATPase (*ATP7a*) gene (Dierick et al. 1995). However, because copper is an important component of COX as well as a cofactor of Cu/Zn-superoxide dismutase (SOD1), disruption in copper metabolism leads to widespread neurological symptoms and severely reduces COX activity (Sparaco et al. 1993; Strausak et al. 2001).

A point mutation in COX subunit insertase *OXA1L* led to mitochondrial encephalopathy and oxidative phosphorylation defect involving both complexes IV and V (Thompson et al. 2018). Compound heterozygous mutations in *COA3* (c.199dupC, which causes a frameshift and a premature stop codon, p.L67Pfs*21, inherited from the father, and c.215A>G, p.Y72C from the mother) in a 34-year old patient were associated with a reduction in the synthesis of COX I subunit, an absence of COX assembly, and isolated COX deficiency in muscle and fibroblasts (Ostergaard et al. 2015). This patient suffered from exercise intolerance, sensorimotor polyneuropathy, obesity, cognitive impairment, and short stature. A homozygous *COA8* defect involving a GACC insertion in exon 2, a change in reading frame, and a premature stop codon in codon 62 (c.170_173 dupGACC; p.P59Tfs*3) was noted in two sisters (52- and 53-year-old) with a multitude of symptoms, including bilateral ptosis, muscle weakness, peripheral neuropathy, hearing loss, and cognitive impairment (Rimoldi et al. 2023). A mutation in the gene of another COX I-specific assembly factor, COX14, in mice resulted in phenotypes like those of patients with COX deficiency (Aich et al. 2024). These include widespread tissue-specific pathologies, such as severe liver inflammation triggered by enhanced ROS production when the COX level is low. A biallelic variant in *COX18* (c.667G>C; p.D223H) was uncovered in a 19-month-old female with a neonatal form of mitochondrial encephalocardiomyopathy and axonal sensory myopathy, failure to thrive, and severe diffuse COX deficiency due to impairment of COX activity and assembly (Ronchi et al. 2023). A founder biallelic loss-of-function variant in the *COX20* gene (c.41A>G; p.K14R) affected eight unrelated families in either a homozygous or compound heterozygous state (Dong et al. 2021). The affected individuals exhibited autosomal recessive sensory neuropathy, impaired balance, and sensory ataxia with a reduction in non-length-dependent sensory potentials. Two mutations in exon 1 of the MR-1S gene (c.66C>T, p.V7A, and c.72C>T, p.V9A) reportedly caused PNKD (paroxysmal nonkinesigenic dyskinesia), an autosomal dominant movement disorder with episodic attacks of dystonia, chorea, and/or athetosis (Wang et al. 2018). A homozygous nonsense mutation in the *PET117* gene (c.172C>T) led to a premature stop codon at position 58 of the protein in two sisters, who manifested neurodevelopmental regression, lesion in the medulla oblongata, reduced COX subunit levels, and isolated COX deficiency (Renkema et al. 2017). A homozygous mutation in *C2ORF64* (*COA5* or *PET191*) was present in two siblings with fatal neonatal cardiomyopathy (Huigsloot et al. 2011). Mutations in *TIMM8A* on the X chromosome caused the neurodegenerative disease, Mohr–Tranebjaerg syndrome or MTS (also known as deafness-dystonia-optic neuropathy or DDON), which affects mainly males, and patients exhibited sensorineural hearing loss, dystonia, and blindness (Kang Y et al. 2019). As for TIM23, homozygous gene knockout in mice is lethal before implantation, whereas heterozygous Tim23 mice have shortened lifespan, early aging, reduced motor coordination, and diminished balance (Ahting et al. 2009).

The association of COX assembly factors in cancers is emerging. COA1, for example, is overexpressed in different cancer types, raising the possibility that its dysregulation may affect the tumor immune microenvironment (Ghosh et al. 2024). COA6 is also overexpressed in lung adenocarcinoma and is associated with reduced immune cell infiltration and overall survival (Liu et al. 2024). Both transcript and protein levels of COX15 are increased in lung cancer cells of patients, and these are positively associated with the overall survival of patients (Zhang et al. 2021). However, the mechanism underlying the possibility that COX15 promotes lung cancer progression is unknown. COX19 is thought to be a target of MACC1 (metastasis-associated with colon cancer protein 1), which promotes the transcription and expression of various tumor-related genes. Overexpressing COX19 increases copper content in the mitochondria, inhibits apoptosis, and promotes tumor growth in mice (Gao et al. 2024). MR-1S is thought to phosphorylate and activate the MEK-ERK-RSK pathway to accelerate cancer growth and enhance metastasis (Wang et al. 2018). However, direct links between COX assembly factors and cancer await further research.

17.4 COX DEFICITS RELATED TO NORMAL AGING AND NEURODEGENERATIVE DISEASES

17.4.1 Oxidative Stress, ROS, and RNS in Aging and Neurodegenerative Diseases

The central tenet of aging, especially accelerated aging, and many neurodegenerative diseases has been excessive accumulation of reactive oxygen species leading to uncompensated oxidative stress or oxidant stress (Nohl 1993; Kadenbach et al. 2009). ROS include superoxide ($^{*}O_2^-$), hydrogen peroxide (H_2O_2), hydroxyl radicals ($^{*}OH$), and singlet oxygen ($^1O_2$), most of which are produced in the mitochondria mainly via complexes I and III as byproducts of oxidative phosphorylation (Reczek and Chandel 2015; Mailloux 2020). Under normal physiological conditions, there is an optimal redox balance in the production of ROS and antioxidants. Antioxidants include superoxide dismutase, catalase, peroxyredoxins, glutathione peroxidase, and glutathione (Redza-Dutordoir and Averill-Bates 2016). Specific oxidants are essential signaling molecules for various biological processes, such as regulation of transcription factors, signaling pathways, programmed cell death, phosphorylation cascade, cellular proliferation, differentiation, hypoxic signal transduction, and the immune response (Reczek and Chandel 2015; Sies and Jones 2020; Averill-Bates 2024). Importantly, ROS signaling is involved in COX subunit composition, redox regulation of copper delivery, and translational control during COX biogenesis (Bourens et al. 2013). However, excessive production of ROS is detrimental, damaging DNA, especially mtDNA, proteins, lipids, enzymes, and membranes (Kadenbach et al. 2000; Auten and Davis 2009). Sirtuin 3 (Sirt3), a histone deacetylase in the mitochondria, deacetylates COX I and counteracts oxidative-stress-induced apoptosis (Tu et al. 2019).

Reactive nitrogen species (RNS), including nitric oxide, nitrogen dioxide (NO_2), peroxynitrite ($ONOO^-$), nitrite (NO_2^-), and nitrate (NO_3^-), also serve the dual roles of signaling molecules and agents for nitrosative stress (Sies et al. 2017). Nitric oxide both stimulates mitochondrial biogenesis and reversibly inhibits COX in competition with oxygen. Excessive RNS irreversibly inhibits multiple sites within the mitochondria, induces ROS production, and can lead to oxidative stress, apoptosis, and necrosis (Brown 2007).

With aging, the efficiency of oxidative metabolism is frequently reduced, resulting in the accumulation of ROS and RNS to different degrees in various tissues and organs. This forms the basis for the Free Radical Theory of Aging, proposed first by Harman (1956) and acknowledged by many subsequent investigators. This theory was later expanded into the Mitochondrial Theory of Aging (Harman 1972) and the association of ROS with mitochondrial diseases (Kirkinezos and Moraes 2001). By the same token, overproduction of ROS and RNS has been linked to several neurodegenerative diseases, including Alzheimer's disease, Parkinson's disease, Huntington's disease (HD), and amyotrophic lateral sclerosis (ALS) (Emerit et al. 2004; Houldsworth 2024). However, free radicals are likely to account for only a part, albeit a significant part, of the mechanism of aging and neurodegenerative diseases.

COX deficiency has been increasingly highlighted in articles on neurodegenerative diseases. As with most human studies, large variations exist among patients, whose genetic background, past medical history, length and extent of illness, and environmental influences are often unknown, making comparisons and conclusions challenging if not impossible. With such a caveat in mind, a few reports are summarized below.

17.4.2 Alzheimer's Disease (AD)

A severe form of presenile dementia was documented by Alois Alzheimer in 1907. He described a 51-year-old female who exhibited severe memory impairment, hallucinations, disorientation, delusions, and psychosocial dysfunction. Following her death, he processed her brain with the Bielschowsky silver stain and observed bundles of abnormal neural fibrils (known as neurofibrillary tangles) and numerous focal lesions (later identified as β-amyloid plaques) in her brain. The disease now bears his name.

One of the earliest reports on COX in AD brains was by Kish et al. (1992), who noted a 16–26% statistically significant decrease in biochemical COX activity in frontal, parietal, and temporal cortices, but not in the occipital cortex and putamen. Surprisingly, the activity in the hippocampus was non-significantly increased by 20%, though a postmortem tissue shrinkage needs to be ruled out. Another study saw a reduction in COX histochemical activity in the AD hippocampus but not the visual cortex (Simonian and Hyman 1995). These same authors reported a decrease in mitochondrial *COX2* mRNA but not nuclear *COX4* transcripts in the AD hippocampus. However, a 50–60% reduction in *COX4* mRNA was detected in other AD brains (Chandrasekaran et al. 1997). Postmortem RNA stability, severity of AD, and disparate phenotypes all deserve attention before definitive conclusions can be made.

Biochemical analyses of COX in different regions of AD brains revealed significant reductions in the prefrontal, parietal, temporal cortices, and hippocampus, where senile plaque densities are high, and in the primary visual and primary motor cortices, where plaque densities are lower yet significantly greater than controls (Wong-Riley et al. 1997b). Various laminae of the prefrontal, parietal, and visual cortices, as well as the hippocampus and lateral geniculate nucleus, also exhibit significantly reduced COX activity and protein, analyzed histochemically and immunohistochemically, respectively, by optical densitometry (Wong-Riley et al. 1997b). These findings confirm a generalized suppression of oxidative metabolism throughout the AD brain, in agreement with PET reports of an overall metabolic rate reduction, which correlates positively with the severity of dementia (Haxby and Rapoport 1986).

Intracellular accumulation of β-amyloid (Aβ) selectively inhibits COX activity (by 50%) and reduces ATP production, but the effects are neutralized by the mitochondrial matrix chaperonin heat-shock protein 60 (Hsp60) (Veereshwarayya et al. 2006). COX-deficient neurons in AD brains harbor elevated levels of mtDNA deletions (Krishnan et al. 2012), further supporting the association of defective mitochondria and COX with neurodegeneration in AD.

Clearly, dysfunctional oxidative metabolism exists in AD that may lead to energy failure and neuronal degeneration. However, correlation is not evidence for causation. There are many known and suspected risk factors for AD, among them are genes for the apolipoprotein E4 (*APOEε4*), amyloid precursor protein (*APP*), presenilin 1 (*PSEN1*), and presenilin 2 (*PSEN2*) (Bekris et al. 2010;

Lambert et al. 2023). Thus far, even though defective energy metabolism has been linked to pathology, it has not been proven to be the primary cause of AD.

17.4.3 Parkinson's Disease (PD)

In 1817, James Parkinson published "An Essay on the Shaking Palsy" or paralysis agitans (Parkinson 2002), which now bears his name. Parkinsonian symptoms include a slow progression of tremor at rest, bradykinesia, hypertonia, rigidity, postural instability, and shuffling gait. Cardinal findings at the cellular level are deaths of dopaminergic neurons in the substantia nigra, pars compacta (SNc) of the midbrain and the deposition of Lewy bodies, which contain aggregated α-synuclein, in the brain stem and cerebral cortex (Spillantini et al. 1997; Olanow and Tatton 1999). Oxidative metabolism of dopamine yields H_2O_2 and other ROS, the abnormal accumulation of which leads to uncompensated oxidative stress and cell death (Olanow and Tatton 1999). Neuromelanin, a byproduct of dopamine metabolism, also avidly binds iron, and excessive iron induces free radical formation, oxidative stress, and eventual neuronal death (Jellinger 1999). Indeed, oxidative stress and mitochondrial defects are prominent features in PD as well as AD (Yan et al. 2013). Depletion of dopamine in the nigrostriatal pathway results in dysregulation of circuits in the basal ganglia and generalized motor dysfunction.

The cause of PD remains elusive, and only 10–20% of PD patients have a family history of the disease, with as many as 90 genes being implicated in monogenic PD (Panicker et al. 2021). These genes include *SNCA/PARK1* (for α-synuclein), *LRRK2/PARK8* (for leucine-rich repeat kinase 2, with both kinase and GTPase domains), *PRKN/PARK2* (for Parkin, an E3 ubiquitin ligase), *PINK1/PARK6* (for PTEN-induced putative kinase 1, a mitochondrial serine/threonine kinase), *PARK7/DJ-1* (for protein deglycase DJ-1, an oxidative stress sensor), *GBA1* (for β-glucocerebrosidase, a lysosomal enzyme), *VPS35* (for vacuolar protein sorting 35, a component of the retromer complex that sorts and transports intracellular protein), *ATP13A2/PARK9* (for ATP13A2, a lysosomal protein that transports polyamines), *MAPT* (for microtubule-associated protein tau), *TMEM230* (for transmembrane protein 230 involved in vesicular recycling and secretion), *LRP10* (for low-density lipoprotein-related protein 10), *NUS1* (for nuclear undecaprenyl pyrophosphate synthase 1, important for protein glycosylation and dolichol synthesis), and *ARSA* (for arylsulfatase A, a lysosomal enzyme) (reviewed in Panicker et al. 2021; Cherian et al. 2023; Croucher and Fleming 2024). A number of these gene products, such as LRRK2, PINK1, parkin, and DJ-1, interact with each other, and their defects lead to mitochondrial dysfunction and oxidative stress, with ultimate neuronal death. Mutations of one or more of these genes thus directly or indirectly induce PD.

Dysfunction of mitochondrial complex I has been implicated in the pathology of PD (Mizuno et al. 1989; Cooper et al. 1992). Animal models of PD have used toxins, such as MPTP and rotenone, to inactivate complex I and cause dopaminergic cell death (Singer and Ramsay 1990; Greenamyre et al. 1999). However, abnormal COX activity has also been reported in PD humans and animals (see below).

As a point of reference, COX activity in the basal ganglia of normal primates and rodents is as follows: dopaminergic neurons in SNc have relatively low levels of COX, whereas GABAergic output neurons in SNr and the internal segment of globus pallidus (GPi), as well as the outer segment of GP (GPe), have high COX activity. The neuropil in these nuclei, however, exhibits low COX labeling, whereas that in the striatum (caudate and putamen), especially in the matrix, is quite high (Wong-Riley 1976; Hevner et al. 1995; and Chapter 2, Section 2.3.2 above). Likewise, 2-deoxyglucose demonstrates that the highest metabolic activity in the striatum is in the matrix rather than in the striosomes in awake behaving rats (Brown et al. 2002). The neuropil of the subthalamic nucleus (STN) also has high COX activity (Hevner et al. 1995). In analyzing pathological samples, then, it is relevant to keep in mind the heterogeneous pattern of metabolic activity among the components of the basal ganglia. For example, if SNc and SNr are not separately analyzed, changes in SNr may be erroneously attributed to those of SNc. With a low baseline level of COX in SNc, it would be difficult to detect any further decrease in enzyme labeling. On the other hand, the STN has a high baseline intensity of COX, and any increase in enzyme activity may also be difficult to identify. Likewise, changes in the striatum often do not distinguish between striosomes and matrix. These possibilities are worth noting when reviewing reports on COX changes in PD.

In humans and MPTP-treated nonhuman primates, increased histochemical COX activity and *COX1* mRNA are noted in the STN, GPi, and SNr, but not the GPe, and chronic L-DOPA treatment reverses such an increase (Vila et al. 1996, 1997). Changes are likely due to reduced inhibition from GPe to STN and increased excitation from STN to output neurons in GPi and SNr. An immunohistochemical and morphometric analysis of COX in four SN nuclei of PD patients failed to identify a clear correlation between reductions in pigmented neurons (due to neuromelanin) and "COX defects" in these neurons, prompting the authors to conclude that "COX defects" in SN neurons was probably related to accelerated aging than to a primary cause of PD (Itoh et al. 1997). The fact that SNc neurons normally have low levels of COX (see above) makes it challenging to distinguish between a true COX defect and normalcy in these neurons. However, high levels of deleted mtDNA are reported in SN neurons, especially those with COX deficiency, in both aging and PD brains (Bender et al. 2006).

In MPTP-induced Parkinsonian monkeys that received grafts of embryonic ventral mesencephalic tissue into the striatum, transplanted dopaminergic neurons exhibit "the highest rate of metabolic activity among all cell types", as revealed by COX histochemistry (Collier et al. 1997). It is not known, however, if such an increase in COX is correlated with functional recovery or if it is associated with GABAergic SNr neurons unintentionally included in the transplant, as dopaminergic SNc neurons normally have low COX activity (see above). A comparable transplant of fetal nigral tissue into the striatum of a PD patient resulted in increased tyrosine hydroxylase and COX labeling in the graft, some synaptic innervation between the graft and the host, and improvement in motor functions; however, the patient died 18 months later from an unrelated cause (Kordower et al. 1996). Using yeast two-hybrid technology, Elkon et al. (2002) found that both normal and mutant α-synuclein interacted with COX in adult human cDNA expression library, and that both wild type and mutant α-synuclein coimmunoprecipitated with COX III. Moreover, the COX inhibitor KCN synergistically augmented the sensitivity of dopaminergic SH-SY5Y neuroblastoma cells to dopamine-induced cell death. Interaction between abnormal aggregates of α-synuclein and COX may contribute to the pathogenesis of PD.

17.4.4 Huntington's Disease (HD)

In 1872, a 22-year-old American physician George Huntington published a paper "On chorea", in which he described a grave hereditary disease characterized by an adult onset, a "hereditary chorea" (as opposed to an ordinary chorea; both consist of involuntary hyperkinetic movements) that progressively worsens to affect every muscle of the body (except the involuntary ones), and a tendency to insanity and suicide. There is neuronal loss in the striatum and the cortex. The disease would later bear his name.

In 1993, an international Huntington's Disease Collaborative Research Group (including researchers such as Nancy S. Wexler) published their findings that identified an expanded and unstable trinucleotide CAG repeat (encoding glutamine) mapped to 4p16.3 on what they referred to as the Huntington's Disease chromosome in multiple HD families, including a large Venezuelan HD pedigree consisting of over 12,000 members. The gene, *IT15*, is also known as *huntingtin* (*HTT*) and the protein huntingtin (Htt). HD is autosomal dominant, with the age of onset determined mainly by the number of CAG/polyQ repeats: 40–55 repeats cause adult onset, and above 55, a juvenile onset (Thompson and Orr 2023).

The exact function of Htt is unclear, but its gene knockout is embryonically lethal in mice (Nasir et al. 1995). It is a large protein of 350-kDa, ubiquitously expressed, and interacts with numerous proteins such that it is implicated in many cellular processes, including vesicular trafficking, axonal transport, signaling, mitochondrial function, selective macroautophagy, cellular scaffolding, synaptic function, transcriptional stimulation of BDNF, and the transport of both BDNF and TrkB (Steffan 2010; Sepers and Raymond 2014; Saudou and Humbert 2016; Vitet et al. 2020; Thompson and Orr 2023). Abnormal accumulation of mutant Htt leads to the disruption of some or all of these processes first in the striatum, where the striatal output neurons of the indirect pathway, the GABAergic medium spiny neurons with dopaminergic D2 receptors, are the most vulnerable. Striatal neurons, and later cortical neurons that project to the striatum, become dysfunctional, exhibit altered synaptic activity, oxidative stress, excitotoxicity, reduced BDNF and TrkB, and ultimate death; clinically, the patients' chorea, gait imbalance, speech and swallowing, and mental abilities worsen with age (reviewed in Sepers and Raymond 2014; Thompson and Orr 2023).

Mitochondrial dysfunction has long been suspected as one of the causes of progressive neuronal atrophy in HD, and defects in complexes II, III, and IV have been reported in the caudate nucleus (Brennan et al. 1985; Gu et al. 1996; Tabrizi et al. 1999). *COX1* mRNAs are markedly reduced in the putamen and GPe of HD brains (Gourfinkel-An et al. 2002). With progressive severity of HD, significant decreases in COX I, COX II, TFAM, and PGC-1α are noted, and mitochondria dynamics are altered in that the fission proteins dynamin-related protein 1 and fission 1 are upregulated and the fusion proteins mitofusin 1 and mitofusin 2 are reduced (Kim et al. 2010; Shirendeb et al. 2011). Mutant Htt increasingly accumulates in both the nucleus and mitochondria of cortical neurons, and the build-up of 8-hydroxy-guanosine in the cortex attests to increasing oxidative DNA damage in HD brains (Shirendeb et al. 2011). In a study of more than 400 German HD patients, Taherzadeh-Fard et al. (2011) found that polymorphism in *NRF-1* and *TFAM* genes, both being downstream transcription factors of PGC-1α, are significantly associated with the age of onset in HD.

17.4.5 Amyotrophic Lateral Sclerosis (ALS)

Amyotrophic lateral sclerosis (ALS or Lou Gehrig's disease) is a progressive neurodegenerative disease characterized by the loss of upper and lower motor neurons that innervate voluntary muscles, leading to severe disability, muscle weakness, paralysis, and eventual death. The disease was first described and named by the French neurologist Jean-Martin Charcot (Charcot 1874) and became well known after the American baseball player Lou Gehrig contracted it in 1939. The majority of ALS cases are sporadic with no known causes, but about 10% are familial with a few identified genes, including a hexanucleotide expansion in *C9orf72*, copper zinc superoxide dismutase (CuZnSOD or *SOD1*), *TDP-43* gene (for transactive response DNA binding protein 43 kDa or TARDBP), and *FUS* (fused in sarcoma) gene (reviewed in Cleveland and Rothstein 2001; Al-Khayri et al. 2024). Currently, there is no known cure for this fatal disease.

In mouse models of ALS caused by *SOD1* mutations, abnormal membrane-bound vacuoles derived from degenerating mitochondria are found in axons and dendrites of motor neurons and are thought to be due to free radical generation (Wong et al. 1995). This is unexpected, as the SOD1 protein is cytosolic, and the mitochondrial SOD2 is not affected in familial ALS. Disrupted mitochondria are also not noted in human ALS (Wong et al. 1995). However, reduced biochemical COX activity is found in the spinal cord of autopsied ALS patients and during early stages of motor neuron degeneration in a mouse model of ALS (Nagata et al. 1998). Decreased biochemical COX activity is also noted in the spinal cord of sporadic ALS patients (Fujita et al. 1996; Delic et al. 2018). Reduced COX activity is further confirmed histochemically in spinal cord motor neurons of ALS patients (Borthwick et al. 1999). In addition to suppressed respiratory chain (complexes I-IV) activity and citrate synthase activity, a significantly higher level of mutant mtDNA and a significantly lower amount of mtDNA are found in the spinal cord of both sporadic and familial ALS patients (Wiedemann et al. 2002), substantiating the basis of mitochondrial dysfunction in this disease.

17.4.6 Summary

A common theme in neurodegenerative diseases is mitochondrial dysfunction that leads to impaired calcium buffering, ROS and RNS generation, oxidative stress, activation of mitochondrial permeability transition, defective oxidative metabolism, secondary excitotoxicity, and eventual neuronal death (Beal 1998; Wallace 2005; Yan et al. 2013; Reiss et al. 2024). Mitochondria are often the first organelle to exhibit morphological and biochemical changes when the cell's environment or function is perturbed, and indeed, they reflect the metabolic integrity of cells and are at the heart of health and disease. A convenient indicator of mitochondrial function is cytochrome *c* oxidase, and COX histochemistry, immunohistochemistry, and/or in situ hybridization of its subunit mRNAs are often used as markers of energy metabolism at the cellular level in human, animal model, or cell line specimens under normal, experimental, or pathological conditions. Thus far, however, mitochondrial dysfunction has been associated with but not proven to be a primary cause of neurodegenerative diseases.

18 Methods for Analyzing COX

Histochemistry, Cytochemistry, Generation of Antibodies, Immunohistochemistry, Biochemistry, and Molecular Biological Techniques

ABSTRACT

Various published protocols from the Wong-Riley laboratory for the detection of COX in neurons are described. They include COX histochemistry and cytochemistry, the generation of COX antibodies and immunohisto-/cytochemistry, optimized biochemical assay for COX in brain tissue homogenates, combined COX histochemistry and immunohistochemistry of other neurochemicals at both the light and EM levels, light and EM in situ hybridization of COX, isolation of COX subunit genes from brain cDNA libraries, and chromosome conformation capture (3C) of interactions among COX subunit genes. Other methods used in the laboratory, including primary neuronal cultures, transcription factor analysis (EMSA, ChIP, promoter mutation, and transfection with shRNA or overexpression vectors), respiratory plethysmographic physiology, and patch-clamp recordings, are universal and will not be described here.

18.1 COX HISTOCHEMISTRY AND CYTOCHEMISTRY

The histochemical method of assessing the relative enzyme activity of COX in neural tissue was published in 1979 (Wong-Riley 1979a). The procedure is more of an enzymatic *reaction* than a simple stain (a term used by others and initially by us) because it only works when the enzyme is *active*. The mere presence of COX, if inactivated, does *not* yield any reaction product. Agents that inactivate the enzyme include prolonged fixation, high heat (such as in paraffin embedding), or drying.

As described in Chapter 2 (Section 2.3.1), COX can withstand a limited amount of fixation necessary to maintain the structural integrity of brain tissues. COX's substrate, cytochrome *c*, is mostly depleted postmortem and is replenished in the incubation medium. 3,3′-diaminobenzidine·4 HCl (DAB) is a chromogen that initiates the reaction by providing electrons to cytochrome *c*, which, in turn, transfers them to COX, and the cycle repeats itself with the accumulation of the brown reaction product, oxidized DAB. The more active COX is, the more reaction product accumulates. For EM localization of enzyme activity, osmium tetroxide is added to convert oxidized DAB to electron-dense osmium black visible along the outer surface of the inner mitochondrial membrane (Wong-Riley 1979a; Wong-Riley and Welt 1980; Wong-Riley et al. 1989a, 1989b). The relative activity of the enzyme derived histochemically and analyzed semi-quantitatively by optical densitometry is proportional to COX activity content and specific activity as analyzed biochemically (Figure 2.2D and E).

DOI: 10.1201/9781003507765-18

Below is the detailed protocol of COX histo/cytochemistry, slightly modified from the original (Wong-Riley 1979a):

1. Good perfusion and fixation promote tissue preservation, eliminate blood elements, and optimize cellular and subcellular localizations of enzyme activity. Perfuse the animal via the left ventricle with an initial warm (37°C) physiological saline solution (colder temperatures coagulate the blood and occlude the blood vessels) to flush out the blood, followed by a cold (4°C) mixture of paraformaldehyde (2–4%) and glutaraldehyde (0.5–1.5%) in 0.1M phosphate or cacodylate buffer, pH 7.35–7.4, and 4% sucrose. For light microscopy, 4% paraformaldehyde with or without 0.5% glutaraldehyde provides sufficient fixation. For EM, the combination of 2.5% paraformaldehyde and 1.5% glutaraldehyde preserves tissue ultrastructure and enzyme activity. If perfusion is not possible, go to step 2.
2. Remove the brain (or other neural tissue, e.g., spinal cord, ganglia, or retina) immediately, dissect the block of tissue containing the region of interest, and postfix it in the original cold fixative for ≤ 1 hour.
3. Rinse the tissue blocks with several changes of the original cold buffer. If the blocks are to be cut with a freezing microtome, first cryoprotect them with increasing concentrations of sucrose (10%, 20%, 30%) or directly into 30% sucrose in 0.1 M phosphate buffer over 1–2 days until the blocks sink. For EM, the blocks can be sectioned in the original cold buffer right away with a vibratome.
4. Cut sections with a freezing microtome, vibratome, or tissue chopper. Thickness can vary from 10 to 60 μm (25 μm is good for light microscopy; 12 μm is possible with a cryostat; thicker sections are cut with a vibratome for EM). If many cryostat sections are mounted on multiple glass slides and cannot be reacted the same day, store them in a moist, cold chamber (e.g., place the slides inside an airtight covered slide box and store overnight at −20°C) before incubating them the next day. Drying inactivates the enzyme.
5. For quantitative comparisons, it is imperative to incubate experimental and control sections under the same conditions at the same time to minimize possible variations under different settings. For frozen sections on a slide, it is optimal to mount experimental and control sections on the *same* slide. For age comparisons, it is ideal to mount samples from different ages on the same slide.

   Incubation medium (100 ml): make up fresh before use in light-sensitive, actinic glass containers:

   DAB (50 mg)
   0.1M phosphate buffer, pH 7.35–7.4 (up to 100 ml)
   Sucrose (4 gm) dissolved in the buffer
   Mix DAB and buffer and filter with #1 Whatman filter paper into another actinic container
   Right before use, add the following:
   Cytochrome C, type III, Sigma-Aldrich (St. Louis, MO; C2506; from equine heart) (20 mg for rats; 30 mg for cats; 50 mg for monkeys)

   Incubate sections at 37°C in an incubator, covered in the dark, with constant agitation on a gentle shaker, for 1–2 h or longer. Check after an hour for the appearance of a light to darker brown reaction product. Vibratome sections take longer to incubate than frozen sections. Stop the incubation when clear differentiation between highly reactive and lightly reactive regions can be discerned. Avoid excessive or inadequate time of incubation.
6. Rinse sections in three changes of 0.1M phosphate-buffered saline (PBS).
7. Free-floating, reacted sections are mounted in filtered (with Whatman #1 filter paper) tap water (avoid distilled water, as it will cause sections to swell) onto glass slides, air dried, and coverslipped. If counterstaining with Nissl is desired, defat the mounted sections for 1 hour in xylene, and stain with cresyl violet.

8. For EM vibratome sections, the area of interest is trimmed in cold buffer after incubation. The pieces are postfixed in 1.3% OsO4 for 1 hour, and either dehydrated and *en bloc* stained with 1% uranyl acetate in absolute alcohol for 1 hour, or in a mixture of 1% OsO4 and 1.5% potassium ferricyanide for 2 h followed by dehydration in an alcohol series. The pieces are then rinsed in propylene oxide and embedded in resin. Under the EM, the reaction product resides along the outer surface of the inner mitochondrial membrane, including the intracristate space.
9. If necessary, use a Kodak #47 dark blue filter to enhance contrast of the brown reaction product at the light microscopic level.
10. Perform optical densitometric readings on groups of cells or the cytoplasm (containing mitochondria) of single neurons. The reference point is the white matter (or a blank area of the slide viewed with a neutral density filter) set to zero for each slide. Data are analyzed with appropriate statistical analyses.

Note: Our semi-quantitative, optical densitometric analyses of COX reaction product in single neurons were done with a Zeiss Zonax MPM 03 microphotometer with a 2–5 µm diameter measuring spot size. It is a direct microdensitometric measurement of the intensity of the COX reaction product and is very accurate. Single neuron measurements are precise and avoid interference by glia or blood elements.

18.2 GENERATION OF COX ANTIBODIES AND IMMUNOHISTOCHEMISTRY/ CYTOCHEMISTRY

Detailed methods of COX purification from bovine brain, generation of COX polyclonal antibodies, and COX immunohistochemistry were published originally by Hevner and Wong-Riley (1989). COX immunocytochemistry was published originally by Liu and Wong-Riley (1994). Below are brief recounts of these methods.

18.2.1 Purification of Brain COX

18.2.1.1 Isolation of Synaptic and Non-Synaptic Mitochondria

1. Obtain a fresh calf brain and pack it in ice. Scrape the gray matter from the cerebral cortex and the caudate nuclei and homogenize them (with a Potter-Elvehjem type homogenizer) within 45 min postmortem. The combined mass is 120–150 g.
2. Proceed with Rendon and Masmoudi's (1985) protocol for purifying synaptic and non-synaptic mitochondria with minor modifications: (a) Wash the nuclear pellet once after the first centrifugation to increase the recovery of mitochondria; (b) resuspend the washed P2 pellet in 120 or 180 ml (rather than 30 ml) of isolation buffer; and (c) lyse the washed synaptosomal fraction in 200 ml (rather than 40 ml) of hypotonic buffer.

18.2.1.2 Solubilization of COX from Isolated Mitochondria

1. Remove cytochrome *c* from isolated mitochondria (50–150 mg protein) according to the method of Jacobs and Sanadi (1960).
2. Prepare partially purified and solubilized COX from the cytochrome *c*-depleted mitochondria according to the method of Jacobs et al. (1966) with some modifications: (a) Use a higher concentration of Triton X-114 (0.006 ml 20% Triton X-l 14/mg protein) in the initial detergent extraction to improve the subsequent purity and stability of COX; and (b) increase the potassium phosphate concentration to 0.5 M in solubilizing with Triton X-100 to improve the solubility and stability of COX.

18.2.1.3 Purification of COX

1. Perform all procedures at 0–4°C. Desalt the solubilized crude brain COX (5–15 mg protein) on a Sephadex G-100 column (Pharmacia, Kalamazoo, MI) pre-equilibrated with 40 mM HEPES buffer, pH 7.4 (adjusted with KOH, pH 7.6 at 4°C) and 0.05% Triton X-100.
2. Pool all fractions with COX activity and apply them (5–10 mg protein) to a 1.5 × 3.7 cm column of yeast cytochrome c-Sepharose 4B (prepared by the method of Broger et al. 1986). (Pre-equilibrate the affinity column with 40 mM HEPES, pH 7.4, and 0.05% Triton X-100.)
3. Wash the sample with pre-equilibration buffer until the A_{280} nm and A_{405} nm bands return to baseline (monitor with an ISCO Type 6 Optical Unit, UA-5 Absorbance Detector, and Model 1133 Multiplexer/Expander).
4. Wash the column with 5–1 0 column volumes of 40 mM HEPES, pH 7.4, and 0.5% Triton X-100, then elute with 25 ml (total) of this buffer containing a linear gradient of 0–100 mM K_3PO_4, pH 7.4 (adjust with KOH). The flow rate of the column should be about 2.1 cm/h during the loading and elution steps and about 4.3 cm/h during the wash.
5. Assay the fractions (2 ml) for COX activity, cytochromes, conductivity, and protein. The last step is immunoblotting and ELISA using the standard methods.

18.2.2 Generation of Polyclonal Antibodies Against Brain COX

1. Immunize New Zealand White rabbits by injecting at multiple intradermal and subcutaneous sites on the animal's back with a primary dose of 100 μg purified (9.9 nmol heme a/mg protein) bovine brain COX emulsified with Difco complete Freund's adjuvant.
2. Ten weeks later, inject intramuscularly in all four limbs a booster dose of 200 μg of purified COX at the same concentration and composition as the primary dose.
3. Two weeks after the booster injection and periodically thereafter, collect blood for antiserum against COX. The titer tends to remain at 1:100,000 for these bleeds until 7 months later.
4. Purify IgG from serum by DEAE-Affi Gel Blue chromatography (Bio-Rad, Hercules, CA) according to the manufacturer's instructions. The anti-COX titer in IgG is ~20-fold less than that in serum because of dilution during chromatography. IgG gives a lower background in COX immunohistochemistry than that of antisera.
5. Store serum and IgG at −20°C, −80°C, or lyophilize and freeze them. Freezing and lyophilization do not affect anti-COX titers.

18.2.3 COX Immunohistochemistry

This method was originally published by Hevner and Wong-Riley (1989). It is briefly described below. All incubations are on a gentle shaker.

1. Cut frozen sections and incubate them in a blocking solution (PBS-nonfat dry milk [PBS-NFDM] containing 1% Triton X-100 and 1–5% normal goat serum (NGS)) for 8–24 h at 4°C, then rinse them 3x for 5 min in PBS.
2. Incubate sections in primary antibodies (anti-COX at 1:2000–8000) diluted in the blocking solution for 4 h at room temperature (RT), then for 8–24 h at 4°C. Rinse sections 3x in PBS.
3. Incubate sections in secondary antibodies (GaR [goat-anti-rabbit]-HRP or fluorescein conjugated GaR IgG) at 1:100 (in PBS-NFDM with 5% NGS) for 6 h at RT or overnight at 4°C. Rinse sections 3x in PBS.
4. For immunofluorescence, incubate sections in tertiary antibody (swine anti-goat IgG, fluorescein conjugate, 1:100) in PBS-NFDM with 5% NGS for 8–12 h at 4°C, rinse, then mount sections in glycerol:PBS (9:1) (PBS is adjusted to pH 9.0 with Na_2CO_3 and contains 10 mg/ml of 1,4-phenylenediamine to reduce fluorescence quenching). Coverslipped sections should be viewed right away. If necessary, they can be stored at 4°C in the dark until viewing.

5. For immunoperoxidase, rinse sections in 0.1 M ammonium phosphate buffer (APB), pH 7.0. Then incubate in 0.05% DAB and 0.004% H_2O_2 in APB for 3–10 min at RT. Stop the reaction by rinsing twice in APB and twice in PBS.
6. Mount sections out of filtered tap water on gelatin-coated slides, air dry, and coverslip.

Note: Controls include preimmune serum and no primary antibodies processed with concentrations equal to or exceeding those of anti-COX antibodies.

18.2.4 COX EM-Immunocytochemistry

Below is a brief account of COX EM-immunocytochemistry originally published by Liu and Wong-Riley (1994) for a precursor protein of COX IV subunit. All reactions are conducted with gentle agitation.

Preembedding

1. Block vibratome sections with the blocking solution (PBS containing 5% NFDM, 5% NGS, and 0.1–0.5% Triton X-100) overnight at 4°C.
2. Incubate sections with purified IgG against COX (or its subunit) diluted at 1:1000 (or empirically tested concentration) in the blocking solution for 48 h at 4°C.
3. Process sections for the immunohistochemical procedure as described in Section 18.2.3, then embed them in Epon-Araldite.

Postembedding (there are variations to this method; see also Section 18.4.2)

1. Osmicate vibratome sections in 1% OsO_4 for 30 min at RT, then dehydrate them in an alcohol series and embed them in Durcupan resin.
2. Cut ultrathin sections and mount them on non-Formvar-coated fine nickel mesh grids.
3. Pretreat sections in 1% periodic acid for 10 min followed by 10% sodium metaperiodate for 15 min, and 1% sodium borohydride for 3 min.
4. Block sections in the blocking solution (5% bovine serum albumin [BSA] in 0.1% Triton X-100, 50 mM Tris, and 0.9% sodium chloride [TTBS]), pH 7.4, for 30 min.
5. Incubate sections in primary antibodies (anti-COX IgG or antisera) at dilutions of 1:100–200 in the blocking solution for 15–40 h at RT.
6. Stop the reaction with TTBS containing 1% BSA for 30 min, then react sections with 10 nm gold-labeled GaR IgG (diluted 1:40 in TTBS containing 0.1% BSA, pH 8.2) for 1 h at RT. Rinse sections in TTBS.
7. Stain the grids with uranyl acetate and lead citrate. Examine samples with an electron microscope.

18.3 OPTIMIZED BIOCHEMICAL ASSAY FOR COX IN BRAIN TISSUE HOMOGENATES

The biochemical assay of COX activity content (tissue wet wt: mU/mg tissue) and specific activity (mU/mg protein) from the whole brain (or separate gray matter and white matter) homogenates was optimized and published originally by Hevner and Wong-Riley (1993). Below is a summary of this protocol for brain tissue homogenates.

1. Animals are deeply anesthetized and decapitated. Remove brains immediately and rinse with cold (4°C) isolation buffer (0.32 M sucrose, 1 mM dipotassium ethylenediaminetetraacetic acid (EDTA), 10 mM Tris, pH 7.4, 4 ml/g tissue), and blot dry with gauze. Dissect regions of interest and freeze them individually in powdered dry ice to preserve enzyme activity. If necessary, samples may be stored at -80°C for assays later.

2. For each sample: quickly weigh and place each sample in a microfuge tube (or whole, small brain in a test tube); immediately add 4 µl/mg tissue (or 4 ml/g whole brain) of cold isolation buffer to yield a 20% homogenate. Set the tubes on ice.
3. Homogenization of large samples is done with ~10 strokes on a Teflon-glass motor-driven homogenizer. For small samples in a microfuge tube, use a motor-driven polypropylene pestle (or manually) for ~1–2 min. The 20% homogenates are stable on ice for hours.
4. Prepare COX assay mixture (0.07% ferrocytochrome c, 10 µM dipotassium EDTA, 0.1 M MES [N-morpholinoethane-sulfonic acid], pH 6.0) from stock solutions and keep it on ice. The stock solutions are 1% ferrocytochrome *c* in 1x MES-EDTA and 10x MES-EDTA. The ferrocytochrome *c* stock solution is prepared by reducing with sodium ascorbate and dialyzing against excess 1x MES-EDTA at 4°C in the dark, with three changes of buffer. A minimum of 8 h is required for equilibration between buffer changes.
5. Set the spectrophotometer wavelength at 550 nm and the temperature at 30°C.
6. Prepare detergent-solubilized homogenates. Mix the appropriate quantities of 20% homogenate, isolation buffer, and deoxycholate (10% stock solution in isolation buffer).

Region of Interest (Rat Brain)	Homogenate (%)	Deoxycholate (%)
Gray matter	0.25	0.5
White matter	1.0	1.0
Retina	0.5	0.75

Note: For other species, these concentrations may be adjusted as necessary.

7. Equilibrate detergent-solubilized homogenates for 4–8 min at RT with gentle, intermittent vortexing. Place samples on ice. Centrifuge in a microfuge at 16,000 × g for 10 s to sediment possible particulate material.
8. Set the spectrophotometer to zero. Use an assay mixture oxidized with 0.1 M potassium ferricyanide (10 µl/ml assay mixture) in both the sample and the reference cuvettes.
9. Assay COX activity in triplicate. Add the sample to the sample cuvette containing the reduced assay mixture. The reference cuvette contains ferricyanide-oxidized assay mixture, and the starting absorbance should be 0.9 – 1.0. Monitor the absorbance change at 550 nm.

   Note: One can use 10 µl detergent-solubilized homogenate in 1 ml of assay mixture at 30°C, with an assay time of 1 min. Assay time and other factors may be modified if necessary.

10. Calculate the enzyme activity in the 10 µl sample as follows:

    COX activity (nmol cyt *c* oxidized/min) = (54.6)· (absorbance change/min)

18.4 COMBINED LABELING OF COX AND OTHER NEUROCHEMICALS

COX histochemistry and/or immunohistochemistry can be combined with immunohistochemical reactions of various neurochemicals at both the light and EM levels. Below are a few published examples.

18.4.1 Double Labeling of COX and GABA at the Light Microscopic Level

This method was originally published by Luo et al. (1989) for double labeling COX and GABA in the CNS of adult rats. Below is a brief account of this approach for combining COX histochemistry or immunohistochemistry with GABA (or other neurochemicals) immunohistochemistry.

18.4.1.1 COX Histochemistry Followed by GABA (or Other Neurochemical) Immunogold-Silver Staining (IGSS) at the Light Microscopic Level (Figure 18.1A and B)

1. Cut 15 μm frozen brain sections and place the free-floating sections in PBS.
2. Incubate the sections for COX histochemistry as described in Section 18.1. Stop the reaction by rinsing the sections with three changes of cold 0.05 M Tris-buffered saline (TBS), pH 7.6.
3. Proceed to immunogold-silver stain (IGSS) for GABA (or other neurochemicals):
   a. Apply Lugol's solution to sections at RT for 5 min.
   (Lugol's solution: Dissolve 1 g iodine and 2 g potassium iodide in 200 ml distilled water [DW])
   b. Rinse thoroughly with 2.5% sodium thiosulfate at RT to remove all traces of iodine.
   c. Rinse with TBS.
   d. Treat sections with 0.05% trypsin in TBS containing 0.1% $CaCl_2$ for 1 min at RT, then rinse with TBS.
   e. Block non-specific background staining with 10% NGS overnight at 4°C, then rinse with TBS.

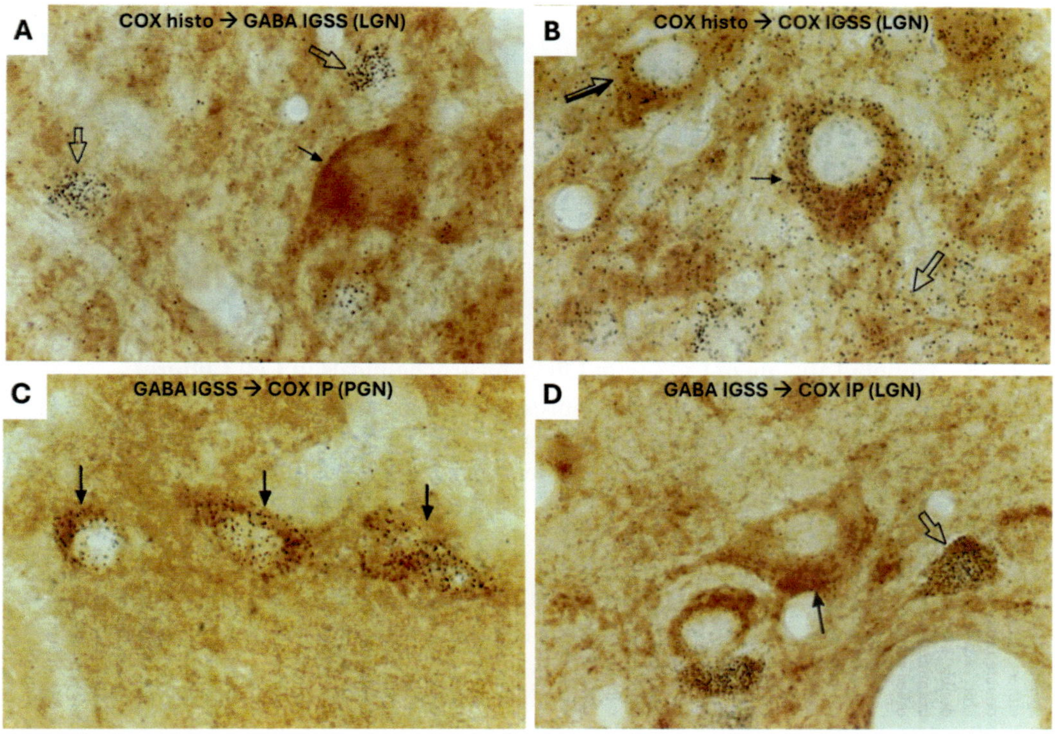

FIGURE 18.1 Double labeling of COX and GABA. (A) LGN neurons were reacted with COX histochemistry followed by GABA IGSS. Note the large neuron darkly reactive for COX with little GABA labeling (solid arrow) and the two small GABA-rich neurons with low COX activity (open arrows). (B) LGN neurons reacted for COX histochemistry followed by COX IGSS. Note the correspondence in relative activity and protein in the dark (solid arrow), moderate (half-solid arrow), and lightly reactive (open arrow) neurons. (C) Perigeniculate neurons (PGN) reacted for GABA IGSS followed by COX immunoperoxidase (IP). Note that these neurons are rich in both GABA and COX. (D) LGN neurons reacted with GABA IGSS followed by COX IP. Here, the large, darkly COX-reactive neuron has no GABA labeling (solid arrow), whereas the small, GABA-rich neuron has low COX labeling (open arrow).

Modified from Luo et al. 1989.

f. Incubate with anti-GABA serum derived from rabbit (Immunonuclear [now ImmunoStar, Hudson, WI]; diluted to 1:100,000 – 1:400,000 in 5% NGS) for 24 h at 4°C, then rinse with TBS.
g. Block again with 10% NGS for 30 min.
h. Incubate with immunogold (Janssen; GaR IgG with 5 nm gold conjugate; 1:100 in TBS) overnight at 4°C,
i. Wash 3 × in TBS for 10 min, then 3 × in DW for 10 min.
j. Immerse in IntenSE M (Janssen, Raritan, NJ) for 13–18 min in the ***dark*** (in the darkroom with a low-intensity red or sodium lamp). The *timing is critical* and is empirically derived from *pretesting* the specific tissue and reagents. (Once the section is exposed to light, it is irreversible).
k. Stop the reaction by rinsing in DW. Check for the presence of silver grains. If the signal is too weak, return sections to IntenSE M for additional time as needed. When done, rinse sections thoroughly in PBS.

4. Mount sections out of filtered tap water on gelatin-coated slides, air dry, and coverslip.

Note: The COX reaction product is brown, and the GABA (or other neurochemical) immunoreaction product is represented by silver grains.

18.4.1.2 GABA (or Other Neurochemical) IGSS Followed by Indirect Immunoperoxidase for COX at the Light Microscopic Level (Figure 18.1C and D)

1. Cut and treat sections with IGSS for GABA as described in Section 18.4.1.1, #3.
2. React the sections for COX immunoperoxidase as described in Section 18.2.3.

Note: GABA (or other neurochemical) immunoreaction product is represented by silver grains, and COX immunoreaction product is brown.

18.4.1.3 Double Labeling of COX with Other Neurochemicals in Cell Cultures

The protocol for combining COX histochemistry with immunohistochemistry of other neurochemicals in primary cultures was described originally by Liu et al. (1990) for neural crest cells and later by Zhang and Wong-Riley (1999) for cortical neurons. These methods are briefly described below.

1. Fix cultures with 4% paraformaldehyde for 1 h on ice, then rinse 3x with PBS.
2. React the cultures for COX histochemistry as described in Section 18.1, #5 (use 0.03% cytochrome *c* and incubate for 2 h). At the end, rinse them 3x with PBS containing 0.1% BSA and block them in this solution for 1 h at RT.
3. Incubate cultures in primary antibodies (against the neurochemical of interest) diluted (at a predetermined optimal concentration) in 0.1% or 0.2% Triton X-100 in PBS overnight at 4°C, followed by three rinses in PBS-BSA, 10 min each.
4. Incubate cultures in secondary antibodies (for single immunofluorescence, use a fluorochrome, such as fluorescein- or rhodamine-conjugated secondary antibodies; for double or triple labeling, use a different fluorochrome conjugate for each neurochemical) for 1 h at RT (in the dark for fluorochromes). At the end, rinse them 3x in PBS-BSA.
5. For IGSS, use a gold-conjugated secondary antibody and proceed with silver staining as described in Section 18.4.1.1.
6. Coat cultures with 50% glycerol containing 1 mg/ml *p*-phenylenediamine, pH 8.5, and coverslip them.

18.4.2 Double Labeling of COX and GABA (or Other Neurochemical) at the EM Level

The protocol was originally published by Nie and Wong-Riley (1995, 1996a). Below is a brief account.

1. React vibratome-cut sections for COX histochemistry and process them for EM as described in Section 18.1.
2. Cut ultrathin sections and place them on Formvar-coated, single-slot nickel grids for postembedding GABA (or other neurochemical) immunocytochemistry.
3. Treat sections with 2% aqueous solution of sodium metaperiodate for 20 min, then rinse 3x in DW.
4. Place sections in Tris (10 mM) -phosphate (10 mM) -buffered isotonic saline, pH 7.6, containing 0.1% Triton X-100 and 1% NGS (TPBS) for 30 min.
5. Transfer grids to drops containing the GABA (or other neurochemical) antisera diluted 1:1,000 (or predetermined optimal concentration) in the same TPBS buffer. Incubate the sections in the primary antibodies for 4 h at RT, then rinse sections 3x in TPBS.
6. Incubate sections in the secondary antibodies (GaR IgG conjugated to 15-nm colloidal gold, Amersham, Chicago, IL) diluted 1:25 in TPBS, pH 8.2, for 2 h at RT. Then rinse sections 3x in TPBS and 3x in DW.
7. Stain sections with alcoholic uranyl acetate and lead citrate, then rinse 3x in DW, air dry, and view by EM.
8. Controls include antibodies preadsorbed with the antigen (GABA or other neurochemicals) or the omission of primary antibodies.

Other examples of double-labeling histochemically reacted COX with immunocytochemically-reacted neurochemicals at the EM level can be found in Kang JJ et al. 2019, 2020.

18.5 MOLECULAR BIOLOGICAL TECHNIQUES FOR ANALYZING COX

18.5.1 Synthesis of Riboprobes for *COX* Subunit Genes

Synthesis of ^{35}S radiolabeled riboprobes for *COX* subunit genes was described by Hevner and Wong-Riley (1991), and that for non-radiolabeled riboprobes was described by Wong-Riley et al. (1997a). Below is a brief account of the digoxigenin-labeled *COX* riboprobes.

1. Linearize purified plasmids of COX subunits by restriction enzyme digestion.
2. Generate antisense and sense riboprobes by in vitro transcription using sets of restriction enzymes and RNA polymerases (e.g., HindIII and T7 for *COX3* antisense, and EcoRI and T3 for *COX3* sense for inserts cloned in pBluescript II KS(+); BamHI and T7 for *COX6a1* antisense, and EcoRV and Sp6 for *COX6a1* sense for inserts cloned in plasmid pCR™II).
3. Label the riboprobes with digoxigenin-11-UTP (Boehringer-Mannheim) according to the manufacturer's instructions. Analyze digoxigenin-labelled antisense and sense riboprobes for size in a 1.2% denaturing agarose gel with formaldehyde, then transfer them to positively charged nylon membrane (Qiabrane Plus, Qiagen) and detect them with Lumi-Phos 530 chemiluminescence according to the manufacturer's instructions (Genius System, Boehringer–Mannheim).

18.5.2 In Situ Hybridization: Light Microscopy

A radioactive riboprobe approach for in situ hybridization of *COX* subunits was originally published by Hevner and Wong-Riley (1991). It is a modification of a procedure developed by Chesselet et al. 1987). As predicted, the mitochondria-derived *COX* subunit mRNAs are distributed in cell bodies and proximal dendrites, but the nuclear-derived ones are located only in cell bodies. It is briefly described below.

With riboprobes, precautions are taken to minimize the presence of RNase. In addition to wearing gloves, water for making solutions should be treated with diethyl pyrocarbonate (DEPC). This is accomplished by adding 1 ml of DEPC to 1 liter of double DW (DDW), shaking vigorously to mix, placing in a 37°C oven overnight, and autoclaving the next day.

For stringent washes, use SSC (sodium chloride – sodium citrate) buffer. Make a 20x stock solution with 3 M NaCl and 0.3 M Na_3 citrate.$2H_2O$. Adjust pH to 7.0 with HCl.

1. Mount unfixed frozen sections on slides, air dry, and bring them to RT under a gentle stream of air. Fix for 5 min in 3% paraformaldehyde with 0.02% DEPC, rinse with 2x SSC, acetylate, and treat with 0.1 M Tris glycine (pH 7.0). Dehydrate sections through a graded ethanol series and air dry.
2. Hybridize the sections for 4 h (for nuclear-encoded COX subunits) or 6 h (for mitochondrial-encoded COX subunits) at 50°C with 20 µl hybridization mix (40% formamide, 8% dextran sulfate, 4x SSC, 1 mg/ml yeast tRNA, 1 mg/ml herring sperm DNA, 10 mM DTT, 1x Denhardt's solution) containing ^{35}S-labeled antisense (sense for controls) riboprobe (250 ng/ml) under coverslips cut from Parafilm (American Can Co., now Amcor, Deerfield, IL).
3. After hybridization, rinse sections with 2x SSC, then wash them for 5 min and again for 20 min in 50% formamide and 2x SSC at 52°C.
4. Digest unhybridized riboprobe with RNAse A (100 µg/ml) in 3x SSC for 30 min at 37°C. Apply a final wash for 5 min in 50% formamide and 2x SSC at 52°C, then rinse the sections overnight in 2x SSC and 0.05% Triton X-100.
5. Dehydrate sections through graded ethanol series, delipidate them in xylene, air dry, then process them for film and emulsion autoradiography (ARG).
6. For film ARG, expose dried sections to Kodak XAR-5 film for 3–5 d at RT. Develop films with a standard processor.
7. For emulsion ARG, dip slides in Kodak NTB-3 emulsion (melted at 43°C with an equal volume of 0.3 M ammonium acetate), dry them in a humid chamber, and expose them in the dark at 4°C for 4 d to 6 wk. Process exposed slides at 15°C for 3.5 min in Kodak Developer D-19, 30 sec rinse in DW, and for 5 min in Kodak Fixer without acid hardener. Dry the developed slides, dehydrate them through a graded ethanol series, clear with xylene, and coverslip with Permount (Fisher, Waltham, MA).
8. Control sections are to be processed under identical conditions as above.

18.5.3 In Situ Hybridization: Electron Microscopy

For EM in situ hybridization of *COX* subunit mRNAs, a non-radioactive, digoxigenin (DIG) riboprobe approach was used on the mouse cerebellum, where the Purkinje cell bodies can be clearly visualized in continuity with their large, primary dendrites in the molecular layer (Wong-Riley et al. 1997a). Mitochondrial-encoded *COX* subunit mRNAs are localized within mitochondria in both cell bodies and dendrites, whereas the nuclear-encoded ones are found exclusively within the cell bodies (associated with the cytoplasmic side of rough endoplasmic reticulum and the outer membrane of the nuclear envelope). The protocol is a modification of

Boehringer–Mannheim's (manufacturer of digoxigenin; Indianapolis, IN) instructions and of Trembleau et al.'s method (1994). It is briefly described below.

1. Cut 40–60 μm sections with a Vibratome and incubate them for 1 h at 37°C in a prehybridization solution (4x SSC, 1x Denhardt's solution, and 10 μg/ml wheat germ tRNA).
2. Immerse sections overnight at 37°C in the hybridization solution (50% formamide, 600 mM NaCl, 80 mM Tris–HCl, pH 7.5, 4 mM EDTA, 0.1% sodium pyrophosphate, 10 μg/ml wheat germ tRNA, and 10 nM of DIG-labeled *COX* subunit riboprobes.
3. Wash sections in 2x SSC followed by 0.1x SSC at 37°C and immerse them in PBS.
4. Incubate sections overnight at 4°C in sheep anti-digoxigenin antibody (Boehringer–Mannheim) diluted (1:1000) in PBS–BSA.
5. Wash sections in PBS, then incubate them in biotinylated anti-sheep secondary antibody (Vector Laboratories, Burlingame, CA; now Newark, CA) (1:200 in PBS–BSA) for 2 h at RT.
6. Incubate sections in avidin–biotin–peroxidase complex (Vector Labs) diluted in PBS (according to the manufacturer's instructions) for 2 h at RT.
7. Rinse sections in PBS and react them in 0.025% DAB and 0.006% H_2O_2 in 50 mM Tris buffer, pH 7.6.
8. Monitor the reaction intensity under a light microscope. When the optimal intensity is reached, stop the reaction by transferring sections to PBS.
9. Osmicate the sections and process them for EM as described in Section 18.1.

18.5.4 Isolation and Subcloning of *COX* Subunit Genes from Brain cDNA Libraries

The original method was published by Wong-Riley et al. (1997a) for *COX3* and *COX6a1* from a rodent brain cDNA library. Using similar methods, *COX6A1* was also isolated from a human brain cDNA library (Wong-Riley et al. 2000). Below is a brief description of the method.

1. Synthesize primers for *COX3* (mitochondrial) based on mouse L cells (Bibb et al. 1981) and for *COX6a* (nuclear) based on the rat liver (Schlerf et al. 1988) or the human liver (Fabrizi et al. 1992). These primers are designed to have a Tm of 58°C and are desalted during the purification process.

   COX3 (rodent): Primer 1: 5'-ACTCATGCATATCACATAGTTA-3'
   Primer 2: 5'-AATGGAGACGTATAGGAAAAG-3'
   COX6a1 (rodent): Primer 1: 5'-CGAGTGGCGCCCACGG-3'
   Primer 2: 5'-ATCTTCATAGCCAGTCGGAA-3'
   COX6A1 (human): Primer 1: 5'-ATGTCGAGTGGCGCCCAT-3'
   Primer 2: 5'-TTCATCTTCGTAGCCAGTTG-3'

2. Perform PCR to isolate and amplify *COX3* and *COX6a* from brain cDNA libraries (Stratagene, La Jolla, CA). Use a hot start and 30 cycles of denaturation (94°C), annealing (53°C), and elongation (72°C).
3. Run the amplified PCR products on a 1.5% low-melting agarose gel. Stain the gel with ethidium bromide to determine the correct sizes and excise the correct bands from the gel.
4. Purify *COX* subunit genes with the Qiaex kit (Qiagen, Chatsworth, CA).
5. Subclone the isolated *COX* subunit cDNAs either into the EcoRV site of the plasmid vector pBluescript II KS(+) (Stratagene) containing the phage T3 and T7 promoters, or into the EcoRI site of the *lac*Z gene in plasmid pCR^{TM}II (TA cloning kit, InVitrogen, San Diego, CA) containing phage SP6 and T7 promoters.

6. Use the ligation reaction to transform OneShot competent cells (InVitrogen). Amplify the white colonies containing plasmids with inserts and purify the plasmid DNAs on a Qiagen plasmid column according to the manufacturer's instructions.
7. Digest plasmid DNAs with EcoRI for confirmation of insert size. Sequence the cDNAs.

18.5.5 Chromosome Conformation Capture (3C) of Interactions Among *COX* Subunit Genes

The 3C method for capturing interactions among the 10 nuclear *COX* subunit genes and the three mitochondrial transcription factor genes (*Tfam*, *Tfb1m*, and *Tfb2m*) responsible for the transcription of the three mitochondrial-encoded *COX* subunit genes was published by Dhar et al. (2009a). The method was a modification of the procedure originally described by Tolhuis et al. (2002) for the β-globin gene. A comparable 3C method was also used to determine interactions between COX subunit genes and glutamatergic receptor subunit genes (Dhar and Wong-Riley 2010). Below is a brief account of the 3C method for *COX*.

1. Harvest N2a cells or primary neurons by trypsinization and resuspension in fresh culture medium with 10% fetal bovine serum.
2. To cross-link chromatin, treat 1×10^7 N2a cells or 5×10^6 primary neurons in 10 ml of cell culture medium with 2% formaldehyde for 10 min at RT. Stop the cross-linking by quenching with 0.125 M glycine for 5 min at RT.
3. Isolate nuclei by incubating cells in a lysis buffer (10 mM Tris, pH 8.0, 10 mM NaCl, 0.2% Nonidet P-40) on ice with agitation for 30 min.
4. Release chromatin by treating isolated nuclei with 0.3% SDS and digesting with 400 units of *Bgl*II (Fermentas, Hanover, MD) overnight.
5. Heat-inactivate the restriction enzyme and dilute the chromatin in T4 ligation buffer to achieve a low genomic DNA concentration of 2.5 ng/μl to facilitate intermolecular re-ligation.
6. Perform chromatin re-ligation by incubating diluted chromatin with 100 Weiss units of T4 ligase (Fermentas) for 4 h at 16 °C.
7. Remove RNA contaminant by RNase A (Roche Applied Science, Madison, WI) digestion, then purify the 3C chromatin by phenol/chloroform extraction.

18.5.5.1 Generation of Control Templates

1. Control PCR templates are necessary to compare with 3C products. To do so, pair all 10 nuclear-encoded *COX* subunit genes as well as pair individual *COX* subunit genes with other genes or gene loci, including *Tfam*, *Tfb1m*, and *Tfb2m* genes. For example, to generate the control template for pairing *COX4i1* and *COX5a*, use PCR to amplify a region of each subunit gene that spans a single *Bgl*II site with 100–150 bp on either side.
2. Digest ~30 μg of each PCR product with 300 units of *Bgl*II overnight at 37°C.
3. Extract the DNA with phenol-chloroform and precipitate it with ethanol.
4. Ligate equimolar amounts of digested product (e.g., *COX4i1* and *COX5a*) at a high concentration (~10 ng/μl for each product) to ensure the formation of intermolecular crosslinked product.
5. Combine one primer of a given restriction fragment (*COX4i1*) with a primer of a second (*COX5a*) in all pairwise combinations of ligated orientations (e.g., forward primer of *COX4i1* with forward primer of *COX5a*, reverse primer of *COX4i1* with forward primer of *COX5a*, etc.). Use the same steps for all other pairings of *COX* genes to ensure all subunits are included.

6. Use the control templates for comparison with 3C samples. Run PCR for both control and 3C at the same time, with all PCR products 150–300 bp and the annealing temperature of 58–60°C. The final chosen control template for each pairing will have the same orientation as the 3C-generated product.
7. A positive 3C control is one that targets an "internal" cross-linking product between two *Bgl*II sites located 1.6 kb apart within the coding region of the calreticulin (*CalR*) gene (+2.8 and +4.4 kb relative to the *CalR* transcriptional start site) (Tolhuis et al. 2002).
8. For negative 3C controls, generate five sets of templates: (*a*) possible cross-linking between the *CalR* 4.4-kb *Bgl*II site and another *Bgl*II site on *COX6a1*; (*b*) possible cross-linking between the *Bgl*II site of citrate synthase (*Cs*) and those of *COX4i1*, *COX6a1*, and *COX8a*, respectively; (*c*) no cross-linking agent (formaldehyde); (*d*) no ligase; and (*e*) possible cross-linking for regions far beyond the *Bgl*II sites for *COX4i1* and *COX8a*.

18.5.5.2 PCR Assays

1. Use very stringent PCR conditions to detect only specific signals. PCR cycles are: denaturation (2 min at 94°C), then 36 cycles of 30 s at 94°C, 15 s at 59°C, and 15 s at 72°C, followed by a final step of 1 min at 72°C. Thirty-six PCR cycles are sufficient to detect the specific PCR fragments generated after digestion and ligation that capture the rare molecular events in the nucleus.
2. Detect cross-linking among *COX* subunit gene loci in 3C samples by using semiquantitative PCR. Specifically, use a pair of primers targeting two different *COX* subunit genes to amplify crosslinking DNA generated from 3C reactions. Each primer for each of the 10 nuclear-encoded *COX* genes and the three mitochondrial transcription factor genes (*Tfam*, *Tfb1m*, and *Tfb2m*) targets a DNA region 100–150 bp from the *Bgl*II site and generates a 200–300-bp amplicon.
3. Determine the linear range of amplification for 3C samples by serial dilution, as described by Miele and Dekker (2009). Add an appropriate amount (150 μg) of 3C DNA within the linear range to each PCR.
4. Compare the 3C products with the corresponding ones from the control templates.
5. Run the PCR product on 2% agarose gel and stain with ethidium bromide. Test all possible interactive pairs among 10 *COX* subunits (*4i1*, *5a*, *5b*, *6a1*, *6b*, *6c*, *7a2*, *7b*, *7c*, and *8a*) by PCRs targeting their possible cross-linked products. Perform all PCRs in triplicate.

Epilogue
Lessons Learned and Future Questions

For the last billion-plus years, cytochrome *c* oxidase has existed, evolved, and assumed vital functions in living organisms from bacteria to humans. Without it, almost all organisms (except for anaerobic bacteria) cannot survive. It is a major contributor to energy generation, and energy supports the functioning of all living organisms.

In neurons, COX serves the added, vital role of sustaining synaptic communications by fueling the energy for membrane repolarizing after not only action potentials but also graded depolarizing potentials. COX reveals heretofore unrecognized properties of neurons: that metabolic activities differ not only at the regional, laminar, and cellular levels, but also at the subcellular level; that the entire neuron is not metabolically homogeneous, but each segment has its microdomain of energy regulation; that dendrites are the major energy consumers of the brain; that metabolic plasticity is still possible in the adult nervous system; and that mature neurons with the highest metabolic demands suffer the most from functional insults.

COX has opened a new field of investigation in the primate visual cortex by its discoveries in labeling metabolically active zones with distinct neurochemical and physiological properties. Unique metabolic domains have been and hopefully will be discovered in other sensory and motor systems and many additional species.

The metallic centers in COX absorb light in the far red to near-infrared range and assist in healing injured neurons exposed to various toxins. More mechanistic questions await investigation.

The bigenomic nature of COX kindles a host of queries: How do the two genomes coordinate with each other? What mediates the crosstalk between the nucleus and the mitochondria? Does one genome dominate, or are there feedforward and feedback loops between them? A great deal has been uncovered regarding the bigenomic regulation of COX in neurons: how they respond to changes in neuronal activity, the key transcription factors involved, and the interactions of all relevant *COX* subunit genes in the same transcription factory. Much more remains to be uncovered: What are the roles of other transcription factors and coactivators? How do they interact with each other? How are the other three bigenomic proteins (complexes I, III, and V) regulated? What are the epigenetic factors? What are the roles of microRNAs?

The coupling of neuronal activity and energy metabolism has been revealed by examining the expressions of COX in relation to those of synaptic neurochemicals at the cellular level and their co-regulation at the molecular level. Many other neurochemicals are likely involved, and the mechanisms of action between and among them deserve rigorous probing.

An unexpected and startling revelation by COX is that the postnatal development in both the brain stem respiratory system and the visual cortex of rats does not follow a smooth, straight path, but rather, is marked by an abrupt though transient episode of reduced metabolic activity induced by suppressed excitation and enhanced inhibition enabled by a downregulation of BDNF/TrkB. This *critical period* of synaptic imbalance and vulnerability is deemed necessary to facilitate inhibitory refinement of neuronal properties, functioning, and connections, especially in response to environmental influences. A host of questions naturally ensue: Are these basic principles applicable to the critical periods of other systems and in other species? What are all the important players in orchestrating coordinated adjustments during this period? What signal(s) trigger(s) the neurochemicals to undergo change? Is there a master regulator or a set of regulators for the critical period? What is/are the mechanism/s of their actions?

DOI: 10.1201/9781003507765-19

Mutations in the mtDNA or the nuclear genome that perturb the synthesis, assembly, and functions of COX or any of the inner mitochondrial membrane complexes will lead to metabolic dysfunction that can negatively affect multiple organs, especially the metabolically most active ones, such as the brain, heart, skeletal muscles, liver, and kidney. Several neurodegenerative diseases exhibit COX deficiency and metabolic distress, but thus far, such deficits can only be attributed to manifestations rather than causes of pathologies. Much more research is necessary to probe the mechanisms underlying these diseases.

Knowledge is never static. Existing knowledge needs to be updated, revised, improved upon, and expanded. In writing this book, my hope is that readers will continue my journey, correct mistakes, and carry the knowledge to the next, higher level. The horizon is never-ending.

Curiosity is the Driver, and Discovery is the Reward. The Road in between can be long, torturous, and disheartening at times. The important focus is to keep the flame of Curiosity blazing. It will not and should not ever be quenched.

References

Abdulhag UN, Soiferman D, Schueler-Furman O, Miller C, Shaag A, Orly Elpeleg O, Edvardson S, Saada A (2015) Mitochondrial complex IV deficiency, caused by mutated COX6B1, is associated with encephalomyopathy, hydrocephalus and cardiomyopathy. *Eur J Hum Genet* 23:159–164. doi: 10.1038/ejhg.2014.85

Achilli A, Iommarini L, Olivieri A, Pala M, Hooshiar Kashani B, Reynier P, La Morgia C, Valentino ML, Liguori R, Pizza F, Barboni P, Sadun F, De Negri AM, Zeviani M, Dollfus H, Moulignier A, Ducos G, Orssaud C, Bonneau D, Procaccio V, Leo-Kottler B, Fauser S, Wissinger B, Amati-Bonneau P, Torroni A, Carelli V (2012) Rare primary mitochondrial DNA mutations and probable synergistic variants in Leber's hereditary optic neuropathy. *PLoS One* 7:e42242. doi: 10.1371/journal.pone.0042242

Acin-Perez R, Gatti DL, Bai Y, Manfredi G (2011) Protein phosphorylation and prevention of cytochrome oxidase inhibition by ATP: coupled mechanisms of energy metabolism regulation. *Cell Metab* 13:712–719. doi: 10.1016/j.cmet.2011.03.024

Ackermann U, Geering K (1990) Mutual dependence of Na,K-ATPase alpha- and beta-subunits for correct posttranslational processing and intracellular transport. *FEBS Lett* 269:105–108. doi: 10.1016/0014-5793(90)81130-g

Afkhami E, Heidari MM, Khatami M, Ghadamyari F, Dianatpour S (2020) Detection of novel mitochondrial mutations in cytochrome C oxidase subunit 1 (COX1) in patients with familial adenomatous polyposis (FAP). *Clin Transl Oncol* 22:908–918. doi: 10.1007/s12094-019-02208-6

Agmon A, Yang LT, Jones EG, O'Dowd DK (1995) Topological precision in the thalamic projection to neonatal mouse barrel cortex. *J Neurosci* 15:549–561. doi: 10.1523/JNEUROSCI.15-01-00549.1995

Ahmed OJ, Mehta MR (2009) The hippocampal rate code: anatomy, physiology and theory. *Trends Neurosci* 32:329–338. doi: 10.1016/j.tins.2009.01.009

Ahmed OM, El-Gareib AW, El-Bakry AM, Abd El-Tawab SM, Ahmed RG (2008) Thyroid hormones states and brain development interactions. *Int J Dev Neurosci* 26:147–209. doi: 10.1016/j.ijdevneu.2007.09.011

Ahting U, Floss T, Uez N, Schneider-Lohmar I, Becker L, Kling E, Iuso A, Bender A, de Angelis MH, Gailus-Durner V, Fuchs H, Meitinger T, Wurst W, Prokisch H, Klopstock T (2009) Neurological phenotype and reduced lifespan in heterozygous Tim23 knockout mice, the first mouse model of defective mitochondrial import. *Biochim Biophys Acta* 1787:371–376. doi: 10.1016/j.bbabio.2008.12.001

Aich A, Boshnakovska A, Witte S, Gall T, Unthan-Fechner K, Yousefi R, Chowdhury A, Dahal D, Methi A, Kaufmann S, Silbern I, Prochazka J, Nichtova Z, Palkova M, Raishbrook M, Koubkova G, Sedlacek R, Tröder SE, Zevnik B, Riedel D, Michanski S, Möbius W, Ströbel P, Lüchtenborg C, Giavalisco P, Urlaub H, Fischer A, Brügger B, Jakobs S, Rehling P (2024) Defective mitochondrial COX1 translation due to loss of COX14 function triggers ROS-induced inflammation in mouse liver. *Nat Commun* 15:6914. doi: 10.1038/s41467-024-51109-y

Aich A, Wang C, Chowdhury A, Ronsör C, Pacheu-Grau D, Richter-Dennerlein R, Dennerlein S, Rehling P (2018) COX16 promotes COX2 metallation and assembly during respiratory complex IV biogenesis. *Elife* 7:e32572. doi: 10.7554/eLife.32572

Aid T, Kazantseva A, Piirsoo M, Palm K, Timmusk T (2007) Mouse and rat BDNF gene structure and expression revisited. *J Neurosci Res* 85:525–535. doi: 10.1002/jnr.21139

Akimoto T, Pohnert SC, Li P, Zhang M, Gumbs C, Rosenberg PB, Williams RS, Yan Z (2005) Exercise stimulates Pgc-1alpha transcription in skeletal muscle through activation of the p38 MAPK pathway. *J Biol Chem* 280:19587–19593. doi: 10.1074/jbc.M408862200

Albert ES, Bec JM, Desmadryl G, Chekroud K, Travo C, Gaboyard S, Bardin F, Marc I, Dumas M, Lenaers G, Hamel C, Muller A, Chabbert C (2012) TRPV4 channels mediate the infrared laser-evoked response in sensory neurons. *J Neurophysiol* 107:3227–3234. doi: 10.1152/jn.00424.2011

Alheid GF, Milsom WK, McCrimmon DR (2004) Pontine influences on breathing: an overview. *Respir Physiol Neurobiol* 143:105–114. doi: 10.1016/j.resp.2004.06.016

Al-Khayri JM, Ravindran M, Banadka A, Vandana CD, Priya K, Nagella P, Kukkemane K (2024) Amyotrophic lateral sclerosis: insights and new prospects in disease pathophysiology, *Biomarkers and Therapies. Pharmaceuticals (Basel)* 17:1391. doi: 10.3390/ph17101391

Alzheimer A (1907) Uber eine eigenartige erkrankung der hirnrinde. *Allgemeine Zeitschrift fur Psychiatrie und Psychisch-Gerichtliche Medizin* 64:146–148.

Amaral DG, Insausti R, Cowan WM (1987) The entorhinal cortex of the monkey: I. Cytoarchitectonic organization. *J Comp Neurol* 264:326–355. doi: 10.1002/cne.902640305

Ames A 3rd, Li YY, Heher EC, Kimble CR (1992) Energy metabolism of rabbit retina as related to function: high cost of Na+ transport. *J Neurosci* 12:840–853. doi: 10.1523/JNEUROSCI.12-03-00840.1992

Anderson AJ, Crameri JJ, Ang CS, Malcolm TR, Kang Y, Baker MJ, Palmer CS, Sharpe AJ, Formosa LE, Ganio K, Baker MJ, McDevitt CA, Ryan MT, Maher MJ, Stojanovski D (2023) Human Tim8a, Tim8b and Tim13 are auxiliary assembly factors of mature Complex IV. *EMBO Rep* 24:e56430. doi: 10.15252/embr.202256430

Anderson S, Bankier AT, Barrell BG, de Bruijn MH, Coulson AR, Drouin J, Eperon IC, Nierlich DP, Roe BA, Sanger F, Schreier PH, Smith AJ, Staden R, Young IG (1981) Sequence and organization of the human mitochondrial genome. *Nature* 290:457–465. doi: 10.1038/290457a0

Andersson U, Scarpulla RC (2001) Pgc-1-related coactivator, a novel, serum-inducible coactivator of nuclear respiratory factor 1-dependent transcription in mammalian cells. *Mol Cell Biol* 21:3738–3749. doi: 10.1128/MCB.21.11.3738-3749.2001

Antonicka H, Leary SC, Guercin GH, Agar JN, Horvath R, Kennaway NG, Harding CO, Jaksch M, Shoubridge EA (2003) Mutations in COX10 result in a defect in mitochondrial heme A biosynthesis and account for multiple, early-onset clinical phenotypes associated with isolated COX deficiency. *Hum Mol Genet* 12:2693–2702. doi: 10.1093/hmg/ddg284

Aqua MS, Bachman NJ, Lomax MI, Grossman LI (1991) Characterization and expression of a cDNA specifying subunit VIIc of bovine cytochrome c oxidase. *Gene* 104:211–217. doi: 10.1016/0378-1119(91)90252-7

Arimura A (1992) Pituitary adenylate cyclase activating polypeptide (PACAP): discovery and current status of research. *Regul Pept* 37:287–303.

Armesto P, Infante C, Cousin X, Ponce M, Manchado M (2015) Molecular and functional characterization of seven Na+/K+-ATPase beta subunit paralogs in Senegalese sole (Solea senegalensis Kaup, 1858). *Comp Biochem Physiol B Biochem Mol Biol* 182:14–26. doi: 10.1016/j.cbpb.2014.11.011

Arnaudo E, Hirano M, Seelan RS, Milatovich A, Hsieh CL, Fabrizi GM, Grossman LI, Francke U, Schon EA (1992) Tissue-specific expression and chromosome assignment of genes specifying two isoforms of subunit VIIa of human cytochrome c oxidase. *Gene* 119:299–305. doi: 10.1016/0378-1119(92)90287-y

Arnold RS, Sun Q, Sun CQ, Richards JC, O'Hearn S, Osunkoya AO, Wallace DC, Petros JA (2013) An inherited heteroplasmic mutation in mitochondrial gene COI in a patient with prostate cancer alters reactive oxygen, reactive nitrogen and proliferation. *Biomed Res Int* 2013:239257. doi: 10.1155/2013/239257

Arnold S, Goglia F, Kadenbach B (1998) 3,5-Diiodothyronine binds to subunit Va of cytochrome c oxidase and abolishes the allosteric inhibition of respiration by ATP. *Eur J Biochem* 252:325–330. doi: 10.1046/j.1432-1327.1998.2520325.x

Assaf Y, Pasternak O (2008) Diffusion tensor imaging (DTI)-based white matter mapping in brain research: a review. *J Mol Neurosci* 34:51–61. doi: 10.1007/s12031-007-0029-0

Attardi G, Schatz G (1988) Biogenesis of mitochondria. *Annu Rev Cell Biol* 4:289–333. doi: 10.1146/annurev.cb.04.110188.001445

Au HC, Scheffler IE (1998) Promoter analysis of the human succinate dehydrogenase iron-protein gene—both nuclear respiratory factors NRF-1 and NRF-2 are required. *Eur J Biochem* 251:164–174. doi: 10.1046/j.1432-1327.1998.2510164.x

Augood SJ, Emson PC (1992) Compartmental distribution of cytochrome oxidase in the striatum of the rat. *Synapse* 10:16–24. doi: 10.1002/syn.890100104

Auten RL, Davis JM (2009) Oxygen toxicity and reactive oxygen species: the devil is in the details. *Pediatr Res* 66:121–127. doi: 10.1203/PDR.0b013e3181a9eafb

Averill-Bates D (2024) Reactive oxygen species and cell signaling. Review. *Biochim Biophys Acta Mol Cell Res* 1871:119573. doi: 10.1016/j.bbamcr.2023.119573

Baar K, Wende AR, Jones TE, Marison M, Nolte LA, Chen M, Kelly DP, Holloszy JO (2002) Adaptations of skeletal muscle to exercise: rapid increase in the transcriptional coactivator PGC-1. *FASEB J* 16:1879–1886. doi: 10.1096/fj.02-0367com

Bachelard HS (1975) Energy utilized by neurotransmitters. In: Ingvar DH, Lassen NA (eds) *Brain Work Alfred Benzon Symposium VIII*. Academic Press, New York, pp 79–81.

Baertling F, Al-Murshedi F, Sánchez-Caballero L, Al-Senaidi K, Joshi NP, Venselaar H, van den Brand MA, Nijtmans LG, Rodenburg RJ (2017) Mutation in mitochondrial complex IV subunit COX5A causes pulmonary arterial hypertension, lactic acidemia, and failure to thrive. *Hum Mutat* 38:692–703. doi: 10.1002/humu.23210

Bai X, Wong-Riley MTT (2003) Neuronal activity regulates protein and gene expression of GluR2 in postnatal rat visual cortical neurons in culture. *J Neurocytol* 32:71–78. doi: 10.1023/a:1027380315902

Baker-Herman TL, Mitchell GS (2002) Phrenic long-term facilitation requires spinal serotonin receptor activation and protein synthesis. *J Neurosci* 22:6239–6246. doi: 10.1523/JNEUROSCI.22-14-06239.2002

Bale G, Elwell CE, Tachtsidis I (2016) From Jobsis to the present day: a review of clinical near-infrared spectroscopy measurements of cerebral cytochrome-c-oxidase. *J Biomed Opt* 21:091307. doi: 10.1117/1.JBO.21.9.091307

Ballanyi K (2004) Neuromodulation of the Perinatal Respiratory Network. *Curr Neuropharmacol* 2:221–243. doi: 10.2174/1570159043476828

Balsa E, Marco R, Perales-Clemente E, Szklarczyk R, Calvo E, Landázuri MO, Enríquez JA (2012) NDUFA4 is a subunit of complex IV of the mammalian electron transport chain. *Cell Metab* 16:378–386. doi: 10.1016/j.cmet.2012.07.015

Bamford OS, Sterni LM, Wasicko MJ, Montrose MH, Carroll JL (1999) Postnatal maturation of carotid body and type I cell chemoreception in the rat. *Am J Physiol* 276:L875–L884. doi: 10.1152/ajplung.1999.276.5.L875

Bandettini PA, Jesmanowicz A, Wong EC, Hyde JS (1993) Processing strategies for time-course data sets in functional MRI of the human brain. *Magn Reason Med* 30:161–173. doi: 10.1002/mrm.1910300204

Barbato A, Gori G, Sacchini M, Pochiero F, Bargiacchi S, Traficante G, Palazzo V, Tiberi L, Bianchini C, Mei D, Parrini E, Pisano T, Procopio E, Guerrini R, Peron A, Stagi S (2024) Endocrinological features and epileptic encephalopathy in COX deficiency due to SCO1 mutations: case series and review of literature. *Endocr Connect* 13:e240221. doi: 10.1530/EC-24-0221

Barde YA, Edgar D, Thoenen H (1982) Purification of a new neurotrophic factor from mammalian brain. *EMBO J* 1:549–553. doi: 10.1002/j.1460-2075.1982.tb01207.x

Bardoni R, Ghirri A, Salio C, Prandini M, Merighi A (2007) BDNF-mediated modulation of GABA and glycine release in dorsal horn lamina II from postnatal rats. *Dev Neurobiol* 67: 960–975. doi: 10.1002/dneu.20401

Bareth B, Dennerlein S, Mick DU, Nikolov M, Urlaub H, Rehling P (2013) The heme a synthase Cox15 associates with cytochrome c oxidase assembly intermediates during Cox1 maturation. *Mol Cell Biol* 33:4128–4137. doi: 10.1128/MCB.00747-13

Barrientos A, Barros MH, Valnot I, Rotig A, Rustin P, Tzagoloff A (2002) Cytochrome oxidase in health and disease. *Gene* 286:53–63. doi: 10.1016/s0378-1119(01)00803-4

Barron MJ, Griffiths P, Turnbull DM, Bates D, Nichols P (2004) The distributions of mitochondria and sodium channels reflect the specific energy requirements and conduction properties of the human optic nerve head. *Br J Ophthalmol* 88:286–290. doi: 10.1136/bjo.2003.027664

Barros MH, Carlson CG, Glerum DM, Tzagoloff A (2001) Involvement of mitochondrial ferredoxin and Cox15p in hydroxylation of heme O. *FEBS Lett* 492:133–138. doi: 10.1016/s0014-5793(01)02249-9

Barros MH, Johnson A, Tzagoloff A (2004) COX23, a homologue of COX17, is required for cytochrome oxidase assembly. *J Biol Chem* 279:31943–31947. doi: 10.1074/jbc.M405014200

Barros MH, Nobrega FG, Tzagoloff A (2002) Mitochondrial ferredoxin is required for heme A synthesis in Saccharomyces cerevisiae. *J Biol Chem* 277:9997–10002. doi: 10.1074/jbc.M112025200

Basu A, Park K, Atchison ML, Carter RS, Avadhani NG (1993) Identification of a transcriptional initiator element in the cytochrome c oxidase subunit Vb promoter which binds to transcription factors NF-E1 (YY-1, delta) and Sp1. *J Biol Chem.* 268:4188–4196.

Bathini M, Raghushaker CR, Mahato KK (2022) The molecular mechanisms of action of photobiomodulation against neurodegenerative diseases: a systematic review. *Cell Mol Neurobiol* 42:955–971. doi: 10.1007/s10571-020-01016-9

Batista G, Hensch TK (2019) Critical period regulation by thyroid hormones: potential mechanisms and sex-specific aspects. *Front Mol Neurosci* 12:77. doi: 10.3389/fnmol.2019.00077

Baty K, Farrugia ME, Hopton S, Falkous G, Schaefer AM, Stewart W, Willison HJ, Reilly MM, Blakely EL, Taylor RW, Ng YS (2021) A novel MT-CO2 variant causing cerebellar ataxia and neuropathy: the role of muscle biopsy in diagnosis and defining pathogenicity. *Neuromuscul Disord* 31:1186–1193. doi: 10.1016/j.nmd.2021.05.014

Bauer MF, Sirrenberg C, Neupert W, Brunner M (1996) Role of Tim23 as voltage sensor and presequence receptor in protein import into mitochondria. *Cell* 87:33–41. doi: 10.1016/s0092-8674(00)81320-3

Bauer N, Mao Q, Vashistha A, Seshadri A, Nancy Du YC, Otterbein L, Tan C, de Caestecker MP, Wang B (2025) Compelling evidence: a critical update on the therapeutic potential of carbon monoxide. *Med Res Rev* 45:1275–1301. doi: 10.1002/med.22116

Bavelier D, Levi DM, Li RW, Dan Y, Hensch TK (2010) Removing brakes on adult brain plasticity: from molecular to behavioral interventions. *J Neurosci* 30:14964–14971. doi: 10.1523/JNEUROSCI.4812-10.2010

Bavis RW, Olson EB Jr, Mitchell GS (2002) Critical developmental period for hyperoxia-induced blunting of hypoxic phrenic responses in rats. *J Appl Physiol* 92:1013–1018. doi: 10.1152/japplphysiol.00859.2001

Beal MF (1998) Mitochondrial dysfunction in neurodegenerative diseases. *Biochim Biophys Acta* 1366:211–223. doi: 10.1016/s0005-2728(98)00114-5

Becker TS, Burgess SM, Amsterdam AH, Allende ML, Hopkins N (1998) *not really finished* is crucial for development of the zebrafish outer retina and encodes a transcription factor highly homologous to human Nuclear Respiratory Factor-1 and avian Initiation Binding Repressor. *Development* 125:4369–4378. doi: 10.1242/dev.125.22.4369

Bekris LM, Yu CE, Bird TD, Tsuang DW (2010) Genetics of Alzheimer disease. *J Geriatr Psychiatry Neurol* 23:213–227. doi: 10.1177/0891988710383571

Belegu R, Hadziefendić S, Dreshaj IA, Haxhiu MA, Martin RJ (1999) CO_2-induced c-fos expression in medullary neurons during early development. *Respir Physiol* 117:13–28. doi: 10.1016/s0034-5687(99)00046-8

Ben-Ari Y, Gaiarsa JL, Tyzio R, Khazipov R (2007) GABA: a pioneer transmitter that excites immature neurons and generates primitive oscillations. *Physiol Rev* 87:1215–1284. doi: 10.1152/physrev.00017.2006

Bender A, Krishnan KJ, Morris CM, Taylor GA, Reeve AK, Perry RH, Jaros E, Hersheson JS, Betts J, Klopstock T, Taylor RW, Turnbull DM (2006) High levels of mitochondrial DNA deletions in substantia nigra neurons in aging and Parkinson disease. *Nat Genet* 38:515–517. doi: 10.1038/ng1769

Bender E, Kadenbach B (2000) The allosteric ATP-inhibition of cytochrome c oxidase activity is reversibly switched on by cAMP-dependent phosphorylation. *FEBS Lett* 466:130–134. doi: 10.1016/s0014-5793(99)01773-1

Bennett MK, Scheller RH (1994) Molecular correlates of synaptic vesicle docking and fusion. *Curr Opin Neurobiol* 4:324–329. doi: 10.1016/0959-4388(94)90092-2

Benson DL, Isackson PJ, Gall CM, Jones EG (1991) Differential effects of monocular deprivation on glutamic acid decarboxylase and type II calcium-calmodulin-dependent protein kinase gene expression in the adult monkey visual cortex. *J Neurosci* 11:31–47. doi: 10.1523/JNEUROSCI.11-01-00031.1991

Berman N, Murphy EH (1981) The critical period for alteration in cortical binocularity resulting from divergent and convergent strabismus. *Dev Brain Res* 2:181–202. doi: 10.1016/0165-3806(81)90031-6

Bernal J (2017) Thyroid hormone regulated genes in cerebral cortex development. *J Endocrinol* 232:R83–R97. doi: 10.1530/JOE-16-0424

Bianchi AL, Denavit-Saubié M, Champagnat J (1995) Central control of breathing in mammals: neuronal circuitry, membrane properties, and neurotransmitters. *Physiol Rev* 75:1–45. doi: 10.1152/physrev.1995.75.1.1

Bibb MJ, Van Etten RA, Wright CT, Walberg MW, Clayton DA (1981) Sequence and gene organization of mouse mitochondrial DNA. *Cell* 26:167–180. doi: 10.1016/0092-8674(81)90300-7

Bisgard GE, Olson EB Jr, Wang ZY, Bavis RW, Fuller DD, Mitchell GS (2003) Adult carotid chemoafferent responses to hypoxia after 1, 2, and 4 wk of postnatal hyperoxia. *J Appl Physiol* 95:946–952. doi: 10.1152/japplphysiol.00985.2002

Blakemore C, Cooper GF (1970) Development of the brain depends on the visual environment. *Nature* 228:477–478. doi: 10.1038/228477a0

Blakemore C, Garey LJ, Vital-Durand F (1978) The physiological effects of monocular deprivation and their reversal in the monkey's visual cortex. *J Physiol* 283:223–262. doi: 10.1113/jphysiol.1978.sp012498

Blanco LN (1995) Mechanisms for the generation of gas-exchange surface area in rat lung. *Am J Physiol* 269:L698–L708. doi: 10.1152/ajplung.1995.269.5.L698

Blood-Siegfried J, Nyska A, Lieder H, Joe M, Vega L, Patterson R, Germolec D (2002) Synergistic effect of influenza a virus on endotoxin-induced mortality in rat pups: a potential model for sudden infant death syndrome. *Pediatr Res* 52:481–490. doi: 10.1203/00006450-200210000-00005

Blot A, Barbour B (2014) Ultra-rapid axon-axon ephaptic inhibition of cerebellar Purkinje cells by the pinceau. *Nat Neurosci* 17:289–295. doi: 10.1038/nn.3624

Bode M, Woellhaf MW, Bohnert M, van der Laan M, Sommer F, Jung M, Zimmermann R, Schroda M, Herrmann JM (2015) Redox-regulated dynamic interplay between Cox19 and the copper-binding protein Cox11 in the intermembrane space of mitochondria facilitates biogenesis of cytochrome c oxidase. *Mol Biol Cell* 26:2385–2401. doi: 10.1091/mbc.E14-11-1526

Bolaños JP, Almeida A, Moncada S (2010) Glycolysis: a bioenergetic or a survival pathway? *Trends Biochem Sci* 35:145–149. doi: 10.1016/j.tibs.2009.10.006

Borthwick GM, Johnson MA, Ince PG, Shaw PJ, Turnbull DM (1999) Mitochondrial enzyme activity in amyotrophic lateral sclerosis: implications for the role of mitochondria in neuronal cell death. *Ann Neurol* 46:787–790. doi: 10.1002/1531-8249(199911)46:5<787::aid-ana17>3.0.co;2-8

Bosman LW, Rosahl TW, Brussaard AB (2002) Neonatal development of the rat visual cortex: synaptic function of GABAA receptor alpha subunits. *J Physiol* 545:169–181. doi: 10.1113/jphysiol.2002.026534

Böttinger L, Guiard B, Oeljeklaus S, Kulawiak B, Zufall N, Wiedemann N, Warscheid B, van der Laan M, Becker T (2013) A complex of Cox4 and mitochondrial Hsp70 plays an important role in the assembly of the cytochrome c oxidase. *Mol Biol Cell* 24:2609–2619. doi: 10.1091/mbc.E13-02-0106

Bourens M, Barrientos A (2017) A CMC1-knockout reveals translation-independent control of human mitochondrial complex IV biogenesis. *EMBO Rep* 18:477–494. doi: 10.15252/embr.201643103

Bourens M, Boulet A, Leary SC, Barrientos A (2014) Human COX20 cooperates with SCO1 and SCO2 to mature COX2 and promote the assembly of cytochrome c oxidase. *Hum Mol Genet* 23:2901–2913. doi: 10.1093/hmg/ddu003

Bourens M, Fontanesi F, Soto IC, Liu J, Barrientos A (2013) Redox and reactive oxygen species regulation of mitochondrial cytochrome c oxidase biogenesis. *Antioxid Redox Signal* 19:1940–1952. doi: 10.1089/ars.2012.4847

Bozzi Y, Pizzorusso T, Cremisi F, Rossi FM, Barsacchi G, Maffei L (1995) Monocular deprivation decreases the expression of messenger RNA for brain-derived neurotrophic factor in the rat visual cortex. *Neuroscience* 69:1133–1144. doi: 10.1016/0306-4522(95)00321-9

Brennan WA Jr, Bird ED, Aprille JR (1985) Regional mitochondrial respiratory activity in Huntington's disease brain. *J Neurochem* 44:1948–1950. doi: 10.1111/j.1471-4159.1985.tb07192.x

Brini M, Carafoli E (2011) The plasma membrane Ca^{2+} ATPase and the plasma membrane sodium calcium exchanger cooperate in the regulation of cell calcium. *Cold Spring Harb Perspect Biol* 3:a004168. doi: 10.1101/cshperspect.a004168

Brischigliaro M, Zeviani M (2021) Cytochrome c oxidase deficiency. *Biochim Biophys Acta Bioenerg* 1862:148335. doi: 10.1016/j.bbabio.2020.148335

Broger C, Bill K, Azzi A (1986) Affinity chromatography purification of cytochrome-c oxidase from bovine heart mitochondria and other sources. *Methods Enzymol* 126:64–72. doi: 10.1016/s0076-6879(86)26008-5

Brown GC (2007) Nitric oxide and mitochondria. *Front Biosci* 12:1024–1033. doi: 10.2741/2122

Brown LL, Feldman SM, Smith DM, Cavanaugh JR, Ackermann RF, Graybiel AM (2002) Differential metabolic activity in the striosome and matrix compartments of the rat striatum during natural behaviors. *J Neurosci* 22:305–314. doi: 10.1523/JNEUROSCI.22-01-00305.2002

Bullier J, Norton TT (1979) Comparison of receptive-field properties of X and Y ganglion cells with X and Y lateral geniculate cells in the cat. *J Neurophysiol* 42:274–291. doi: 10.1152/jn.1979.42.1.274

Bundschuh FA, Hannappel A, Anderka O, Ludwig B (2009) Surf1, associated with Leigh syndrome in humans, is a heme-binding protein in bacterial oxidase biogenesis. *J Biol Chem* 284:25735–25741. doi: 10.1074/jbc.M109.040295

Burchfiel JL, Duffy FH (1981) Role of intracortical inhibition in deprivation amblyopia: reversal by microiontophoretic bicuculline. *Brain Res* 206:479–484. doi: 10.1016/0006-8993(81)90551-5

Burkhalter J, Fiumelli H, Allaman I, Chatton JY, Martin JL (2003) Brain-derived neurotrophic factor stimulates energy metabolism in developing cortical neurons. *J Neurosci* 23:8212–8220. doi: 10.1523/JNEUROSCI.23-23-08212.2003

Burnashev N, Monyer H, Seeburg PH, Sakmann B (1992) Divalent ion permeability of AMPA receptor channels is dominated by the edited form of a single subunit. *Neuron* 8:189–198. doi: 10.1016/0896-6273(92)90120-3

Burri PH, Dbaly J, Weibel ER (1974) The postnatal growth of the rat lung. 1. Morphometry. *Anat Rec* 178:711–730. doi: 10.1002/ar.1091780405

Campos Y, García-Redondo A, Fernández-Moreno MA, Martínez-Pardo M, Goda G, Rubio JC, Martín MA, del Hoyo P, Cabello A, Bornstein B, Garesse R, Arenas J (2001) Early-onset multisystem mitochondrial disorder caused by a nonsense mutation in the mitochondrial DNA cytochrome C oxidase II gene. *Ann Neurol* 50:409–413. doi: 10.1002/ana.1141

Cao W, Daniel KW, Robidoux J, Puigserver P, Medvedev AV, Bai X, Floering LM, Spiegelman BM, Collins S (2004) p38 mitogen-activated protein kinase is the central regulator of cyclic AMP-dependent transcription of the brown fat uncoupling protein 1 gene. *Mol Cell Biol* 24:3057–3067. doi: 10.1128/MCB.24.7.3057-3067.2004

Carder RK, Hendry SH (1994) Neuronal characterization, compartmental distribution, and activity-dependent regulation of glutamate immunoreactivity in adult monkey striate cortex. *J Neurosci* 14:242–262. doi: 10.1523/JNEUROSCI.14-01-00242.1994

Caron-Godon CA, Della Vecchia S, Romano A, Doccini S, Dal Canto F, Pasquariello R, Rubegni A, Battini R, Santorelli FM, Glerum DM, Nesti C (2023) Novel COX11 mutations associated with mitochondrial disorder: functional characterization in patient fibroblasts and Saccharomyces cerevisiae. *Int J Mol Sci* 24:16636. doi: 10.3390/ijms242316636

Carpenter MB (1976) *Human Neuroanatomy*. 7th ed. The Williams & Wilkins Co, Baltimore.

Carroll E, Wong-Riley M (1982) Light and E.M. analysis of cytochrome oxidase-rich zones in the striate cortex of squirrel monkeys. *Soc Neurosci Abstr* 8:706.

Carroll EW, Wong-Riley MTT (1984) Quantitative light and electron microscopic analysis of cytochrome oxidase-rich zones in the striate cortex of the squirrel monkey. *J Comp Neurol* 222:1–17. doi: 10.1002/cne.902220102

Carroll J, Fearnley IM, Skehel JM, Shannon RJ, Hirst J, Walker JE (2006) Bovine complex I is a complex of 45 different subunits. *J Biol Chem* 281:32724–32727. doi: 10.1074/jbc.M607135200

Carroll JL (2003) Developmental plasticity in respiratory control. *J Appl Physiol* 94:375–389. doi: 10.1152/japplphysiol.00809.2002

Carter RS, Avadhani NG (1994) Cooperative binding of GA-binding protein transcription factors to duplicated transcription initiation region repeats of the cytochrome c oxidase subunit IV gene. *J Biol Chem* 269:4381–4387.

Carter RS, Bhat NK, Basu A, Avadhani NG (1992) The basal promoter elements of murine cytochrome c oxidase subunit IV gene consist of tandemly duplicated Ets motifs that bind to GABP-related transcription factors. *J Biol Chem* 267:23418–23426.

Carvalho AL, Caldeira MV, Santos SD, Duarte CB (2008) Role of the brain-derived neurotrophic factor at glutamatergic synapses. *Br J Pharmacol* 153 (Suppl 1):S310–S324. doi: 10.1038/sj.bjp.0707509

Castora FJ, Conyers BL, Gershon BS, Kerns KA, Campbell R, Simsek-Duran F (2019) The T9861C mutation in the mtDNA-encoded cytochrome C oxidase subunit III gene occurs in high frequency but with unequal distribution in the Alzheimer's disease brain. *J Alzheimers Dis* 72:257–269. doi: 10.3233/JAD-190176

Castresana J, Lübben M, Saraste M, Higgins DG (1994) Evolution of cytochrome oxidase, an enzyme older than atmospheric oxygen. *EMBO J* 13:2516–2525. doi: 10.1002/j.1460-2075.1994.tb06541.x

Cavara NA, Orth A, Hicking G, Seebohm G, Hollmann M (2010) Residues at the tip of the pore loop of NR3B-containing NMDA receptors determine Ca^{2+} permeability and Mg^{2+} block. *BMC Neurosci* 11:133. doi: 10.1186/1471-2202-11-133

Celio MR, Schärer L, Morrison JH, Norman AW, Bloom FE (1986) Calbindin immunoreactivity alternates with cytochrome c-oxidase-rich zones in some layers of the primate visual cortex. *Nature* 323:715–717. doi: 10.1038/323715a0

Chacinska A, Lind M, Frazier AE, Dudek J, Meisinger C, Geissler A, Sickmann A, Meyer HE, Truscott KN, Guiard B, Pfanner N, Rehling P (2005) Mitochondrial presequence translocase: switching between TOM tethering and motor recruitment involves Tim21 and Tim17. *Cell* 120:817–829. doi: 10.1016/j.cell.2005.01.011

Chan JY, Han XL, Kan YW (1993) Cloning of Nrf1, an NF-E2-related transcription factor, by genetic selection in yeast. *Proc Natl Acad Sci USA* 90(23):11371–11375. doi: 10.1073/pnas.90.23.11371

Chance B, Williams GR (1955) Respiratory enzymes in oxidative phosphorylation. III. The steady state. *J Biol Chem* 217:409–427.

Chance B, Williams GR (1956) The respiratory chain and oxidative phosphorylation. In Nord FF (ed) *Advances in Enzymology and Related Subjects of Biochemistry*, vol 17. Interscience Publishers Inc, New York, pp 65–134.

Chandrasekaran K, Hatanpää K, Rapoport SI, Brady DR (1997) Decreased expression of nuclear and mitochondrial DNA-encoded genes of oxidative phosphorylation in association neocortex in Alzheimer disease. *Brain Res Mol Brain Res* 44:99–104. doi: 10.1016/s0169-328x(96)00191-x

Charcot J (1874) Sclerose laterale amytrophique. *Oeuvres Compltes Bureauxdu Proges Medical* 2:249–266.

Chau CM, Evans MJ, Scarpulla RC (1992) Nuclear respiratory factor 1 activation sites in genes encoding the gamma-subunit of ATP synthase, eukaryotic initiation factor 2 alpha, and tyrosine aminotransferase. Specific interaction of purified NRF-1 with multiple target genes. *J Biol Chem* 267:6999–7006.

Chen N, Luo T, Raymond LA (1999) Subtype-dependence of NMDA receptor channel open probability. *J Neurosci* 19:6844–6854. doi: 10.1523/JNEUROSCI.19-16-06844.1999

Cheng Q, Yeh HH (2003) Brain-derived neurotrophic factor attenuates mouse cerebellar granule cell GABA(A) receptor-mediated response via postsynaptic mechanisms. *J. Physiol* 548:711–721. doi:10.1113/jphysiol.2002.037846

Cherian A, Divya KP, Vijayaraghavan A (2023) Parkinson's disease - genetic cause. *Curr Opin Neurol* 36:292–301. doi: 10.1097/WCO.0000000000001167

Chesselet MF, Weiss L, Wuenschell C, Tobin AJ, Affolter HU (1987) Comparative distribution of mRNAs for glutamic acid decarboxylase, tyrosine hydroxylase, and tachykinins in the basal ganglia: an in situ hybridization study in the rodent brain. *J Comp Neurol* 262:125–140. doi: 10.1002/cne.902620110

Choi DW (1988) Glutamate neurotoxicity and diseases of the nervous system. *Neuron* 1:623–634. doi: 10.1016/0896-6273(88)90162-6

Chu PJ, Rivera JF, Arnold DB (2006) A role for Kif17 in transport of Kv4.2. *J Biol Chem* 281:365–373. doi: 10.1074/jbc.M508897200

Church C, Chapon C, Poyton RO (1996) Cloning and characterization of PET100, a gene required for the assembly of yeast cytochrome c oxidase. *J Biol Chem* 271:18499–18507. doi: 10.1074/jbc.271.31.18499

Clark DD, Sokoloff, L (1999) Circulation and energy metabolism of the brain. In: Siegel GJ, Agranoff BW, Albers RW, Fisher SK, Uhler MD (eds) *Basic Neurochemistry: Molecular, Cellular and Medical Aspects*. Lippincott, Philadelphia, pp 637–670.

Clausen MV, Hilbers F, Poulsen H (2017) The structure and function of the Na,K-ATPase isoforms in health and disease. *Front Physiol* 8:371. doi: 10.3389/fphys.2017.00371

Clayton DA (2000) Transcription and replication of mitochondrial DNA. *Hum Reprod* 15:11–17. doi: 10.1093/humrep/15.suppl_2.11

Clemente P, Peralta S, Cruz-Bermudez A, Echevarría L, Fontanesi F, Barrientos A, Fernandez-Moreno MA, Garesse R (2013) hCOA3 stabilizes cytochrome c oxidase 1 (COX1) and promotes cytochrome c oxidase assembly in human mitochondria. *J Biol Chem* 288:8321–8331. doi: 10.1074/jbc.M112.422220

Cleveland DW, Rothstein JD (2001) From Charcot to Lou Gehrig: deciphering selective motor neuron death in ALS. *Nat Rev Neurosci* 2:806–819. doi: 10.1038/35097565

Cohen D (1972) Magnetoencephalography: detection of the brain's electrical activity with a superconducting magnetometer. *Science* 175:664–666. doi: 10.1126/science.175.4022.664

Cole EG, Gaston K (1997) A functional YY1 binding site is necessary and sufficient to activate Surf-1 promoter activity in response to serum growth factors. *Nucleic Acids Res* 25:3705–3711. doi: 10.1093/nar/25.18.3705

Collier TJ, Redmond DE Jr, Roth RH, Elsworth JD, Taylor JR, Sladek JR Jr (1997) Metabolic energy capacity of dopaminergic grafts and the implanted striatum in parkinsonian nonhuman primates as visualized with cytochrome oxidase histochemistry. *Cell Transplant* 6:135–140. doi: 10.1177/096368979700600207

Comi GP, Bordoni A, Salani S, Franceschina L, Sciacco M, Prelle A, Fortunato F, Zeviani M, Napoli L, Bresolin N, Moggio M, Ausenda CD, Taanman JW, Scarlato G (1998) Cytochrome c oxidase subunit I microdeletion in a patient with motor neuron disease. *Ann. Neurol* 43:110–116. doi: 10.1002/ana.410430119

Conlan MJ, Rapley JW, Cobb CM (1996) Biostimulation of wound healing by low-energy laser irradiation. A review. *J Clin Periodontol* 23:492–496. doi: 10.1111/j.1600-051x.1996.tb00580.x

Connelly CA, Ellenberger HH, Feldman JL (1989) Are there serotonergic projections from raphe and retrotrapezoid nuclei to the ventral respiratory group in the rat? *Neurosci Lett* 105:34–40. doi: 10.1016/0304-3940(89)90007-4

Conner JM, Lauterborn JC, Yan Q, Gall CM, Varon S (1997) Distribution of brain-derived neurotrophic factor (BDNF) protein and mRNA in the normal adult rat CNS: evidence for anterograde axonal transport. *J Neurosci* 17:2295–2313. doi: 10.1523/JNEUROSCI.17-07-02295.1997

Contreras RG, Torres-Carrillo A, Flores-Maldonado C, Shoshani L, Ponce A (2024) Na(+)/K(+)-ATPase: more than an electrogenic pump. *Int J Mol Sci* 25:6122. doi: 10.3390/ijms25116122

Cooper CE (1990) The steady-state kinetics of cytochrome c oxidation by cytochrome oxidase. *Biochim Biophys Acta* 1017:187–203. doi: 10.1016/0005-2728(90)90184-6

Cooper CE (2002) Nitric oxide and cytochrome oxidase: substrate, inhibitor or effector? *Trends Biochem Sci* 27:33–39. doi: 10.1016/s0968-0004(01)02035-7

Cooper CE, Brown GC (2008) The inhibition of mitochondrial cytochrome oxidase by the gases carbon monoxide, nitric oxide, hydrogen cyanide and hydrogen sulfide: chemical mechanism and physiological significance. *J Bioenerg Biomembr* 40:533–539. doi: 10.1007/s10863-008-9166-6

Cooper CE, Cope M, Springett R, Amess PN, Penrice J, Tyszczuk L, Punwani S, Ordidge R, Wyatt J, Delpy DT (1999) Use of mitochondrial inhibitors to demonstrate that cytochrome oxidase near-infrared spectroscopy can measure mitochondrial dysfunction noninvasively in the brain. *J Cereb Blood Flow Metab* 19:27–38. doi: 10.1097/00004647-199901000-00003

Cooper CE, Springett R (1997) Measurement of cytochrome oxidase and mitochondrial energetics by near-infrared spectroscopy. *Philos Trans R Soc Lond B Biol Sci* 352:669–676. doi: 10.1098/rstb.1997.0048

Cooper JM, Mann VM, Krige D, Schapira AH (1992) Human mitochondrial complex I dysfunction. *Biochim Biophys Acta* 1101:198–203. doi: 10.1016/s0005-2728(05)80019-2

Crawford ML, Harwerth RS, Smith EL, von Noorden GK (1993) Keeping an eye on the brain: the role of visual experience in monkeys and children. *J Gen Psychol* 120:7–19. doi: 10.1080/00221309.1993.9917858

Crawford ML, Marc RE (1976) Light transmission of cat and monkey eyelids. *Vision Res* 16:323–324. doi: 10.1016/0042-6989(76)90118-8

Creutzfeldt OD (1975) Neurophysiological correlates of different functional states of the brain. In: Ingvar DH, Lassen NA (eds) *Brain Work, Alfred Benzon Symposium VIII*. Academic Press, New York, pp 21–46.

Crittenden JR, Graybiel AM (2011) Basal Ganglia disorders associated with imbalances in the striatal striosome and matrix compartments. *Front Neuroanat* 5:59. doi: 10.3389/fnana.2011.00059

Croucher KM, Fleming SM (2024) ATP13A2 (PARK9) and basal ganglia function. *Front Neurol* 14:1252400. doi: 10.3389/fneur.2023.1252400

Cull-Candy S, Brickley S, Farrant M (2001) NMDA receptor subunits: diversity, development and disease. *Curr Opin Neurobiol* 11:327–335. doi: 10.1016/s0959-4388(00)00215-4

Cummings KJ, Leiter JC, Trachtenberg FL, Okaty BW, Darnall RA, Haas EA, Harper RM, Nattie EE, Krous HF, Mena OJ, Richerson GB, Dymecki SM, Kinney HC, Haynes RL (2024) Altered 5-HT2A/C receptor binding in the medulla oblongata in the sudden infant death syndrome (SIDS): part II. Age-associated alterations in serotonin receptor binding profiles within medullary nuclei supporting cardiorespiratory homeostasis. *J Neuropathol Exp Neurol* 83:144–160. doi: 10.1093/jnen/nlae004

Cynader M (1983) Prolonged sensitivity to monocular deprivation in dark-reared cats: effects of age and visual exposure. *Brain Res* 284:155–164. doi: 10.1016/0165-3806(83)90002-0

Dahlstroem A, Fuxe K (1964) Evidence for the existence of monoamine-containing neurons in the central nervous system. I. Demonstration of monoamines in the cell bodies of brain stem neurons. *Acta Physiol Scand Suppl* 232:1–55.

Dalmonte ME, Forte E, Genova ML, Giuffrè A, Sarti P, Lenaz G (2009) Control of respiration by cytochrome c oxidase in intact cells: role of the membrane potential. *J Biol Chem* 284:32331–32335. doi: 10.1074/jbc.M109.050146

D'Aurelio M, Pallotti F, Barrientos A, Gajewski CD, Kwong JQ, Bruno C, Beal MF, Manfredi G (2001) In vivo regulation of oxidative phosphorylation in cells harboring a stop-codon mutation in mitochondrial DNA-encoded cytochrome c oxidase subunit I. *J Biol Chem* 276:46925–46932. doi: 10.1074/jbc.M106429200

Davis MF, Figueroa Velez DX, Guevarra RP, Yang MC, Habeeb M, Carathedathu MC, Gandhi SP (2015) Inhibitory neuron transplantation into adult visual cortex creates a new critical period that rescues impaired vision. *Neuron* 86:1055–1066. doi: 10.1016/j.neuron.2015.03.062

Davis SE, Solhied G, Castillo M, Dwinell M, Brozoski D, Forster HV, (2006) Postnatal developmental changes in CO_2 sensitivity in rats. *J Appl Physiol* 101:1097–1103. doi: 10.1152/japplphysiol.00378.2006

Dawson TM, Dawson VL (1996) Nitric oxide synthase: role as a transmitter/mediator in the brain and endocrine system. *Annu Rev Med* 47:219–227. doi: 10.1146/annurev.med.47.1.219

Decoster E, Simon M, Hatat D, Faye G (1990) The MSS51 gene product is required for the translation of the COX1 mRNA in yeast mitochondria. *Mol Gen Genet* 224:111–118. doi: 10.1007/BF00259457

Dekker J, Rippe K, Dekker M, Kleckner N (2002) Capturing chromosome conformation. *Science* 295:1306–1311. doi: 10.1126/science.1067799

Delic V, Kurien C, Cruz J, Zivkovic S, Barretta J, Thomson A, Hennessey D, Joseph J, Ehrhart J, Willing AE, Bradshaw P, Garbuzova-Davis S (2018) Discrete mitochondrial aberrations in the spinal cord of sporadic ALS patients. *J Neurosci Res* 96:1353–1366. doi: 10.1002/jnr.24249

Delpire E (2000) Cation-chloride cotransporters in neuronal communication. *News Physiol Sci* 309–312. doi: 10.1152/physiologyonline.2000.15.6.309

den Brave F, Schulte U, Fakler B, Pfanner N, Becker T (2024) Mitochondrial complexome and import network. *Trends Cell Biol* 34:578–594. doi: 10.1016/j.tcb.2023.10.004

del Río E, McLaughlin M, Downes CP, Nicholls DG (1999) Differential coupling of G-protein-linked receptors to Ca^{2+} mobilization through inositol(1,4,5)trisphosphate or ryanodine receptors in cerebellar granule cells in primary culture. *Eur J Neurosci* 11:3015–3022. doi: 10.1046/j.1460-9568.1999.00714.x

Dennerlein S, Oeljeklaus S, Jans D, Hellwig C, Bareth B, Jakobs S, Deckers M, Warscheid B, Rehling P (2015) MITRAC7 acts as a COX1-specific chaperone and reveals a checkpoint during cytochrome c oxidase assembly. *Cell Rep* 12:1644–1655. doi: 10.1016/j.celrep.2015.08.009

de Oliveira RC, Dos Reis SP, Cavalcante GC (2023) Mutations in structural genes of the mitochondrial complex IV may influence breast cancer. *Genes (Basel)* 14:1465. doi: 10.3390/genes14071465

DeYoe EA, Felleman DJ, Van Essen DC, McClendon E (1994) Multiple processing streams in occipitotemporal visual cortex. *Nature* 371:151–154. doi: 10.1038/371151a0

DeYoe EA, Trusk TC, Wong-Riley MTT (1995) Activity correlates of cytochrome oxidase-defined compartments in granular and supragranular layers of primary visual cortex of the macaque monkey. *Vis Neurosci* 12:629–639. doi: 10.1017/s0952523800008920

DeYoe EA, Van Essen DC (1985) Segregation of efferent connections and receptive field properties in visual area V2 of the macaque. *Nature* 317:58–61. doi: 10.1038/317058a0

DeYoe EA, Van Essen DC (1988) Concurrent processing streams in monkey visual cortex. *Trends Neurosci* 11:219–226. doi: 10.1016/0166-2236(88)90130-0

Dhar SS, Johar K, Wong-Riley MTT (2013) Bigenomic transcriptional regulation of all thirteen cytochrome c oxidase subunit genes by specificity protein 1. *Open Biol* 3:120176. doi: 10.1098/rsob.120176

Dhar SS, Liang HL, Wong-Riley MTT (2009b) Nuclear respiratory factor 1 co-regulates AMPA glutamate receptor subunit 2 and cytochrome c oxidase: tight coupling of glutamatergic transmission and energy metabolism in neurons. *J Neurochem* 108:1595–1606. doi: 10.1111/j.1471-4159.2009.05929.x

Dhar SS, Liang HL, Wong-Riley MTT (2009c) Transcriptional coupling of synaptic transmission and energy metabolism: role of nuclear respiratory factor 1 in co-regulating neuronal nitric oxide synthase and cytochrome c oxidase genes in neurons. *Biochim Biophys Acta Mol Cell Res* 1793:1604–1613. doi: 10.1016/j.bbamcr.2009.07.001

Dhar SS, Ongwijitwat S, Wong-Riley MTT (2008) Nuclear respiratory factor 1 regulates all ten nuclear-encoded subunits of cytochrome c oxidase in neurons. *J Biol Chem* 283:3120–3129. doi: 10.1074/jbc.M707587200

Dhar SS, Ongwijitwat S, Wong-Riley MTT (2009a) Chromosome conformation capture of all 13 genomic loci in the transcriptional regulation of the multisubunit bigenomic cytochrome c oxidase in neurons. *J Biol Chem* 284:18644–18650. doi: 10.1074/jbc.M109.019976

Dhar SS, Wong-Riley MTT (2009) Coupling of energy metabolism and synaptic transmission at the transcriptional level: role of nuclear respiratory factor 1 in regulating both cytochrome c oxidase and NMDA glutamate receptor subunit genes. *J Neurosci* 29:483–492. doi: 10.1523/JNEUROSCI.3704-08.2009

Dhar SS, Wong-Riley MTT (2010) Chromosome conformation capture of transcriptional interactions between cytochrome c oxidase genes and genes of glutamatergic synaptic transmission in neurons. *J Neurochem* 115:676–683. doi: 10.1111/j.1471-4159.2010.06956.x

Dhar SS, Wong-Riley MTT (2011) The kinesin superfamily protein KIF17 is regulated by the same transcription factor (NRF-1) as its cargo NR2B in neurons. *Biochim Biophys Acta* 1813:403–411. doi: 10.1016/j.bbamcr.2010.12.013

Di Pasquale E, Lindsay A, Feldman J, Monteau R, Hilaire G (1997) Serotonergic inhibition of phrenic motoneuron activity: an in vitro study in neonatal rat. *Neurosci Lett* 230:29–32. doi: 10.1016/s0304-3940(97)00469-2

Dierick HA, Ambrosini L, Spencer J, Glover TW, Mercer JF (1995) Molecular structure of the Menkes disease gene (ATP7A). *Genomics* 28:462–469. doi: 10.1006/geno.1995.1175

Diering GH, Huganir RL (2018) The AMPA receptor code of synaptic plasticity. *Neuron* 100:314–329. doi: 10.1016/j.neuron.2018.10.018

Difiglia M, Graveland GA, Schiff L (1987) Cytochrome oxidase activity in the rat caudate nucleus: light and electron microscopic observations. *J Comp Neurol* 255:137–145. doi: 10.1002/cne.902550111

Dillingham CM, Milczarek MM, Perry JC, Vann SD (2021) Time to put the mammillothalamic pathway into context. *Neurosci Biobehav Rev* 121:60–74. doi: 10.1016/j.neubiorev.2020.11.031

DiMauro S, Mendell JR, Sahenk Z, Bachman D, Scarpa A, Scofield RM, Reiner C (1980) Fatal infantile mitochondrial myopathy and renal dysfunction due to cytochrome c oxidase deficiency. *Neurol* 30:795–804. doi: 10.1212/wnl.30.8.795

DiMauro S, Nicholson JF, Hays AP, Eastwood AB, Papadimitriou A, Koenigsberger R, DeVivo DC (1983) Benign infantile mitochondrial myopathy due to reversible cytochrome c oxidase deficiency. *Ann Neurol* 14:226–234. doi: 10.1002/ana.410140209

DiMauro S, Zeviani M, Rizzuto R, Lombes A, Nakase H, Bonilla E, Miranda A, Schon E (1988) Molecular defects in cytochrome oxidase in mitochondrial diseases. *J Bioenerg Biomembr* 20:353–364. doi: 10.1007/BF00769637

Dong HL, Ma Y, Yu H, Wei Q, Li JQ, Liu GL, Li HF, Chen L, Chen DF, Bai G, Wu ZY (2021) Bi-allelic loss of function variants in COX20 gene cause autosomal recessive sensory neuronopathy. *Brain* 144:2457–2470. doi: 10.1093/brain/awab135

Doupe AJ, Kuhl PK (1999) Birdsong and human speech: common themes and mechanisms. *Annu Rev Neurosci* 22:567–631. doi: 10.1146/annurev.neuro.22.1.567

Duffin J, Ezure K, Lipski J (1995) Breathing rhythm generation: focus on the rostral ventrolateral medulla. *News Physiol Sci* 10:133–140. doi: 10.1152/physiologyonline.1995.10.3.133

Duffy FH, Snodgrass SR, Burchfiel JL, Conway JL (1976) Bicuculline reversal of deprivation amblyopia in the cat. *Nature* 260:256–257. doi: 10.1038/260256a0

Duffy KR, Murphy KM, Frosch MP, Livingstone MS (2007) Cytochrome oxidase and neurofilament reactivity in monocularly deprived human primary visual cortex. *Cereb Cortex* 17:1283–1291. doi: 10.1093/cercor/bhl038

Dumont S, Le Pennec S, Donnart A, Teusan R, Steenman M, Chevalier C, Houlgatte R, Savagner F (2018) Transcriptional orchestration of mitochondrial homeostasis in a cellular model of PGC-1-related coactivator-dependent thyroid tumor. *Oncotarget* 9:15883–15894. doi: 10.18632/oncotarget.24633

Dutertre S, Becker CM, Betz H (2012) Inhibitory glycine receptors: an update. *J Biol Chem* 287:40216–40223. doi: 10.1074/jbc.R112.408229

Economides JR, Sincich LC, Adams DL, Horton JC (2011) Orientation tuning of cytochrome oxidase patches in macaque primary visual cortex. *Nat Neurosci* 14:1574–1580. doi: 10.1038/nn.2958

Edwards DP, Purpura KP, Kaplan E (1995) Contrast sensitivity and spatial frequency response of primate cortical neurons in and around the cytochrome oxidase blobs. *Vision Res* 35:1501–1523. doi: 10.1016/0042-6989(94)00253-i

Edwards IJ, Bruce G, Lawrenson C, Howe L, Clapcote SJ, Deuchars SA, Deuchars J (2013) Na+/K+ ATPase alpha1 and alpha3 isoforms are differentially expressed in alpha- and gamma-motoneurons. *J Neurosci* 33:9913–9919. doi: 10.1523/JNEUROSCI.5584-12.2013

Eells JT, Henry MM, Lewandowski MF, Seme MT, Murray TG (2000) Development and characterization of a rodent model of methanol-induced retinal and optic nerve toxicity. *Neurotoxicology* 21:321–330.

Eells JT, Henry MM, Summerfelt P, Wong-Riley MT, Buchmann EV, Kane M, Whelan NT, Whelan HT (2003) Therapeutic photobiomodulation for methanol-induced retinal toxicity. *Proc Natl Acad Sci USA* 100:3439–3444. doi: 10.1073/pnas.0534746100

Egger A, Samardzija M, Sothilingam V, Tanimoto N, Lange C, Salatino S, Fang L, Garcia-Garrido M, Beck S, Okoniewski MJ, Neutzner A, Seeliger MW, Grimm C, Handschin C (2012) PGC-1α determines light damage susceptibility of the murine retina. *PLoS One* 7:e31272. doi: 10.1371/journal.pone.0031272

Einarsdottir O, Caughey WS (1985) Bovine heart cytochrome c oxidase preparations contain high affinity binding sites for magnesium as well as for zinc, copper, and heme. *Biochem Biophys Res Commun* 129:840–847. doi: 10.1016/0006-291x(85)91968-0

Elkon H, Don J, Melamed E, Ziv I, Shirvan A, Offen D (2002) Mutant and wild-type alpha-synuclein interact with mitochondrial cytochrome C oxidase. *J Mol Neurosci* 18:229–238. doi: 10.1385/JMN:18:3:229

Emerit J, Edeas M, Bricaire F (2004) Neurodegenerative diseases and oxidative stress. *Biomed Pharmacother* 58:39–46. doi: 10.1016/j.biopha.2003.11.004

Engelhardt JK, Silveira V, Morales FR, Pose I, Chase MH (2010) Serotoninergic control of glycinergic inhibitory postsynaptic currents in rat hypoglossal motoneurons. *Brain Res* 1345:1–8. doi: 10.1016/j.brainres.2010.05.009

Enwemeka CS, Bumah VV, Masson-Meyers DS (2020) Light as a potential treatment for pandemic coronavirus infections: a perspective. *J Photochem Photobiol B* 207:111891. doi: 10.1016/j.jphotobiol.2020.111891

Erickson JT, Conover JC, Borday V, Champagnat J, Barbacid M, Yancopoulos G, Katz DM (1996) Mice lacking brain-derived neurotrophic factor exhibit visceral sensory neuron losses distinct from mice lacking NT4 and display a severe developmental deficit in control of breathing. *J Neurosci* 16:5361–5371. doi: 10.1523/JNEUROSCI.16-17-05361.1996

Espinosa JS, Stryker MP (2012) Development and plasticity of the primary visual cortex. *Neuron* 75:230–249. doi: 10.1016/j.neuron.2012.06.009

Evans MJ, Scarpulla RC (1989) Interaction of nuclear factors with multiple sites in the somatic cytochrome c promoter. Characterization of upstream NRF-1, ATF, and intron Sp1 recognition sequences. *J Biol Chem* 264:14361–14368.

Evans MJ, Scarpulla RC (1990) NRF-1: a trans-activator of nuclear-encoded respiratory genes in animal cells. *Genes Dev* 4:1023–1034. doi: 10.1101/gad.4.6.1023

Ezure K, Tanaka I, Saito Y (2003) Activity of brainstem respiratory neurones just before the expiration-inspiration transition in the rat. *J Physiol* 547:629–640. doi: 10.1113/jphysiol.2002.032805

Fabrizi GM, Rizzuto R, Nakase H, Mita S, Kadenbach B, Schon EA (1989) Sequence of a cDNA specifying subunit VIa of human cytochrome c oxidase. *Nucleic Acids Res* 17:6409. doi: 10.1093/nar/17.15.6409

Fabrizi GM, Sadlock J, Hirano M, Mita S, Koga Y, Rizzuto R, Zeviani M, Schon EA (1992) Differential expression of genes specifying two isoforms of subunit VIa of human cytochrome c oxidase. *Gene* 119:307–312. doi: 10.1016/0378-1119(92)90288-z

Fagiolini M, Fritschy JM, Löw K, Möhler H, Rudolph U, Hensch TK (2004) Specific GABAA circuits for visual cortical plasticity. *Science* 303:1681–1683. doi: 10.1126/science.1091032

Fagiolini M, Hensch TK (2000) Inhibitory threshold for critical-period activation in primary visual cortex. *Nature* 404:183–186. doi: 10.1038/35004582

Fagiolini M, Pizzorusso T, Berardi N, Domenici L, Maffei L (1994) Functional postnatal development of the rat primary visual cortex and the role of visual experience: dark rearing and monocular deprivation. *Vision Res* 34:709–720. doi: 10.1016/0042-6989(94)90210-0

Falkenberg M, Gaspari M, Rantanen A, Trifunovic A, Larsson NG, Gustafsson CM (2002) Mitochondrial transcription factors B1 and B2 activate transcription of human mtDNA. *Nat Genet* 31:289–294. doi: 10.1038/ng909

Farias MF, Ungerleider LG, Pereira SS, Amorim AKJ, Soares JGM, Gattass R (2019) Time course of cytochrome oxidase blob plasticity in the primary visual cortex of adult monkeys after retinal laser lesions. *J Comp Neurol* 527:600–613. doi: 10.1002/cne.24434

Feger J, Deniau JM, Hammond-Le Guyader C, Ohye C (1976) Connections from the basal ganglia to the thalamus. *Appl Neurophysiol* 39:272–284. doi: 10.1159/000102504

Feldman JL, Mitchell GS, Nattie EE (2003) Breathing: rhythmicity, plasticity, chemosensitivity. *Annu Rev Neurosci* 26:239–266. doi: 10.1146/annurev.neuro.26.041002.131103

Felleman DJ, Van Essen DC (1991) Distributed hierarchical processing in the primate cerebral cortex. *Cereb Cortex* 1:1–47. doi: 10.1093/cercor/1.1.1-a

Fernández-Vizarra E, Tiranti V, Zeviani M (2009) Assembly of the oxidative phosphorylation system in humans: what we have learned by studying its defects. *Biochim Biophy Acta* 1793:200–211. doi: 10.1016/j.bbamcr.2008.05.028

Ferrari M, Hanley DF, Wilson DA, Traystman RJ (1990) Redox changes in cat brain cytochrome-c oxidase after blood-fluorocarbon exchange. *Am J Physiol* 258:H1706–H1713. doi: 10.1152/ajpheart.1990.258.6.H1706

Filiano JJ, Kinney HC (1994) A perspective on neuropathologic findings in victims of the sudden infant death syndrome: the triple-risk model. *Biol Neonate* 65:194–197. doi: 10.1159/000244052

Finsen NR (1895) The red light treatment of small-pox. *Br Med J* 2:1412–1414. doi: 10.1136/bmj.2.1823.1412-a

Fitzpatrick D, Itoh K, Diamond IT (1983) The laminar organization of the lateral geniculate body and the striate cortex in the squirrel monkey (Saimiri sciureus). *J Neurosci* 3:673–702. doi: 10.1523/JNEUROSCI.03-04-00673.1983

Fontanesi F, Clemente P, Barrientos A (2011) Cox25 teams up with Mss51, Ssc1, and Cox14 to regulate mitochondrial cytochrome c oxidase subunit 1 expression and assembly in Saccharomyces cerevisiae. *J Biol Chem* 286:555–566. doi: 10.1074/jbc.M110.188805

Formosa LE, Maghool S, Sharpe AJ, Reljic B, Muellner-Wong L, Stroud DA, Ryan MT, Maher MJ (2022) Mitochondrial COA7 is a heme-binding protein with disulfide reductase activity, which acts in the early stages of complex IV assembly. *Proc Natl Acad Sci USA* 119:e2110357119. doi: 10.1073/pnas.2110357119

Frappell P, Lanthier C, Baudinette RV, Mortola JP (1992) Metabolism and ventilation in acute hypoxia: a comparative analysis in small mammalian species. *Am J Physiol Regul Integr Comp Physiol* 262:R1040–R1046. doi: 10.1152/ajpregu.1992.262.6.R1040

Frenkel MY, Bear MF (2004) How monocular deprivation shifts ocular dominance in visual cortex of young mice. *Neuron* 44:917–923. doi: 10.1016/j.neuron.2004.12.003

Fujita K, Yamauchi M, Shibayama K, Ando M, Honda M, Nagata Y (1996) Decreased cytochrome c oxidase activity but unchanged superoxide dismutase and glutathione peroxidase activities in the spinal cords of patients with amyotrophic lateral sclerosis. *J Neurosci Res* 45:276–281. doi: 10.1002/(SICI)1097-4547(19960801)45:3<276::AID-JNR9>3.0.CO;2-A

Fukuda R, Zhang H, Kim JW, Shimoda L, Dang CV, Semenza GL (2007) HIF-1 regulates cytochrome oxidase subunits to optimize efficiency of respiration in hypoxic cells. *Cell* 129:111–122. doi: 10.1016/j.cell.2007.01.047

Gaisne M, Bonnefoy N (2006) The COX18 gene, involved in mitochondrial biogenesis, is functionally conserved and tightly regulated in humans and fission yeast. *FEMS Yeast Res* 6:869–882. doi: 10.1111/j.1567-1364.2006.00083.x

Gallardo ME, Moreno-Loshuertos R, López C, Casqueiro M, Silva J, Bonilla F, Rodríguez de Córdoba S, Enríquez JA (2006) m.6267G>A: a recurrent mutation in the human mitochondrial DNA that reduces cytochrome c oxidase activity and is associated with tumors. *Hum Mutat* 27:575–582. doi: 10.1002/humu.20338

Galvão de Oliveira M, Tengan C, Micheletti C, Ramos de Macedo P, Soares Pinho Cernach MC, Cavole TR, de França Basto M, Filho JS, Virmond LA, Milanezi F, Nakano V, Falconi A, Perrone E (2021) A novel variant in the COX15 gene causing a fatal infantile cardioencephalomyopathy: a case report with clinical and molecular review. *Eur J Med Genet* 64:104195. doi: 10.1016/j.ejmg.2021.104195

Gangfuß A, Hentschel A, Rademacher N, Sickmann A, Stüve B, Horvath R, Gross C, Kohlschmidt N, Förster F, Abicht A, Schänzer A, Schara-Schmidt U, Roos A, Della Marina A (2022) Identification of a novel homozygous synthesis of cytochrome c oxidase 2 variant in siblings with early-onset axonal Charcot-Marie-Tooth disease. *Hum Mutat* 43:477–486. doi: 10.1002/humu.24338

Gao S, Zhang H, Zhang X, Wang J, Bai W, Jiang B (2024) COX19 is a new target of MACC1 and promotes colorectal cancer progression by regulating copper transport in mitochondria. *J Nutr* 154:381–394. doi: 10.1016/j.tjnut.2023.12.032

Gao XP, Liu Q, Nair B, Wong-Riley MTT (2014) Reduced levels of brain-derived neurotrophic factor contribute to synaptic imbalance during the critical period of respiratory development in rats. *Eur J Neurosci* 40:2183–2195. doi: 10.1111/ejn.12568

Gao XP, Liu QS, Liu Q, Wong-Riley MTT (2011) Excitatory-inhibitory imbalance in hypoglossal neurons during the critical period of postnatal development in the rat. *J Physiol* 589.8:1991–2006. doi: 10.1113/jphysiol.2010.198945

Gao XP. Zhang H, Wong-Riley M (2015) Role of brain-derived neurotrophic factor in the excitatory-inhibitory imbalance during the critical period of postnatal respiratory development in the rat. *Physiol Rep* 3:e12631. doi: 10.14814/phy2.12631

Garg AK, Li P, Rashid MS, Callaway EM (2019) Color and orientation are jointly coded and spatially organized in primate primary visual cortex. *Science* 364:1275–1279. doi: 10.1126/science.aaw5868

Garthwaite J (1991) Glutamate, nitric oxide and cell-cell signalling in the nervous system. *Trends Neurosci* 14:60–67. doi: 10.1016/0166-2236(91)90022-m

Garthwaite J (2008) Concepts of neural nitric oxide-mediated transmission. *Eur J Neurosci* 27:2783–2802. doi: 10.1111/j.1460-9568.2008.06285.x

Geering K (2008) Functional roles of Na,K-ATPase subunits. *Curr Opin Nephrol Hypertens* 17:526–532. doi: 10.1097/MNH.0b013e3283036cbf

Gerlai R, Henderson JT, Roder JC, Jia Z (1998) Multiple behavioral anomalies in GluR2 mutant mice exhibiting enhanced LTP. *Behav Brain Res* 95:37–45. doi: 10.1016/s0166-4328(98)00002-3

Ghit A, Assal D, Al-Shami AS, Hussein DEE (2021) GABA(A) receptors: structure, function, pharmacology, and related disorders. *J Genet Eng Biotechnol* 19:123. doi: 10.1186/s43141-021-00224-0

Ghosh S, Goswami D, Dutta R, Ghatak D, De R (2024) A comprehensive pan-cancer analysis of cytochrome C oxidase assembly factor 1 (COA1) reveals instrumental role of mitochondrial protein assembly in cancer that modulates disease progression and prognostic outcome. *Cell Biochem Biophys* 82:2533–2555. doi: 10.1007/s12013-024-01366-x

Gianfranceschi L, Siciliano R, Walls J, Morales B, Kirkwood A, Huang ZJ, Tonegawa S, Maffei L (2003) Visual cortex is rescued from the effects of dark rearing by overexpression of BDNF. *Proc Natl Acad Sci USA* 100(21):12486–12491. doi: 10.1073/pnas.1934836100

Gidoni D, Dynan WS, Tjian R (1984) Multiple specific contacts between a mammalian transcription factor and its cognate promoters. *Nature* 312:409–413. doi: 10.1038/312409a0

Glerum DM, Tzagoloff A (1994) Isolation of a human cDNA for heme A: farnesyltransferase by functional complementation of a yeast cox10 mutant. *Proc Natl Acad Sci USA* 91:8452–8456. doi: 10.1073/pnas.91.18.8452

Gleyzer N, Scarpulla RC (2013) Activation of a PGC-1-related coactivator (PRC)-dependent inflammatory stress program linked to apoptosis and premature senescence. *J Biol Chem* 288:8004–8015. doi: 10.1074/jbc.M112.426841

Gleyzer N, Vercauteren K, Scarpulla RC (2005) Control of mitochondrial transcription specificity factors (TFB1M and TFB2M) by nuclear respiratory factors (NRF-1 and NRF-2) and PGC-1 family coactivators. *Mol Cell Biol* 25:1354–1366. doi: 10.1128/MCB.25.4.1354-1366.2005

Goebel DJ, Poosch MS (1999) NMDA receptor subunit gene expression in the rat brain: a quantitative analysis of endogenous mRNA levels of NR1Com, NR2A, NR2B, NR2C, NR2D and NR3A. *Brain Res Mol Brain Res* 69:164–170. doi: 10.1016/s0169-328x(99)00100-x

Goldberg A, Wildman DE, Schmidt TR, Hüttemann M, Goodman M, Weiss ML, Grossman LI (2003) Adaptive evolution of cytochrome c oxidase subunit VIII in anthropoid primates. *Proc Natl Acad Sci USA* 100:5873–5878. doi: 10.1073/pnas.0931463100

Golshani P, Truong H, Jones EG (1997) Developmental expression of GABA(A) receptor subunit and GAD genes in mouse somatosensory barrel cortex. *J Comp Neurol* 383:199–219. doi: 10.1002/(sici)1096-9861(19970630)383:2<199::aid-cne7>3.0.co;2-w

Gonzalez-Lima F (2021) Neuroprotection and neurocognitive augmentation by photobiomodulation. In: Opris I, Lebedev MA, Casanova MF (eds) *Modern Approaches to Augmentation of Brain Function*. Contemporary Clinical Neuroscience. Springer Nature, Switzerland, pp 165–207. doi: 10.1007/978-3-030-54564-2_9

Gopalakrishnan L, Scarpulla RC (1995) Structure, expression, and chromosomal assignment of the human gene encoding nuclear respiratory factor 1. *J Biol Chem* 270:18019–18025. doi: 10.1074/jbc.270.30.18019

Gordon JA, Stryker MP (1996) Experience-dependent plasticity of binocular responses in the primary visual cortex of the mouse. *J Neurosci* 16:3274–3286. doi: 10.1523/JNEUROSCI.16-10-03274.1996

Goto Y, Nonaka I, Horai S (1990) A mutation in the tRNA(Leu)(UUR) gene associated with the MELAS subgroup of mitochondrial encephalomyopathies. *Nature* 348:651–653. doi: 10.1038/348651a0

Gourfinkel-An I, Vila M, Faucheux B, Duyckaerts C, Viallet F, Hauw JJ, Brice A, Agid Y, Hirsch EC (2002) Metabolic changes in the basal ganglia of patients with Huntington's disease: an in situ hybridization study of cytochrome oxidase subunit I mRNA. *J Neurochem* 80:466–476. doi: 10.1046/j.0022-3042.2001.00734.x

Gray SL, Cummings KJ, Jirik FR, Sherwood NM (2001) Targeted disruption of the pituitary adenylate cyclase-activating polypeptide gene results in early postnatal death associated with dysfunction of lipid and carbohydrate metabolism. *Mol Endocrinol* 15:1739–1747. doi: 10.1210/mend.15.10.0705

Graybiel AM (2000) The basal ganglia. *Curr Biol* 10:R509–R511. doi: 10.1016/s0960-9822(00)00593-5

Greenamyre JT, MacKenzie G, Peng TI, Stephans SE (1999) Mitochondrial dysfunction in Parkinson's disease. *Biochem Soc Symp* 66:85–97. doi: 10.1042/bss0660085

Grevel A, Pfanner N, Becker T (2019) Coupling of import and assembly pathways in mitochondrial protein biogenesis. *Biol Chem* 401:117–129. doi: 10.1515/hsz-2019-0310

Griffiths DE, Wharton DC (1961) Studies of the electron transport system. XXXV. Purification and properties of cytochrome oxidase. *J Biol Chem* 236:1850–1856.

Grossman LI, Lomax M I (1997) Nuclear genes for cytochrome c oxidase [review]. *Biochim Biophys Acta* 1352:174–192. doi: 10.1016/s0167-4781(97)00025-0

Grossman LI, Purandare N, Arshad R, Gladyck S, Somayajulu M, Hüttemann M, Aras S (2017) MNRR1, a biorganellar regulator of mitochondria. *Oxid Med Cell Longev* 2017:6739236. doi: 10.1155/2017/6739236

Gu M, Gash MT, Mann VM, Javoy-Agid F, Cooper JM, Schapira AH (1996) Mitochondrial defect in Huntington's disease caudate nucleus. *Ann Neurol* 39:385–389. doi: 10.1002/ana.410390317

Gugneja S, Virbasius CM, Scarpulla RC (1996) Nuclear respiratory factors 1 and 2 utilize similar glutamine-containing clusters of hydrophobic residues to activate transcription. *Mol Cell Biol* 16:5708–5716. doi: 10.1128/MCB.16.10.5708

Gugneja S, Virbasius JV, Scarpulla RC (1995) Four structurally distinct, non-DNA-binding subunits of human nuclear respiratory factor 2 share a conserved transcriptional activation domain. *Mol Cell Biol* 15:102–111. doi: 10.1128/MCB.15.1.102

Guillaud L, Setou M, Hirokawa N (2003) KIF17 dynamics and regulation of NR2B trafficking in hippocampal neurons. *J Neurosci* 23:131–140. doi: 10.1523/JNEUROSCI.23-01-00131.2003

Guillery RW (1969) The organization of synaptic interconnections in the laminae of the dorsal lateral geniculate nucleus of the cat. *Z Zellforsch Mikrosk Anat* 96:1–38. doi: 10.1007/BF00321474

Guillery RW (1971) Survival of large cells in the dorsal lateral geniculate laminae after interruption of retino-geniculate afferents. *Brain Res* 28:541–544. doi: 10.1016/0006-8993(71)90062-x

Guire ES, Lickey ME, Gordon B (1999) Critical period for the monocular deprivation effect in rats: assessment with sweep visually evoked potentials. *J Neurophysiol* 81:121–128. doi: 10.1152/jn.1999.81.1.121

Guo AL, Nie F, Wong-Riley M (2000) Human nuclear respiratory factor 2α subunit cDNA: isolation, subcloning, sequencing, and in situ hybridization of transcripts in normal and monocularly deprived macaque visual system. *J Comp Neurol* 417:221–232. doi: 10.1002/(sici)1096-9861(20000207)417:2<221::aid-cne7>3.0.co;2-4

Habiba A, Blanco G, Mercer RW (2000) Expression, activity and distribution of Na,K-ATPase subunits during in vitro neuronal induction. *Brain Res* 875:1–13. doi: 10.1016/s0006-8993(00)02541-5

Hagen G, Dennig J, Preiss A, Beato M, Suske G (1995) Functional analyses of the transcription factor Sp4 reveal properties distinct from Sp1 and Sp3. *J Biol Chem* 270:24989–24994. doi: 10.1074/jbc.270.42.24989

Hall AV, Antoniou H, Wang Y, Cheung AH, Arbus AM, Olson SL, Lu WC, Kau CL, Marsden PA (1994) Structural organization of the human neuronal nitric oxide synthase gene (NOS1). *J Biol Chem* 269:33082–33090.

Hall J, Moubarak A, O'Brien P, Pan LP, Cho I., Millett F (1988) Topological studies of monomeric and dimeric cytochrome c oxidase and identification of the copper A site using fluorescence probe. *J Biol Chem* 263:8142–8149.

Hallén S, Brzezinski P (1994) Light-induced structural changes in cytochrome c oxidase: implication for the mechanism of electron and proton gating. *Biochim Biophys Acta* 1184:207–218. doi: 10.1016/0005-2728(94)90225-9

Hallmann K, Kudin AP, Zsurka G, Kornblum C, Reimann J, Stüve B, Waltz S, Hattingen E, Thiele H, Nürnberg P, Rüb C, Voos W, Kopatz J, Neumann H, Kunz WS (2016) Loss of the smallest subunit of cytochrome c oxidase, COX8A, causes Leigh-like syndrome and epilepsy. *Brain* 139:338–345. doi: 10.1093/brain/awv357

Halperin D, Drabkin M, Wormser O, Yogev Y, Dolgin V, Shorer Z, Gradstein L, Shelef I, Flusser H, Birk OS (2020) Phenotypic variability and mutation hotspot in COX15-related Leigh syndrome. *Am J Med Genet A* 182:1506–1512. doi: 10.1002/ajmg.a.61577

Hamblin MR (2019) Photobiomodulation for Alzheimer's disease: has the light dawned? *Photonics* 6:77. doi: 10.3390/photonics6030077

Hamblin MR, de Sousa MVP, Agrawal T (eds) (2017) *Handbook of Low-Level Laser Therapy*. Pan Stanford Publishing, Singapore.

Handy DE, Loscalzo J (2012) Redox regulation of mitochondrial function. *Antioxid Redox Signal* 16:1323–1367. doi: 10.1089/ars.2011.4123

Hannibal J, Møller M, Ottersen OP, Fahrenkrug J (2000) PACAP and glutamate are co-stored in the retinohypothalamic tract. *J Comp Neurol* 418:147–155.

Hanover JL, Huang ZJ, Tonegawa S, Stryker MP (1999) Brain-derived neurotrophic factor overexpression induces precocious critical period in mouse visual cortex. *J Neurosci* 19:RC40. doi: 10.1523/JNEUROSCI.19-22-j0003.1999

Harauzov A, Spolidoro M, DiCristo G, De Pasquale R, Cancedda L, Pizzorusso T, Viegi A, Berardi N, Maffei L (2010) Reducing intracortical inhibition in the adult visual cortex promotes ocular dominance plasticity. *J Neurosci* 30:361–371. doi: 10.1523/JNEUROSCI.2233-09.2010

Harman D (1956) Aging: a theory based on free radical and radiation chemistry. *J Gerontol* 11:298–300. doi: 10.1093/geronj/11.3.298

Harman D (1972) The biologic clock: the mitochondria? *J Am Geriatr Soc* 20:145–147. doi: 10.1111/j.1532-5415.1972.tb00787.x

Harper RM, Gozal D, Bandler R, Spriggs D, Lee J, Alger J (1998) Regional brain activation in humans during respiratory and blood pressure challenges. *Clin Exp Pharmacol Physiol* 25:483–486. doi: 10.1111/j.1440-1681.1998.tb02240.x

Haseltine EC, DeBruyn EJ, Casagrande VA (1979) Demonstration of ocular dominance columns in Nissl-stained sections of monkey visual cortex following enucleation. *Brain Res* 176:153–158. doi: 10.1016/0006-8993(79)90878-3

Hassler R (1966) Comparative anatomy of the central visual system in day- and night-active primates. In: Hassler R, Stephan H (eds) *Evolution of the Forebrain*. Springer, Boston, pp. 419–434. doi: 10.1007/978-1-4899-6527-1_40

Hata Y, Stryker MP (1994) Control of thalamocortical afferent rearrangement by postsynaptic activity in developing visual cortex. *Science* 265:1732–1735. doi: 10.1126/science.8085163

Hatta S, Kumagami T, Qian J, Thornton M, Smith EL 3rd, Chino YM (1998) Nasotemporal directional bias of V1 neurons in young infant monkeys. *Invest Ophthalmol Vis Sci* 39:2259–2267.

Haxby JV, Rapoport SI (1986) Abnormalities of regional brain metabolism in Alzheimer's disease and their relation to functional impairment. *Prog Neuropsychopharmacol Biol Psychiatry* 10:427–438. doi: 10.1016/0278-5846(86)90016-3

Haynes WM (2011) *CRC Handbook of Chemistry and Physics*, 92nd ed. CRC Press, Hoboken.

Heinze L, Harvey RJ, Haverkamp S, Wässle H (2007) Diversity of glycine receptors in the mouse retina: localization of the alpha4 subunit. *J Comp Neurol* 500:693–707. doi: 10.1002/cne.21201

Hejzlarová K, Kaplanová V, Nůsková H, Kovářová N, Ješina P, Drahota Z, Mráček T, Seneca S, Houštěk J (2015) Alteration of structure and function of ATP synthase and cytochrome c oxidase by lack of Fo-a and Cox3 subunits caused by mitochondrial DNA 9205delTA mutation. *Biochem J* 466:601–611. doi: 10.1042/BJ20141462

Hell K, Tzagoloff A, Neupert W, Stuart RA (2000) Identification of Cox20p, a novel protein involved in the maturation and assembly of cytochrome oxidase subunit 2. *J Biol Chem* 275:4571–4578. doi: 10.1074/jbc.275.7.4571

Hellwig P, Soulimane T, Buse G, Mantele W (1999) Electrochemical, FTIR, and UV/VIS spectroscopic properties of the ba3 oxidase from *Thermus thermophilus*. *Biochemistry* 38:9648–9658. doi: 10.1021/bi9903401

Hendrickson AE (1985) Dots, stripes and columns in monkey visual cortex. *Trends Neurosci* 8:406–410. doi: 10.1016/0166-2236(85)90143-2

Hendrickson AE, Hunt SP, Wu JY (1981) Immunocytochemical localization of glutamic acid decarboxylase in monkey striate cortex. *Nature* 292:605–607. doi: 10.1038/292605a0

Hendrickson AE, Tigges M (1985) Enucleation demonstrates ocular dominance columns in Old World macaque but not in New World squirrel monkey visual cortex. *Brain Res* 333:340–344. doi: 10.1016/0006-8993(85)91589-6

Hendry SH, Bhandari MA (1992) Neuronal organization and plasticity in adult monkey visual cortex: immunoreactivity for microtubule-associated protein 2. *Vis Neurosci* 9:445–459. doi: 10.1017/s0952523800011251

Hendry SH, Fuchs J, deBlas AL, Jones EG (1990) Distribution and plasticity of immunocytochemically localized GABAA receptors in adult monkey visual cortex. *J Neurosci* 10:2438–2450. doi: 10.1523/JNEUROSCI.10-07-02438.1990

Hendry SH, Jones EG, Burstein N (1988) Activity-dependent regulation of tachykinin-like immunoreactivity in neurons of monkey visual cortex. *J Neurosci* 8:1225–1238. doi: 10.1523/JNEUROSCI.08-04-01225.1988

Henneberger C, Jüttner R, Rothe T, Grantyn R (2002) Postsynaptic action of BDNF on GABAergic synaptic transmission in the superficial layers of the mouse superior colliculus. *J Neurophysiol* 88:595–603. doi:10.1152/jn.2002.88.2.595

Hensch TK, Fagiolini M (2005) Excitatory-inhibitory balance and critical period plasticity in developing visual cortex. *Prog Brain Res* 147:115–124. doi: 10.1016/S0079-6123(04)47009-5

Hensch TK, Fagiolini M, Mataga N, Stryker MP, Baekkeskov S, Kash SF (1998) Local GABA circuit control of experience-dependent plasticity in developing visual cortex. *Science* 282:1504–1508. doi: 10.1126/science.282.5393.1504

Hensch TK, Quinlan EM (2018) Critical periods in amblyopia. *Vis Neurosci* 35:E014. doi: 10.1017/S0952523817000219

Hensch TK, Stryker MP (2004) Columnar architecture sculpted by GABA circuits in developing cat visual cortex. *Science* 303:1678–1681. doi: 10.1126/science.1091031

Hevner RF, Duff RS, Wong-Riley MTT (1992) Coordination of ATP production and consumption in brain: parallel regulation of cytochrome oxidase and Na^+,K^+-ATPase. *Neurosci Lett* 138:188–192. doi: 10.1016/0304-3940(92)90502-x

Hevner RF, Liu S, Wong-Riley MTT (1993) An optimized method for determining cytochrome oxidase activity in brain tissue homogenates. *J Neurosci Methods* 50:309–319. doi: 10.1016/0165-0270(93)90038-s

Hevner RF, Liu S, Wong-Riley MTT (1995) A metabolic map of cytochrome oxidase in the rat brain: histochemical, densitometric and biochemical studies. *Neuroscience* 65:313–342. doi: 10.1016/0306-4522(94)00514-6

Hevner RF, Wong-Riley MTT (1989) Brain cytochrome oxidase: purification, antibody generation, and immunohistochemical/histochemical correlations in the CNS. *J Neurosci* 9:3884–3898. doi: 10.1523/JNEUROSCI.09-11-03884.1989

Hevner RF, Wong-Riley MTT (1990) Regulation of cytochrome oxidase protein levels by functional activity in the macaque monkey visual system. *J Neurosci* 10:1331–1340. doi: 10.1523/JNEUROSCI.10-04-01331.1990

Hevner RF, Wong-Riley MTT (1991) Neuronal expression of nuclear and mitochondrial genes for cytochrome oxidase (CO) subunits analyzed by in situ hybridization: Comparison with CO activity and protein. *J Neurosci* 11:1942–1958. doi: 10.1523/JNEUROSCI.11-07-01942.1991

Hevner RF, Wong-Riley MTT (1992) Entorhinal cortex of the human, monkey, and rat: metabolic map as revealed by cytochrome oxidase. *J Comp Neurol* 326:451–469. doi: 10.1002/cne.903260310

Hevner RF, Wong-Riley MTT (1993) Mitochondrial and nuclear gene expression for cytochrome oxidase subunits are disproportionately regulated by functional activity in neurons. *J Neurosci* 13:1805–1819. doi: 10.1523/JNEUROSCI.13-05-01805.1993

HGNC Database (2025) HUGO Gene Nomenclature Committee (HGNC), Department of Haematology, Long Road, Cambridge CB2 0PT, United Kingdom. https://www.genenames.org.

Higuchi Y, Okunushi R, Hara T, Hashiguchi A, Yuan J, Yoshimura A, Murayama K, Ohtake A, Ando M, Hiramatsu Y, Ishihara S, Tanabe H, Okamoto Y, Matsuura E, Ueda T, Toda T, Yamashita S, Yamada K, Koide T, Yaguchi H, Mitsui J, Ishiura H, Yoshimura J, Doi K, Morishita S, Sato K, Nakagawa M, Yamaguchi M, Tsuji S, Takashima H (2018) Mutations in COA7 cause spinocerebellar ataxia with axonal neuropathy. *Brain* 141:1622–1636. doi: 10.1093/brain/awy104

Hikosaka O (2007) GABAergic output of the basal ganglia. *Prog Brain Res* 160:209–226. doi: 10.1016/S0079-6123(06)60012-5

Hilaire G, Bou C, Monteau R (1997) Serotonergic modulation of central respiratory activity in the neonatal mouse: an in vitro study. *Eur J Pharmacol* 329:115–120. doi: 10.1016/S0014-2999(97)89173-X

Hilaire G, Duron B (1999) Maturation of the mammalian respiratory system. *Physiol Rev* 79:325–360. doi: 10.1152/physrev.1999.79.2.325

Hilaire G, Monteau R, Gauthier P, Rega P, Morin D (1990) Functional significance of the dorsal respiratory group in adult and newborn rats: in vivo and in vitro studies. *Neurosci Lett* 111:133–138. doi: 10.1016/0304-3940(90)90357-f

Hill BC (1991) The reaction of the electrostatic cytochrome c-cytochrome oxidase complex with oxygen. *J Biol Chem* 266:2219–2226.

Hirokawa N (1998) Kinesin and dynein superfamily proteins and the mechanism of organelle transport. *Science* 279:519–526. doi: 10.1126/science.279.5350.519

Hirokawa N, Nitta R, Okada Y (2009) The mechanisms of kinesin motor motility: lessons from the monomeric motor KIF1A. *Nat Rev Mol Cell Biol* 10:877–884. doi: 10.1038/nrm2807

Hirokawa N, Takemura R (2004) Kinesin superfamily proteins and their various functions and dynamics. *Exp Cell Res* 301:50–59. doi: 10.1016/j.yexcr.2004.08.010

Hodges MR, Richerson GB (2010) The role of medullary serotonin (5-HT) neurons in respiratory control: contributions to eupneic ventilation, CO2 chemoreception, and thermoregulation. *J Appl Physiol* 108:1425–1432. doi: 10.1152/japplphysiol.01270.2009

Hodges MR, Wehner M, Aungst J, Smith JC, Richerson GB (2009) Transgenic mice lacking serotonin neurons have severe apnea and high mortality during development. *J Neurosci* 29:10341–10349. doi: 10.1523/JNEUROSCI.1963-09.2009

Hofer MM, Barde YA (1988) Brain-derived neurotrophic factor prevents neuronal death in vivo. *Nature* 331:261–262. doi: 10.1038/331261a0

Hollmann M, Heinemann S (1994) Cloned glutamate receptors. *Annu Rev Neurosci* 17:31–108. doi: 10.1146/annurev.ne.17.030194.000335

Holtman JR Jr, Norman WP, Skirboll L, Dretchen KL, Cuello C, Visser TJ, Hökfelt T, Gillis RA (1984) Evidence for 5-hydroxytryptamine, substance P, and thyrotropin-releasing hormone in neurons innervating the phrenic motor nucleus. *J Neurosci* 4:1064–1071. doi: 10.1523/JNEUROSCI.04-04-01064.1984

Horn D, Al-Ali H, Barrientos A (2008) Cmc1p is a conserved mitochondrial twin CX9C protein involved in cytochrome c oxidase biogenesis. *Mol Cell Biol* 28:4354–4364. doi: 10.1128/MCB.01920-07

Horner RL (2007) Respiratory motor activity: influence of neuromodulators and implications for sleep disordered breathing. *Can J Physiol Pharmacol* 85:155–165. doi: 10.1139/y06-089

Horton JC (1984) Cytochrome oxidase patches: a new cytoarchitectonic feature of monkey visual cortex. *Philos Trans R Soc Lond B Biol Sci* 304:199–253. doi: 10.1098/rstb.1984.0021

Horton JC, Hedley-Whyte ET (1984) Mapping of cytochrome oxidase patches and ocular dominance columns in human visual cortex. *Philos Trans R Soc Lond B Biol Sci* 304:255–272. doi: 10.1098/rstb.1984.0022

Horton JC, Hocking DR (1996) Intrinsic variability of ocular dominance column periodicity in normal macaque monkeys. *J Neurosci* 16:7228–7239. doi: 10.1523/JNEUROSCI.16-22-07228.1996

Horton JC, Hocking DR (1997) Timing of the critical period for plasticity of ocular dominance columns in macaque striate cortex. *J Neurosci* 17:3684–3709. doi: 10.1523/JNEUROSCI.17-10-03684.1997

Horton JC, Hocking DR (1998) Monocular core zones and binocular border strips in primate striate cortex revealed by the contrasting effects of enucleation, eyelid suture, and retinal laser lesions on cytochrome oxidase activity. *J Neurosci* 18:5433–5455. doi: 10.1523/JNEUROSCI.18-14-05433.1998

Horton JC, Hubel DH (1981) Regular patchy distribution of cytochrome oxidase staining in primary visual cortex of macaque monkey. *Nature* 292:762–764. doi: 10.1038/292762a0

Horvat S, Beyer C, Arnold S (2006) Effect of hypoxia on the transcription pattern of subunit isoforms and the kinetics of cytochrome c oxidase in cortical astrocytes and cerebellar neurons. *J Neurochem* 99:937–951. doi: 10.1111/j.1471-4159.2006.04134.x

Houldsworth A (2024) Role of oxidative stress in neurodegenerative disorders: a review of reactive oxygen species and prevention by antioxidants. *Brain Commun* 6:fcad356. doi: 10.1093/braincomms/fcad356

Houreld NN, Masha RT, Abrahamse H (2012) Low-intensity laser irradiation at 660 nm stimulates cytochrome c oxidase in stressed fibroblast cells. *Lasers Surg Med* 44:429–434. doi: 10.1002/lsm.22027

Hoy JL, Farrow K (2025) The superior colliculus. *Curr Biol* 35:R164–R168. doi: 10.1016/j.cub.2025.01.022

Hoyer D, Clarke DE, Fozard JR, Hartig PR, Martin GR, Mylecharane EJ, Saxena PR, Humphrey PP (1994) International Union of Pharmacology classification of receptors for 5-hydroxytryptamine (Serotonin). *Pharmacol Rev* 46:157–203.

Huang J, Waters KA, Machaalani R (2017a) Hypoxia and nicotine effects on pituitary adenylate cyclase activating polypeptide (PACAP) and its receptor 1 (PAC1) in the developing piglet brainstem. *Neurotoxicology* 62:30–38. doi: 10.1016/j.neuro.2017.05.005

Huang J, Waters KA, Machaalani R (2017b) Pituitary adenylate cyclase activating polypeptide (PACAP) and its receptor 1 (PAC1) in the human infant brain and changes in the Sudden Infant Death Syndrome (SIDS). *Neurobiol Dis* 103:70–77. doi: 10.1016/j.nbd.2017.04.002

Huang ZJ, Kirkwood A, Pizzorusso T, Porciatti V, Morales B, Bear MF, Maffei L, Tonegawa S (1999) BDNF regulates the maturation of inhibition and the critical period of plasticity in mouse visual cortex. *Cell* 98:739–755. doi: 10.1016/s0092-8674(00)81509-3

Hubel DH (1982) Exploration of the primary visual cortex, 1955–78. *Nature* 299:515–524. doi: 10.1038/299515a0

Hubel DH, Livingstone MS (1985) Complex-unoriented cells in a subregion of primate area 18. *Nature* 315:325–327. doi: 10.1038/315325a0

Hubel DH, Livingstone MS (1987) Segregation of form, color, and stereopsis in primate area 18. *J Neurosci* 7:3378–3415. doi: 10.1523/JNEUROSCI.07-11-03378.1987

Hubel DH, Wiesel TN (1968) Receptive fields and functional architecture of monkey striate cortex. *J Physiol* 195:215–243. doi: 10.1113/jphysiol.1968.sp008455

Hubel DH, Wiesel TN (1970) The period of susceptibility to the physiological effects of unilateral eye closure in kittens. *J Physiol* 206:419–436. doi: 10.1113/jphysiol.1970.sp009022

Hubel DH, Wiesel TN (1972) Laminar and columnar distribution of geniculo-cortical fibers in the macaque monkey. *J Comp Neurol* 146:421–450. doi: 10.1002/cne.901460402

Hubel DH, Wiesel TN (1974) Sequence regularity and geometry of orientation columns in the monkey striate cortex. *J Comp Neurol* 158:267–293. doi: 10.1002/cne.901580304

Hubel DH, Wiesel TN (1977) Ferrier lecture. Functional architecture of macaque monkey visual cortex. *Proc R Soc Lond B Biol Sci* 198:1–59. doi: 10.1098/rspb.1977.0085

Hubel DH, Wiesel TN, LeVay S (1977) Plasticity of ocular dominance columns in monkey striate cortex. *Philos Trans R Soc Lond B Biol Sci* 278:377–409. doi: 10.1098/rstb.1977.0050

Hudgson P, Bradley WG, Jenkison M (1972) Familial "mitochondrial" myopathy. A myopathy associated with disordered oxidative metabolism in muscle fibres. 1. Clinical, electrophysiological and pathological findings. *J Neurol Sci* 16:343–370. doi: 10.1016/0022-510x(72)90197-9

Huigsloot M, Nijtmans LG, Szklarczyk R, Baars MJ, van den Brand MA, Hendriksfranssen MG, van den Heuvel LP, Smeitink JA, Huynen MA, Rodenburg RJ (2011) A mutation in C2orf64 causes impaired cytochrome c oxidase assembly and mitochondrial cardiomyopathy. *Am J Hum Genet* 88:488–493. doi: 10.1016/j.ajhg.2011.03.002

Humphrey AL, Hendrickson AE (1983) Background and stimulus-induced patterns of high metabolic activity in the visual cortex (area 17) of the squirrel and macaque monkey. *J Neurosci* 3:345–358. doi: 10.1523/JNEUROSCI.03-02-00345.1983

Hunt JA (1974) Half-life and rate of synthesis of globin messenger ribonucleic acid. Determination of half-life of messenger ribonucleic acid and its relative synthetic rate in erythroid cells. *Biochem J* 138:487–498. doi: 10.1042/bj1380487

Huntington G (1872) On chorea. *Med Surg Reporter Philadelphia* 26:317–321.

Huntsman MM, Isackson PJ, Jones EG (1994) Lamina-specific expression and activity-dependent regulation of seven GABAA receptor subunit mRNAs in monkey visual cortex. *J Neurosci* 14:2236–2259. doi: 10.1523/JNEUROSCI.14-04-02236.1994

Huo L, Scarpulla RC (2001) Mitochondrial DNA instability and peri-implantation lethality associated with targeted disruption of nuclear respiratory factor 1 in mice. *Mol Cell Biol* 21:644–654. doi: 10.1128/MCB.21.2.644-654.2001

Hüttemann M, Jaradat S, Grossman LI (2003a) Cytochrome c oxidase of mammals contains a testes-specific isoform of subunit VIb--the counterpart to testes-specific cytochrome C? *Mol Reprod Dev* 66:8–16. doi: 10.1002/mrd.10327

Hüttemann M, Kadenbach B, Grossman LI (2001) Mammalian subunit IV isoforms of cytochrome c oxidase. *Gene* 267:111–123. doi: 10.1016/s0378-1119(01)00385-7

Hüttemann M, Klewer S, Lee I, Pecinova A, Pecina P, Liu J, Lee M, Doan JW, Larson D, Slack E, Maghsoodi B, Erickson RP, Grossman LI (2012) Mice deleted for heart-type cytochrome c oxidase subunit 7a1 develop dilated cardiomyopathy. *Mitochondrion* 12:294–304. doi: 10.1016/j.mito.2011.11.002

Hüttemann M, Mühlenbein N, Schmidt TR, Grossman LI, Kadenbach B (2000) Isolation and sequence of the human cytochrome c oxidase subunit VIIaL gene. *Biochim Biophys Acta* 1492:252–258. doi: 10.1016/s0167-4781(00)00087-7

Hüttemann M, Schmidt TR, Grossman LI (2003b) A third isoform of cytochrome c oxidase subunit VIII is present in mammals. *Gene* 312:95–102. doi: 10.1016/s0378-1119(03)00604-8

Indrieri A, van Rahden VA, Tiranti V, Morleo M, Iaconis D, Tammaro R, D'Amato I, Conte I, Maystadt I, Demuth S, Zvulunov A, Kutsche K, Zeviani M, Franco B (2012) Mutations in COX7B cause microphthalmia with linear skin lesions, an unconventional mitochondrial disease. *Am J Hum Genet.* 91(5):942–949. doi: 10.1016/j.ajhg.2012.09.016

Insinna C, Humby M, Sedmak T, Wolfrum U, Besharse JC (2009) Different roles for KIF17 and kinesin II in photoreceptor development and maintenance. *Dev Dyn* 238:2211–2222. doi: 10.1002/dvdy.21956

Irrcher I, Adhihetty PJ, Sheehan T, Joseph AM, Hood DA (2003) PPARgamma coactivator-1alpha expression during thyroid hormone- and contractile activity-induced mitochondrial adaptations. *Am J Physiol Cell Physiol* 284:C1669–C1677. doi: 10.1152/ajpcell.00409.2002

Isaac JT, Ashby MC, McBain CJ (2007) The role of the GluR2 subunit in AMPA receptor function and synaptic plasticity. *Neuron* 54:859–871. doi: 10.1016/j.neuron.2007.06.001

Ishimaru N, Tabuchi A, Hara D, Hayashi H, Sugimoto T, Yasuhara M, Shiota J, Tsuda M (2007) Regulation of neurotrophin-3 gene transcription by Sp3 and Sp4 in neurons. *J Neurochem* 100:520–531. doi: 10.1111/j.1471-4159.2006.04216.x

Itoh K, Weis S, Mehraein P, Müller-Höcker J (1997) Defects of cytochrome c oxidase in the substantia nigra of Parkinson's disease: and immunohistochemical and morphometric study. *Mov Disord* 12:9–16. doi: 10.1002/mds.870120104

Jackson DA, Iborra FJ, Manders EM, Cook PR (1998) Numbers and organization of RNA polymerases, nascent transcripts, and transcription units in HeLa nuclei. *Mol Biol Cell* 9:1523–1536. doi: 10.1091/mbc.9.6.1523

Jacobs BL, Fornal CA (1997) Serotonin and motor activity. *Curr Opin Neurobiol* 7:820–825. doi: 10.1016/s0959-4388(97)80141-9

Jacobs EE, Andrews EC, Cunningham W, Crane FL (1966) Membraneous oxidase purification, properties and reaction characteristics. *Biochem Biophys Res Commun* 25:87–95. doi: 10.1016/0006-291x(66)90644-9

Jacobs EE, Sanadi DR (1960) The reversible removal of cytochrome c from mitochondria. *J Biol Chem* 235:531–534.

Jaradat SA, Ko MS, Grossman LI (1998) Tissue-specific expression and mapping of the Cox7ah gene in mouse. *Genomics* 49:363–370. doi: 10.1006/geno.1998.5279

Jaynes J (1957) Imprinting: the interaction of learned and innate behavior: II. The critical period. *J Comp Physiol Psychol* 50:6–10. doi:10.1037/h0044716

Jędrzejewska-Szmek J, Dorman DB, Blackwell KT (2023) Making time and space for calcium control of neuron activity. *Curr Opin Neurobiol* 83:102804. doi: 10.1016/j.conb.2023.102804

Jellinger KA (1999) The role of iron in neurodegeneration: prospects for pharmacotherapy of Parkinson's disease. *Drugs Aging* 14:115–140. doi: 10.2165/00002512-199914020-00004

Jenks KR, Tsimring K, Ip JPK, Zepeda JC, Sur M (2021) Heterosynaptic plasticity and the experience-dependent refinement of developing neuronal circuits. *Front Neural Circuits* 15:803401. doi: 10.3389/fncir.2021.803401

Jewell EA, Lingrel JB (1991) Comparison of the substrate dependence properties of the rat Na,K-ATPase alpha 1, alpha 2, and alpha 3 isoforms expressed in HeLa cells. *J Biol Chem* 266:16925–16930.

Jia L, Dienhart M, Schramp M, McCauley M, Hell K, Stuart RA (2003) Yeast Oxa1 interacts with mitochondrial ribosomes: the importance of the C-terminal region of Oxa1. *EMBO J* 22:6438–6447. doi: 10.1093/emboj/cdg624

Jia Z, Agopyan N, Miu P, Xiong Z, Henderson J, Gerlai R, Taverna FA, Velumian A, MacDonald J, Carlen P, Abramow-Newerly W, Roder J (1996) Enhanced LTP in mice deficient in the AMPA receptor GluR2. *Neuron* 17:945–956. doi: 10.1016/s0896-6273(00)80225-1

Jöbsis FF (1977) Noninvasive, infrared monitoring of cerebral and myocardial oxygen sufficiency and circulatory parameters. *Science* 198:1264–1267. doi: 10.1126/science.929199

Johar K, Priya A, Dhar S, Liu Q, Wong-Riley MTT (2013) Neuron-specific specificity protein 4 bigenomically regulates the transcription of all mitochondria- and nucleus-encoded cytochrome c oxidase subunit genes in neurons. *J Neurochem* 127:496–508. doi: 10.1111/jnc.12433

Johar K, Priya A, Wong-Riley MTT (2012) Regulation of Na^+/K^+-ATPase by nuclear respiratory factor 1: implication in the tight coupling of neuronal activity, energy generation, and energy consumption. *J Biol Chem* 287:40381–40390. doi: 10.1074/jbc.M112.414573

Johar K, Priya A, Wong-Riley MTT (2014) Regulation of Na^+/K^+-ATPase by neuron-specific transcription factor Sp4: implication in the tight coupling of energy production, neuronal activity and energy consumption in neurons. *Eur J Neurosci* 39:566–578. doi: 10.1111/ejn.12415

Johnson EN, Hawken MJ, Shapley R (2008) The orientation selectivity of color-responsive neurons in macaque V1. *J Neurosci* 28:8096–8106. doi: 10.1523/JNEUROSCI.1404-08.2008

Johnson GC, May V, Parsons RL, Hammack SE (2019) Parallel signaling pathways of pituitary adenylate cyclase activating polypeptide (PACAP) regulate several intrinsic ion channels. *Ann N Y Acad Sci* 1455:105–112. doi: 10.1111/nyas.14116

Jordan D (2001) Central nervous pathways and control of the airways. *Respir Physiol* 125:67–81. doi: 10.1016/s0034-5687(00)00205-x

Kaas JH, Guillery RW, Allman JM (1972) Some principles of organization in the dorsal lateral geniculate nucleus. *Brain Behav Evol* 6:253–299. doi: 10.1159/000123713

Kadenbach B (2017) Regulation of mammalian 13-subunit cytochrome c oxidase and binding of other proteins: role of NDUFA4. *Trends Endocrinol Metab* 28:761–770. doi: 10.1016/j.tem.2017.09.003

Kadenbach B (2020) Regulation of cytochrome c oxidase contributes to health and optimal life. *World J Biol Chem* 11:52–61. doi: 10.4331/wjbc.v11.i2.52

Kadenbach B, Frank V, Rieger T, Napiwotzki J (1997) Regulation of respiration and energy transduction in cytochrome c oxidase isozymes by allosteric effectors. *Mol Cell Biochem* 174:131–135.

Kadenbach B, Hüttemann M (2015) The subunit composition and function of mammalian cytochrome c oxidase. *Mitochondrion* 24:64–76. doi: 10.1016/j.mito.2015.07.002

Kadenbach B, Hüttemann M, Arnold S, Lee I, Bender E (2000) Mitochondrial energy metabolism is regulated via nuclear-coded subunits of cytochrome c oxidase. *Free Radic Biol Med* 29:211–221. doi: 10.1016/s0891-5849(00)00305-1

Kadenbach B, Jarausch J, Hartmann R, Merle P (1983) Separation of mammalian cytochrome c oxidase into 13 polypeptides by a sodium dodecyl sulfate-gel electrophoretic procedure. *Anal Biochem* 129:517–521. doi: 10.1016/0003-2697(83)90586-9

Kadenbach B, Ramzan R, Vogt S (2009) Degenerative diseases, oxidative stress and cytochrome c oxidase function. *Trends Mol Med* 15:139–147. doi: 10.1016/j.molmed.2009.02.004

Kadenbach B, Ramzan R, Vogt S (2013) High efficiency versus maximal performance – the cause of oxidative stress in eukaryotes. A hypothesis. *Mitochondrion* 13:1–6. doi: 10.1016/j.mito.2012.11.005

Kadenbach B, Stroh A, Ungibauer M, Kuhn-Nentwig L, Büge U, Jarausch J (1986) Isozymes of cytochrome-c oxidase: characterization and isolation from different tissues. *Methods Enzymol* 126:32–45. doi: 10.1016/s0076-6879(86)26006-1

Kageyama G, Wong-Riley M (1987) Light and EM localization of cytochrome oxidase at several reported sites of active ion transport and transport-ATPase in neural/nonneural tissues in cat and fish. *Anat Rec* 218:70A.

Kageyama GH, Wong-Riley M (1985) An analysis of the cellular localization of cytochrome oxidase in the lateral geniculate nucleus of the adult cat. *J Comp Neurol* 242:338–357. doi: 10.1002/cne.902420304

Kageyama GH, Wong-Riley MTT (1982) Histochemical localization of cytochrome oxidase in the hippocampus: correlation with specific neuronal types and afferent pathways. *Neuroscience* 7:2337–2361. doi: 10.1016/0306-4522(82)90199-3

Kageyama GH, Wong-Riley MTT (1984) The histochemical localization of cytochrome oxidase in the retina and lateral geniculate nucleus of the ferret, cat, and monkey, with particular reference to retinal mosaics and ON/OFF-center visual channels. *J Neurosci* 4:2445–2459. doi: 10.1523/JNEUROSCI.04-10-02445.1984

Kanehisa M, Furumichi M, Sato Y, Matsuura Y, Ishiguro-Watanabe M (2025) KEGG: biological systems database as a model of the real world. *Nucleic Acids Res* 53(D1):D672–D677. doi: 10.1093/nar/gkae909

Kang JJ, Fung ML, Lam CS, Wu SX, Yang SJ, Huang XF, Wong-Riley MTT, Liu YY (2020) Chronic intermittent hypoxia alters dendritic mitochondrial structure and activity in the pre-Bötzinger complex of rats. *FASEB J* 34:14588–14601, 2020. doi: 10.1096/fj.201902141R

Kang JJ, Guo B, Liang WH, Lam CS, Wu SX, Huang XF, Wong-Riley MTT, Fung ML, Liu YY (2019) Daily acute intermittent hypoxia induced dynamic changes in dendritic mitochondrial ultrastructure and cytochrome oxidase activity in the pre-Bötzinger complex of rats. *Ex. Neurol* 313:124–134. doi: 10.1016/j.expneurol.2018.12.008

Kang Y, Anderson AJ, Jackson TD, Palmer CS, De Souza DP, Fujihara KM, Stait T, Frazier AE, Clemons NJ, Tull D, Thorburn DR, McConville MJ, Ryan MT, Stroud DA, Stojanovski D (2019) Function of hTim8a in complex IV assembly in neuronal cells provides insight into pathomechanism underlying Mohr-Tranebjaerg syndrome. *Elife* 8:e48828. doi: 10.7554/eLife.48828

Kaplan JH (2002) Biochemistry of Na,K-ATPase. *Annu Rev Biochem* 71:511–535. doi: 10.1146/annurev.biochem.71.102201.141218

Karim FD, Urness LD, Thummel CS, Klemsz MJ, McKercher SR, Celada A, Van Beveren C, Maki RA, Gunther CV, Nye JA, Graves BJ (1990) The ETS-domain: a new DNA-binding motif that recognizes a purine-rich core DNA sequence. *Genes Dev* 4:1451–1453. doi: 10.1101/gad.4.9.1451

Karu T (1999) Primary and secondary mechanisms of action of visible to near-IR radiation on cells. *J Photochem Photobiol B* 49:1–17. doi: 10.1016/S1011-1344(98)00219-X

Karu TI, Pyatibrat LV, Kolyakov SF, Afanasyeva NI (2005) Absorption measurements of a cell monolayer relevant to phototherapy: reduction of cytochrome c oxidase under near IR radiation. *J Photochem Photobiol B* 81:98–106. doi: 10.1016/j.jphotobiol.2005.07.002

Karu TI, Pyatibrat LV, Ryabykh TP (1997) Nonmonotonic behavior of the dose dependence of the radiation effect on cells in vitro exposed to pulsed laser radiation at lambda = 820 nm. *Lasers Surg Med* 21:485–492. doi: 10.1002/(sici)1096-9101(1997)21:5<485::aid-lsm11>3.0.co;2-8

Katoh-Semba R, Takeuchi IK, Semba R, Kato K (1997) Distribution of brain-derived neurotrophic factor in rats and its changes with development in the brain. *J Neurochem* 69:34–42. doi: 10.1046/j.1471-4159.1997.69010034.x

Katz DM (2005) Regulation of respiratory neuron development by neurotrophic and transcriptional signaling mechanisms. *Respir Physiol Neurobiol* 149:99–109. doi: 10.1016/j.resp.2005.02.007

Katz DM (2014) Brain-derived neurotrophic factor and Rett syndrome. *Handb Exp Pharmacol* 220:481–495. doi: 10.1007/978-3-642-45106-5_18

Kausz M, Murai Z, Arimura A, Köves K (1999) Distribution of pituitary adenylate cyclase activating polypeptide (PACAP) immunoreactive elements in the brain stem of rats studied by immunohistochemistry. *Neurobiology (Bp)* 7:19–31.

Kay AR, Blaustein MP (2019) Evolution of our understanding of cell volume regulation by the pump-leak mechanism. *J Gen Physiol* 151:407–416. doi: 10.1085/jgp.201812274

Kayadjanian N, Lee HS, Piña-Crespo J, Heinemann SF (2007) Localization of glutamate receptors to distal dendrites depends on subunit composition and the kinesin motor protein KIF17. *Mol Cell Neurosci* 34:219–230. doi: 10.1016/j.mcn.2006.11.001

Keilin D (1925) On cytochrome, a respiratory pigment, common to animals, yeast, and higher plants. *Proc R Soc Lond B Biol Sci* 98:312–339.

Keilin D, Hartree EF (1938) Cytochrome oxidase. *Proc R Soc Lond B Biol Sci* 125:171–186.

Kennedy C, Des Rosiers MH, Jehle JW, Reivich M, Sharpe F, Sokoloff L (1977) Mapping of functional neural pathways by autoradiographic survey of local metabolic rate with (14C)deoxyglucose. *Science* 187:850–853. doi: 10.1126/science.1114332

Kenyon KA, Bushong EA, Mauer AS, Strehler EE, Weinberg RJ, Burette AC (2010) Cellular and subcellular localization of the neuron-specific plasma membrane calcium ATPase PMCA1a in the rat brain. *J Comp Neurol* 518:3169–3183. doi: 10.1002/cne.22409

Khalimonchuk O, Kim H, Watts T, Perez-Martinez X, Winge DR (2012) Oligomerization of heme o synthase in cytochrome oxidase biogenesis is mediated by cytochrome oxidase assembly factor Coa2. *J Biol Chem* 287:26715–26726. doi: 10.1074/jbc.M112.377200

Khalimonchuk O, Rigby K, Bestwick M, Pierrel F, Cobine PA, Winge DR (2008) Pet191 is a cytochrome c oxidase assembly factor in Saccharomyces cerevisiae. *Eukaryot Cell* 7:1427–1431. doi: 10.1128/EC.00132-08

Kholwadwala D, Donnelly DF (1992) Maturation of carotid chemoreceptor sensitivity to hypoxia: in vitro studies in the newborn rat. *J Physiol* 453:461–473. doi: 10.1113/jphysiol.1992.sp019239

Kim J, Moody JP, Edgerly CK, Bordiuk OL, Cormier K, Smith K, Beal MF, Ferrante RJ (2010) Mitochondrial loss, dysfunction and altered dynamics in Huntington's disease. *Hum Mol Genet* 19:3919–3935. doi: 10.1093/hmg/ddq306. Epub 2010 Jul 21.

Kinney HC (2005) Abnormalities of the brainstem serotonergic system in the sudden infant death syndrome: a review. *Pediatr Dev Pathol* 8:507–524. doi: 10.1007/s10024-005-0067-y

Kirkinezos IG, Moraes CT (2001) Reactive oxygen species and mitochondrial diseases. *Semin Cell Dev Biol* 12:449–457. doi: 10.1006/scdb.2001.0282

Kirsch J (2006) Glycinergic transmission. *Cell Tissue Res* 326:535–540. doi: 10.1007/s00441-006-0261-x

Kish SJ, Bergeron C, Rajput A, Dozic S, Mastrogiacomo F, Chang LJ, Wilson JM, DiStefano LM, Nobrega JN (1992) Brain cytochrome oxidase in Alzheimer's disease. *J Neurochem* 59:776–779. doi: 10.1111/j.1471-4159.1992.tb09439.x

Kitamura K, Kano M (2013) Dendritic calcium signaling in cerebellar Purkinje cell. *Neural Netw* 47:11–17. doi: 10.1016/j.neunet.2012.08.001

Knudsen EI, Knudsen PF (1990) Sensitive and critical periods for visual calibration of sound localization by barn owls. *J Neurosci* 10:222–232. doi:10.1523/JNEUROSCI.10-01-00222.1990

Knutti D, Kressler D, Kralli A (2001) Regulation of the transcriptional coactivator PGC-1 via MAPK-sensitive interaction with a repressor. *Proc Natl Acad Sci U S A* 98:9713–9718. doi: 10.1073/pnas.171184698

Koludarova L, Battersby BJ (2024) Mitochondrial protein synthesis quality control. *Hum Mol Genet* 33(R1):R53–R60. doi: 10.1093/hmg/ddae012

Kordower JH, Rosenstein JM, Collier TJ, Burke MA, Chen EY, Li JM, Martel L, Levey AE, Mufson EJ, Freeman TB, Olanow CW (1996) Functional fetal nigral grafts in a patient with Parkinson's disease: chemoanatomic, ultrastructural, and metabolic studies. *J Comp Neurol* 370:203–230. doi: 10.1002/(SICI)1096-9861(19960624)370:2<203::AID-CNE6>3.0.CO;2-6

Kreile AK, Bonhoeffer T, Hübener M (2011) Altered visual experience induces instructive changes of orientation preference in mouse visual cortex. *J Neurosci* 31:13911–13920. doi: 10.1523/JNEUROSCI.2143-11.2011

Krishnan KJ, Ratnaike TE, De Gruyter HL, Jaros E, Turnbull DM (2012) Mitochondrial DNA deletions cause the biochemical defect observed in Alzheimer's disease. *Neurobiol Aging* 33:2210–2214. doi: 10.1016/j.neurobiolaging.2011.08.009

Krnjević K (1970) Glutamate and gamma-aminobutyric acid in brain. *Nature* 228:119–124. doi: 10.1038/228119a0

Krubitzer LA, Kaas JH (1990) Cortical connections of MT in four species of primates: areal, modular, and retinotopic patterns. *Vis Neurosci* 5:165–204. doi: 10.1017/s0952523800000213

Kuhn-Nentwig L, Kadenbach B (1985) Isolation and properties of cytochrome c oxidase from rat liver and quantification of immunological differences between isozymes from various rat tissues with subunit-specific antisera. *Eur J Biochem* 149:147–158. doi: 10.1111/j.1432-1033.1985.tb08905.x

Kuljis RO, Rakic P (1989) Neuropeptide Y-containing neurons are situated predominantly outside cytochrome oxidase puffs in macaque visual cortex. *Vis Neurosci* 2:57–62. doi: 10.1017/s0952523800004326

Kwong KK, Belliveau JW, Chesler DA, Goldberg IE, Weisskoff RM, Poncelet BP, Kennedy DN, Hoppel BE, Cohen MS, Turner R, Cheng H-M, Brady TJ, Rosen BR (1992) Dynamic magnetic resonance imaging of human brain activity during primary sensory stimulation. *Proc Natl Acad Sci USA* 89:5675–5679. doi: 10.1073/pnas.89.12.5675

Lachica EA, Casagrande VA (1992) Direct W-like geniculate projections to the cytochrome oxidase (CO) blobs in primate visual cortex: axon morphology. *J Comp Neurol* 319:141–158. doi: 10.1002/cne.903190112

LaManna JC, Kutina-Nelson KL, Hritz MA, Huang Z, Wong-Riley MT (1996) Decreased rat brain cytochrome oxidase activity after prolonged hypoxia. *Brain Res* 720:1–6. doi: 10.1016/0006-8993(95)01495-0

LaManna JC, Vendel LM, Farrell RM (1992) Brain adaptation to chronic hypobaric hypoxia in rats. *J Appl Physiol* (1985) 72:2238–2243. doi: 10.1152/jappl.1992.72.6.2238

LaMarco K, Thompson CC, Byers BP, Walton EM, McKnight SL (1991) Identification of Ets- and notch-related subunits in GA binding protein. *Science* 253:789–792. doi: 10.1126/science.1876836

Lambert JC, Ramirez A, Grenier-Boley B, Bellenguez C (2023) Step by step: towards a better understanding of the genetic architecture of Alzheimer's disease. *Mol Psychiatry* 28:2716–2727. doi: 10.1038/s41380-023-02076-1

Lamperti C, Diodato D, Lamantea E, Carrara F, Ghezzi D, Mereghetti P, Rizzi R, Zeviani M (2012) MELAS-like encephalomyopathy caused by a new pathogenic mutation in the mitochondrial DNA encoded cytochrome c oxidase subunit I. *Neuromuscul Disord* 22:990–994. doi: 10.1016/j.nmd.2012.06.003

Lania L, Majello B, De Luca P (1997) Transcriptional regulation by the Sp family proteins. *Int J Biochem Cell Biol* 29:1313–1323. doi: 10.1016/s1357-2725(97)00094-0

Lapuente-Brun E, Moreno-Loshuertos R, Acín-Pérez R, Latorre-Pellicer A, Colás C, Balsa E, Perales-Clemente E, Quirós PM, Calvo E, Rodríguez-Hernández MA, Navas P, Cruz R, Carracedo Á, López-Otín C, Pérez-Martos A, Fernández-Silva P, Fernández-Vizarra E, Enríquez JA (2013) Supercomplex assembly determines electron flux in the mitochondrial electron transport chain. *Science* 340:1567–1570. doi: 10.1126/science.1230381

Larsson NG, Barsh GS, Clayton DA (1997) Structure and chromosomal localization of the mouse mitochondrial transcription factor A gene (Tfam). *Mamm Genome* 8:139–140. doi: 10.1007/s003359900373

Lau CG, Takeuchi K, Rodenas-Ruano A, Takayasu Y, Murphy J, Bennett MV, Zukin RS (2009) Regulation of NMDA receptor Ca^{2+} signalling and synaptic plasticity. *Biochem Soc Trans* 37:1369–1374. doi: 10.1042/BST0371369

Lauer SM, Schneeweiß U, Brecht M, Ray S (2018) Visualization of cortical modules in flattened mammalian cortices. *J Vis Exp* 131:e56992. doi: 10.3791/56992

Laurie DJ, Wisden W, Seeburg PH (1992) The distribution of thirteen GABAA receptor subunit mRNAs in the rat brain. III. Embryonic and postnatal development. *J Neurosci* 12:4151–4172. doi: 10.1523/JNEUROSCI.12-11-04151.1992

Lavezzi AM, Corna M, Mingrone R, Matturri L (2010) Study of the human hypoglossal nucleus: normal development and morpho-functional alterations in sudden unexplained late fetal and infant death. *Brain Dev* 32:275–284. doi: 10.1016/j.braindev.2009.05.006

Leary SC, Cobine PA, Kaufman BA, Guercin GH, Mattman A, Palaty J, Lockitch G, Winge DR, Rustin P, Horvath R, Shoubridge EA (2007) The human cytochrome c oxidase assembly factors SCO1 and SCO2 have regulatory roles in the maintenance of cellular copper homeostasis. *Cell Metab* 5:9–20. doi: 10.1016/j.cmet.2006.12.001

Leary SC, Kaufman BA, Pellecchia G, Guercin GH, Mattman A, Jaksch M, Shoubridge EA (2004) Human SCO1 and SCO2 have independent, cooperative functions in copper delivery to cytochrome c oxidase. *Hum Mol Genet* 13:1839–1848. doi: 10.1093/hmg/ddh197

Leclerc N, Doré L, Parent A, Hawkes R (1990) The compartmentalization of the monkey and rat cerebellar cortex: zebrin I and cytochrome oxidase. *Brain Res* 506:70–78. doi: 10.1016/0006-8993(90)91200-z

Lee I, Salomon AR, Ficarro S, Mathes I, Lottspeich F, Grossman LI, Hüttemann M (2005) cAMP-dependent tyrosine phosphorylation of subunit I inhibits cytochrome c oxidase activity. *J Biol Chem* 280:6094–6100. doi: 10.1074/jbc.M411335200

Légrádi G, Shioda S, Arimura A (1994) Pituitary adenylate cyclase-activating polypeptide-like immunoreactivity in autonomic regulatory areas of the rat medulla oblongata. *Neurosci Lett* 193–196. doi: 10.1016/0304-3940(94)90080-9

Leigh D (1951) Subacute necrotizing encephalomyelopathy in an infant. *J Neurol Neurosurg Psychiatry* 14:216–221. doi: 10.1136/jnnp.14.3.216

Lessmann V (1998) Neurotrophin-dependent modulation of glutamatergic synaptic transmission in the mammalian CNS. *Gen Pharmacol* 31:667–674. doi: 10.1016/s0306-3623(98)00190-6

Letovsky J, Dynan WS (1989) Measurement of the binding of transcription factor Sp1 to a single GC box recognition sequence. *Nucleic Acids Res* 17:2639–2653. doi: 10.1093/nar/17.7.2639

Leung MC, Lo SC, Siu FK, So KF (2002) Treatment of experimentally induced transient cerebral ischemia with low energy laser inhibits nitric oxide synthase activity and up-regulates the expression of transforming growth factor-beta 1. *Lasers Surg Med* 31:283–288. doi: 10.1002/lsm.10096

Levelt CN, Hübener M (2012) Critical-period plasticity in the visual cortex. *Annu Rev Neurosci* 35:309–330. doi: 10.1146/annurev-neuro-061010-113813

Levine ES, Dreyfus CF, Black IB and Plummer MR (1995) Brain-derived neurotrophic factor rapidly enhances synaptic transmission in hippocampal neurons via postsynaptic tyrosine kinase receptors. *Proc Natl Acad Sci USA* 92:8074–8077. doi: 10.1073/pnas.92.17.8074

Levitt JB, Yoshioka T, Lund JS (1994) Intrinsic cortical connections in macaque visual area V2: evidence for interaction between different functional streams. *J Comp Neurol* 342:551–570. doi: 10.1002/cne.903420405

Li Y, D'Aurelio M, Deng JH, Park JS, Manfredi G, Hu P, Lu J, Bai Y (2007) An assembled complex IV maintains the stability and activity of complex I in mammalian mitochondria. *J Biol Chem* 282:17557–17562. doi: 10.1074/jbc.M701056200

Li Y, Park JS, Deng JH, Bai Y (2006) Cytochrome c oxidase subunit IV is essential for assembly and respiratory function of the enzyme complex. *J Bioenerg Biomembr* 38:283–291. doi: 10.1007/s10863-006-9052-z

Liang HL, Dhar SS, Wong-Riley MTT (2010) p38 mitogen-activated protein kinase and calcium channels mediate signaling in depolarization-induced activation of peroxisome proliferators-activated receptor gamma coactivator-1alpha in neurons. *J Neurosci Res* 88:640–649. doi: 10.1002/jnr.22222

Liang HL, Ongwijitwat S, Wong-Riley MTT (2006b) Bigenomic functional regulation of all 13 cytochrome c oxidase subunit transcripts in rat neurons in vitro and in vivo. *Neuroscience* 140:177–190. doi: 10.1016/j.neuroscience.2006.01.056

Liang HL, Whelan H, Eells J, Meng H, Buchmann E, Lerch-Gaggl A, Wong-Riley M (2006a) Photobiomodulation partially rescues visual cortical neurons from cyanide-induced apoptosis. *Neuroscience* 139:639–649. doi: 10.1016/j.neuroscience.2005.12.047

Liang HL, Whelan HT, Eells JT, Wong-Riley MTT (2008) Near-infrared light via light-emitting diode treatment is therapeutic against rotenone- and MPP^+-induced neurotoxicity. *Neuroscience* 153:963–974. doi: 10.1016/j.neuroscience.2008.03.042

Liang HL, Wong-Riley MTT (2006) Activity-dependent regulation of nuclear respiratory factor-1, nuclear respiratory factor-2, and peroxisome proliferator-activated receptor gamma coactivator-1 in neurons. *NeuroReport* 17:401–405. doi: 10.1097/01.wnr.0000204980.98876.11

Liang T, Bernal B, Qi W, Ikeno Y, Salmon AB, Kwok D, Michalek J, Zhou M, Bai Y (2025) Robust assembly of electron transfer chain in the brain of naked mole-rats during aging. *Organelle* 1. doi: 10.61747/0ifp.202408002

Lightowlers R, Takamiya S, Wessling R, Lindorfer M, Capaldi RA (1989) Cloning and sequencing of the cDNA for a 13th different subunit (IHQ) of beef heart cytochrome c oxidase. *J Biol Chem* 264:16858–16861.

Lim SC, Smith KR, Stroud DA, Compton AG, Tucker EJ, Dasvarma A, Gandolfo LC, Marum JE, McKenzie M, Peters HL, Mowat D, Procopis PG, Wilcken B, Christodoulou J, Brown GK, Ryan MT, Bahlo M, Thorburn DR (2014) A founder mutation in PET100 causes isolated complex IV deficiency in Lebanese individuals with Leigh syndrome. *Am J Hum Genet* 94:209–222. doi: 10.1016/j.ajhg.2013.12.015

Lima PLV, Pereira CV, Nissanka N, Arguello T, Gavini G, Maranduba CMDC, Diaz F, Moraes CT (2019) Photobiomodulation enhancement of cell proliferation at 660 nm does not require cytochrome c oxidase. *J Photochem Photobiol B* 194:71–75. doi: 10.1016/j.jphotobiol.2019.03.015

Lin J, Puigserver P, Donovan J, Tarr P, Spiegelman BM (2002) Peroxisome proliferator-activated receptor gamma coactivator 1beta (PGC-1beta), a novel PGC-1-related transcription coactivator associated with host cell factor. *J Biol Chem* 277:1645–1648. doi: 10.1074/jbc.C100631200

Linden DC, Guillery RW, Cucchiaro J (1981) The dorsal lateral geniculate nucleus of the normal ferret and its postnatal development. *J Comp Neurol* 203:189–211. doi: 10.1002/cne.902030204

Lindsay AD, Feldman JL (1993) Modulation of respiratory activity of neonatal rat phrenic motoneurones by serotonin. *J Physiol* 461:213–233. doi: 10.1113/jphysiol.1993.sp019510

Ling JQ, Li T, Hu JF, Vu TH, Chen HL, Qiu XW, Cherry AM, Hoffman AR (2006) CTCF mediates interchromosomal colocalization between Igf2/H19 and Wsb1/Nf1. *Science* 312:269–272. doi: 10.1126/science.1123191

Lithgow T (2000) Targeting of proteins to mitochondria. *FEBS Lett* 476:22–26. doi: 10.1016/s0014-5793(00)01663-x

Liu DM, Cuevas J, Adams DJ (2000) VIP and PACAP potentiation of nicotinic ACh-evoked currents in rat parasympathetic neurons is mediated by G-protein activation. *Eur J Neurosci* 12:2243–2251. doi: 10.1046/j.1460-9568.2000.00116.x

Liu J, Wei X, Zhao C, Hu S, Duan J, Ju G, Wong-Riley MTT, Liu Y (2011) 5-HT induces enhanced phrenic nerve activity via 5-HT_{2A} receptor/PKC mechanism in anesthetized rats. *Eur J Pharmacol* 657:67–75. doi: 10.1016/j.ejphar.2011.01.048

Liu Q, Fehring C, Lowry TF, Wong-Riley MTT (2009) Postnatal development of metabolic rate during normoxia and acute hypoxia in rats: implication for a sensitive period. *J Appl Physiol* 106:1212–1222. doi: 10.1152/japplphysiol.90949.2008

Liu Q, Kim J, Cinotte J, Homolka P, Wong-Riley MTT (2003) Carotid body denervation effect on cytochrome oxidase activity in pre- Bötzinger complex of developing rats. *J Appl Physiol* 94:1115–1121. doi: 10.1152/japplphysiol.00765.2002

Liu Q, Lowry T, Wong-Riley MTT (2006) Postnatal changes in ventilation during normoxia and acute hypoxia in the rat: implication for a sensitive period. *J Physiol* 577.3:957–970. doi: 10.1113/jphysiol.2006.121970

Liu Q, Wong-Riley MTT (2002) Postnatal expression of neurotransmitters, receptors, and cytochrome oxidase in the rat pre-Bötzinger complex. *J Appl Physiol* 92:923–934. doi: 10.1152/japplphysiol.00977.2001

Liu Q, Wong-Riley MTT (2003) Postnatal changes in cytochrome oxidase expressions in brain stem nuclei of rats: implications for sensitive periods. *J Appl Physiol* 95:2285–2291. doi: 10.1152/japplphysiol.00638.2003

Liu Q, Wong-Riley MTT (2004) Developmental changes in the expression of $GABA_A$ receptor subunits α1, α2, and α3 in the rat pre-Bötzinger complex. *J Appl Physiol* 96:1825–1831. doi: 10.1152/japplphysiol.01264.2003

Liu Q, Wong-Riley MTT (2005) Postnatal developmental expressions of neurotransmitters and receptors in various brain stem nuclei of rats. *J Appl Physiol* 98:1442–1457. doi: 10.1152/japplphysiol.01301.2004

Liu Q, Wong-Riley MTT (2006) Developmental changes in the expression of $GABA_A$ receptor subunits α1, α2, and α3 in brain stem nuclei of rats. *Brain Res* 1098:129–138. doi: 10.1016/j.brainres.2006.05.001

Liu Q, Wong-Riley MTT (2008) Postnatal changes in the expression of serotonin 2A receptors in various brain stem nuclei of the rat. *J Appl Physiol* 104:1801–1808. doi: 10.1152/japplphysiol.00057.2008

Liu Q, Wong-Riley MTT (2010a) Postnatal changes in the expressions of serotonin 1A, 1B, and 2A receptors in ten brain stem nuclei of the rat: implication for a sensitive period. *Neuroscience* 165:61–78. doi: 10.1016/j.neuroscience.2009.09.078

Liu Q, Wong-Riley MTT (2010b) Postnatal changes in tryptophan hydroxylase and serotonin transporter immunoreactivity in multiple brainstem nuclei of the rat: implications for a sensitive period. *J Comp Neurol* 518:1082–1097. doi: 10.1002/cne.22265

Liu Q, Wong-Riley MTT (2010c) Postnatal development of N-methyl-D-aspartate receptor subunits 2A, 2B, 2C, 2D, and 3B immunoreactivity in brain stem respiratory nuclei of the rat. *Neuroscience* 171:637–654. doi: 10.1016/j.neuroscience.2010.09.055

Liu Q, Wong-Riley MTT (2012) Postnatal development of Na^+-K^+-$2Cl^-$ co-transporter 1 and K^+-Cl^- co-transporter 2 immunoreactivity in multiple brain stem respiratory nuclei of the rat. *Neuroscience* 210: 1–20. doi: 10.1016/j.neuroscience.2012.03.018

Liu Q, Wong-Riley MTT (2013a) Postnatal development of glycine receptor subunits α1, α2, α3, and β immunoreactivity in multiple brain stem respiratory-related nuclei of the rat. *Brain Res* 1538:1–16. doi: 10.1016/j.brainres.2013.09.028

Liu Q, Wong-Riley MTT (2013b) Postnatal development of brain-derived neurotrophic factor (BDNF) and tyrosine protein kinase B (TrkB) receptor immunoreactivity in multiple brain stem respiratory-related nuclei of the rat. *J Comp Neurol* 521:109–129. doi: 10.1002/cne.23164

Liu Q, Wong-Riley MTT (2013c) Gender considerations in ventilatory and metabolic development in rats: special emphasis on the critical period. *Resp Physiol Neurobiol* 188:200–207. doi: 10.1016/j.resp.2013.06.013

Liu Q, Wong-Riley MTT (2019) Pituitary adenylate cyclase-activating polypeptide: postnatal development in multiple brain stem respiratory-related nuclei in the rat. *Respir Physiol Neurobiol* 259:149–155. doi: 10.1016/j.resp.2018.10.005

Liu QR, Walther D, Drgon T, Polesskaya O, Lesnick TG, Strain KJ, de Andrade M, Bower JH, Maraganore DM, Uhl GR (2005) Human brain derived neurotrophic factor (BDNF) genes, splicing patterns, and assessments of associations with substance abuse and Parkinson's Disease. *Am J Med Genet B Neuropsychiatr Genet* 134B:93–103. doi: 10.1002/ajmg.b.30109

Liu S, Wilcox DA, Sieber-Blum M, Wong-Riley M (1990) Developing neural crest cells in culture: correlation of cytochrome oxidase activity with SSEA-1 and dopamine-beta-hydroxylase immunoreactivity. *Brain Res* 535:271–280. doi: 10.1016/0006-8993(90)91610-s

Liu S, Wong-Riley MTT (1994) Nuclear-encoded mitochondrial precursor protein: intramitochondrial delivery to dendrites and axon terminals of neurons and regulation by neuronal activity. *J Neurosci* 14:5338–5351. doi: 10.1523/JNEUROSCI.14-09-05338.1994

Liu SJ, Zukin RS (2007) Ca2+-permeable AMPA receptors in synaptic plasticity and neuronal death. *Trends Neurosci* 30:126–134. doi: 10.1016/j.tins.2007.01.006

Liu W, Jiang Y, Li G, Huang D, Qin T (2024) Oxidative phosphorylation related gene COA6 is a novel indicator for the prognosis and immune response in lung adenocarcinoma. *Sci Rep* 14:25970. doi: 10.1038/s41598-024-77775-y

Liu YY, Wong-Riley MTT (2001) Developmental study of cytochrome oxidase activity in the brain stem respiratory nuclei of postnatal rats. *J Appl Physiol* 90:685–694. doi: 10.1152/jappl.2001.90.2.685

Livingstone M, Hubel D (1988) Segregation of form, color, movement, and depth: anatomy, physiology, and perception. *Science* 240:740–749. doi: 10.1126/science.3283936

Livingstone MS, Hubel DH (1982) Thalamic inputs to cytochrome oxidase-rich regions in monkey visual cortex. *Proc Natl Acad Sci USA* 79:6098–6101. doi: 10.1073/pnas.79.19.6098

Livingstone MS, Hubel DH (1984a) Anatomy and physiology of a color system in the primate visual cortex. *J Neurosci* 4:309–356. doi: 10.1523/JNEUROSCI.04-01-00309.1984

Livingstone MS, Hubel DH (1984b) Specificity of intrinsic connections in primate primary visual cortex. *J Neurosci* 4:2830–2835. doi: 10.1523/JNEUROSCI.04-11-02830.1984

Lomax MI, Bachman NJ, Nasoff MS, Caruthers MH, Grossman LI (1984) Isolation and characterization of a cDNA clone for bovine cytochrome c oxidase subunit IV. *Proc Natl Acad Sci USA* 81:6295–6299. doi: 10.1073/pnas.81.20.6295

Lomax MI, Riggs PK, Womack JE (1995) Structure and chromosomal location of the bovine gene for the heart muscle isoform of cytochrome c oxidase subunit VIII. *Mamm Genome* 6:118–122. doi: 10.1007/BF00303255

Lomvardas S, Barnea G, Pisapia DJ, Mendelsohn M, Kirkland J, Axel R (2006) Interchromosomal interactions and olfactory receptor choice. *Cell* 126:403–413. doi: 10.1016/j.cell.2006.06.035

Lowry OH (1975) Energy metabolism in brain and its control. In: Ingvar DH, Lassen NA (eds) *Brain Work, Alfred Benzon Symposium VIII*. Academic Press, New York, pp 48–64.

Ludwig B (1980) Heme aa3-type cytochrome c oxidase from bacteria. *Biochim Biophys Acta* 594:177–189. doi: 10.1016/0304-4173(80)90008-7

Ludwig B, Bender E, Arnold S, Hüttemann M, Lee I, Kadenbach B (2001) Cytochrome c oxidase and the regulation of oxidative phosphorylation. *ChemBioChem.* 2:392–403. doi: 10.1002/1439-7633(20010601)2:6<392::AID-CBIC392>3.0.CO;2-N

Lund JS (1988) Anatomical organization of macaque monkey striate visual cortex. *Annu Rev Neurosci* 11:253–288. doi: 10.1146/annurev.ne.11.030188.001345

Lund JS, Angelucci A, Bressloff PC (2003) Anatomical substrates for functional columns in macaque monkey primary visual cortex. *Cereb Cortex* 13:15–24. doi: 10.1093/cercor/13.1.15

Lund JS, Lund RD, Hendrickson AE, Bunt AH, Fuchs AF (1975) The origin of efferent pathways from the primary visual cortex, area 17, of the macaque monkey as shown by retrograde transport of horseradish peroxidase. *J Comp Neurol* 164:287–303. doi: 10.1002/cne.901640303

Lund RD (1965) Uncrossed visual pathways of Hooded and Albino Rats. *Science* 149:1506–1507. doi: 10.1126/science.149.3691.1506

Luo XG, Hevner RF, Wong-Riley MTT (1989) Double labeling of cytochrome oxidase and gamma-aminobutyric acid in central nervous system neurons of adult cats. *J Neurosci Methods* 30:189–195. doi: 10.1016/0165-0270(89)90129-5

Lyon DC, Kaas JH (2002) Evidence from V1 connections for both dorsal and ventral subdivisions of V3 in three species of New World monkeys. *J Comp Neurol* 449:281–297. doi: 10.1002/cne.10297

Macdonald DS, Weerapura M, Beazely MA, Martin L, Czerwinski W, Roder JC, Orser BA, MacDonald JF (2005) Modulation of NMDA receptors by pituitary adenylate cyclase activating peptide in CA1 neurons requires G alpha q, protein kinase C, and activation of Src. *J Neurosci* 25:11374–11384. doi: 10.1523/JNEUROSCI.3871-05.2005

Mailloux RJ (2020) An update on mitochondrial reactive oxygen species production. *Antioxidants (Basel)* 9:472. doi: 10.3390/antiox9060472

Manita S, Miyakawa H, Kitamura K, Murayama M (2017) Dendritic spikes in sensory perception. *Front Cell Neurosci* 11:29. doi: 10.3389/fncel.2017.00029

Marcondes M, Farias MF, Pary LF, Fiorani M, Lima B, Amorim AKJ, Gattass R (2024) Impact of monocular retinal lesions on blob size in adult human V1. *J Cogn Neurosci* 36:2652–2666. doi: 10.1162/jocn_a_02133

Marcondes M, Rosa MGP, Fiorani M, Lima B, Gattass R (2019) Distribution of cytochrome oxidase-rich patches in human primary visual cortex. *J Comp Neurol* 527:614–624. doi: 10.1002/cne.24435

Markham A, Cameron I, Franklin P, Spedding M (2004) BDNF increases rat brain mitochondrial respiratory coupling at complex I, but not complex II. *Eur J Neurosci* 20:1189–1196. doi: 10.1111/j.1460-9568.2004.03578.x

Martin CT, Scholes CP, Chan SI (1988) On the nature of cysteine coordination to CuA in cytochrome c oxidase. *J Biol Chem* 263:8420–8429.

Martin JL, Finsterwald C (2011) Cooperation between BDNF and glutamate in the regulation of synaptic transmission and neuronal development. *Commun Integr Biol* 4:14–16. doi: 10.4161/cib.13761

Martinez Lyons A, Ardissone A, Reyes A, Robinson AJ, Moroni I, Ghezzi D, Fernandez-Vizarra E, Zeviani M (2016) COA7 (C1orf163/RESA1) mutations associated with mitochondrial leukoencephalopathy and cytochrome c oxidase deficiency. *J Med Genet* 53:846–849. doi: 10.1136/jmedgenet-2016-104194

Mason MG, Nicholls P, Cooper CE (2014) Re-evaluation of the near infrared spectra of mitochondrial cytochrome c oxidase: implications for non invasive in vivo monitoring of tissues. *Biochim Biophys Acta* 1837:1882–1891. doi: 10.1016/j.bbabio.2014.08.005

Massa V, Fernandez-Vizarra E, Alshahwan S, Bakhsh E, Goffrini P, Ferrero I, Mereghetti P, D'Adamo P, Gasparini P, Zeviani M (2008) Severe infantile encephalomyopathy caused by a mutation in COX6B1, a nucleus-encoded subunit of cytochrome c oxidase. *Am J Hum Genet* 82:1281–1289. doi: 10.1016/j.ajhg.2008.05.002

Maunsell JH, Newsome WT (1987) Visual processing in monkey extrastriate cortex. *Annu Rev Neurosci* 10:363–401. doi: 10.1146/annurev.ne.10.030187.002051

May PJ (2006) The mammalian superior colliculus: laminar structure and connections. *Prog Brain Res* 151:321–378. doi: 10.1016/S0079-6123(05)51011-2

Mayer CA, Di Fiore JM, Martin RJ, MacFarlane PM (2014) Vulnerability of neonatal respiratory neural control to sustained hypoxia during a uniquely sensitive window of development. *J Appl Physiol* 116:514–521. doi: 10.1152/japplphysiol.00976.2013

McAllister AK, Lo DC, Katz LC (1995) Neurotrophins regulate dendritic growth in developing visual cortex. *Neuron* 15:791–803. doi: 10.1016/0896-6273(95)90171-x

McEntee WJ, Crook TH (1993) Glutamate: its role in learning, memory, and the aging brain. *Psychopharmacology (Berl)* 111:391–401. doi: 10.1007/BF02253527

McEwen JE, Hong KH, Park S, Preciado GT (1993) Sequence and chromosomal localization of two PET genes required for cytochrome c oxidase assembly in Saccharomyces cerevisiae. *Curr Genet* 23:9–14. doi: 10.1007/BF00336742

McGinty DJ, Harper RM (1976) Dorsal raphe neurons: depression of firing during sleep in cats. *Brain Res* 101:569–575. doi: 10.1016/0006-8993(76)90480-7

McStay GP, Su CH, Thomas SM, Xu JT, Tzagoloff A (2013) Characterization of assembly intermediates containing subunit 1 of yeast cytochrome oxidase. *J Biol Chem* 288:26546–26556. doi: 10.1074/jbc.M113.498592

Meng H, Liang HL, Wong-Riley MTT (2007) Quantitative immuno-electron microscopic analysis of depolarization-induced expression of PGC-1α in cultured rat visual cortical neurons. *Brain Res* 1175:10–16. doi: 10.1016/j.brainres.2007.07.063

Merante F, Petrova-Benedict R, MacKay N, Mitchell G, Morin C, DeBraekeleer M, Laframboise R, Gagne R, Robinson BH (1993) A biochemically distinct form of cytochrome oxidase (COX) deficiency in the Saguenay-Lac-Saint-Jean region of Québec. *Am J Hum Genet* 53:481–487.

Messier ML, Li A, Nattie EE (2004) Inhibition of medullary raphe serotonergic neurons has age-dependent effects on the CO2 response in newborn piglets. *J Appl Physiol* 96:1909–1919. doi: 10.1152/japplphysiol.00805.2003

Mester E, Nagylucskay S, Tisza S, Mester A (1978) Stimulation of wound healing by means of laser rays. Part III--Investigation of the effect on immune competent cells. *Acta Chir Acad Sci Hung* 19:163–170.

Mester E, Spiry T, Szende B, Tota JG (1971) Effect of laser rays on wound healing. *Am J Surg* 122:532–535. doi: 10.1016/0002-9610(71)90482-x

Mick DU, Dennerlein S, Wiese H, Reinhold R, Pacheu-Grau D, Lorenzi I, Sasarman F, Weraarpachai W, Shoubridge EA, Warscheid B, Rehling P (2012) MITRAC links mitochondrial protein translocation to respiratory-chain assembly and translational regulation. *Cell* 151:1528–1541. doi: 10.1016/j.cell.2012.11.053

Mick DU, Vukotic M, Piechura H, Meyer HE, Warscheid B, Deckers M, Rehling P (2010) Coa3 and Cox14 are essential for negative feedback regulation of COX1 translation in mitochondria. *J Cell Biol* 191:141–154. doi: 10.1083/jcb.201007026

Miele A, Dekker J (2009) Mapping cis- and trans- chromatin interaction networks using chromosome conformation capture (3C). *Methods Mol Biol* 464:105–121. doi: 10.1007/978-1-60327-461-6_7

Milenkovic D, Misic J, Hevler JF, Molinié T, Chung I, Atanassov I, Li X, Filograna R, Mesaros A, Mourier A, Heck AJR, Hirst J, Larsson NG (2023) Preserved respiratory chain capacity and physiology in mice with profoundly reduced levels of mitochondrial respirasomes. *Cell Metab* 35:1799–1813.e7. doi: 10.1016/j.cmet.2023.07.015

Misceo D, Strømme P, Bitarafan F, Chawla MS, Sheng Y, Bach de Courtade SM, Eide L, Frengen E (2024) Biallelic NDUFA4 deletion causes mitochondrial complex IV deficiency in a patient with leigh syndrome. *Genes (Basel)* 15:500. doi: 10.3390/genes15040500

Mishmar D, Ruiz-Pesini E, Mondragon-Palomino M, Procaccio V, Gaut B, Wallace DC (2006) Adaptive selection of mitochondrial complex I subunits during primate radiation. *Gene* 378:11–18. doi: 10.1016/j.gene.2006.03.015

Mitchell GS, Johnson SM (2003) Neuroplasticity in respiratory motor control. *J Appl Physiol* 94:358–374. doi: 10.1152/japplphysiol.00523.2002

Mitchell P (1961) Coupling of phosphorylation to electron and hydrogen transfer by a chemi-osmotic type of mechanism. *Nature* 191:144–148. doi: 10.1038/191144a0

Miyata A, Arimura A, Dahl RR, Minamino N, Uehara A, Jiang L, Culler MD, Coy DH (1989) Isolation of a novel 38 residue-hypothalamic polypeptide which stimulates adenylate cyclase in pituitary cells. *Biochem Biophys Res Commun* 164:567–754. doi: 10.1016/0006-291x(89)91757-9

Mizuno Y, Ohta S, Tanaka M, Takamiya S, Suzuki K, Sato T, Oya H, Ozawa T, Kagawa Y (1989) Deficiencies in complex I subunits of the respiratory chain in Parkinson's disease. *Biochem Biophys Res Commun* 163:1450–1455. doi: 10.1016/0006-291x(89)91141-8

Mjaatvedt AE, Wong-Riley MTT (1988) Relationship between synaptogenesis and cytochrome oxidase activity in Purkinje cells of the developing rat cerebellum. *J Comp Neurol* 277:155–182. doi: 10.1002/cne.902770202

Möhler H (2006) GABA(A) receptor diversity and pharmacology. *Cell Tissue Res* 326:505–516. doi: 10.1007/s00441-006-0284-3

Moi P, Chan K, Asunis I, Cao A, Kan YW (1994) Isolation of NF-E2-related factor 2 (Nrf2), a NF-E2-like basic leucine zipper transcriptional activator that binds to the tandem NF-E2/AP1 repeat of the beta-globin locus control region. *Proc Natl Acad Sci USA* 91:9926–9930. doi: 10.1073/pnas.91.21.9926

Monyer H, Burnashev N, Laurie DJ, Sakmann B, Seeburg PH (1994) Developmental and regional expression in the rat brain and functional properties of four NMDA receptors. *Neuron* 12:529–540. doi: 10.1016/0896-6273(94)90210-0

Mori H, Mishina M (1995) Structure and function of the NMDA receptor channel. *Neuropharmacology* 34:1219–1237. doi: 10.1016/0028-3908(95)00109-j

Mortola JP (1987) Dynamics of breathing in newborn mammals. *Physiol Rev* 67:187–243. doi: 10.1152/physrev.1987.67.1.187

Mortola JP (2004) Implications of hypoxic hypometabolism during mammalian ontogenesis. *Respir Physiol Neurobiol* 141:345–356. doi: 10.1016/j.resp.2004.01.011

Moser MB, Rowland DC, Moser EI (2015) Place cells, grid cells, and memory. *Cold Spring Harb Perspect Biol* 7:a021808. doi: 10.1101/cshperspect.a021808

Mouradian GC Jr, Alvarez-Argote S, Gorzek R, Thuku G, Michkalkiewicz T, Wong-Riley MTT, Konduri GG, Hodges MR. (2019) Acute and chronic changes in the control of breathing in a rat model of bronchopulmonary dysplasia. *Am J Physiol Lung Cell Mol Physiol* 316:L506–L518. doi: 10.1152/ajplung.00086.2018

Mu L, Xia D, Michalkiewicz T, Hodges M, Mouradian G, Konduri GG, Wong-Riley MTT (2018) Effects of neonatal hyperoxia on the critical period of postnatal development of neurochemical expressions in brain stem respiratory-related nuclei in the rat. *Physiol Rep* 6:e13627. doi: 10.14814/phy2.13627

Mullur R, Liu YY, Brent GA (2014) Thyroid hormone regulation of metabolism. *Physiol Rev* 94:355–382. doi: 10.1152/physrev.00030.2013

Murata K, Kinoshita T, Ishikawa T, Kuroda K, Hoshi M, Fukazawa Y (2020) Region- and neuronal-subtype-specific expression of Na,K-ATPase alpha and beta subunit isoforms in the mouse brain. *J Comp Neurol* 528:2654–2678. doi: 10.1002/cne.24924

Murphy KM, Beston BR, Boley PM, Jones DG (2005) Development of human visual cortex: a balance between excitatory and inhibitory plasticity mechanisms. *Dev Psychobiol* 46:209–221. doi: 10.1002/dev.20053

Murphy KM, Jones DG, Van Sluyters RC (1995) Cytochrome-oxidase blobs in cat primary visual cortex. *J Neurosci* 15:4196–4208. doi: 10.1523/JNEUROSCI.15-06-04196.1995

Myers SJ, Peters J, Huang Y, Comer MB, Barthel F, Dingledine R (1998) Transcriptional regulation of the GluR2 gene: neural-specific expression, multiple promoters, and regulatory elements. *J Neurosci* 18:6723–6739. doi: 10.1523/JNEUROSCI.18-17-06723.1998

Nagata Y, Fujita K, Yamauchi M, Kato T, Ando M, Honda M (1998) Neurochemical changes in the spinal cord in degenerative motor neuron diseases. *Mol Chem Neuropathol* 33:237–247. doi: 10.1007/BF02815185

Nair B, Johar K, Priya A, Wong-Riley MTT (2016) Specificity protein 4 (Sp4) transcriptionally regulates inhibitory GABAergic receptors in neurons. *Biochim Biophys Acta Mol Cell Res* 1863:1–9. doi: 10.1016/j.bbamcr.2015.10.005

Nair B, Wong-Riley MTT (2016) Transcriptional regulation of brain-derived neurotrophic factor coding exon IX: role of nuclear respiratory factor 2. *J Biol Chem* 291:22583–22593. doi: 10.1074/jbc.M116.742304

Nairuz T, Sangwoo-Cho, Lee JH (2024) Photobiomodulation therapy on brain: pioneering an innovative approach to revolutionize cognitive dynamics. *Cells* 13:966. doi: 10.3390/cells13110966

Nakamura H, Gattass R, Desimone R, Ungerleider LG (1993) The modular organization of projections from areas V1 and V2 to areas V4 and TEO in macaques. *J Neurosci* 13:3681–3691. doi: 10.1523/JNEUROSCI.13-09-03681.1993

Nakamura H, Le WR, Wakita M, Mikami A, Itoh K (2004) Projections from the cytochrome oxidase modules of visual area V2 to the ventral posterior area in the macaque. *Exp Brain Res* 155:102–110. doi: 10.1007/s00221-003-1698-8

Nakanishi S (1992) Molecular diversity of glutamate receptors and implications for brain function. *Science* 258:597–603. doi: 10.1126/science.1329206

Namslauer I, Brzezinski P (2009) A mitochondrial DNA mutation linked to colon cancer results in proton leaks in cytochrome c oxidase. *Proc Natl Acad Sci USA* 106:3402–3407. doi: 10.1073/pnas.0811450106

Napiwotzki J, Kadenbach B (1998) Extramitochondrial ATP/ADP ratios regulate cytochrome c oxidase activity via binding to the cytosolic domain of subunit IV. *Biol Chem* 379:335–339. doi: 10.1515/bchm.1998.379.3.335

Nascimento-Silva S, Gattass R, Fiorani M Jr, Sousa AP (2003) Three streams of visual information processing in V2 of Cebus monkey. *J Comp Neurol* 466:104–118. doi: 10.1002/cne.10878

Nasir J, Floresco SB, O'Kusky JR, Diewert VM, Richman JM, Zeisler J, Borowski A, Marth JD, Phillips AG, Hayden MR (1995) Targeted disruption of the Huntington's disease gene results in embryonic lethality and behavioral and morphological changes in heterozygotes. *Cell* 81:811–823. doi: 10.1016/0092-8674(95)90542-1

Nassi JJ, Callaway EM (2009) Parallel processing strategies of the primate visual system. *Nat Rev Neurosci* 10:360–372. doi: 10.1038/nrn2619

Nataraj K, Le Roux N, Nahmani M, Lefort S, Turrigiano G (2010) Visual deprivation suppresses L5 pyramidal neuron excitability by preventing the induction of intrinsic plasticity. *Neuron* 68:750–762. doi: 10.1016/j.neuron.2010.09.033

Nathan C, Xie QW (1994) Nitric oxide synthases: roles, tolls, and controls. *Cell* 78:915–918. doi: 10.1016/0092-8674(94)90266-6

Nattie EE (2001) Central chemosensitivity, sleep, and wakefulness. *Respir Physiol* 129:257–268. doi: 10.1016/s0034-5687(01)00295-x

Neupert W, Herrmann JM (2007) Translocation of proteins into mitochondria. *Annu Rev Biochem* 76:723–749. doi: 10.1146/annurev.biochem.76.052705.163409

Nicholls P (1975) Formate as an inhibitor of cytochrome c oxidase. *Biochem Biophys Res Commun* 67:610–616. doi: 10.1016/0006-291x(75)90856-6

Nie F, Wong-Riley M (1999) Nuclear respiratory factor-2 subunit protein: correlation with cytochrome oxidase and regulation by functional activity in the monkey primary visual cortex. *J Comp Neurol* 404:310–320. doi: 10.1002/(sici)1096-9861(19990215)404:3<310::aid-cne3>3.0.co;2-4

Nie F, Wong-Riley MTT (1995) Double labeling of GABA and cytochrome oxidase in the macaque visual cortex: quantitative EM analysis. *J Comp Neurol* 356:115–131. doi: 10.1002/cne.903560108

Nie F, Wong-Riley MTT (1996a) Differential glutamatergic innervation in cytochrome oxidase-rich and -poor regions of the macaque striate cortex: quantitative EM analysis of neurons and neuropil. *J Comp Neurol* 369:571–590. doi: 10.1002/(SICI)1096-9861(19960610)369:4<571::AID-CNE7>3.0.CO;2-1

Nie F, Wong-Riley MTT (1996b) Metabolic and neurochemical plasticity of gamma-aminobutyric acid-immunoreactive neurons in the adult macaque striate cortex following monocular impulse blockade: quantitative electron microscopic analysis. *J Comp Neurol* 370:350–366. doi: 10.1002/(SICI)1096-9861(19960701)370:3<350::AID-CNE6>3.0.CO;2-3

Nijtmans LG, Taanman JW, Muijsers AO, Speijer D, Van den Bogert C (1998) Assembly of cytochrome-c oxidase in cultured human cells. *Eur J Biochem* 254:389–394. doi: 10.1046/j.1432-1327.1998.2540389.x

Nohl H (1993) Involvement of free radicals in ageing: a consequence or cause of senescence. *Br Med Bull* 49:653–667. doi: 10.1093/oxfordjournals.bmb.a072638

Novack GV, Galeano P, Castaño EM, Morelli L (2020) Mitochondrial supercomplexes: physiological organization and dysregulation in age-related neurodegenerative disorders. *Front Endocrinol (Lausanne)* 11:600. doi: 10.3389/fendo.2020.00600

Núñez-Abades PA, Pásaro R, Bianchi AL (1992) Study of the topographical distribution of different populations of motoneurons within rat's nucleus ambiguus, by means of four different fluorochromes. *Neurosci Lett* 135:103–107. doi: 10.1016/0304-3940(92)90146-x

Nusser Z, Sieghart W, Benke D, Fritschy JM, Somogyi P (1996) Differential synaptic localization of two major gamma-aminobutyric acid type A receptor alpha subunits on hippocampal pyramidal cells. *Proc Natl Acad Sci USA* 93(21):11939–11944. doi: 10.1073/pnas.93.21.11939

Nývltová E, Dietz JV, Seravalli J, Khalimonchuk O, Barrientos A (2022) Coordination of metal center biogenesis in human cytochrome c oxidase. *Nat Commun* 13:3615. doi: 10.1038/s41467-022-31413-1

Oefele M, Hau M, Ruuskanen S, Casagrande S (2024) Mitochondrial function is enhanced by thyroid hormones during zebra finch development. *R Soc Open Sci* 11:240417. doi: 10.1098/rsos.240417

Ogawa S, Tank DW, Menon R, Ellermann JM, Kim SG, Merkle H, Ugurbil K (1992) Intrinsic signal changes accompanying sensory stimulation: functional brain mapping with magnetic resonance imaging. *Proc Natl Acad Sci USA* 89:5951–5955. doi: 10.1073/pnas.89.13.5951

Ogier M, Wang H, Hong E, Wang Q, Greenberg ME, Katz DM (2007) Brain-derived neurotrophic factor expression and respiratory function improve after ampakine treatment in a mouse model of Rett syndrome. *J Neurosci* 27:10912–10917. doi: 10.1523/JNEUROSCI.1869-07.2007

Oishi H, Berezovskii VK, Livingstone MS, Weiner KS, Arcaro MJ (2024) Inferotemporal face patches are histo-architectonically distinct. *Cell Rep* 43:114732. doi: 10.1016/j.celrep.2024.114732

Okada M, Onodera K, Van Renterghem C, Sieghart W, Takahashi T (2000) Functional correlation of GABA(A) receptor alpha subunits expression with the properties of IPSCs in the developing thalamus. *J Neurosci* 20:2202–2208. doi: 10.1523/JNEUROSCI.20-06-02202.2000

Okada Y, Chen Z, Kuwana S (2001) Cytoarchitecture of central chemoreceptors in the mammalian ventral medulla. *Respir Physiol* 129:13–23. doi: 10.1016/s0034-5687(01)00279-1

Okaty BW, Commons KG, Dymecki SM (2019) Embracing diversity in the 5-HT neuronal system. *Nat Rev Neurosci* 20:397–424. doi: 10.1038/s41583-019-0151-3

O'Keefe J, Dostrovsky J (1971) The hippocampus as a spatial map. Preliminary evidence from unit activity in the freely-moving rat. *Brain Res* 34:171–175. doi: 10.1016/0006-8993(71)90358-1

O'Keefe LP, Levitt JB, Kiper DC, Shapley RM, Movshon JA (1998) Functional organization of owl monkey lateral geniculate nucleus and visual cortex. *J Neurophysiol* 80:594–609. doi: 10.1152/jn.1998.80.2.594

Olanow CW, Tatton WG (1999) Etiology and pathogenesis of Parkinson's disease. *Annu Rev Neurosci* 22:123–144. doi: 10.1146/annurev.neuro.22.1.123

Olavarria JF, Qi H, Takahata T, Kaas JH (2022) Overall patterns of eye-specific retino-geniculo-cortical projections to layers III, IV, and VI in primary visual cortex of the greater galago (Otolemur crassicudatus), and correlation with cytochrome oxidase blobs. *Vis Neurosci* 39:E007. doi: 10.1017/S0952523822000062

Olsen RW, Bureau M, Ransom RW, Deng L, Dilber A, Smith G, Krestchatisky M, Tobin AJ (1988) The GABA receptor-chloride ion channel protein complex. *Adv Exp Med Biol* 236:1–14. doi: 10.1007/978-1-4757-5971-6_1

Olsen RW, Tobin AJ (1990) Molecular biology of GABAA receptors. *FASEB J* 4:1469–1480. doi: 10.1096/fasebj.4.5.2155149

Olson EJ, Simon PM (1996) Sleep-wake cycles and the management of respiratory failure. *Curr Opin Pulm Med* 2:500–506.

Ongwijitwat S, Liang HL, Graboyes EM, Wong-Riley MTT (2006) Nuclear respiratory factor 2 senses changing cellular energy demands and its silencing down-regulates cytochrome oxidase and other target gene mRNAs. *Gene* 374:39–49. doi: 10.1016/j.gene.2006.01.009

Ongwijitwat S, Wong-Riley MTT (2004) Functional analysis of the rat cytochrome c oxidase subunit 6A1 promoter in primary neurons. *Gene* 337:163–171. doi: 10.1016/j.gene.2004.04.024

Ongwijitwat S, Wong-Riley MTT (2005) Is nuclear respiratory factor 2 a master transcriptional coordinator for all ten nuclear-encoded cytochrome c oxidase subunits in neurons? *Gene* 360:65–77. doi: 10.1016/j.gene.2005.06.015

Onimaru H, Arata A, Homma I (1988) Primary respiratory rhythm generator in the medulla of brainstem-spinal cord preparation from newborn rat. *Brain Res* 445:314–324. doi: 10.1016/0006-8993(88)91194-8

Onimaru H, Homma I (2003) A novel functional neuron group for respiratory rhythm generation in the ventral medulla. *J Neurosci* 23:1478–1486. doi: 10.1523/JNEUROSCI.23-04-01478.2003

Osborne CS, Chakalova L, Brown KE, Carter D, Horton A, Debrand E, Goyenechea B, Mitchell JA, Lopes S, Reik W, Fraser P (2004) Active genes dynamically colocalize to shared sites of ongoing transcription. *Nat Genet* 36:1065–1071. doi: 10.1038/ng1423

Ostergaard E, Weraarpachai W, Ravn K, Born AP, Jønson L, Duno M, Wibrand F, Shoubridge EA, Vissing J (2015) Mutations in COA3 cause isolated complex IV deficiency associated with neuropathy, exercise intolerance, obesity, and short stature. *J Med Genet* 52:203–207. doi: 10.1136/jmedgenet-2014-102914

Otsuka M, Mizuno Y, Yoshida M, Kagawa Y, Ohta S (1988) Nucleotide sequence of cDNA encoding human cytochrome c oxidase subunit VIc. *Nucleic Acids Res* 16:10916. doi: 10.1093/nar/16.22.10916

Ouchi S, Ishii K, Kosaki K, Suzuki H, Yamada M, Takenouchi T, Tamaoka A (2023) Parkinsonism in spinocerebellar ataxia with axonal neuropathy caused by adult-onset COA7 variants: a case report. *BMC Neurol* 23:211. doi: 10.1186/s12883-023-03202-w

Pacheu-Grau D, Bareth B, Dudek J, Juris L, Vögtle FN, Wissel M, Leary SC, Dennerlein S, Rehling P, Deckers M (2015) Cooperation between COA6 and SCO2 in COX2 maturation during cytochrome c oxidase assembly links two mitochondrial cardiomyopathies. *Cell Metab* 21:823–833. doi: 10.1016/j.cmet.2015.04.012

Paharkova V, Alvarez G, Nakamura H, Cohen P, Lee KW (2015) Rat Humanin is encoded and translated in mitochondria and is localized to the mitochondrial compartment where it regulates ROS production. *Mol Cell Endocrinol* 413:96–100. doi: 10.1016/j.mce.2015.06.015

Palmer CS, Lou J, Kouskousis B, Pandzic E, Anderson AJ, Kang Y, Hinde E, Stojanovski D (2021) Super-resolution microscopy reveals the arrangement of inner membrane protein complexes in mammalian mitochondria. *J Cell Sci* 134:jcs252197. doi: 10.1242/jcs.252197

Palumaa P, Kangur L, Voronova A, Sillard R (2004) Metal-binding mechanism of Cox17, a copper chaperone for cytochrome c oxidase. *Biochem J* 382:307–314. doi: 10.1042/BJ20040360

Pan LP, He Q, Chan SI (1991b) The nature of zinc in cytochrome c oxidase. *J Biol Chem* 266:19109–19112.

Pan LP, Li Z, Larsen R, Chan SI (1991a) The nature of Cux in cytochrome c oxidase. *J Biol Chem* 266:1367–1370.

Panicker N, Ge P, Dawson VL, Dawson TM (2021) The cell biology of Parkinson's disease. *J Cell Biol* 220:e202012095. doi: 10.1083/jcb.202012095

Parkinson J (2002) An essay on the shaking palsy. 1817. *J Neuropsychiatry Clin Neurosci* 14:223–236; discussion 222. doi: 10.1176/jnp.14.2.223

Pastore D, Greco M, Passarella S (2000) Specific helium-neon laser sensitivity of the purified cytochrome c oxidase. *Int J Radiat Biol* 76:863–870. doi: 10.1080/09553000050029020

Patterson SL, Abel T, Deuel TA, Martin KC, Rose JC, Kandel ER (1996) Recombinant BDNF rescues deficits in basal synaptic transmission and hippocampal LTP in BDNF knockout mice. *Neuron* 16:1137–1145. doi: 10.1016/s0896-6273(00)80140-3

Payne JA, Stevenson TJ, Donaldson LF (1996) Molecular characterization of a putative K-Cl cotransporter in rat brain. A neuronal-specific isoform. *J Biol Chem* 271:16245–16252. doi: 10.1074/jbc.271.27.16245

Pellegri G, Magistretti PJ, Martin JL (1998) VIP and PACAP potentiate the action of glutamate on BDNF expression in mouse cortical neurones. *Eur J Neurosci* 10:272–280. doi: 10.1046/j.1460-9568.1998.00052.x

Peña F, Ramirez JM (2002) Endogenous activation of serotonin-2A receptors is required for respiratory rhythm generation in vitro. *J Neurosci* 22:11055–11064. doi: 10.1523/JNEUROSCI.22-24-11055.2002

Penn RD, Hagins WA (1969) Signal transmission along retinal rods and the origin of the electroretinographic a-wave. *Nature* 223:201–204. doi: 10.1038/223201a0

Perez-Martinez X, Broadley SA, Fox TD (2003) Mss51p promotes mitochondrial Cox1p synthesis and interacts with newly synthesized Cox1p. *EMBO J* 22:5951–5961. doi: 10.1093/emboj/cdg566

Perkins AT 4th, Teyler TJ (1988) A critical period for long-term potentiation in the developing rat visual cortex. *Brain Res* 439:222–229. doi: 10.1016/0006-8993(88)91478-3

Petros JA, Baumann AK, Ruiz-Pesini E, Amin MB, Sun CQ, Hall J, Lim S, Issa MM, Flanders WD, Hosseini SH, Marshall FF, Wallace DC (2005) mtDNA mutations increase tumorigenicity in prostate cancer. *Proc Natl Acad Sci USA* 102:719–724. doi: 10.1073/pnas.0408894102

Pfeffer S, Woellhaf MW, Herrmann JM, Förster F (2015) Organization of the mitochondrial translation machinery studied in situ by cryoelectron tomography. *Nat Commun* 6:6019. doi: 10.1038/ncomms7019

Pham L, Arroum T, Wan J, Pavelich L, Bell J, Morse PT, Lee I, Grossman LI, Sanderson TH, Malek MH, Hüttemann M (2024) Regulation of mitochondrial oxidative phosphorylation through tight control of cytochrome c oxidase in health and disease - Implications for ischemia/reperfusion injury, inflammatory diseases, diabetes, and cancer. *Redox Biol* 78:103426. doi: 10.1016/j.redox.2024.103426

Pillai NR, AlDhaheri NS, Ghosh R, Lim J, Streff H, Nayak A, Graham BH, Hanchard NA, Elsea SH, Scaglia F (2019) Biallelic variants in COX4I1 associated with a novel phenotype resembling Leigh syndrome with developmental regression, intellectual disability, and seizures. *Am J Med Genet A* 179:2138–2143. doi: 10.1002/ajmg.a.61288

Pinti P, Tachtsidis I, Hamilton A, Hirsch J, Aichelburg C, Gilbert S, Burgess PW (2020) The present and future use of functional near-infrared spectroscopy (fNIRS) for cognitive neuroscience. *Ann N Y Acad Sci* 1464:5–29. doi: 10.1111/nyas.13948

Pitceathly RD, Rahman S, Wedatilake Y, Polke JM, Cirak S, Foley AR, Sailer A, Hurles ME, Stalker J, Hargreaves I, Woodward CE, Sweeney MG, Muntoni F, Houlden H, Taanman JW, Hanna MG; UK10K Consortium (2013) NDUFA4 mutations underlie dysfunction of a cytochrome c oxidase subunit linked to human neurological disease. *Cell Rep* 3:1795–1805. doi: 10.1016/j.celrep.2013.05.005

Pitceathly RDS, Taanman JW (2018) NDUFA4 (renamed COXFA4) is a cytochrome-c oxidase subunit. *Trends Endocrinol Metab* 29:452–454. doi: 10.1016/j.tem.2018.03.009

Pizzorusso T, Medini P, Berardi N, Chierzi S, Fawcett JW, Maffei L (2002) Reactivation of ocular dominance plasticity in the adult visual cortex. *Science* 298:1248–1251. doi: 10.1126/science.1072699

Plotkin MD, Snyder EY, Hebert SC, Delpire E (1997) Expression of the Na-K-2Cl cotransporter is developmentally regulated in postnatal rat brains: a possible mechanism underlying GABA's excitatory role in immature brain. *J Neurobiol* 33:781–795. doi: 10.1002/(sici)1097-4695(19971120)33:6<781::aid-neu6>3.0.co;2-5

Poo MM (2001) Neurotrophins as synaptic modulators. *Nat Rev Neurosci* 2:24–32. doi: 10.1038/35049004

Poole OV, Everett CM, Gandhi S, Marino S, Bugiardini E, Woodward C, Lam A, Quinlivan R, Hanna MG, Pitceathly RDS (2019) Adult-onset Leigh syndrome linked to the novel stop codon mutation m.6579G>A in MT-CO1. *Mitochondrion* 47:294–297. doi: 10.1016/j.mito.2019.02.004

Portnow LH, Vaillancourt DE, Okun MS (2013) The history of cerebral PET scanning: from physiology to cutting-edge technology. *Neurology* 80:952–956. doi: 10.1212/WNL.0b013e318285c135

Poyton RO, Ball KA (2011) Therapeutic photobiomodulation: nitric oxide and a novel function of mitochondrial cytochrome c oxidase. *Discov Med* 11:154–159.

Presson J, Gordon B (1979) Critical period and minimum exposure required for the effects of alternating monocular occlusion in cat visual cortex. *Vision Res* 19:807–811. doi: 10.1016/0042-6989(79)90157-3

Priya A, Johar K, Nair B, Wong-Riley MTT (2014a) Specificity protein 4 (Sp4) regulates the transcription of AMPA receptor subunit GluA2 (Gria2). *Biochim Biophys Acta Mol Cell Res* 1843:1196–1206. doi: 10.1016/j.bbamcr.2014.02.008

Priya A, Johar K, Nair B, Wong-Riley MTT (2014b) Nuclear respiratory factor 2 regulates the transcription of AMPA receptor subunit GluA2 (Gria2). *Biochim Biophys Acta Mol Cell Res* 1843:3018–3028. doi: 10.1016/j.bbamcr.2014.09.006

Priya A, Johar K, Wong-Riley MTT (2013b) Specificity protein 4 functionally regulates the transcription of NMDA receptor subunits GluN1, GluN2A, and GluN2B. *Biochim Biophys Acta Mol Cell Res* 1833:2745–2756. doi: 10.1016/j.bbamcr.2013.07.002

Priya A. Johar K, Wong-Riley MTT (2013a) Nuclear respiratory factor 2 regulates the expression of the same NMDA receptor subunit genes as NRF-1: both factors act by a concurrent and parallel mechanism to couple energy metabolism and synaptic transmission. *Biochim Biophys Acta Mol Cell Res* 1833:48–58. doi: 10.1016/j.bbamcr.2012.10.014

Priya R, Rakela B, Kaneko M, Spatazza J, Larimer P, Hoseini MS, Hasenstaub AR, Alvarez-Buylla A, Stryker MP (2019) Vesicular GABA transporter is necessary for transplant-induced critical period plasticity in mouse visual cortex. *J Neurosci* 39:2635–2648. doi: 10.1523/JNEUROSCI.1253-18.2019

Pruunsild P, Kazantseva A, Aid T, Palm K, Timmusk T (2007) Dissecting the human BDNF locus: bidirectional transcription, complex splicing, and multiple promoters. *Genomics* 90:397–406. doi: 10.1016/j.ygeno.2007.05.004

Puigserver P, Spiegelman BM (2003) Peroxisome proliferator-activated receptor-gamma coactivator 1 alpha (PGC-1 alpha): transcriptional coactivator and metabolic regulator. *Endocr Rev* 24:78–90. doi: 10.1210/er.2002-0012

Puigserver P, Wu Z, Park CW, Graves R, Wright M, Spiegelman BM (1998) A cold-inducible coactivator of nuclear receptors linked to adaptive thermogenesis. *Cell* 92:829–839. doi: 10.1016/s0092-8674(00)81410-5

Putnam RW (2001) Intracellular pH regulation of neurons in chemosensitive and nonchemosensitive areas of brain slices. *Respir Physiol* 129:37–56. doi: 10.1016/s0034-5687(01)00281-x

Quast KB, Reh RK, Caiati MD, Kopell N, McCarthy MM, Hensch TK (2023) Rapid synaptic and gamma rhythm signature of mouse critical period plasticity. *Proc Natl Acad Sci USA* 120:e2123182120. doi: 10.1073/pnas.2123182120

Quirk B, Whelan HT (2021) Effect of red-to-near infrared light and a nitric oxide donor on the oxygen consumption of isolated cytochrome c oxidase. *Photobiomodul Photomed Laser Surg* 39:463–470. doi: 10.1089/photob.2020.4978

Quirk BJ, Whelan HT (2020) What lies at the heart of photobiomodulation: light, cytochrome C oxidase, and nitric oxide-review of the evidence. *Photobiomodul Photomed Laser Surg* 38:527–530. doi: 10.1089/photob.2020.4905

Radius RL (1987) Anatomy of the optic nerve head and glaucomatous optic neuropathy. *Surv Ophthalmol* 32:35–44. doi: 10.1016/0039-6257(87)90072-5

Rahman S, Thorburn D (2020) Nuclear gene-encoded leigh syndrome spectrum overview. In: Adam MP, Feldman J, Mirzaa GM, Pagon RA, Wallace SE, Amemiya A (eds) *GeneReviews® [Internet]*. University of Washington, Seattle, 1993–2024.

Raisman G, Cowan WM, Powell TPS (1965) The extrinsic afferent, commissural and association fibers of the hippocampus. *Brain* 88:963–996. doi: 10.1093/brain/88.5.963

Ramón y Cajal S (1909) *Histologie du Système Nerveux de L'homme et des Vertébres*. A Maloine, Paris.

Ramzan R, Kadenbach B, Vogt S (2021) Multiple mechanisms regulate eukaryotic cytochrome C oxidase. *Cells* 10:514. doi: 10.3390/cells10030514

Rebelo AP, Saade D, Pereira CV, Farooq A, Huff TC, Abreu L, Moraes CT, Mnatsakanova D, Mathews K, Yang H, Schon EA, Zuchner S, Shy ME (2018) SCO2 mutations cause early-onset axonal Charcot-Marie-Tooth disease associated with cellular copper deficiency. *Brain* 141:662–672. doi: 10.1093/brain/awx369

Reczek CR, Chandel NS (2015) ROS-dependent signal transduction. *Curr Opin Cell Biol* 33:8–13. doi: 10.1016/j.ceb.2014.09.010

Redza-Dutordoir M, Averill-Bates DA (2016) Activation of apoptosis signalling pathways by reactive oxygen species. *Biochim Biophys Acta* 1863:2977–2992. doi: 10.1016/j.bbamcr.2016.09.012

Reh RK, Dias BG, Nelson CA 3rd, Kaufer D, Werker JF, Kolb B, Levine JD, Hensch TK (2020) Critical period regulation across multiple timescales. *Proc Natl Acad Sci U S A* 117:23242–23251. doi: 10.1073/pnas.1820836117

Reiss AB, Gulkarov S, Jacob B, Srivastava A, Pinkhasov A, Gomolin IH, Stecker MM, Wisniewski T, De Leon J (2024) Mitochondria in Alzheimer's disease pathogenesis. *Life (Basel)* 14:196. doi: 10.3390/life14020196

Reiter HO, Stryker MP (1988) Neural plasticity without postsynaptic action potentials: less-active inputs become dominant when kitten visual cortical cells are pharmacologically inhibited. *Proc Natl Acad Sci USA* 85:3623–3627. doi: 10.1073/pnas.85.10.3623

Rekling JC, Feldman JL (1998) PreBötzinger complex and pacemaker neurons: hypothesized site and kernel for respiratory rhythm generation. *Annu Rev Physiol* 60:385–405. doi: 10.1146/annurev.physiol.60.1.385

Remmers JE, deGroot WJ, Sauerland EK, Anch AM (1978) Pathogenesis of upper airway occlusion during sleep. *J Appl Physiol Respir Environ Exerc Physiol* 44:931–938. doi: 10.1152/jappl.1978.44.6.931

Rendon A, Masmoudi A (1985) Purification of non-synaptic and synaptic mitochondria and plasma membranes from rat brain by a rapid Percoll gradient procedure. *J Neurosci Methods* 14:41–51. doi: 10.1016/0165-0270(85)90113-x

Renkema GH, Visser G, Baertling F, Wintjes LT, Wolters VM, van Montfrans J, de Kort GAP, Nikkels PGJ, van Hasselt PM, van der Crabben SN, Rodenburg RJT (2017) Mutated PET117 causes complex IV deficiency and is associated with neurodevelopmental regression and medulla oblongata lesions. *Hum Genet* 136:759–769. doi: 10.1007/s00439-017-1794-7

Richter-Dennerlein R, Oeljeklaus S, Lorenzi I, Ronsör C, Bareth B, Schendzielorz AB, Wang C, Warscheid B, Rehling P, Dennerlein S (2016) Mitochondrial protein synthesis adapts to influx of nuclear-encoded protein. *Cell* 167:471–483.e10. doi: 10.1016/j.cell.2016.09.003

Rimoldi M, Magri F, Antognozzi S, Ripolone M, Salani S, Piga D, Bertolasi L, Zanotti S, Ciscato P, Fortunato F, Moggio M, Corti S, Comi GP, Ronchi D (2023) Prominent muscle involvement in a familial form of mitochondrial disease due to a COA8 variant. *Front Genet* 14:1278572. doi: 10.3389/fgene.2023.1278572

Ristevski S, O'Leary DA, Thornell AP, Owen MJ, Kola I, Hertzog PJ (2004) The ETS transcription factor GABPalpha is essential for early embryogenesis. *Mol Cell Biol* 24:5844–5849. doi: 10.1128/MCB.24.13.5844-5849.2004

Rivera C, Li H, Thomas-Crusells J, Lahtinen H, Viitanen T, Nanobashvili A, Kokaia Z, Airaksinen MS, Voipio J, Kaila K, Saarma M (2002) BDNF-induced TrkB activation down-regulates the K+-Cl- cotransporter KCC2 and impairs neuronal Cl- extrusion. *J Cell Biol* 159:747–752. doi: 10.1083/jcb.200209011

Rivnyak A, Kiss P, Tamas A, Balogh D, Reglodi D (2018) Review on PACAP-induced transcriptomic and proteomic changes in neuronal development and repair. *Int J Mol Sci* 19:1020. doi: 10.3390/ijms19041020

Rizzuto R, Nakase H, Darras B, Francke U, Fabrizi GM, Mengel T, Walsh F, Kadenbach B, DiMauro S, Schon EA (1989) A gene specifying subunit VIII of human cytochrome c oxidase is localized to chromosome 11 and is expressed in both muscle and non-muscle tissues. *J Biol Chem* 264:10595–10600.

Rizzuto R, Nakase H, Zeviani M, DiMauro S, Schon EA (1988) Subunit Va of human and bovine cytochrome c oxidase is highly conserved. *Gene* 69:245–256. doi: 10.1016/0378-1119(88)90435-0

Robinson BH (2000) Human cytochrome oxidase deficiency. *Pediatr Res* 48:581–585. doi: 10.1203/00006450-200011000-00004

Rochefort NL, Narushima M, Grienberger C, Marandi N, Hill DN, Konnerth A (2011) Development of direction selectivity in mouse cortical neurons. *Neuron* 71:425–432. doi: 10.1016/j.neuron.2011.06.013

Rockland KS, Lund JS (1983) Intrinsic laminar lattice connections in primate visual cortex. *J Comp Neurol* 216:303–318. doi: 10.1002/cne.902160307

Rockoff EC, Balaram P, Kaas JH (2014) Patchy distributions of myelin and vesicular glutamate transporter 2 align with cytochrome oxidase blobs and interblobs in the superficial layers of the primary visual cortex. *Eye Brain* 6(Suppl 1):19–27. doi: 10.2147/EB.S59797

Ronchi D, Garbellini M, Magri F, Menni F, Meneri M, Bedeschi MF, Dilena R, Cecchetti V, Picciolli I, Furlan F, Polimeni V, Salani S, Pezzoli L, Fortunato F, Bellini M, Piga D, Ripolone M, Zanotti S, Napoli L, Ciscato P, Sciacco M, Mangili G, Mosca F, Corti S, Iascone M, Comi GP (2023) A biallelic variant in COX18 cause isolated Complex IV deficiency associated with neonatal encephalo-cardio-myopathy and axonal sensory neuropathy. *Eur J Hum Genet* 31:1414–1420. doi: 10.1038/s41431-023-01433-6

Rossi FM, Bozzi Y, Pizzorusso T, Maffei L (1999) Monocular deprivation decreases brain-derived neurotrophic factor immunoreactivity in the rat visual cortex. *Neuroscience* 90:363–368. doi: 10.1016/s0306-4522(98)00463-1

Rothblat LA, Schwartz ML (1979) The effect of monocular deprivation on dendritic spines in visual cortex of young and adult albino rats: evidence for a sensitive period. *Brain Res* 161:156–161. doi: 10.1016/0006-8993(79)90203-8

Rothman SM, Olney JW (1995) Excitotoxicity and the NMDA receptor--still lethal after eight years. *Trends Neurosci* 18:57–58. doi: 10.1016/0166-2236(95)93869-y

Rötig A, Parfait B, Heidet L, Dujardin G, Rustin P, Munnich A (1997) Sequence and structure of the human OXA1L gene and its upstream elements. *Biochim Biophys Acta* 1361:6–10. doi: 10.1016/s0925-4439(97)00031-8

Row BW, Kheirandish L, Neville JJ, Gozal D (2002) Impaired spatial learning and hyperactivity in developing rats exposed to intermittent hypoxia. *Pediatr Res* 52:449–453. doi: 10.1203/00006450-200209000-00024

Roy CS, Sherrington CS (1890) On the regulation of the blood-supply of the brain. *J Physiol* 11:85–108. doi: 10.1113/jphysiol.1890.sp000321

Rudecki AP, Gray SL (2016) PACAP in the defense of energy homeostasis. *Trends Endocrinol Metab* 27:620–632. doi: 10.1016/j.tem.2016.04.008

Rudolph U, Crestani F, Benke D, Brünig I, Benson JA, Fritschy JM, Martin JR, Bluethmann H, Möhler H (1999) Benzodiazepine actions mediated by specific gamma-aminobutyric acid(A) receptor subtypes. *Nature* 401:796–800. doi: 10.1038/44579

Ruigrok TJ (1997) Cerebellar nuclei: the olivary connection. *Prog Brain Res* 114:167–192. doi: 10.1016/s0079-6123(08)63364-6

Runcie MJ, Ulman LG, Potter EK (1995) Effects of pituitary adenylate cyclase-activating polypeptide on cardiovascular and respiratory responses in anaesthetised dogs. *Regul Pept* 60:193–200. doi: 10.1016/0167-0115(95)00131-x

Sadlock JE, Lightowlers RN, Capaldi RA, Schon EA (1993) Isolation of a cDNA specifying subunit VIIb of human cytochrome c oxidase. *Biochim Biophys Acta* 1172: 223–225. doi: 10.1016/0167-4781(93)90301-s

Salehpour F, Mahmoudi J, Kamari F, Sadigh-Eteghad S, Rasta SH, Hamblin MR (2018) Brain photobiomodulation therapy: a narrative review. *Mol Neurobiol* 55:6601–6636. doi: 10.1007/s12035-017-0852-4

Salviati L, Sacconi S, Rasalan MM, Kronn DF, Braun A, Canoll P, Davidson M, Shanske S, Bonilla E, Hays AP, Schon EA, DiMauro S (2002) Cytochrome c oxidase deficiency due to a novel SCO2 mutation mimics Werdnig-Hoffmann disease. *Arch Neurol* 59:862–865. doi: 10.1001/archneur.59.5.862

Sanderson KJ (1974) Lamination of the dorsal lateral geniculate nucleus in carnivores of the weasel (Mustelidae), raccoon (Procyonidae) and fox (Canidae) families. *J Comp Neurol* 153:238–266. doi: 10.1002/cne.901530303

Sanderson TH, Wider JM, Lee I, Reynolds CA, Liu J, Lepore B, Tousignant R, Bukowski MJ, Johnston H, Fite A, Raghunayakula S, Kamholz J, Grossman LI, Przyklenk K, Hüttemann M (2018) Inhibitory modulation of cytochrome c oxidase activity with specific near-infrared light wavelengths attenuates brain ischemia/reperfusion injury. *Sci Rep* 8:3481. doi: 10.1038/s41598-018-21869-x

Santana-Blank L, Rodriguez-Santana E, Santana-Rodriguez JA, Santana-Rodriguez KE, Reyes-Barrios H (2017) Water as a photoacceptor, energy transducer, and rechargeable electrolytic bio-battery in photobiomodulation. In: Hamblin MR, de Sousa MVP, Agrawal T (eds) *Handbook of Low-Level Laser Therapy*. Pan Stanford Publishing, Singapore, pp 119–139.

Sant'Anna GM, Mortola JP (2002) Thermal and respiratory control in young rats with altered caloric intake during postnatal development. *Respir Physiol Neurobiol* 133:215–227. doi: 10.1016/s1569-9048(02)00148-9

Sanz-Morello B, Pfisterer U, Winther Hansen N, Demharter S, Thakur A, Fujii K, Levitskii SA, Montalant A, Korshunova I, Mammen PP, Kamenski P, Noguchi S, Aldana BI, Hougaard KS, Perrier JF, Khodosevich K (2020) Complex IV subunit isoform COX6A2 protects fast-spiking interneurons from oxidative stress and supports their function. *EMBO J* 39:e105759. doi: 10.15252/embj.2020105759

Saudou F, Humbert S (2016) The biology of huntingtin. *Neuron* 89:910–926. doi: 10.1016/j.neuron.2016.02.003

Sawa C, Goto M, Suzuki F, Watanabe H, Sawada J, Handa H (1996) Functional domains of transcription factor hGABP beta1/E4TF1-53 required for nuclear localization and transcription activation. *Nucleic Acids Res* 24:4954–4961. doi: 10.1093/nar/24.24.4954

Scarpulla RC (1997) Nuclear control of respiratory chain expression in mammalian cells. *J Bioenerg Biomembr* 29:109–119. doi: 10.1023/a:1022681828846

Scarpulla RC (2002) Transcriptional activators and coactivators in the nuclear control of mitochondrial function in mammalian cells. *Gene* 286:81–89. doi: 10.1016/s0378-1119(01)00809-5

Scarpulla RC, Vega RB, Kelly DP (2012) Transcriptional integration of mitochondrial biogenesis. *Trends Endocrinol Metab* 23:459–466. doi: 10.1016/j.tem.2012.06.006

Schägger H, Pfeiffer K (2000) Supercomplexes in the respiratory chains of yeast and mammalian mitochondria. *EMBO J* 19:1777–1783. doi: 10.1093/emboj/19.8.1777

Schägger H, Pfeiffer K (2001) The ratio of oxidative phosphorylation complexes I-V in bovine heart mitochondria and the composition of respiratory chain supercomplexes. *J Biol Chem* 276:37861–37867. doi: 10.1074/jbc.M106474200

Schinder AF, Berninger B & Poo M (2000) Postsynaptic target specificity of neurotrophin-induced presynaptic potentiation. *Neuron* 25:151–163. doi: 10.1016/s0896-6273(00)80879-x

Schlaggar BL, Fox K, O'Leary DD (1993) Postsynaptic control of plasticity in developing somatosensory cortex. *Nature* 364:623–626. doi: 10.1038/364623a0

Schlerf A, Droste M, Winter M, Kadenbach B (1988) Characterization of two different genes (cDNA) for cytochrome c oxidase subunit VIa from heart and liver of the rat. *EMBO J* 7:2387–2391. doi: 10.1002/j.1460-2075.1988.tb03083.x

Schmidt N, Ferger B (2001) Neurochemical findings in the MPTP model of Parkinson's disease. *J Neural Transm (Vienna)* 108:1263–1282. doi: 10.1007/s007020100004

Schon EA, Rizzuto R, Moraes CT, Nakase H, Zeviani M, DiMauro S (1989) A direct repeat is a hotspot for large-scale deletion of human mitochondrial DNA. *Science* 244:346–349. doi: 10.1126/science.2711184

Schotanus SM, Chergui K (2008) Long-term potentiation in the nucleus accumbens requires both NR2A- and NR2B-containing N-methyl-D-aspartate receptors. *Eur J Neurosci* 27:1957–1964. doi: 10.1111/j.1460-9568.2008.06173.x

Schreiber SN, Emter R, Hock MB, Knutti D, Cardenas J, Podvinec M, Oakeley EJ, Kralli A (2004) The estrogen-related receptor alpha (ERRalpha) functions in PPARgamma coactivator 1alpha (PGC-1alpha)-

induced mitochondrial biogenesis. *Proc Natl Acad Sci USA* 101:6472–6477. doi: 10.1073/pnas.0308686101

Schwarzacher SW, Pestean A, Günther S, Ballanyi K (2002) Serotonergic modulation of respiratory motoneurons and interneurons in brainstem slices of perinatal rats. *Neuroscience* 115:1247–1259. doi: 10.1016/s0306-4522(02)00540-7

Schweitzer P, Fortin G, Beloeil JC, Champagnat J (1992) In vitro study of newborn rat brain maturation: implication for sudden infant death syndrome. *Neurochem Int* 20:109–112. doi: 10.1016/0197-0186(92)90133-c

Seeburg PH (1993) The TINS/TiPS lecture. The molecular biology of mammalian glutamate receptor channels. *Trends Neurosci* 16:359–365. doi: 10.1016/0166-2236(93)90093-2

Seelan RS, Gopalakrishnan L, Scarpulla RC, Grossman LI (1996) Cytochrome c oxidase subunit VIIa liver isoform. Characterization and identification of promoter elements in the bovine gene. *J Biol Chem* 271:2112–2120. doi: 10.1074/jbc.271.4.2112

Seelan RS, Grossman LI (1992) Structure and organization of the heart isoform gene for bovine cytochrome c oxidase subunit VIIa. *Biochemistry* 31:4696–4704. doi: 10.1021/bi00134a024

Seelan RS, Grossman LI (1997) Structural organization and promoter analysis of the bovine cytochrome c oxidase subunit VIIc gene. A functional role for YY1. *J Biol Chem* 272:10175–10181. doi: 10.1074/jbc.272.15.10175

Segade F, Hurlé B, Claudio E, Ramos S, Lazo PS (1996) Identification of an additional member of the cytochrome c oxidase subunit VIIa family of proteins. *J Biol Chem* 271:12343–12349. doi: 10.1074/jbc.271.21.12343

Sepers MD, Raymond LA (2014) Mechanisms of synaptic dysfunction and excitotoxicity in Huntington's disease. *Drug Discov Today* 19:990–996. doi: 10.1016/j.drudis.2014.02.006

Seress L, Abrahám H, Hajnal A, Lin H, Totterdell S (2005) NOS-positive local circuit neurons are exclusively axo-dendritic cells both in the neo- and archi-cortex of the rat brain. *Brain Res* 1056:183–190. doi: 10.1016/j.brainres.2005.07.034

Serra A, Brozoski D, Hedin N, Franciosi R, Forster HV (2001) Mortality after carotid body denervation in rats. *J Appl Physiol* 91:1298–1306. doi: 10.1152/jappl.2001.91.3.1298

Setou M, Nakagawa T, Seog DH, Hirokawa N (2000) Kinesin superfamily motor protein KIF17 and mLin-10 in NMDA receptor-containing vesicle transport. *Science* 288:1796–1802. doi: 10.1126/science.288.5472.1796

Sharma V, Ala-Vannesluoma P, Vattulainen I, Wikström M, Róg T (2015) Role of subunit III and its lipids in the molecular mechanism of cytochrome c oxidase. *Biochim Biophys Acta* 1847:690–697. doi: 10.1016/j.bbabio.2015.04.007

Sharp T, Barnes NM (2020) Central 5-HT receptors and their function; present and future. *Neuropharmacology* 177:108155. doi: 10.1016/j.neuropharm.2020.108155

Shaye H, Stauch B, Gati C, Cherezov V (2021) Molecular mechanisms of metabotropic GABA(B) receptor function. *Sci Adv* 7:eabg3362. doi: 10.1126/sciadv.abg3362

Sheedy D, Lara A, Garrick T, Harper C (1999) Size of mamillary bodies in health and disease: useful measurements in neuroradiological diagnosis of Wernicke's encephalopathy. *Alcohol Clin Exp Res* 23:1624–1628.

Shen L, Duffin J (2002) Caudal expiratory neurones in the rat. *Pflugers Arch* 444:405–410. doi: 10.1007/s00424-002-0817-x

Sherer TB, Betarbet R, Testa CM, Seo BB, Richardson JR, Kim JH, Miller GW, Yagi T, Matsuno-Yagi A, Greenamyre JT (2003) Mechanism of toxicity in rotenone models of Parkinson's disease. *J Neurosci* 23:10756–10764. doi: 10.1523/JNEUROSCI.23-34-10756.2003

Sherman SM, Spear PD (1982) Organization of visual pathways in normal and visually deprived cats. *Physiol Rev* 62:738–855. doi: 10.1152/physrev.1982.62.2.738

Shimada S, Shinzawa-Itoh K, Baba J, Aoe S, Shimada A, Yamashita E, Kang J, Tateno M, Yoshikawa S, Tsukihara T (2017) Complex structure of cytochrome c-cytochrome c oxidase reveals a novel protein-protein interaction mode. *EMBO J* 36:291–300. doi: 10.15252/embj.201695021

Shimokata K, Katayama Y, Murayama H, Suematsu M, Tsukihara T, Muramoto K, Aoyama H, Yoshikawa S, Shimada H (2007) The proton pumping pathway of bovine heart cytochrome c oxidase. *Proc Natl Acad Sci USA* 104:4200–4205. doi: 10.1073/pnas.0611627104

Shinde SS, Sharma S, Teekas L, Sharma A, Vijay N (2021) Recurrent erosion of COA1/MITRAC15 exemplifies conditional gene dispensability in oxidative phosphorylation. *Sci Rep* 11:24437. doi: 10.1038/s41598-021-04077-y

Shipp S, Zeki S (1985) Segregation of pathways leading from area V2 to areas V4 and V5 of macaque monkey visual cortex. *Nature* 315:322–325. doi: 10.1038/315322a0

Shirendeb U, Reddy AP, Manczak M, Calkins MJ, Mao P, Tagle DA, Reddy PH (2011) Abnormal mitochondrial dynamics, mitochondrial loss and mutant huntingtin oligomers in Huntington's disease: implications for selective neuronal damage. *Hum Mol Genet* 20:1438–1455. doi: 10.1093/hmg/ddr024

Shoham D, Hübener M, Schulze S, Grinvald A, Bonhoeffer T (1997) Spatio-temporal frequency domains and their relation to cytochrome oxidase staining in cat visual cortex. *Nature* 385:529–533. doi: 10.1038/385529a0

Shrivastava AN, Triller A, Melki R (2020) Cell biology and dynamics of Neuronal Na(+)/K(+)-ATPase in health and diseases. *Neuropharmacology* 169:107461. doi: 10.1016/j.neuropharm.2018.12.008

Shteyer E, Saada A, Shaag A, Al-Hijawi FA, Kidess R, Revel-Vilk S, Elpeleg O (2009) Exocrine pancreatic insufficiency, dyserythropoeitic anemia, and calvarial hyperostosis are caused by a mutation in the COX4I2 gene. *Am J Hum Genet* 84:412–417. doi: 10.1016/j.ajhg.2009.02.006

Shulga A, Blaesse A, Kysenius K, Huttunen HJ, Tanhuanpää K, Saarma M, Rivera C (2009) Thyroxin regulates BDNF expression to promote survival of injured neurons. *Mol Cell Neurosci* 42:408–418. doi: 10.1016/j.mcn.2009.09.002

Sies H, Berndt C, Jones DP (2017) Oxidative stress. *Annu Rev Biochem* 86:715–748. doi: 10.1146/annurev-biochem-061516-045037

Sies H, Jones DP (2020) Reactive oxygen species (ROS) as pleiotropic physiological signalling agents. *Nat Rev Mol Cell Biol* 1:363–383. doi: 10.1038/s41580-020-0230-3

Signes A, Cerutti R, Dickson AS, Benincá C, Hinchy EC, Ghezzi D, Carrozzo R, Bertini E, Murphy MP, Nathan JA, Viscomi C, Fernandez-Vizarra E, Zeviani M (2019) APOPT1/COA8 assists COX assembly and is oppositely regulated by UPS and ROS. *EMBO Mol Med* 11:e9582. doi: 10.15252/emmm.201809582

Sillito AM, Kemp JA, Blakemore C (1981) The role of GABAergic inhibition in the cortical effects of monocular deprivation. *Nature* 291:318–320. doi: 10.1038/291318a0

Silverman MA, Leroux MR (2009) Intraflagellar transport and the generation of dynamic, structurally and functionally diverse cilia. *Trends Cell Biol* 19:306–316. doi: 10.1016/j.tcb.2009.04.002

Silverman MS, Grosof DH, De Valois RL, Elfar SD (1989) Spatial-frequency organization in primate striate cortex. *Proc Natl Acad Sci USA* 86:7I1–7I5. doi: 10.1073/pnas.86.2.711

Simonian NA, Hyman BT (1995) Functional alterations in neural circuits in Alzheimer's disease. *Neurobiol Aging* 16:305–309. doi: 10.1016/0197-4580(95)00034-c

Sincich LC, Horton JC (2002) Divided by cytochrome oxidase: a map of the projections from V1 to V2 in macaques. *Science* 295:1734–1737. doi: 10.1126/science.1067902

Sincich LC, Jocson CM, Horton JC (2010) V1 interpatch projections to V2 thick stripes and pale stripes. *J Neurosci* 30:6963–6974. doi: 10.1523/JNEUROSCI.5506-09.2010

Singer JH, Bellingham MC, Berger AJ (1996) Presynaptic inhibition of glutamatergic synaptic transmission to rat motoneurons by serotonin. *J Neurophysiol* 76:799–807. doi: 10.1152/jn.1996.76.2.799

Singer JH, Talley EM, Bayliss DA, Berger AJ (1998) Development of glycinergic synaptic transmission to rat brain stem motoneurons. *J Neurophysiol* 80:2608–2620. doi: 10.1152/jn.1998.80.5.2608

Singer TP, Ramsay RR (1990) Mechanism of the neurotoxicity of MPTP. *An update. FEBS Lett* 274:1–8. doi: 10.1016/0014-5793(90)81315-f

Sinkler CA, Kalpage H, Shay J, Lee I, Malek MH, Grossman LI, Hüttermann M (2017) Tissue- and condition-specific isoforms of mammalian cytochrome c oxidase subunits: from function to human disease. *Oxid Med Cell Longev* 2017:1534056. doi: 10.1155/2017/1534056

Skoglösa Y, Takei N, Lindholm D (1999) Distribution of pituitary adenylate cyclase activating polypeptide mRNA in the developing rat brain. *Brain Res Mol Brain Res* 65:1–13. doi: 10.1016/s0169-328x(98)00294-0

Skou JC (1957) The influence of some cations on an adenosine triphosphatase from peripheral nerves. *Biochim Biophys Acta* 23:394–401. doi: 10.1016/0006-3002(57)90343-8

Skou JC (1998) The Identification of the Sodium-Potassium Pump (Nobel Lecture). *Angew Chem Int Ed Engl* 37:2320–2328. doi: 10.1002/(SICI)1521-3773(19980918)37:17<2320::AID-ANIE2320>3.0.CO;2-2

Smale ST, Baltimore D (1989) The "initiator" as a transcription control element. *Cell* 57:103–113. doi: 10.1016/0092-8674(89)90176-1

Smith JC, Ellenberger HH, Ballanyi K, Richter DW, Feldman JL (1991) Pre-Botzinger complex: a brainstem region that may generate respiratory rhythm in mammals. *Science* 254:726–729. doi: 10.1126/science.1683005

Snyder SH, Jaffrey SR, Zakhary R (1998) Nitric oxide and carbon monoxide: parallel roles as neural messengers. *Brain Res Brain Res Rev* 26:167–175. doi: 10.1016/s0165-0173(97)00032-5

Sokoloff L (1981) Localization of functional activity in the central nervous system by measurement of glucose utilization with radioactive deoxyglucose. *J Cereb Blood Flow Metab* 1:7–36. doi: 10.1038/jcbfm.1981.4

Sokoloff L, Reivich M, Kennedy C, Des Rosiers MH, Patlak CS, Pettigrew KD, Sakurada O, Shinohara M (1977) The [14C]deoxyglucose method for the measurement of local cerebral glucose utilization: theory, procedure, and normal values in the conscious and anesthetized albino rat. *J Neurochem* 28:897–916. doi: 10.1111/j.1471-4159.1977.tb10649.x

Solodkin A, Van Hoesen GW (1996) Entorhinal cortex modules of the human brain. *J Comp Neurol* 365:610–617. doi: 10.1002/(SICI)1096-9861(19960219)365:4<610::AID-CNE8>3.0.CO;2-7

Sommer AP, Schemmer P, Pavláth AE, Försterling HD, Mester ÁR, Trelles MA (2020) Quantum biology in low level light therapy: death of a dogma. *Ann Transl Med* 8:440. doi: 10.21037/atm.2020.03.159

Sommer B, Köhler M, Sprengel R, Seeburg PH (1991) RNA editing in brain controls a determinant of ion flow in glutamate-gated channels. *Cell* 67:11–19. doi: 10.1016/0092-8674(91)90568-j

Sommer N, Hüttemann M, Pak O, Scheibe S, Knoepp F, Sinkler C, Malczyk M, Gierhardt M, Esfandiary A, Kraut S, Jonas F, Veith C, Aras S, Sydykov A, Alebrahimdehkordi N, Giehl K, Hecker M, Brandes RP, Seeger W, Grimminger F, Ghofrani HA, Schermuly RT, Grossman LI, Weissmann N (2017) Mitochondrial complex IV subunit 4 isoform 2 is essential for acute pulmonary oxygen sensing. *Circ Res* 121:424–438. doi: 10.1161/CIRCRESAHA.116.310482

Southan AP, Robertson B (1998) Patch-clamp recordings from cerebellar basket cell bodies and their presynaptic terminals reveal an asymmetric distribution of voltage-gated potassium channels. *J Neurosci* 18:948–955. doi: 10.1523/JNEUROSCI.18-03-00948.1998

Southwell DG, Froemke RC, Alvarez-Buylla A, Stryker MP, Gandhi SP (2010) Cortical plasticity induced by inhibitory neuron transplantation. *Science* 327:1145–1148. doi: 10.1126/science.1183962

Sparaco M, Schon EA, DiMauro S, Bonilla E (1995) Myoclonic epilepsy with ragged-red fibers (MERRF): an immunohistochemical study of the brain. *Brain Pathol* 5:125–133. doi: 10.1111/j.1750-3639.1995.tb00586.x

Sparaco M., Hirano A., Hirano M, DiMauro S, Bonilla E (1993) Cytochrome C oxidase deficiency and neuronal involvement in Menkes' kinky hair disease: immunohistochemical study. *Brain Pathol* 3:349–354. doi: 10.1111/j.1750-3639.1993.tb00762.x

Spiering BA, Weakley J, Mujika I (2023) Effects of Bed Rest on Physical Performance in Athletes: A Systematic and Narrative Review. *Sports Med* 53:2135–2146. doi: 10.1007/s40279-023-01889-y

Spilianakis CG, Lalioti MD, Town T, Lee GR, Flavell RA (2005) Interchromosomal associations between alternatively expressed loci. *Nature* 435:637–645. doi: 10.1038/nature03574

Spillantini MG, Schmidt ML, Lee VM, Trojanowski JQ, Jakes R, Goedert M (1997) Alpha-synuclein in Lewy bodies. *Nature* 388:839–840. doi: 10.1038/42166

Stafford CA (1984) Critical period plasticity for visual function: definition in monocularly deprived rats using visually evoked potentials. *Ophthalmic Physiol Opt* 4:95–100.

Steffan JS (2010) Does Huntingtin play a role in selective macroautophagy? *Cell Cycle* 9:3401–3413. doi: 10.4161/cc.9.17.12718

Steward O, Scoville SA (1976) Cells of origin of entorhinal cortical afferents to the hippocampus and fascia dentata of the rat. *J Comp Neurol* 169:347–370. doi: 10.1002/cne.901690306

Stiburek L, Vesela K, Hansikova H, Hulkova H, Zeman J (2009) Loss of function of Sco1 and its interaction with cytochrome c oxidase. *Am J Physiol Cell Physiol* 296:C1218–C1226. doi: 10.1152/ajpcell.00564.2008

Stiburek L, Vesela K, Hansikova H, Pecina P, Tesarova M, Cerna L, Houstek J, Zeman J (2005) Tissue-specific cytochrome c oxidase assembly defects due to mutations in SCO2 and SURF1. *Biochem J* 392:625–632. doi: 10.1042/BJ20050807

Stirling CE, Lee A (1980) [3H]ouabain autoradiography of frog retina. *J Cell Biol* 85:313–324. doi: 10.1083/jcb.85.2.313

St-John WM (1998) Neurogenesis of patterns of automatic ventilatory activity. *Prog Neurobiol* 56:97–117. doi: 10.1016/s0301-0082(98)00031-8

St-Pierre J, Drori S, Uldry M, Silvaggi JM, Rhee J, Jäger S, Handschin C, Zheng K, Lin J, Yang W, Simon DK, Bachoo R, Spiegelman BM (2006) Suppression of reactive oxygen species and neurodegeneration by the PGC-1 transcriptional coactivators. *Cell* 127:397–408. doi: 10.1016/j.cell.2006.09.024

Strausak D, Mercer JF, Dieter HH, Stremmel W, Multhaup G (2001) Copper in disorders with neurological symptoms: Alzheimer's, Menkes, and Wilson diseases. *Brain Res Bull* 55:175–185. doi: 10.1016/s0361-9230(01)00454-3

Stryer L (1975) *Biochmistry*, 2nd ed. WH Freeman & Co, San Francisco.

Stryker MP, Sherk H (1975) Modification of cortical orientation selectivity in the cat by restricted visual experience: a reexamination. *Science* 190:904–906. doi: 10.1126/science.1188372

Stryker MP, Zahs KR (1983) On and off sublaminae in the lateral geniculate nucleus of the ferret. *J Neurosci* 3:1943–1951. doi: 10.1523/JNEUROSCI.03-10-01943.1983

Stunden CE, Filosa JA, Garcia AJ, Dean JB, Putnam RW (2001) Development of in vivo ventilatory and single chemosensitive neuron responses to hypercapnia in rats. *Respir Physiol* 127:135–155. doi: 10.1016/s0034-5687(01)00242-0

Sucharov C, Basu A, Carter RS, Avadhani NG (1995) A novel transcriptional initiator activity of the GABP factor binding ets sequence repeat from the murine cytochrome c oxidase Vb gene. *Gene Expr* 5:93–111.

Sugiyama H, Ito I, Hirono C (1987) A new type of glutamate receptor linked to inositol phospholipid metabolism. *Nature* 325:531–533. doi: 10.1038/325531a0

Supp DM, Witte DP, Branford WW, Smith EP, Potter SS (1996) Sp4, a member of the Sp1-family of zinc finger transcription factors, is required for normal murine growth, viability, and male fertility. *Dev Biol* 176:284–299. doi: 10.1006/dbio.1996.0134

Suske G (1999) The Sp-family of transcription factors. *Gene* 238:291–300. doi: 10.1016/s0378-1119(99)00357-1

Suske G, Mengel T, Cordingley M, Kadenbach B (1987) Molecular cloning and further characterization of cDNAs for rat nuclear-encoded cytochrome c oxidase subunits VIc and VIII. *Eur J Biochem* 168:233–237. doi: 10.1111/j.1432-1033.1987.tb13410.x

Sutor B, Luhmann HJ (1995) Development of excitatory and inhibitory postsynaptic potentials in the rat neocortex. *Perspect Dev Neurobiol* 2:409–419.

Suzuki-Yagawa Y, Kawakami K, Nagano K (1992) Housekeeping Na,K-ATPase alpha 1 subunit gene promoter is composed of multiple cis elements to which common and cell type-specific factors bind. *Mol Cell Biol* 12:4046–4055. doi: 10.1128/mcb.12.9.4046-4055.1992

Swaminathan AB, Soma S, Vicary AC, Zulkifli M, Kaur H, Gohil VM (2022) A yeast suppressor screen links Coa4 to the mitochondrial copper delivery pathway for cytochrome c oxidase. *Genetics* 221:iyac090. doi: 10.1093/genetics/iyac090

Sweadner KJ (1989) Isozymes of the Na+/K+-ATPase. *Biochim Biophys Acta* 988:185–220. doi: 10.1016/0304-4157(89)90019-1

Szklarczyk R, Wanschers BF, Nijtmans LG, Rodenburg RJ, Zschocke J, Dikow N, van den Brand MA, Hendriks-Franssen MG, Gilissen C, Veltman JA, Nooteboom M, Koopman WJ, Willems PH, Smeitink JA, Huynen MA, van den Heuvel LP (2013) A mutation in the FAM36A gene, the human ortholog of COX20, impairs cytochrome c oxidase assembly and is associated with ataxia and muscle hypotonia. *Hum Mol Genet* 22:656–667. doi: 10.1093/hmg/dds473

Szyrach G, Ott M, Bonnefoy N, Neupert W, Herrmann JM (2003) Ribosome binding to the Oxa1 complex facilitates co-translational protein insertion in mitochondria. *EMBO J* 22:6448–6457. doi: 10.1093/emboj/cdg623

Taanman JW, Schrage C, Ponne NJ, Das AT, Bolhuis PA, de Vries H, Agsteribbe E (1990) Isolation of cDNAs encoding subunit VIb of cytochrome c oxidase and steady-state levels of coxVIb mRNA in different tissues. *Gene* 93:285–291. doi: 10.1016/0378-1119(90)90237-l

Tabrizi SJ, Cleeter MW, Xuereb J, Taanman JW, Cooper JM, Schapira AH (1999) Biochemical abnormalities and excitotoxicity in Huntington's disease brain. *Ann Neurol* 45:25–32. doi: 10.1002/1531-8249(199901)45:1<25::aid-art6>3.0.co;2-e

Taherzadeh-Fard E, Saft C, Akkad DA, Wieczorek S, Haghikia A, Chan A, Epplen JT, Arning L (2011) PGC-1alpha downstream transcription factors NRF-1 and TFAM are genetic modifiers of Huntington disease. *Mol Neurodegener* 6:32. doi: 10.1186/1750-1326-6-32

Takahata T, Patel NB, Balaram P, Chino YM, Kaas JH (2018) Long-term histological changes in the macaque primary visual cortex and the lateral geniculate nucleus after monocular deprivation produced by early restricted retinal lesions and diffuser induced form deprivation. *J Comp Neurol* 526:2955–2972. doi: 10.1002/cne.24494

Takesian AE, Hensch TK (2013) Balancing plasticity/stability across brain development. *Prog Brain Res* 207:3–34. doi: 10.1016/B978-0-444-63327-9.00001-1

Tamiya G, Makino S, Hayashi M, Abe A, Numakura C, Ueki M, Tanaka A, Ito C, Toshimori K, Ogawa N, Terashima T, Maegawa H, Yanagisawa D, Tooyama I, Tada M, Onodera O, Hayasaka K (2014) A mutation of COX6A1 causes a recessive axonal or mixed form of Charcot-Marie-Tooth disease. *Am J Hum Genet* 95:294–300. doi: 10.1016/j.ajhg.2014.07.013

Tan L, Ringach DL, Trachtenberg JT (2022) The development of receptive field tuning properties in mouse binocular primary visual cortex. *J Neurosci* 42:3546–3556. doi: 10.1523/JNEUROSCI.1702-21.2022

Tanaka H, Grooms SY, Bennett MV, Zukin RS (2000) The AMPAR subunit GluR2: still front and center-stage. *Brain Res* 886:190–207. doi: 10.1016/s0006-8993(00)02951-6

Tanaka T, Saito H & Matsuki N (1997) Inhibition of GABAA synaptic responses by brain-derived neurotrophic factor(BDNF)inrathippocampus.*JNeurosci*17:2959–2966.doi:10.1523/JNEUROSCI.17-09-02959.1997

Tang-Siegel GG, Maughan DW, Frownfelter MB, Light AR (2024) Mitochondrial DNA missense mutations ChrMT: 8981A > G and ChrMT: 6268C > T identified in a Caucasian female with myalgic encephalomyelitis/chronic fatigue syndrome (ME/CFS) triggered by the Epstein-Barr virus. *Case Rep Genet* 2024:6475425. doi: 10.1155/2024/6475425

Tao X, West AE, Chen WG, Corfas G, Greenberg ME (2002) A calcium-responsive transcription factor, CaRF, that regulates neuronal activity-dependent expression of BDNF. *Neuron* 33:383–395. doi: 10.1016/s0896-6273(01)00561-x

Taylor NG, Swenson S, Harris NJ, Germany EM, Fox JL, Khalimonchuk O (2017) The assembly factor Pet117 couples heme a synthase activity to cytochrome oxidase assembly. *J Biol Chem* 292:1815–1825. doi: 10.1074/jbc.M116.766980

Tello D, Balsa E, Acosta-Iborra B, Fuertes-Yebra E, Elorza A, Ordóñez Á, Corral-Escariz M, Soro I, López-Bernardo E, Perales-Clemente E, Martínez-Ruiz A, Enríquez JA, Aragonés J, Cadenas S, Landázuri MO (2011) Induction of the mitochondrial NDUFA4L2 protein by HIF-1alpha decreases oxygen consumption by inhibiting complex I activity. *Cell Metab* 14:768–779. doi: 10.1016/j.cmet.2011.10.008

Tephly TR (1991) The toxicity of methanol. *Life Sci* 48:1031–1041. doi: 10.1016/0024-3205(91)90504-5

The Huntington's Disease Collaborative Research Group (1993) A novel gene containing a trinucleotide repeat that is expanded and unstable on Huntington's disease chromosomes. *Cell* 72:971–983. doi: 10.1016/0092-8674(93)90585-e

Thoenen H (1995) Neurotrophins and neuronal plasticity. *Science* 270:593–598. doi: 10.1126/science.270.5236.593

Thompson CC, Brown TA, McKnight SL (1991) Convergence of Ets- and notch-related structural motifs in a heteromeric DNA binding complex. *Science* 253:762–768. doi: 10.1126/science.1876833

Thompson K, Mai N, Oláhová M, Scialó F, Formosa LE, Stroud DA, Garrett M, Lax NZ, Robertson FM, Jou C, Nascimento A, Ortez C, Jimenez-Mallebrera C, Hardy SA, He L, Brown GK, Marttinen P, McFarland R, Sanz A, Battersby BJ, Bonnen PE, Ryan MT, Chrzanowska-Lightowlers ZM, Lightowlers RN, Taylor RW (2018) OXA1L mutations cause mitochondrial encephalopathy and a combined oxidative phosphorylation defect. *EMBO Mol Med* 10:e9060. doi: 10.15252/emmm.201809060

Thompson LM, Orr HT (2023) HD and SCA1: tales from two 30-year journeys since gene discovery. *Neuron* 111:3517–3530. doi: 10.1016/j.neuron.2023.09.036

Thörnström PE, Brzezinski P, Fredriksson PO, Malmström BG (1988) Cytochrome c oxidase as an electron-transport-driven proton pump: pH dependence of the reduction levels of the redox centers during turnover. *Biochemistry* 27:5441–5447. doi: 10.1021/bi00415a009

Tighilet B, Hashikawa T, Jones EG (1998) Cell- and lamina-specific expression and activity-dependent regulation of type II calcium/calmodulin-dependent protein kinase isoforms in monkey visual cortex. *J Neurosci* 18:2129–2146. doi: 10.1523/JNEUROSCI.18-06-02129.1998

Timney B (1983) The effects of early and late monocular deprivation on binocular depth perception in cats. *Brain Res* 283:235–243. doi: 10.1016/0165-3806(83)90180-3

Timón-Gómez A, Garlich J, Stuart RA, Ugalde C, Barrientos A (2020) Distinct roles of mitochondrial HIGD1A and HIGD2A in respiratory complex and supercomplex biogenesis. *Cell Rep* 31:107607. doi: 10.1016/j.celrep.2020.107607

Timón-Gómez A, Nývltová E, Abriata LA, Vila AJ, Hosler J, Barrientos A (2018) Mitochondrial cytochrome c oxidase biogenesis: recent developments. *Semin Cell Dev Biol* 76:163–178. doi: 10.1016/j.semcdb.2017.08.055

Tiranti V, Hoertnagel K, Carrozzo R, Galimberti C, Munaro M, Granatiero M, Zelante L, Gasparini P, Marzella R, Rocchi M, Bayona-Bafaluy MP, Enriquez JA, Uziel G, Bertini E, Dionisi-Vici C, Franco B, Meitinger T, Zeviani M (1998) Mutations of SURF-1 in Leigh disease associated with cytochrome c oxidase deficiency. *Am J Hum Genet* 63:1609–1621. doi: 10.1086/302150

Tolhuis B, Palstra RJ, Splinter E, Grosveld F, de Laat W (2002) Looping and interaction between hypersensitive sites in the active beta-globin locus. *Mol Cell* 10:1453–1465. doi: 10.1016/s1097-2765(02)00781-5

Tootell RB, Hamilton SL, Silverman MS (1985) Topography of cytochrome oxidase activity in owl monkey cortex. *J Neurosci* 5:2786–2800. doi: 10.1523/JNEUROSCI.05-10-02786.1985

Tootell RBH, Hamilton SL, Silverman MS, Switkes E (1988a) Functional anatomy of macaque striate cortex. I. Ocular dominance, binocular interactions, and baseline conditions. *J Neurosci* 8:1500–1530. doi: 10.1523/JNEUROSCI.08-05-01500.1988

Tootell RBH, Hamilton SL, Switkes E (1988d) Functional anatomy of macaque striate cortex. IV. Contrast and magno-parvo streams. *J Neurosci* 8:1594–1609. doi: 10.1523/JNEUROSCI.08-05-01594.1988

Tootell RBH, Silverman MS, De Valois RL, Jacobs GH (1983) Functional organization of the second cortical visual area in primates. *Science* 220:737–739. doi: 10.1126/science.6301017

Tootell RBH, Silverman MS, Hamilton SL, De Valois RL, Switkes E (1988c) Functional anatomy of macaque striate cortex. III. Color. *J Neurosci* 8:1569–1593. doi: 10.1523/JNEUROSCI.08-05-01569.1988

Tootell RBH, Silverman MS, Hamilton SL, Switkes E, De Valois RL (1988e) Functional anatomy of macaque striate cortex. V. Spatial frequency. *J Neurosci* 8:1610–1624. doi: 10.1523/JNEUROSCI.08-05-01610.1988

Tootell RBH, Switkes E, Silverman MS, Hamilton SL (1988b) Functional anatomy of macaque striate cortex. II. Retinotopic organization. *J Neurosci* 8:1531–1568. doi: 10.1523/JNEUROSCI.08-05-01531.1988

Tóth D, Simon G, Reglődi D (2023) Pituitary adenylate cyclase-activating polypeptide (PACAP) and sudden infant death syndrome: a potential model for investigation. *Int J Mol Sci.* 24:15063. doi: 10.3390/ijms242015063

Traub RD, Wong RK (1982) Cellular mechanism of neuronal synchronization in epilepsy. *Science* 216:745–747. doi: 10.1126/science.7079735

Traynelis SF, Wollmuth LP, McBain CJ, Menniti FS, Vance KM, Ogden KK, Hansen KB, Yuan H, Myers SJ, Dingledine R (2010) Glutamate receptor ion channels: structure, regulation, and function. *Pharmacol Rev* 62:405–496. doi: 10.1124/pr.109.002451

Trembleau A, Morales M, Bloom FE (1994) Aggregation of vasopressin mRNA in a subset of axonal swellings of the median eminence and posterior pituitary: light and electron microscopic evidence. *J Neurosci* 14:39–53. doi: 10.1523/JNEUROSCI.14-01-00039.1994

Trusk TC, Kaboord WS, Wong-Riley MTT (1990) Effects of monocular enucleation, tetrodotoxin, and lid suture on cytochrome oxidase reactivity in supragranular puffs of adult macaque striated cortex. *Vis Neurosci* 4:185–204. doi: 10.1017/s0952523800003345

Tryba AK, Peña F, Ramirez JM (2006) Gasping activity in vitro: a rhythm dependent on 5-HT2A receptors. *J Neurosci* 26:2623–2634. doi: 10.1523/JNEUROSCI.4186-05.2006

Ts'o DY, Frostig RD, Lieke EE, Grinvald A (1990) Functional organization of primate visual cortex revealed by high resolution optical imaging. *Science* 249:417–420. doi: 10.1126/science.2165630

Ts'o DY, Gilbert CD (1988) The organization of chromatic and spatial interactions in the primate striate cortex. *J Neurosci* 8:1712–1727. doi: 10.1523/JNEUROSCI.08-05-01712.1988

Tsukihara T, Aoyama H, Yamashita E, Tomizaki T, Yamaguchi H, Shinzawa-Itoh K, Nakashima R, Yaono R, Yoshikawa S (1995) Structures of metal sites of oxidized bovine heart cytochrome c oxidase at 2.8 A. *Science* 269:1069–1074. doi: 10.1126/science.7652554

Tsukihara T, Aoyama H, Yamashita E, Tomizaki T, Yamaguchi H, Shinzawa-Itoh K, Nakashima R, Yaono R, Yoshikawa S (1996) The whole structure of the 13-subunit oxidized cytochrome c oxidase at 2.8 Å. *Science* 272:1136–1144. doi: 10.1126/science.272.5265.1136

Tu LF, Cao LF, Zhang YH, Guo YL, Zhou YF, Lu WQ, Zhang TZ, Zhang T, Zhang GX, Kurihara H, Li YF, He RR (2019) Sirt3-dependent deacetylation of COX-1 counteracts oxidative stress-induced cell apoptosis. *FASEB J* 33:14118–14128. doi: 10.1096/fj.201900708R

Tuthill MC, Wada RK, Arimoto JM, Sugino CN, Kanemaru KK, Takeuchi KK, Sidell N (2003) N-myc oncogene expression in neuroblastoma is driven by Sp1 and Sp3. *Mol Genet Metab* 80:272–280. doi: 10.1016/s1096-7192(03)00133-1

Uddman R, Luts A, Arimura A, Sundler F (1991) Pituitary adenylate cyclase-activating peptide (PACAP), a new vasoactive intestinal peptide (VIP)-like peptide in the respiratory tract. *Cell Tissue Res* 265:197–201. doi: 10.1007/BF00318155

Ungerleider LG, Mishkin M (1982) Two cortical visual systems. In: Dingle DJ, Goodale MA, Mansfield RJW (eds) *Analysis of Visual Behavior*. MIT Press, Cambridge, MA, pp. 549–586.

Uusimaa J, Finnilä S, Vainionpää L, Kärppä M, Herva R, Rantala H, Hassinen IE, Majamaa K (2003) A mutation in mitochondrial DNA-encoded cytochrome c oxidase II gene in a child with Alpers-Huttenlocher-like disease. *Pediatrics* 111:e262–268. doi: 10.1542/peds.111.3.e262

Valic M, Pecotic R, Dogas Z (2008) Phrenic nerve activity is enhanced by 5-HT1A receptor agonist 8-OH-DPAT in spontaneously breathing anesthetized rats. *J Physiol Pharmacol* 59:17–25.

Valnot I, von Kleist-Retzow JC, Barrientos A, Gorbatyuk M, Taanman JW, Mehaye B, Rustin P, Tzagoloff A, Munnich A, Rötig A (2000) A mutation in the human heme A: farnesyltransferase gene (COX10) causes cytochrome c oxidase deficiency. *Hum Mol Genet* 9:1245–1249. doi: 10.1093/hmg/9.8.1245

Van Essen DC (2005) Corticocortical and thalamocortical information flow in the primate visual system. *Prog Brain Res* 149:173–185. doi: 10.1016/S0079-6123(05)49013-5

Van Hoesen GW, Augustinack JC, Dierking J, Redman SJ, Thangavel R (2000) The parahippocampal gyrus in Alzheimer's disease. Clinical and preclinical neuroanatomical correlates. *Ann N Y Acad Sci* 911:254–274. doi: 10.1111/j.1749-6632.2000.tb06731.x

van Rahden VA, Fernandez-Vizarra E, Alawi M, Brand K, Fellmann F, Horn D, Zeviani M, Kutsche K (2015) Mutations in NDUFB11, encoding a complex I component of the mitochondrial respiratory chain, cause microphthalmia with linear skin defects syndrome. *Am J Hum Genet* 96:640–650. doi: 10.1016/j.ajhg.2015.02.002

Vanneste WH (1966) The stoichiometry and absorption spectra of components a and a-3 in cytochrome c oxidase. *Biochemistry* 5:838–848. doi: 10.1021/bi00867a005

Vanneste WH, Ysebaert-Vanneste M, Mason HS (1974) The decline of molecular activity of cytochrome oxidase during purification. *J Biol Chem* 249:7390–7401.

Varlamov DA, Kudin AP, Vielhaber S, Schröder R, Sassen R, Becker A, Kunz D, Haug K, Rebstock J, Heils A, Elger CE, Kunz WS (2002) Metabolic consequences of a novel missense mutation of the mtDNA CO I gene. *Hum Mol Genet* 11:1797–1805. doi: 10.1093/hmg/11.16.1797

Vaudry D, Falluel-Morel A, Bourgault S, Basille M, Burel D, Wurtz O, Fournier A, Chow BKC, Hashimoto H, Galas L, Vaudry H (2009) Pituitary adenylate cyclase-activating polypeptide and its receptors: 20 years after the discovery. *Pharmacol Rev* 61:283–357. doi: 10.1124/pr.109.001370

Veereshwarayya V, Kumar P, Rosen KM, Mestril R, Querfurth HW (2006) Differential effects of mitochondrial heat shock protein 60 and related molecular chaperones to prevent intracellular beta-amyloid-induced inhibition of complex IV and limit apoptosis. *J Biol Chem* 281:29468–29478. doi: 10.1074/jbc.M602533200

Venugopal R, Jaiswal AK (1996) Nrf1 and Nrf2 positively and c-Fos and Fra1 negatively regulate the human antioxidant response element-mediated expression of NAD(P)H:quinone oxidoreductase1 gene. *Proc Natl Acad Sci USA* 93(25):14960–14965. doi: 10.1073/pnas.93.25.14960

Vercauteren K, Gleyzer N, Scarpulla RC (2008) PGC-1-related coactivator complexes with HCF-1 and NRF-2beta in mediating NRF-2(GABP)-dependent respiratory gene expression. *J Biol Chem* 283:12102–12111. doi: 10.1074/jbc.M710150200

Verdoorn TA (1994) Formation of heteromeric gamma-aminobutyric acid type A receptors containing two different alpha subunits. *Mol Pharmacol* 45:475–480.

Vidoni S, Harbour ME, Guerrero-Castillo S, Signes A, Ding S, Fearnley IM, Taylor RW, Tiranti V, Arnold S, Fernandez-Vizarra E, Zeviani M (2017) MR-1S Interacts with PET100 and PET117 in Module-Based Assembly of Human Cytochrome c Oxidase. *Cell Rep* 18:1727–1738. doi: 10.1016/j.celrep.2017.01.044

Vila M, Levy R, Herrero MT, Faucheux B, Obeso JA, Agid Y, Hirsch EC (1996) Metabolic activity of the basal ganglia in parkinsonian syndromes in human and non-human primates: a cytochrome oxidase histochemistry study. *Neuroscience* 71:903–912. doi: 10.1016/0306-4522(95)00549-8

Vila M, Levy R, Herrero MT, Ruberg M, Faucheux B, Obeso JA, Agid Y, Hirsch EC (1997) Consequences of nigrostriatal denervation on the functioning of the basal ganglia in human and nonhuman primates: an in situ hybridization study of cytochrome oxidase subunit I mRNA. *J Neurosci* 17:765–773. doi: 10.1523/JNEUROSCI.17-02-00765.1997

Villani G, Greco M, Papa S, Attardi G (1998) Low reserve of cytochrome c oxidase capacity in vivo in the respiratory chain of a variety of human cell types. *J Biol Chem* 273:31829–31836. doi: 10.1074/jbc.273.48.31829

Villringer A, Chance B (1997) Non-invasive optical spectroscopy and imaging of human brain function. *Trends Neurosci* 20:435–442. doi: 10.1016/s0166-2236(97)01132-6

Vinogradova OS (1975) Functional organization of the limbic system in the process of registration of information: facts and hypotheses. In: Isaacson RL, Pribram KH (eds) *The Hippocampus*, vol 2, Neurophysiology and behavior. Plenum Press, New York, pp. 3–69.

Virbasius CA, Virbasius JV, Scarpulla RC (1993) NRF-1, an activator involved in nuclear-mitochondrial interactions, utilizes a new DNA-binding domain conserved in a family of developmental regulators. *Genes Dev* 7:2431–2445. doi: 10.1101/gad.7.12a.2431

Virbasius JV, Scarpulla RC (1991) Transcriptional activation through ETS domain binding sites in the cytochrome c oxidase subunit IV gene. *Mol Cell Biol* 11:5631–5638. doi: 10.1128/mcb.11.11.5631-5638.1991

Virbasius JV, Scarpulla RC (1994) Activation of the human mitochondrial transcription factor A gene by nuclear respiratory factors: a potential regulatory link between nuclear and mitochondrial gene expression in organelle biogenesis. *Proc Natl Acad Sci USA* 91:1309–1313. doi: 10.1073/pnas.91.4.1309

Virbasius JV, Virbasius CA, Scarpulla RC (1993) Identity of GABP with NRF-2, a multisubunit activator of cytochrome oxidase expression, reveals a cellular role for an ETS domain activator of viral promoters. *Genes Dev* 7:380–392. doi: 10.1101/gad.7.3.380

Vitet H, Brandt V, Saudou F (2020) Traffic signaling: new functions of huntingtin and axonal transport in neurological disease. *Curr Opin Neurobiol* 63:122–130. doi: 10.1016/j.conb.2020.04.001

Vogt S, Ramzan R, Grossman LI, Singh KK, Ferguson-Miller S, Yoshikawa S, Lee I, Hüttemann M (2021) Mitochondrial respiration is controlled by allostery, subunit composition and phosphorylation sites of cytochrome c oxidase: a trailblazer's tale - Bernhard Kadenbach. *Mitochondrion* 60:228–233. doi: 10.1016/j.mito.2021.08.015

Voss MD, De Castro D, Lipski J, Pilowsky PM, Jiang C (1990) Serotonin immunoreactive boutons form close appositions with respiratory neurons of the dorsal respiratory group in the cat. *J Comp Neurol* 295:208–218. doi: 10.1002/cne.902950205

Wallace DC (2005) A mitochondrial paradigm of metabolic and degenerative diseases, aging, and cancer: a dawn for evolutionary medicine. *Annu Rev Genet* 39:359–407. doi: 10.1146/annurev.genet.39.110304.095751

Wallace DC, Singh G, Lott MT, Hodge JA, Schurr TG, Lezza AM, Elsas LJ 2nd, Nikoskelainen EK (1988) Mitochondrial DNA mutation associated with Leber's hereditary optic neuropathy neuropathy. *Science* 242:1427–1430. doi: 10.1126/science.3201231

Wallace MN (1988) Lattices of high histochemical activity occur in the human, monkey, and cat superior colliculus. *Neuroscience* 25:569–583. doi: 10.1016/0306-4522(88)90259-x

Wang J, Zhao W, Liu H, He H, Shao R (2018) Myofibrillogenesis regulator 1 (MR-1): a potential therapeutic target for cancer and PNKD. *J Drug Target* 26:643–648. doi: 10.1080/1061186X.2017.1401077

Wang R, Dong Y, Lu Y, Zhang W, Brann DW, Zhang Q (2019) Photobiomodulation for global cerebral ischemia: targeting mitochondrial dynamics and functions. *Mol Neurobiol* 56:1852–1869. doi: 10.1007/s12035-018-1191-9

Wang W, Sun Y, Lin Y, Xu X, Zhao D, Ji K, Li W, Zhao Y, Yan C (2021) A novel nonsense variant in MT-CO3 causes MELAS syndrome. *Neuromuscul Disord* 31:558–565. doi: 10.1016/j.nmd.2021.02.020

Wang X, Tian F, Reddy DD, Nalawade SS, Barrett DW, Gonzalez-Lima F, Liu H (2017) Up-regulation of cerebral cytochrome-c-oxidase and hemodynamics by transcranial infrared laser stimulation: a broadband near-infrared spectroscopy study. *J Cereb Blood Flow Metab* 37:3789–3802. doi: 10.1177/0271678X17691783

Warburg O (1924) Über Eisen, den sauerstoffübertragenden Bestandteil des Atmungsferments. *Biochem Z* 152:479.

Warburg O (1931) The oxygen-transferring ferment of respiration. Nobel Lecture for Nobel Prize in Physiology or Medicine, December 10, 1931. NobedlPrize.Org.

Wardle RA, Poo MM (2003) Brain-derived neurotrophic factor modulation of GABAergic synapses by postsynaptic regulation of chloride transport. *J Neurosci* 23:8722–8732. doi: 10.1523/JNEUROSCI.23-25-08722.2003

Wareski P, Vaarmann A, Choubey V, Safiulina D, Liiv J, Kuum M, Kaasik A (2009) PGC-1{alpha} and PGC-1{beta} regulate mitochondrial density in neurons. *J Biol Chem* 284:21379–21385. doi: 10.1074/jbc.M109.018911

Watson SA, McStay GP (2020) Functions of cytochrome c oxidase assembly factors. *Int J Mol Sci* 21:7254. doi: 10.3390/ijms21197254

Weber B, Keller AL, Reichold J, Logothetis NK (2008) The microvascular system of the striate and extrastriate visual cortex of the macaque. *Cereb Cortex* 18:2318–2330. doi: 10.1093/cercor/bhm259

Weidner VR, Barnes PY, Eckerle KL (1977) A wavelength standard for the near infrared based on the reflectance of rare-earth oxides. *J Res Natl Bur Stand* 91:243–253. doi: 10.6028/jres.091.030

Weishaupt A, Kadenbach B (1992) Selective removal of subunit VIb increases the activity of cytochrome c oxidase. *Biochemistry* 31:11477–11481. doi: 10.1021/bi00161a028

Weller RE, Kaas JH (1983) Retinotopic patterns of connections of area 17 with visual areas V-II and MT in macaque monkeys. *J Comp Neurol* 220:253–279. doi: 10.1002/cne.902200302

Weraarpachai W, Antonicka H, Sasarman F, Seeger J, Schrank B, Kolesar JE, Lochmüller H, Chevrette M, Kaufman BA, Horvath R, Shoubridge EA (2009) Mutation in TACO1, encoding a translational activator of COX I, results in cytochrome c oxidase deficiency and late-onset Leigh syndrome. *Nat Genet* 41:833–837. doi: 10.1038/ng.390

Westerholz S, de Lima AD, Voigt T (2013) Thyroid hormone-dependent development of early cortical networks: temporal specificity and the contribution of trkB and mTOR pathways. *Front Cell Neurosci* 7:121. doi: 10.3389/fncel.2013.00121

Wharton DC, Tzagoloff A (1964) Studies on the electron transfer system. LVII. The near infrared absorption band of cytochrome oxidase. *J Biol Chem* 239:2036–2041.

Whelan HT, Smits RL Jr, Buchman EV, Whelan NT, Turner SG, Margolis DA, Cevenini V, Stinson H, Ignatius R, Martin T, Cwiklinski J, Philippi AF, Graf WR, Hodgson B, Gould L, Kane M, Chen G, Caviness J (2001) Effect of NASA light-emitting diode irradiation on wound healing. *J Clin Laser Med Surg* 19:305–314. doi: 10.1089/104454701753342758

Wickström R, Hökfelt T, Lagercrantz H (2002) Development of CO(2)-response in the early newborn period in rat. *Respir Physiol Neurobiol* 132:145–158. doi: 10.1016/s1569-9048(02)00076-9

Wiedemann FR, Manfredi G, Mawrin C, Beal MF, Schon EA (2002) Mitochondrial DNA and respiratory chain function in spinal cords of ALS patients. *J Neurochem* 80:616–625. doi: 10.1046/j.0022-3042.2001.00731.x

Wierstra I (2008) Sp1: emerging roles--beyond constitutive activation of TATA-less housekeeping genes. *Biochem Biophys Res Commun* 372:1–13. doi: 10.1016/j.bbrc.2008.03.074

Wiesel TN, Hubel DH (1963a) Effects of visual deprivation on morphology and physiology of cells in the cats lateral geniculate body. *J Neurophysiol* 26:978–993. doi: 10.1152/jn.1963.26.6.978

Wiesel TN, Hubel DH (1963b) Single-cell responses in striate cortex of kittens deprived of vision in one eye. *J Neurophysiol* 26:1003–1017. doi: 10.1152/jn.1963.26.6.1003

Wikström M, Krab K, Saraste M (1981) *Cytochrome Oxidase. A Synthesis*. Academic Press, New York.

Willems JL, Monnens LA, Trijbels JM, Veerkamp JH, Meyer AE, van Dam K, van Haelst U (1977) Leigh's encephalomyelopathy in a patient with cytochrome c oxidase deficiency in muscle tissue. *Pediatrics* 60:850–857.

Williams SL, Valnot I, Rustin P, Taanman JW (2004) Cytochrome c oxidase subassemblies in fibroblast cultures from patients carrying mutations in COX10, SCO1, or SURF1. *J Biol Chem* 279:7462–7469. doi: 10.1074/jbc.M309232200

Wilson PD, Rowe MH, Stone J (1976) Properties of relay cells in cat's lateral geniculate nucleus: a comparison of W-cells with X- and Y-cells. *J Neurophysiol* 39:1193–1209. doi: 10.1152/jn.1976.39.6.1193

Wilson RJ, Cummings KJ (2008) Pituitary adenylate cyclase-activating polypeptide is vital for neonatal survival and the neuronal control of breathing. *Respir Physiol Neurobiol* 164:168–178. doi: 10.1016/j.resp.2008.06.003

Winter DB, Bruyninckx WJ, Foulke FG, Grinich NP, Mason HS (1980) Location of heme a on subunit I and II and copper on subunit II of cytochrome c oxidase. *J Bid Chem* 255:11408–11414.

Wintjes LTM, Kava M, van den Brandt FA, van den Brand MAM, Lapina O, Bliksrud YT, Kulseth MA, Amundsen SS, Selberg TR, Ybema-Antoine M, Tutakhel OAZ, Greed L, Thorburn DR, Tangeraas T, Balasubramaniam S, Rodenburg RJT (2021) A novel variant in COX16 causes cytochrome c oxidase deficiency, severe fatal neonatal lactic acidosis, encephalopathy, cardiomyopathy, and liver dysfunction. *Hum Mutat* 42:135–141. doi: 10.1002/humu.24137

Witter MP, Amaral DG (1991) Entorhinal cortex of the monkey: V. Projections to the dentate gyrus, hippocampus, and subicular complex. *J Comp Neurol* 307:437–459. doi: 10.1002/cne.903070308

Wolf W, Hicks TP, Albus K (1986) The contribution of GABA-mediated inhibitory mechanisms to visual response properties of neurons in the kitten's striate cortex. *J Neurosci* 6:2779–2795. doi: 10.1523/JNEUROSCI.06-10-02779.1986

Wong PC, Pardo CA, Borchelt DR, Lee MK, Copeland NG, Jenkins NA, Sisodia SS, Cleveland DW, Price DL (1995) An adverse property of a familial ALS-linked SOD1 mutation causes motor neuron disease characterized by vacuolar degeneration of mitochondria. *Neuron* 14:1105–1116. doi: 10.1016/0896-6273(95)90259-7

Wong RW, Setou M, Teng J, Takei Y, Hirokawa N (2002) Overexpression of motor protein KIF17 enhances spatial and working memory in transgenic mice. *Proc Natl Acad Sci USA* 99:14500–14505. doi: 10.1073/pnas.222371099

Wong-Riley M (1979a) Changes in the visual system of monocularly sutured or enucleated cats demonstrable with cytochrome oxidase histochemistry. *Brain Res* 171:11–28. doi: 10.1016/0006-8993(79)90728-5

Wong-Riley M (1979b) Columnar cortico-cortical interconnections within the visual system of the squirrel and macaque monkeys. *Brain Res* 162:201–217. doi: 10.1016/0006-8993(79)90284-1

Wong-Riley M, Anderson B, Liebl W, Huang Z (1998a) Neurochemical organization of the macaque striate cortex: correlation of cytochrome oxidase with Na^+K^+ATPase, NADPH-diaphorase, nitric oxide synthase, and N-methyl-D-aspartate receptor subunit 1. *Neuroscience* 83:1025–1045. doi: 10.1016/s0306-4522(97)00432-6

Wong-Riley M, Antuono P, Ho KC, Egan R, Hevner R, Liebl W, Huang Z, Rachel R, Jones J (1997b) Cytochrome oxidase in Alzheimer's disease: biochemical, histochemical, and immunohistochemical analyses of the visual and other systems. *Vision Res* 37:3593–3608. doi: 10.1016/S0042-6989(96)00210-6

Wong-Riley M, Carroll EW (1984a) Effect of impulse blockage on cytochrome oxidase activity in monkey visual system. *Nature* 307:262–264. doi: 10.1038/307262a0

Wong-Riley M, Guo A, Bachman NJ, Lomax MI (2000) Human COX6A1 gene: promoter analysis, cDNA isolation and expression in the monkey brain. *Gene* 247:63–75. doi: 10.1016/s0378-1119(00)00121-9

Wong-Riley M, Riley DA (1983) The effect of impulse blockage on cytochrome oxidase activity in the cat visual system. *Brain Res* 261:185–193. doi: 10.1016/0006-8993(83)90622-4

Wong-Riley MTT (1974a) Demonstration of geniculocortical and callosal projection neurons in the squirrel monkey by means of retrograde axonal transport of horseradish peroxidase. *Brain Res* 79:267–272. doi: 10.1016/0006-8993(74)90415-6

Wong-Riley MTT (1974b) Synaptic organization of the inner plexiform layer in the retina of the tiger salamander. *J Neurocytol* 3:1–33. doi: 10.1007/BF01111929

Wong-Riley MTT (1976) Endogenous peroxidatic activity in brain stem neurons as demonstrated by their staining with diaminobenzidine in normal squirrel monkeys. *Brain Res* 108:257–277. doi: 10.1016/0006-8993(76)90185-2

Wong-Riley MTT (1989) Cytochrome oxidase: an endogenous metabolic marker for neuronal activity. *Trends Neurosci* 12:94–101. doi: 10.1016/0166-2236(89)90165-3

Wong-Riley MTT (1993) Cytochrome oxidase studies on the human visual cortex. In: Gulyas B, Ottoson D, Roland PE (eds) *Functional Organization of the Human Visual Cortex*. Plenum Press, New York, pp 165–180.

Wong-Riley MTT (1994) Primate visual cortex: dynamic metabolic organization and plasticity revealed by cytochrome oxidase. In: Peters A, Rockland K (eds) *Cerebral Cortex*, vol 10. Primary visual cortex in primates. Plenum Press, New York, pp 141–200.

Wong-Riley MTT (2004) Striate cortex. In: Craighead WE, Nemeroff CB (eds) *The Concise Corsini Encyclopedia of Psychology and Behavioral Science*. 3rd ed. John Wiley & Sons, New Jersey, pp 947–949.

Wong-Riley MTT (2012) Bigenomic regulation of cytochrome c oxidase in neurons and the tight coupling between neuronal activity and energy metabolism. *Adv Exp Med Biol* 748:283–304. doi: 10.1007/978-1-4614-3573-0_12

Wong-Riley MTT (2021) The critical period: neurochemical and synaptic mechanisms shared by the visual cortex and the brain stem respiratory system. *Proc Biol Sci* 288:20211025. doi: 10.1098/rspb.2021.1025

Wong-Riley MTT, Bai X, Buchmann E, Whelan HT (2001) Light-emitting diode treatment reverses the effect of TTX on cytochrome oxidase in neurons. *NeuroReport* 12:3033–3037. doi: 10.1097/00001756-200110080-00011

Wong-Riley MTT, Besharse JC (2012) The kinesin superfamily protein KIF17: one protein with many functions. *Invited review. Biomol Concepts* 3:267–282. doi: 10.1515/bmc-2011-0064

Wong-Riley MTT, Carroll EW (1984b) Quantitative light and electron microscopic analysis of cytochrome oxidase-rich zones in V II prestriate cortex of the squirrel monkey. *J Comp Neurol* 222:18–37. doi: 10.1002/cne.902220103

Wong-Riley MTT, Hevner RF, Cutlan R, Earnest M, Egan R, Frost J, Nguyen T (1993) Cytochrome oxidase in the human visual cortex: distribution in the developing and the adult brain. *Vis Neurosci* 10:41–58. doi: 10.1017/s0952523800003217

Wong-Riley MTT, Huang Z, Liebl W, Nie F, Xu H, Zhang C (1998b) Neurochemical organization of the macaque retina: effect of TTX on levels and gene expression of cytochrome oxidase and nitric oxide synthase and on the immunoreactivity of $Na^{+}K^{+}$ATPase and NMDA receptor subunit 1. *Vision Res* 38:1455–1477. doi: 10.1016/s0042-6989(98)00001-7

Wong-Riley MTT, Jacobs P (2002) AMPA glutamate receptor subunit 2 in normal and visually deprived macaque visual cortex. *Vis Neurosci* 19:563–573. doi: 10.1017/s0952523802195022

Wong-Riley MTT, Kageyama GH (1986) Localization of cytochrome oxidase in the mammalian spinal cord and dorsal root ganglia, with quantitative analysis of ventral horn cells in monkeys. *J Comp Neurol* 245:41–61. doi: 10.1002/cne.902450104

Wong-Riley MTT, Liang HL, Eells JT, Chance B, Henry MM, Buchmann E, Kane M, Whelan HT (2005a) Photobiomodulation directly benefits primary neurons functionally inactivated by toxins: role of cytochrome c oxidase. *J Biol Chem* 280:4761–4771. doi: 10.1074/jbc.M409650200

Wong-Riley MTT, Liang HL, Ongwijitwat S (2008) Activity-dependent bigenomic transcriptional regulation of cytochrome c oxidase in neurons. In: Dudek SM (ed) *Transcriptional Regulation by Neuronal Activity: To the Nucleus and Back*. Springer Science, New York, Chapter 11, pp 209–228.

Wong-Riley MTT, Liu Q (2005) Neurochemical development of brain stem nuclei involved in the control of respiration. Invited review. In: Bavis R, Carroll J (eds) Development of Respiratory Control. *Respir Physiol Neurobiol* 149:83–98. doi: 10.1016/j.resp.2005.01.011

Wong-Riley MTT, Liu Q (2008) Neurochemical and physiological correlates of a critical period of respiratory development in the rat. Invited review. In: McCrimmon D, Mitchell G, Alheid G (eds) Neurochemistry of Respiratory Control. *Respir Physiol Neurobiol* 164:28–37. doi: 10.1016/j.resp.2008.04.014

Wong-Riley MTT, Liu Q, Gao X (2019) Mechanisms underlying a critical period of respiratory development in the rat. Invited review. In: MacFarlane PM, Bavis RW (eds) Developmental Diversity in the Control of Breathing. *Respir Physiol Neurobiol* 264:40–50. doi: 10.1016/j.resp.2019.04.006

Wong-Riley MTT, Liu Q, Gao XP (2013) Peripheral-central chemoreceptor interaction and the significance of a critical period in the development of respiratory control. Invited review. In: Carroll JL, Donnelly D, Bairm A (eds) Development of the Carotid Body. *Respir Physiol Neurobiol* 185:156–169. doi: 10.1016/j.resp.2012.05.026

Wong-Riley MTT, Merzenich MM, Leake PA (1978) Changes in endogenous enzymatic reactivity to DAB induced by neuronal inactivity. *Brain Res* 141:185–192. doi: 10.1016/0006-8993(78)90629-7

Wong-Riley MTT, Mullen MA, Huang Z, Guyer C (1997a) Brain cytochrome oxidase subunit complementary DNAs: isolation, subcloning, sequencing, light and electron microscopic in situ hybridization of transcripts, and regulation by neuronal activity. *Neuroscience* 76:1035–1055. doi: 10.1016/s0306-4522(96)00410-1

Wong-Riley MTT, Nie F, Hevner RF, Liu S (1998c) Brain cytochrome oxidase: functional significance and bigenomic regulation in the CNS. In: Gonzalez-Lima F (ed) *Cytochrome Oxidase in Neuronal Metabolism and Alzheimer's Disease*. Plenum Press, New York, pp 1–53.

Wong-Riley MTT, Tripathi SC, Trusk TC, Hoppe DA (1989a) Effect of retinal impulse blockage on cytochrome oxidase-rich zones in the macaque striate cortex. I. Quantitative electron-microscopic (EM) analysis of neurons. *Vis Neurosci* 2:483–497. doi: 10.1017/s0952523800012384

Wong-Riley MTT, Trusk TC, Kaboord W, Huang Z (1994) Effect of retinal impulse blockage on cytochrome oxidase-poor interpuffs in the macaque striate cortex: quantitative EM analysis of neurons. *J Neurocytol* 23:533–553. doi: 10.1007/BF01262055

Wong-Riley MTT, Trusk TC, Tripathi SC, Hoppe DA (1989b) Effect of retinal impulse blockage on cytochrome oxidase-rich zones in the macaque striate cortex. II. Quantitative electron-microscopic (EM) analysis of neuropil. *Vis Neurosci* 2:499–514. doi: 10.1017/s0952523800012396

Wong-Riley MTT, Walsh SM, Leake-Jones PA, Merzenich MM (1981) Maintenance of neuronal activity by electrical stimulation of unilaterally deafened cats demonstrable with cytochrome oxidase technique. *Ann Otol Rhinol Laryngol Suppl* 90:30–32. doi: 10.1177/00034894810902s211

Wong-Riley MTT, Welt C (1980) Histochemical changes in cytochrome oxidase of cortical barrels after vibrissal removal in neonatal and adult mice. *Proc Natl Acad Sci USA* 77:2333–2337. doi: 10.1073/pnas.77.4.2333

Wong-Riley MTT, Yang SJ, Liang HL, Ning G, Jacobs P (2005b) Quantitative immuno-electron microscopic analysis of nuclear respiratory factor 2 alpha and beta subunits: normal distribution and activity-dependent regulation in mammalian visual cortex. *Vis Neurosci* 22:1–18. doi: 10.1017/S0952523805221016

Woolsey TA, Van der Loos H (1970) The structural organization of layer IV in the somatosensory region (SI) of mouse cerebral cortex. The description of a cortical field composed of discrete cytoarchitectonic units. *Brain Res* 17:205–242. doi: 10.1016/0006-8993(70)90079-x

Woolsey TA, Wann JR (1976) Areal changes in mouse cortical barrels following vibrissal damage at different postnatal ages. *J Comp Neurol* 170:53–66. doi: 10.1002/cne.901700105

Wray S, Cope M, Delpy DT, Wyatt JS, Reynolds EO (1988) Characterization of the near infrared absorption spectra of cytochrome aa3 and haemoglobin for the non-invasive monitoring of cerebral oxygenation. *Biochim Biophys Acta* 933:184–192. doi: 10.1016/0005-2728(88)90069-2

Wu Z, Puigserver P, Andersson U, Zhang C, Adelmant G, Mootha V, Troy A, Cinti S, Lowell B, Scarpulla RC, Spiegelman BM 1999) Mechanisms controlling mitochondrial biogenesis and respiration through the thermogenic coactivator PGC-1. *Cell* 98:115–124. doi: 10.1016/S0092-8674(00)80611-X

Yamada M, Amuro N, Goto Y, Okazaki T (1990) Structural organization of the rat cytochrome c oxidase subunit IV gene. *J Biol Chem* 265:7687–7692.

Yan MH, Wang X, Zhu X (2013) Mitochondrial defects and oxidative stress in Alzheimer's disease and Parkinson disease. *Free Radic Biol Med* 62:90–101. doi: 10.1016/j.freeradbiomed.2012.11.014

Yang SJ, Liang HL, Ning G, Wong-Riley MTT (2004) Ultrastructural study of depolarization-induced translocation of NRF-2 transcription factor in cultured rat visual cortical neurons. *Eur J Neurosci* 19:1153–1162. doi: 10.1111/j.1460-9568.2004.03250.x

Yang SJ, Liang HL, Wong-Riley MTT (2006) Activity-dependent transcriptional regulation of nuclear respiratory factor-1 in cultured rat visual cortical neurons. *Neurosci* 141:1181–1192. doi: 10.1016/j.neuroscience.2006.04.063

Yang WL, Iacono L, Tang WM, Chin KV (1998) Novel function of the regulatory subunit of protein kinase A: regulation of cytochrome c oxidase activity and cytochrome c release. *Biochemistry* 37:14175–14180. doi: 10.1021/bi981402a

Yen K, Miller B, Kumagai H, Silverstein A, Cohen P (2025) Mitochondrial-derived microproteins: from discovery to function. *Trends Genet* 41:132–145. doi: 10.1016/j.tig.2024.11.010

Yoon JC, Puigserver P, Chen G, Donovan J, Wu Z, Rhee J, Adelmant G, Stafford J, Kahn CR, Granner DK, Newgard CB, Spiegelman BM (2001) Control of hepatic gluconeogenesis through the transcriptional coactivator PGC-1. *Nature* 413:131–138. doi: 10.1038/35093050

Yoshii A, Constantine-Paton M (2010) Postsynaptic BDNF-TrkB signaling in synapse maturation, plasticity, and disease. *Dev Neurobiol* 70:304–322. doi: 10.1002/dneu.20765

Yoshikawa S, Muramoto K, Shinzawa-Itoh K (2011) Proton-pumping mechanism of cytochrome C oxidase. *Annu Rev Biophys* 40:205–223. doi: 10.1146/annurev-biophys-042910-155341

Yoshioka T, Dow BM (1996) Color, orientation and cytochrome oxidase reactivity in areas V1, V2 and V4 of macaque monkey visual cortex. *Behav Brain Res* 76:71–88. doi: 10.1016/0166-4328(95)00184-0

Yoshioka T, Levitt JB, Lund JS (1994) Independence and merger of thalamocortical channels within macaque monkey primary visual cortex: anatomy of interlaminar projections. *Vis Neurosci* 11:467489. doi: 10.1017/s0952523800002406

Yu W, Naim JO, McGowan M, Ippolito K, Lanzafame RJ (1997) Photomodulation of oxidative metabolism and electron chain enzymes in rat liver mitochondria. *Photochem Photobiol* 66:866–871. doi: 10.1111/j.1751-1097.1997.tb03239.x

Yu ZY, Wang W, Fritschy JM, Witte OW, Redecker C (2006) Changes in neocortical and hippocampal GABAA receptor subunit distribution during brain maturation and aging. *Brain Res* 1099:73–81. doi: 10.1016/j.brainres.2006.04.118

Zaninello M, Baptista P, Duarte FV (2024) Mitochondrial dynamics and mRNA translation: A local synaptic tale. *Biology (Basel)* 13:746. doi: 10.3390/biology13090746

Zeki S, Shipp S (1988) The functional logic of cortical connections. *Nature* 335:311–317. doi: 10.1038/335311a0

Zeviani M, Moraes CT, DiMauro S, Nakase H, Bonilla E, Schon EA, Rowland LP (1988b) Deletions of mitochondrial DNA in Kearns-Sayre syndrome. *Neurology* 38:1339–1346. doi: 10.1212/wnl.38.9.1339

Zeviani M, Nakagawa M, Herbert J, Lomax MI, Grossman LI, Sherbany AA, Miranda AF, DiMauro S, Schon EA (1987) Isolation of a cDNA clone encoding subunit IV of human cytochrome c oxidase. *Gene* 55(2–3):205–217. doi: 10.1016/0378-1119(87)90281-2

Zeviani M, Sakoda S, Sherbany AA, Nakase H, Rizzuto R, Samitt CE, DiMauro S, Schon EA (1988a) Sequence of cDNAs encoding subunit Vb of human and bovine cytochrome c oxidase. *Gene* 65:1–11. doi: 10.1016/0378-1119(88)90411-8

Zhang B, Bi H, Sakai E, Maruko I, Zheng J, Smith EL 3rd, Chino YM (2005) Rapid plasticity of binocular connections in developing monkey visual cortex (V1). *Proc Natl Acad Sci U S A* 102:9026–9031. doi: 10.1073/pnas.0500280102

Zhang C, Li N, Liu YY, Yuan T, Yang S, Wang XP (2021) Cox15 is a novel oncogene that required for lung cancer cell proliferation. *Biochem Biophys Res Commun* 578:70–76. doi: 10.1016/j.bbrc.2021.09.010

Zhang C, Wong-Riley M (1999) Expression and regulation of NMDA receptor subunit R1 and neuronal nitric oxide synthase in cortical neuronal cultures: correlation with cytochrome oxidase. *J Neurocytol* 28:525–539. doi: 10.1023/a:1007053204929

Zhang C, Wong-Riley MTT (2000a) Synthesis and degradation of cytochrome oxidase subunit mRNAs in neurons: differential bigenomic regulation by neuronal activity. *J Neurosci Res* 60:338–344. doi: 10.1002/(SICI)1097-4547(20000501)60:3<338::AID-JNR8>3.0.CO;2-1

Zhang C, Wong-Riley MTT (2000b) Depolarizing stimulation upregulates GA-binding protein in neurons: a transcription factor involved in the bigenomic expression of cytochrome oxidase subunits. *Eur J Neurosci* 12:1013–1023. doi: 10.1046/j.1460-9568.2000.00997.x

Zhang H, Mu L, Wang D, Xia D, Salmon A, Liu Q, Wong-Riley MTT (2018) Uncovering a critical period of synaptic imbalance during postnatal development of the rat visual cortex: role of brain-derived neurotrophic factor. *J Physiol* 596.18:4511–4536. doi: 10.1113/JP275814

Zhang K, Wang G, Zhang X, Hüttemann PP, Qiu Y, Liu J, Mitchell A, Lee I, Zhang C, Lee JS, Pecina P, Wu G, Yang ZQ, Hüttemann M, Grossman LI (2016) COX7AR is a stress-inducible mitochondrial COX Subunit that promotes breast cancer malignancy. *Sci Rep* 6:31742. doi: 10.1038/srep31742

Zheng F, Zhou X, Moon C, Wang H (2012) Regulation of brain-derived neurotrophic factor expression in neurons. *Int J Physiol Pathophysiol Pharmacol* 4:188–200.

Zhu Z, Yao J, Johns T, Fu K, De Bie I, Macmillan C, Cuthbert AP, Newbold RF, Wang J, Chevrette M, Brown GK, Brown RM, Shoubridge EA (1998) SURF1, encoding a factor involved in the biogenesis of cytochrome c oxidase, is mutated in Leigh syndrome. *Nat Genet* 20:337–343. doi: 10.1038/3804

Zifa E, Fillion G (1992) 5-Hydroxytryptamine receptors. *Pharmacol Rev* 44(3):401–458.

Zimmerman AL, Yamanaka G, Eckstein F, Baylor DA, Stryer L (1985) Interaction of hydrolysis-resistant analogs of cyclic GMP with the phosphodiesterase and light-sensitive channel of retinal rod outer segments. *Proc Natl Acad Sci USA* 82:8813–8817. doi: 10.1073/pnas.82.24.8813

Zong S, Wu M, Gu J, Liu T, Guo R, Yang M (2018) Structure of the intact 14-subunit human cytochrome c oxidase. *Cell Res* 28:1026–1034. doi: 10.1038/s41422-018-0071-1

Zuckerman R (1973) Ionic analysis of photoreceptor membrane currents. *J Physiol* 235:333–354. doi: 10.1113/jphysiol.1973.sp010390

Author Index

Pages in *italics* refer to figures and pages in **bold** refer to tables.
(Only first authors and single co-authors are listed)

F

G

H

M

N

O

T

Subject Index

Pages in *italics* refer to figures and pages in **bold** refer to tables.

D

E

F

G

H

I

K

L

W

X

Y

For Product Safety Concerns and Information please contact our
EU representative GPSR@taylorandfrancis.com Taylor & Francis
Verlag GmbH, Kaufingerstraße 24, 80331 München, Germany